# OPTICAL PROPERTIES OF LOW-DIMENSIONAL MATERIALS

# OPTICAL PROPERTIES OF LOW-DIMENSIONAL MATERIALS

Editors

**T. Ogawa**

*Osaka City University*

**Y. Kanemitsu**

*University of Tsukuba*

World Scientific

NEW JERSEY · LONDON · SINGAPORE · BEIJING · SHANGHAI · HONG KONG · TAIPEI · CHENNAI

*Published by*

World Scientific Publishing Co. Pte. Ltd.

5 Toh Tuck Link, Singapore 596224

*USA office:* 27 Warren Street, Suite 401-402, Hackensack, NJ 07601

*UK office:* 57 Shelton Street, Covent Garden, London WC2H 9HE

**British Library Cataloguing-in-Publication Data**
A catalogue record for this book is available from the British Library.

**OPTICAL PROPERTIES OF LOW–DIMENSIONAL MATERIALS**



ISBN-13 978-981-02-2231-4
ISBN-10 981-02-2231-9

# PREFACE

In contemporary condensed-matter physics and materials science, a great deal of interest is focused on low-dimensional materials. The purpose of the study on low-dimensional systems is threefold: (i) the simplified description of several phenomena which take place in three-dimensional (3D) materials, (ii) the discovery of novel phenomena peculiar to low-dimensional materials, and (iii) the creation of artificial novel 3D materials composed of low-dimensional elements. The first two purposes can be summarized as "a trend from 3D to low-dimensional systems," while the last one is "a trend from low-dimensional to new 3D systems."

From the standpoint of (i), low-dimensional systems are treated as expedients for understanding complicated properties of 3D materials in our real world. An exact solution of a special low-dimensional problem is often applied to higher-dimensional problems to find a guide for understanding them. The viewpoint of (ii), on the other hand, often comes from actual synthesis and experimental observation of low-dimensional materials. Experiments reveal new phenomena in such materials, which never be expected in 3D systems. Hence novel concepts and paradigms are required and created for full understanding, which, in turn, induce new experiments and synthesis of novel materials. These material-leading and paradigm-leading researches have affected each other and have been main motive powers for developing and activating the research field on low-dimensional materials.

We here stress that another viewpoint of (iii) is of special importance in the present and future study on low-dimensional materials. In this view, a low-dimensional material is treated as a constituting element for creating an artificial novel 3D system (denoted here as a "3′D system") by its rearrangement. Not only the characteristics of the low-dimensional elements themselves but also the global connection and 3D networking among them play crucial roles in yielding a variety of phenomena and characteristics of 3′D materials. For example, electronic and structural phase transitions and/or cooperative phenomena depend sensitively on the way of their rearrangement in the 3D space, hence a possibility for artificial control of such phenomena will be expected. This will bring also a wide application.

We should note that above three aspects of the research purpose are inseparably related to each other. Characteristics of both the low-dimensional material itself and its assembly should be clarified, then we have a guiding principle for designing and synthesizing novel artificial materials with appropriate and desired properties. This is a final goal of this field. In this respect, the purposes of (ii) and (iii) are like the two wheels of a cart in this research field.

This book surveys, from the (ii) and (iii) viewpoints, recent theoretical and experimental studies on optical and electronic properties of low-dimensional materials, including some quite novel materials: artificial structures of inorganic and organic semiconductors, silicon nanocrystals, $C_{60}$-related molecular materials, and artificial crystals in zeolites. Remarkable progress in the research of low-dimensional materials has been achieved. In this respect we think that it is a good timing to publish a collection of excellent review articles on such materials. The detailed critical reviews are written by up-and-coming leading researchers, all of which cover a wide range of subjects starting from fundamental theories to device applications, focusing especially on optical properties. All the materials introduced and elucidated in this book yield new optical phenomena originating from their mesoscopic and low-dimensional characters, which phenomena contribute to a new research field of condensed-matter and optical physics.

Since low-dimensional materials often possess their own characteristic structures, there are a variety of electronic and optical properties unique to that material. Therefore, almost all portions of this book are devoted to introduction and discussion on characteristic features of low-dimensional materials which attract recent interest. However, the limited volume of this book compelled us to select the issues to which we devote more attention. The subject, therefore, has by no means been exhausted, nor is it treated absolutely objectively. This book is not intended as a catalog of low-dimensional materials; it does not contain their full characteristics, either. It is intended to help all readers who want to become acquainted with the latest problems of low-dimensional condensed-matter physics, to benefit from its achievements.

As a tool for investigating electron, exciton, and phonon structures of materials, optical measurements and laser spectroscopic techniques are greatly developed and now almost established. These are powerful also for the study of low-dimensional materials. Their low dimensionality is well reflected in their optical responses. In turn, the optical properties are good measures to explore the intrinsic natures of low-dimensional systems. We stress that nanostructures (nanometer-scaled structures) are attracting much interest not only from the viewpoints of fundamental condensed-matter physics but also from the expectation of their potential applications to various optical and electronic devices. Therefore, a technological use of the optical properties of low-dimensional materials is also eagerly looked for at an applied side of material engineering. Thus the optical properties of the materials are not only a key or a clue for exploring their electronic structures but also related directly to optical device applications. This book is written from both viewpoints to show readers a close link between novel optical properties and the low dimensionality. Controllability of the optical properties is also examined often in this book.

The readers descend from generals to particulars in this book. This book consists of eight chapters, each of which is intended to be read individually without referring other chapters. Chapters 1 and 2 are devoted to the theoretical descriptions of material-independent features of low-dimensional systems. In Chapters 3–8, six contemporary low-dimensional materials are elucidated in detail from experimental viewpoints. Two-dimensional inorganic semiconductor compounds, II–VI and III–V artificial quantum wells and superlattices, are revisited in Chapters 3 and 4 from more fundamental physical viewpoints to reveal their peculiar optical phenomena. Chapter 4 is related to the device application of blue-green laser diodes. The dimensionality of materials is tried to be controlled in Chapters 5 and 6 in Si- and Pb-related compounds. The readers find that their optical properties are strongly related to the material dimensionality, and that the Coulomb interaction between two charged particles can be controlled in low-dimensional materials. Materials which recently attract a great deal of interest are examined and explained in the last two Chapters 7 and 8. There, $C_{60}$, a natural quasi-zero-dimensional molecular material, and artificial 3D metal- and semiconductor-cluster lattices in zeolites are selected. These chapters demonstrate the recreation of novel macroscopic crystals ("3′D systems") from microscopic small clusters of carbon, metal, or semiconductor atoms. Organization of this book is illustrated in the next page, where one finds that which dimensional (3D–0D and 3′D) systems are treated in each chapter.

This book is intended for graduate students, researchers, and technologists working on condensed-matter physics, materials science, and optoelectronics. It is well-illustrated with a wealth of diagrams and numerous data of low-dimensional materials. We hope that it fulfills its purpose of providing the readers with valuable insight and prospective into this rapidly developing and potentially important field.

In editing this book, we took advantage both of original and review works, references to which are given throughout the book. We are deeply grateful to contributors. All the contributors wish to thank all of their colleagues, who have kindly supplied with the necessary materials even before publication on some occasions, for collaboration and discussions at various stages of their researches on low-dimensional materials. Some figures in this book were reproduced from several journals, owing to the kind permission granted by authors and publishers. We express our sincere gratitude and deep appreciation to the following journal publishers: Akademie Verlag, The American Association for the Advancement Science, The American Institute of Physics, The American Physical Society, Elsevier Science Ltd, The Japan Society of Applied Physics, Macmillan Magazines Ltd, Narosa Publishing House, North-Holland Publishing Co., Pergamon Press Ltd, The Physical Society of Japan, and World Scientific Publishing Co.

Osaka and Tsukuba
September 1995

*Tetsuo Ogawa*
*Yoshihiko Kanemitsu*

# ORGANIZATION OF THIS BOOK

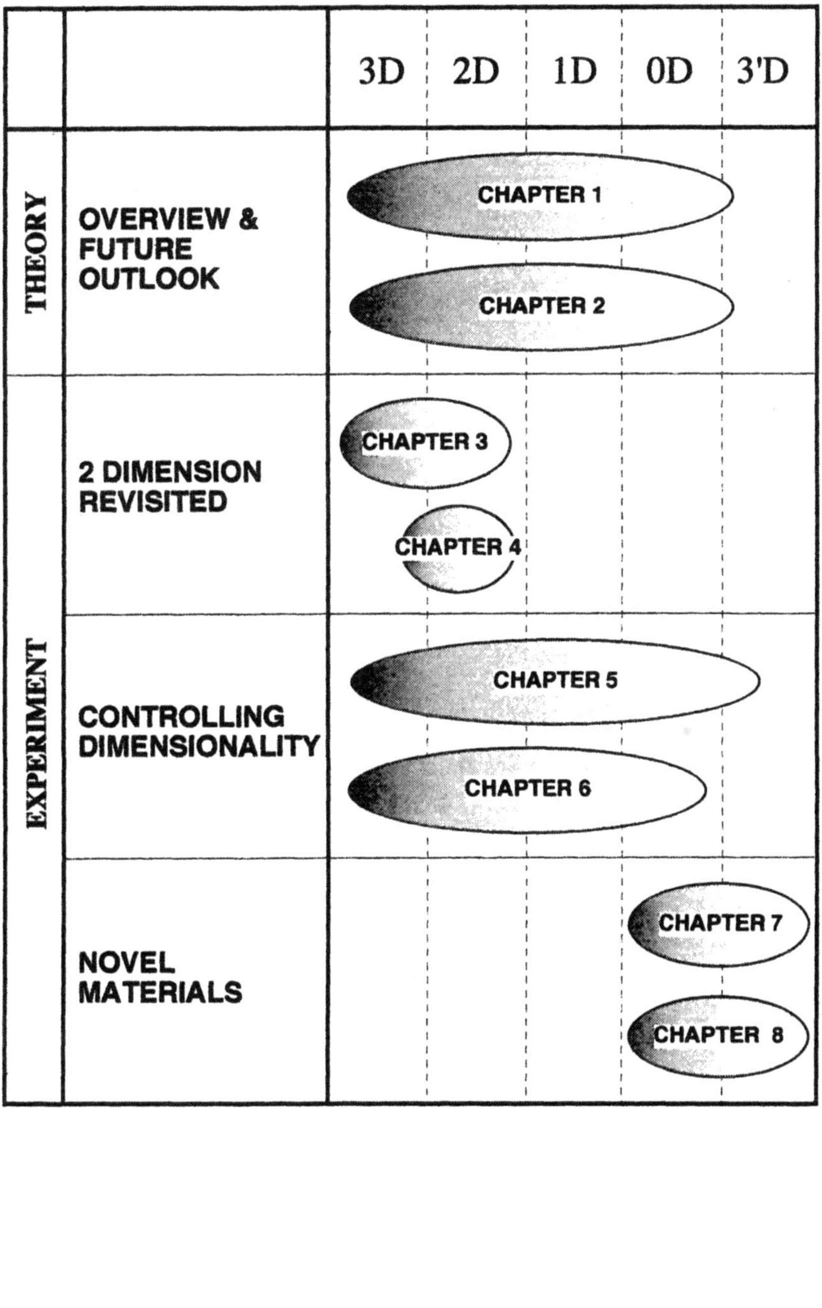

# CONTRIBUTORS

**Tetsuo Ogawa** (Chapter 1, p. 1–89)
*Department of Applied Physics, Osaka City University, 3–3–138 Sugimoto, Sumiyoshi-ku, Osaka 558, Japan*
Electronic-mail: ogawa@a-phys.eng.osaka-cu.ac.jp

**Takashi Nakayama** (Chapter 2, p. 90–146)
*Department of Physics, Chiba University, 1–33 Yayoi, Inage-ku, Chiba 263, Japan*
Electronic-mail: miho@tansei.cc.u-tokyo.ac.jp

**Masaaki Nakayama** (Chapter 3, p. 147–201)
*Department of Applied Physics, Osaka City University, 3–3–138 Sugimoto, Sumiyoshi-ku, Osaka 558, Japan*
Electronic-mail: nakayama@a-phys.eng.osaka-cu.ac.jp

**Yoichi Yamada** (Chapter 4, p. 202–239)
*Department of Electrical and Electronic Engineering, Yamaguchi University, 2557 Tokiwadai, Ube, Yamaguchi 755, Japan*
Electronic-mail: yamada@po.cc.yamaguchi-u.ac.jp

**Yoshihiko Kanemitsu** (Chapter 5, p. 240–287)
*Institute of Physics, University of Tsukuba, Tsukuba, Ibaraki 305, Japan*
Electronic-mail: sunyu@sakura.cc.tsukuba.ac.jp

**Teruya Ishihara** (Chapter 6, p. 288–339)
*Department of Physical Electronics, Hiroshima University, 1–4–1 Kagamiyama, Higashi-Hiroshima 724, Japan*
Electronic-mail: terry@hws.ipc.hiroshima-u.ac.jp

**Yoshihiro Iwasa** (Chapter 7, p. 340–386)
*Japan Advanced Institute of Science and Technology, School of Materials Science, Ishikawa 923–12, Japan*
Electronic-mail: iwasa@jaist.ac.jp

**Yasuo Nozue** (Chapter 8, p. 387–414)
*Department of Physics, Graduate School of Science, Tohoku University, Aramaki, Aoba-ku, Sendai, Miyagi 980–77, Japan*
Electronic-mail: nozue-jn@msp.phys.tohoku.ac.jp

# CONTENTS

*CHAPTER 1*

# DIMENSIONALITY AND OPTICAL RESPONSES OF MATERIALS

Tetsuo Ogawa
*Department of Applied Physics, Osaka City University*

## 1.1. INTRODUCTION

In modern condensed-matter physics, main interests have been focused especially on "low dimensionality," "strong electronic correlations," and "strong electron-lattice coupling." Actually, in lower-dimensional systems, the electronic correlation and the electron-lattice interaction become of more importance, as will be shown in this chapter, then variety of phenomena have been observed, which never be expected in bulk systems. Simultaneously, theoretically novel concepts peculiar to the low-dimensional systems are discovered and developed [1]. In this chapter, we pay attention to the low dimensionality of materials in terms of not only their electronic structure but also their optical properties. Thus here we pay our attention to the general and universal features of low-dimensional systems [2, 3]. Material-dependent characteristics are left for the other chapters.

At the same time, the advent of a new crystal-growth techniques, *e.g.*, molecular beam epitaxy, opens the way to the artificially-controlled growth of an atomic layer on an atomic layer, owing to progress in crystal availability and control, basic understanding of low-dimensional systems, and applicability of heterostructure concepts. The recent years have seen the emergence of a wide family of structures and devices. We will be interested in (quasi-)$d$-dimensional[1] properties; the free motion of the carriers occurs in only $d$ directions, the motion in the residual $3 - d$ directions being restricted to a well-defined portion of space by momentum, energy, and wavefunction quantizations.

There are several aspects of "low dimensionality" of materials. Therefore, before we discuss the electronic and optical properties of materials, what the "low dimensionality" means should be defined clearly. In the following, we shall study six different aspects of the low dimensionality to clarify how the optical responses are affected by the dimensionality.

The plan of this chapter[2] is as follows. In Sec. 1.2, dimensionality of the single-electron states is discussed in terms of the density of states of electrons, zone-folding and band-mixing mechanisms, and the topological nature. Wannier excitons in semiconductors are also one of the most interesting quasiparticles which exhibit strong dependence on the material dimension. This is studied in detail in Sec. 1.3 and the low-dimensional exciton effects on linear and nonlinear optical processes are clarified.

[1] We use the word "quasi" to mark the difference with exact $d$-dimensional ($d$D) systems in which the wavefunction is exactly confined in a $d$D space, with no extension outside the space.

[2] We take $\hbar = k_B = 1$ throughout this chapter.

The two-photon absorption experiment in a semiconductor wire structure is reviewed as an effective probe for dimensionality of Wannier exciton wavefunctions. Related studies on the low-dimensional Wannier excitons and excitonic molecules are given there. Dimensionality of the Coulomb interaction can be controlled with the use of the dielectric confinement effects (image charge effects), which is examined in Sec. 1.4. In Sec. 1.5, we study low-dimensional conductors (metals) and the Fermi-surface effects. In 1D conductors, a concept of the Tomonaga-Luttinger (TL) liquid is introduced and is applied to the study of the optical Fermi-edge singularity, which is one of the most striking optical phenomena due to the Fermi-surface effects. Similar to the dimensionality of the Anderson localization of an electron in random media, the self-trapping phenomena of excitons and electrons are examined in Sec. 1.6 and its dependence on the system dimension is clarified. In Sec. 1.7, dynamics of photoinduced structure changes, observed in linear chain systems, are considered from two viewpoints of the nucleation picture and the phase-transition picture.

We note here that nanostructures (nanometer-scaled structures) are attracting much interest not only from the viewpoints of fundamental condensed-matter physics but also from the expectation of their potential applications to various optical and electronic devices. In low-dimensional materials, effects of their geometrical restrictions are well reflected in their optical responses. In turn, the optical properties are good measures to explore the intrinsic natures of low-dimensional systems.

## 1.2. DIMENSIONALITY IN THE SINGLE-ELECTRON STATES

First of all, we focus our attention to the single-electron states in perfect and rigid crystals. Artificial low-dimensional structures such as superlattices (SLs), quantum wells (QWs), and quantum wires can be fabricated recently. These structures can undergo drastic changes in their material properties, some of which have been observed also in natural low-dimensional materials [4, 5]. Crystal symmetry modulation results in the zone folding effect, *e.g.*, in SLs, which is a direct consequence of the synthetically induced periodicity. Energy-level quantization in QWs and wires, resulting from carrier confinement in an artificial electrostatic potential, is also based on this modulation. In addition, the mixing of different materials yields the band mixing effect, which is achieved using ternary or quarternary alloys such as AlGaAs and InGaAsP.

Dimensionality of the single-electron states is determined by a length scale of an electron wavefunction, *i.e.*, the de Broglie length of electrons, $\lambda_e$, in semiconductors and insulators or the Fermi wavelength, $\lambda_F$, in metals and conductors. When a geometrical-confinement length of the system is smaller than $\lambda_e$ or $\lambda_F$, the electron motion along the direction is confined geometrically and the system dimension decreases by one. In this section, we first overview the dimensionality of the single-electron Bloch states in terms of the density of states. Second, an example of the zone folding and the band mixing effects is given in the electronic structure of Si quantum wires comparing with that of bulk Si. Lastly, we comment on the topology of artificial quantum structures in terms of the control of electronic states.

### 1.2.1. *Density of States of Bloch Electrons*

Consider a quasi-$d$-dimensional (Q$d$D) semiconductor, which consists of a Q$d$D well region surrounded by barrier regions: a bulk in $d = 3$, a QW in $d = 2$, a quantum wire in $d = 1$, and a quantum dot (box) in $d = 0$. The well width is assumed to be $L_{\perp z}$ for $d = 2$, $L_{\perp y}$ and $L_{\perp z}$ for $d = 1$, and $L_{\perp x}$, $L_{\perp y}$ and $L_{\perp z}$ for $d = 0$ systems. These confinement lengths are assumed to be smaller than $\lambda_e$ or $\lambda_F$. Thus the confinement direction(s) is chosen throughout this chapter to be $z$ direction for $d = 2$, $y$ and $z$ directions for $d = 1$, and all the directions for $d = 0$ systems.

The dimensionality in the density of states (DOS) spectrum of the Bloch electrons is one of the familiar properties. The number of quantum states of an electron whose energy lies between $E$ and $E + dE$ is given as $\rho^{(d)}(E)dE$ par a unit volume in $d$ dimension. Here $\rho^{(d)}(E)$ is the DOS in a $d$D crystal, which is explicitly given as

$$\rho^{(3)}(E) = \frac{1}{2\pi^2}(2m^*)^{3/2}E^{1/2}, \tag{1.1}$$

$$\rho^{(2)}(E) = \sum_{n_z}\frac{m^*}{\pi}\Theta(E - E_{n_z}), \tag{1.2}$$

$$\rho^{(1)}(E) = \sum_{n_y}\sum_{n_z}\frac{(2m^*)^{1/2}}{\pi}\left(E - E_{n_y} - E_{n_z}\right)^{-1/2}, \tag{1.3}$$

$$\rho^{(0)}(E) = 2\sum_{n_x}\sum_{n_y}\sum_{n_z}\delta\left(E - E_{n_x} - E_{n_y} - E_{n_z}\right), \tag{1.4}$$

where $m^*$ is an effective mass of the band and the parabolic approximation is used. Here $\Theta(x)$ is the step function and $E_{n_i}$ $(i = x, y, z)$ is an energy of the sublevel along the $i$th confinement direction labeled by the quantum number $n_i = 1, 2, 3, \cdots$.

We find that the DOS in 3D crystals increases by $\sqrt{E}$ as a function of the energy $E$. In the 2D case, step-like DOS appears. The DOS diverges at the band-edge energy $(E = E_{n_y} + E_{n_z})$ in the 1D case, and decreases in proportion with $1/\sqrt{E}$. In the 0D case, the density of states consists of a series of the delta functions. It is important to note that the DOS is finite even at the bottom of the 2D level, whereas it tends towards zero in the 3D system. More drastically, the DOS diverges to infinity at the bottom of the 1D level.

### 1.2.2. *Band Structures of a Quasi-One-Dimensional System: Comparison between a Bulk Crystal and a Wire-Shaped Crystal of Silicon*

Here we study the zone folding and the band mixing effects in a Q1D system to demonstrate drastic change of its electronic, hence optical properties, with the use of an *ab initio* band structure calculation. Consider a crystalline Si wire along the (100) direction, hereafter denoted the $x$-axis (see an inset of Fig. 1.1) [6]. Low-dimensional silicon is one of the most important systems to yield dramatic changes of their electronic and optical properties depending on their dimension. This problem is closely related to the recent topics on "visible light emission from silicon" [7, 8, 9],

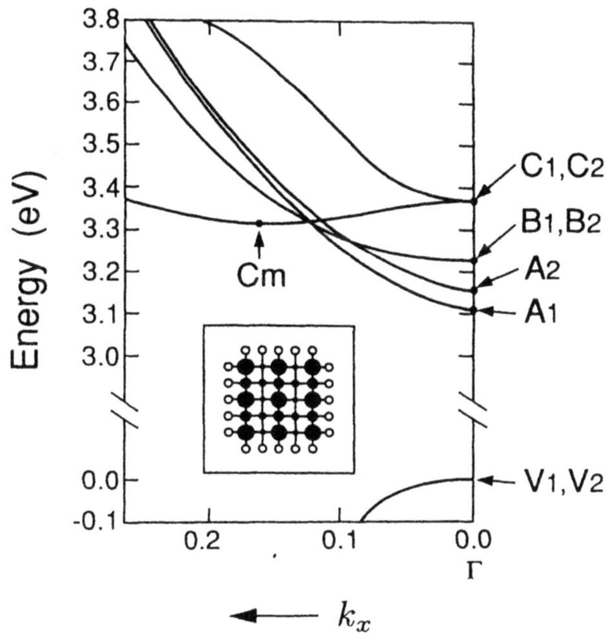


Figure 1.1. Electronic structure near the band gap region for a 3 × 3 Si wire. A plane view of the atomic structure of the 3×3 Si wire is illustrated in the inset. The open (solid) circles describe the hydrogen (silicon) atoms. The electronic structures for the 4 × 4 Si wires have almost the same characteristics as those of the 3 × 3 wire. (After [6]).

which will be extensively discussed in Chapter 5. The Si wires with 3 × 3 and 4 × 4 structures are considered, which have the characteristic diameter $L_\perp$ of 7.7 and 11.5 Å, respectively. Ohno, Shiraishi, and Ogawa [6] evaluated two quantities: the band-gap energy $E_{\rm gap}^{\alpha\beta}$, which is the difference in energy between a conduction band (whose index is $\alpha$) and a valence band $\beta$, and the transition strength $|d_{\alpha\beta}(\vec{\varepsilon})|^2$. Here $d_{\alpha\beta}(\vec{\varepsilon}) \equiv \langle\psi_\alpha|\vec{\varepsilon}\cdot\vec{p}|\psi_\beta\rangle$ is the interband-transition dipole matrix element with polarization unit-vector $\vec{\varepsilon}$ ($\vec{\varepsilon} = \hat{x}, \hat{y}, \hat{z}$) and momentum operator $\vec{p}$, and $|\psi\rangle$ is the band wavefunction. They employ the local-density-functional formalism with *ab initio* norm-conserving pseudopotentials of the Kleinman-Bylander form [10].

Figure 1.1 shows the calculated band structure of the 3×3 Si wire. One of the most important features is that the band gap is direct at the zone center, which is in striking contrast to an indirect gap of a bulk Si. These structures are well explained in terms of the mixing and the zone folding of the bulk Si bands. The four lowest conduction band states, $A_1$, $A_2$, $B_1$, and $B_2$, originate from the four transverse valleys of bulk Si. The conduction band state $C_1$ reaches the minimum at point $C_{\rm m}$, which corresponds to the longitudinal valley of bulk Si. The four transverse valleys of the Si wire are

Table 1.1. The band energy $E_{\text{gap}}^{\alpha\beta}$ (eV) measured from the valence-band-maximum, the effective mass $m^*$ in the unit of free electron mass, and the interband matrix element $\tilde{d}_{\alpha\beta}(\vec{\varepsilon}) \equiv |d_{\alpha\beta}(\vec{\varepsilon})/d_{\text{GaAs}}|$ (where $\vec{\varepsilon} = \hat{x}, \hat{y}, \hat{z}$) relative to the direct transition in bulk GaAs are listed for the 3 × 3 and 4 × 4 Si wires. The wavefunction symmetry is also presented: the $\sigma_0$ symmetry is the identity, $\sigma_y$ ($\sigma_z$) is the mirror symmetry with respect to the $z$–$x$ ($x$–$y$) plane, and $\sigma_{yz}$ equals $\sigma_y\sigma_z$. In the Table, $\hat{\nu}$ stands for $\hat{y}$ or $\hat{z}$.

| Si wire | | 3 × 3 | | | | 4 × 4 | | | |
|---|---|---|---|---|---|---|---|---|---|
| $L_\perp$ | | 7.7 Å | | | | 11.5 Å | | | |
| $\alpha, \beta$ | | $E_{\text{gap}}^{\alpha\beta}$ | $m^*$ | $\tilde{d}_{\alpha\beta}(\hat{x})$ | $\tilde{d}_{\alpha\beta}(\hat{\nu})$ | $E_{\text{gap}}^{\alpha\beta}$ | $m^*$ | $\tilde{d}_{\alpha\beta}(\hat{x})$ | $\tilde{d}_{\alpha\beta}(\hat{\nu})$ |
| $A_1$ | $\sigma_0$ | 3.11 | 0.3 | 0 | 0.01 | 2.47 | 0.3 | 0 | 0.003 |
| $A_2$ | $\sigma_{yz}$ | 3.16 | 0.3 | 0 | 0.01 | 2.52 | 0.3 | 0 | 0.003 |
| $B_1$ | $\sigma_y$ | 3.23 | 1.5 | 0.3 | 0 | 2.54 | 0.6 | 0.09 | 0 |
| $B_2$ | $\sigma_z$ | 3.23 | 1.5 | 0.3 | 0 | 2.54 | 0.6 | 0.09 | 0 |
| $C_m$ | | 3.31 | | | | 2.62 | | | |
| $V_1$ | $\sigma_y$ | 0 | 1.1 | | | 0 | 1.0 | | |
| $V_2$ | $\sigma_z$ | 0 | 1.1 | | | 0 | 1.0 | | |

lower in energy than the two longitudinal valleys, making the band gap direct. This direct band-gap characteristic results from the anisotropic effective mass of a bulk Si. For the conduction band valley of a bulk Si, the longitudinal effective mass is about four times as large as the transverse mass. Therefore the folded transverse valley of the Si wire has a larger effective mass in the lateral direction than the folded longitudinal valley. The difference in effective mass gives the order of the conduction band minima, and then gives the direct band gap of the Si wire, since bands with a larger effective mass exhibit smaller energy shifts due to carrier confinement.

Another important feature of the band structure is that the direct band gap is optically allowed. This is because the quantum confinement and surface scattering in the Si wire result in a finite matrix element $d_{\alpha\beta}(\vec{\varepsilon})$ between the zone-center hole state and the electron state folded from near the zone edge of bulk Si. When the Si wire decreases in width, these effects are significantly enhanced. Thus the enlargement in $d_{\alpha\beta}(\vec{\varepsilon})$ depends strongly on the wire width. The band-gap energy is also enlarged by 1.3 and 1.9 eV with decreasing Si wire width from 11.5 to 7.7 Å. The calculated results of band energy, effective mass, and transition matrix element for Si wires with 3 × 3 and 4 × 4 structures are tabulated in Table 1.1. We should note also that the anisotropy of the transition matrix element appears clearly. This characteristics depends on the crystal direction of the Si wire.[3] As well as the low dimensionality of the Bloch electron states, excitonic effects [11] also play important roles in visible light emission from Si low-dimensional structures (see Sec. 1.3 and Chapter 5).

[3] In the limit of small cross section, the (100) wire structure corresponds to the gauche-helix-type polysilane chain.

### 1.2.3. *Topological Features of Artificial Low-Dimensional Structures*

In designing artificial structures, there is a peculiar guiding principle, called the topology, introduced by Kumagai and Ohno [12]. By topology they mean the topology of the space where the Schrödinger equation is defined. They clarified how the topology of the range of the Schrödinger equation relates to its solutions, *i.e.*, single-electron energy eigenstates. Along their argument, we shall discuss the difference in the energy eigenstates of two types of quantum structures with different topologies. Using the 3D Schrödinger equation, we consider a torus-shaped quantum wire structure with a finite width and thickness, which is referred as $\mathbf{S}_\upsilon$ space, and a straight quantum wire structure with length $L_x$, finite width, and thickness, denoted as $\mathbf{R}_\upsilon$ space.

$\mathbf{R}_\upsilon$ *space* — The Schrödinger equation in $\mathbf{R}_\upsilon$ space can be separated into three independent eigenvalue equations for the rectangular coordinates, $x$, $y$, and $z$. The quantum wire geometry is assumed to be characterized by a wire length $L_x$, a width $L_{\perp y}$, and a thickness $L_{\perp z}$ with an infinite potential barrier. The boundary conditions for the single-electron wavefunction $\psi(x, y, z)$ are

$$\psi(0, y, z) = \psi(L_x, y, z) = 0, \tag{1.5}$$
$$\psi(x, 0, z) = \psi(x, L_{\perp y}, z) = 0, \tag{1.6}$$
$$\psi(x, y, 0) = \psi(x, y, L_{\perp z}) = 0. \tag{1.7}$$

The energy eigenfunction and eigenvalue are then expressed as

$$\psi_{n_x n_y n_z}(x, y, z) = \left(\frac{8}{L_x L_{\perp y} L_{\perp z}}\right)^{1/2} \sin\left(\frac{\pi n_x x}{L_x}\right) \sin\left(\frac{\pi n_y y}{L_{\perp y}}\right) \sin\left(\frac{\pi n_z z}{L_{\perp z}}\right), \tag{1.8}$$
$$E_{n_x n_y n_z} = \frac{1}{2m}\left[\left(\frac{\pi n_x}{L_x}\right)^2 + \left(\frac{\pi n_y}{L_{\perp y}}\right)^2 + \left(\frac{\pi n_z}{L_{\perp z}}\right)^2\right], \tag{1.9}$$

where $n_x$, $n_y$, and $n_z$ are quantum numbers taking positive nonzero integers.

$\mathbf{S}_\upsilon$ *space* — Here we employ the circular cylindrical coordinates, $\rho$, $\phi$, and $z$. The $\mathbf{S}_\upsilon$ space is characterized as a circle with finite width and thickness, defined by two concentric circles of radii $R_1$, $R_2$ ($R_2 > R_1$), and thickness $L_{\perp z}$. The boundary conditions become

$$\psi(R_1, \phi, z) = \psi(R_2, \phi, z) = 0, \tag{1.10}$$
$$\psi(\rho, \phi, z) = \psi(\rho, \phi + 2\pi, z), \tag{1.11}$$
$$\psi(\rho, \phi, 0) = \psi(\rho, \phi, L_{\perp z}) = 0. \tag{1.12}$$

Then the energy eigenfunction and eigenvalue are given as

$$\psi_{n,s,n_z}(\rho, \phi, z) = C\left[J_n(\beta_{ns}\rho) - N_n(\beta_{ns}\rho)\frac{J_n(\beta_{ns}R_2)}{N_n(\beta_{ns}R_2)}\right] e^{in\phi} \sin\left(\frac{\pi n_z z}{L_{\perp z}}\right), \tag{1.13}$$
$$E_{n,s,n_z} = \frac{1}{2m}\left[\beta_{ns}^2 + \left(\frac{\pi n_z}{L_{\perp z}}\right)^2\right], \tag{1.14}$$

where $n = 0, 1, 2, \cdots$, $s = 1, 2, 3, \cdots$, and $n_z = 1, 2, 3, \cdots$ are quantum numbers, $J_n$ and $N_n$ are the $n$th order Bessel functions of the first and second kind, and $\beta_{ns}$ is the $s$th root of the equation: $J_n(\beta R_1)N_n(\beta R_2) - J_n(\beta R_2)N_n(\beta R_1) = 0$. When $R_0 \equiv \sqrt{R_1 R_2} \gg W \equiv R_2 - R_1$,

$$E_{n,s,n_z} \simeq \frac{1}{2m}\left[\frac{n^2}{R_0^2} + \left(\frac{\pi s}{W} - \frac{W}{8\pi s R_0^2}\right)^2 + \left(\frac{nW}{2\pi s R_0^2}\right)^2\left(n^2 - \frac{1}{2}\right) + \left(\frac{\pi n_z}{L_{\perp z}}\right)^2\right]. \quad (1.15)$$

This consists of four terms. The first and second terms are the $n$- and $s$-dependent quantization energies, and the third contains both $n$ and $s$. The fourth term only involves the quantum number $n_z$.

*Comparison between* $\mathbf{R}_v$ *and* $\mathbf{S}_v$ *spaces* — The difference between $\mathbf{R}_v$ and $\mathbf{S}_v$ stems from two factors; one is the fact that $\mathbf{R}_v$ is fundamentally a straight line, whereas $\mathbf{S}_v$ is curved. The other is the connection of the space. $\mathbf{S}_v$ can be made by bending and connecting the ends of $\mathbf{R}_v$. The curving relates to the Euclidean geometric character, while the connection affects the topological character. The topological nature of the artificial structure is related to its boundary condition. There are four main effects of the topological difference between $\mathbf{R}_v$ and $\mathbf{S}_v$.

*(a) The spatial distribution of electrons.* — The $\mathbf{S}_v$ topology yields the eigenfunctions of the form $e^{ikx}$, thus the spatial distribution of electrons is uniform and constant. On the other hand, the $\mathbf{R}_v$ topology leads $\sin(kx)$-type eigenfunctions, so the spatial distribution is $x$-dependent.

*(b) The frozen ground state in* $\mathbf{S}_v$ *space.* — The $e^{ikx}$ eigenfunction becomes a constant when $k = 0$, while the $\sin(kx)$ eigenfunction never become constant. The $k = 0$ state electron in the $\mathbf{S}_v$ topology has no quantization energy. Such a state is called the "frozen state." Thus the ground state of S-type topology becomes the frozen state, which is free from the structural restriction induced zero-point energy.

*(c) The degeneracy of eigenstates.* — All the eigenstates, except $n = 0$, are 2-fold degenerate in $\mathbf{S}_v$, while there is no degeneracy in $\mathbf{R}_v$.

*(d) The energy separation of each eigenstates.* — The energy separation is larger in $\mathbf{S}_v$ case than in $\mathbf{R}_v$ case. This comes from the fact that the $m$th eigenfunction of **S** corresponds to the $2m$th eigenfunction of **R**. Thus $\mathbf{S}_v$ has no eigenstates corresponding to the odd eigenstates of $\mathbf{R}_v$.

As we mentioned, the topological characteristics of artificial low-dimensional structures affects their single-particle electronic structure. Note that conventional periodic boundary conditions in **R** space yield an **S**-type topology, which leads 2-fold degenerate states except for the ground state. Also, **S**-type topology induces a frozen state.

## 1.3. DIMENSIONALITY IN THE WANNIER EXCITON STATES

We turn to problems of the Wannier exciton confined in an arbitrary dimension $d$ under the weak excitation condition.[4] There are two limiting types of exciton: the Wannier exciton and the Frenkel exciton. We here pay attention only to the former, which is a weakly bound state of an electron and a hole. Because spatial extent of the Wannier exciton is much larger than a lattice constant, its wavefunction is more sensitively affected by a spatial geometry than that of the Frenkel excitons. The Wannier exciton has three degrees of freedom: a center-of-mass motion, an electron-hole ($e$-$h$) relative motion, and a spin configuration. The latter two are internal degrees of freedom. In perfect rigid crystals in 3D, the center-of-mass motion is well described by a plane wave with a 3D wave vector $\vec{K}^{(3)} = (K_x, K_y, K_z)$. Wavefunctions of an $e$-$h$ relative motion, which are sensitive to the geometry of the semiconductor structures, determine the optical properties.[5] To simplify the problem, we employ the effective-mass approximation and also the envelope function approximation. Exchange and spin-orbit interactions are neglected here, then only the spin-singlet exciton is considered.

There is an additional length scale characterizing a Coulomb system, called the exciton effective Bohr radius. Hence the dimensionality of exciton systems depends on the ratio of the exciton Bohr radius and the geometrical confinement length; when $L_{\perp i} < a_{\mathrm{B}}^{(d+1)}$ with an effective Bohr radius of Q$(d+1)$D exciton $a_{\mathrm{B}}^{(d+1)}$, the exciton becomes a Q$d$D one. We also assume a semiconductor with a one-photon-allowed direct band gap $E_{\mathrm{gap}}^{(d)}$, which consists of a single conduction band and a single valence band. First, we review fundamental natures of the $d$D Wannier excitons and the one-photon absorption rates. Next, nonlinear optical quantities (*i.e.*, two-photon absorption spectra) are studied in terms of a probe for determining the exciton's dimensionality. An experimental study and other geometrical effects on exciton states are also reviewed in this section.

### 1.3.1. *Excitonic Effects on One-Photon Absorption Processes in d Dimension*

We discuss the $d$D excitonic effects on the one-photon absorption (OPA) spectrum near the band edge. The coupling between electromagnetic radiation and electron

---

[4] In a strong excitation regime where many excitons are created, the exciton-exciton interaction becomes crucial and the Bose-Einstein condensation of excitons is expected at low temperature. Possibility of the exciton condensation in 1D structure will be discussed theoretically afterwards (Sec. 1.5.3.4).

[5] Because an $e$-$h$ relative motion of Frenkel excitons is almost suppressed, dimensionality and geometrical effects are reflected only in the center-of-mass motion, *i.e.*, resonant transfer of the Frenkel exciton. For Wannier excitons, on the other hand, both the $e$-$h$ relative motion and the center-of-mass motion can be affected by the spatial geometry, *i.e.*, the dimensionality. Optical responses of the Wannier excitons are mainly determined by the $e$-$h$ relative motion. Thus we pay special attention to the $e$-$h$ relative motion of Wannier excitons confined in a $d$D space. Geometry and dimensionality of the center-of-mass motion will be discussed in Sec. 1.3.5.

systems is given by the term:

$$\sum_j \left[ -\frac{e}{mc}\vec{p}_j \cdot \vec{A}(\vec{r}_j) + \frac{e^2}{2mc^2}A^2(\vec{r}_j) \right]. \tag{1.16}$$

To describe one-photon processes we only need the term linear in the vector potential $\vec{A}$. We write $\vec{A}$ in terms of photon operators $\hat{a}_{\vec{k}\lambda}$:

$$\vec{A}(\vec{r}) = \sum_{\vec{k}} \sum_{\lambda} \left( \frac{2\pi c}{k\Omega} \right)^{1/2} \left[ e^{i\vec{k}\cdot\vec{r}} \vec{\varepsilon}_{\vec{k}\lambda} \hat{a}_{\vec{k}\lambda} + e^{-i\vec{k}\cdot\vec{r}} \vec{\varepsilon}^{\,*}_{\vec{k}\lambda} \hat{a}^{\dagger}_{\vec{k}\lambda} \right], \tag{1.17}$$

where the sum over $\lambda$ covers the two transverse polarization ($\vec{k}\cdot\vec{\varepsilon}_{\vec{k}\lambda} = 0$), and $\Omega$ is the volume of the quantization box. The probability for a one-photon process is given by the Golden Rule expression:

$$W = 2\pi \sum_f |\langle \Phi_f | \hat{V} | \Phi_i \rangle|^2 \delta(E_f - E_i), \tag{1.18}$$

where $\hat{V}$ is the first term in Eq. (1.16). Consider an absorption process. The initial state $|\Phi_i\rangle$ then contains a photon and can be written as $|\Phi_i\rangle = \hat{a}^{\dagger}_{\vec{k}\lambda}|\Psi_0\rangle$, where $|\Psi_0\rangle$ is the initial state for the system of electrons and nuclei involved. The initial energy is the sum of the photon energy and the energy of the system (electronic and vibrational), $E_i = \omega_k + E_0$. The photon part in Eq. (1.18) can immediately be sorted out, and we have the OPA rate as

$$\begin{aligned} W_{\text{OPA}} &= \frac{2\pi e^2}{m^2c^2} \sum_n \left| \langle \Psi_n | \sum_j \left( \frac{2\pi c}{k\Omega} \right)^{1/2} e^{i\vec{k}\cdot\vec{r}_j} \vec{\varepsilon}_{\vec{k}\lambda} \cdot \vec{p}_j | \Psi_0 \rangle \right|^2 \delta(E_n - E_0 - \omega_k) \\ &= \left( \frac{2\pi e}{m} \right)^2 \frac{1}{\omega_k \Omega} \sum_n \left| \langle \Psi_n | \sum_j e^{i\vec{k}\cdot\vec{r}_j} \vec{\varepsilon}_{\vec{k}\lambda} \cdot \vec{p}_j | \Psi_0 \rangle \right|^2 \delta(E_n - E_0 - \omega_k). \end{aligned} \tag{1.19}$$

Here $E_n$ is the eigenenergy of the state $|\Psi_n\rangle$, and the vibrational part (phonon) is neglected. When we take the envelope function approximation to take into account exciton effects and the subband quantization, for the one-photon-allowed direct transition, the OPA rate is proportional to

$$W^{(d)}_{\text{OPA}} \propto |\langle c|\vec{\varepsilon}\cdot\vec{p}|v\rangle|^2 \sum_i \sum_{\alpha_i,\beta_i} |\langle \phi_{\alpha_i} | \phi_{\beta_i} \rangle|^2 \sum_{\nu} \left| \psi^{(d)}_{\nu;\{\alpha,\beta\}}(\vec{r}_{\parallel} = 0) \right|^2, \tag{1.20}$$

where the first term, $d^{(d)}_{cv}(\vec{\varepsilon}) \equiv \langle c|\vec{\varepsilon}\cdot\vec{p}|v\rangle \neq 0$, is the interband dipole transition matrix element with the momentum operator $\vec{p}$ and the Bloch function of the conduction (valence) band $|c\rangle$ ($|v\rangle$) in $d$ dimension. Here $\vec{\varepsilon}$ is the polarization vector of incident light. The second term, $|\langle \phi_{\alpha_i} | \phi_{\beta_i} \rangle|$, is the subband overlap integral appeared only in the $d \leq 2$ case. The subbands along the $i$th confinement direction in the conduction (valence) band are labeled by an index $\alpha_i$ ($\beta_i$). This overlap integral leads the subband selection rule: $\alpha_i - \beta_i$ = even $\simeq 0$. The $d$D excitonic effects are described by the third term, $|\psi^{(d)}_{\nu;\{\alpha,\beta\}}(\vec{r}_\parallel = 0)|^2$, which is the probability for finding an electron and a hole at the same position. Here $\vec{r}_\parallel$ is the relative-coordinate vector along unconfinement directions. The exciton states where the wavefunction has a node at the origin, *i.e.*, $\psi^{(d)}_{\nu;\{\alpha,\beta\}}(\vec{r}_\parallel = 0) = 0$, are forbidden for one-photon processes. We note that an indirect transition rate is also affected by the exciton wavefunction $|\psi^{(d)}_\nu(\vec{r}_\parallel = 0)|^2$.

### *1.3.1.1. The three-dimensional case*

Total energy of Wannier excitons in $d = 3$ is given by $E^{(3)}_{\vec{K}^{(3)}} + E^{(3)}_\nu$, where $E^{(3)}_{\vec{K}^{(3)}} = |\vec{K}^{(3)}|^2/2(m_e^* + m_h^*)$ is the kinetic energy of the center-of-mass motion and $E^{(3)}_\nu$ is that of the $e$-$h$ relative motion. The latter is an eigenvalue of the Schrödinger equation for wavefunction $\psi^{(3)}_\nu$ of an $e$-$h$ relative motion [13], *i.e.*,

$$\triangle_3 \psi^{(3)}_\nu(\vec{r}) + 2\mu \left[E^{(3)}_\nu - V^{(3)}(\vec{r})\right] \psi^{(3)}_\nu(\vec{r}) = 0, \tag{1.21}$$

with an index $\nu$. Here, $\vec{r} = (x, y, z) \equiv (x_e - x_h, y_e - y_h, z_e - z_h)$ is a relative coordinate of an $e$-$h$ pair, $\mu \equiv m_e^* m_h^*/(m_e^* + m_h^*)$ is the $e$-$h$ reduced mass, $\triangle_d$ is the Laplacian for the relative coordinate in $d$ dimension, and

$$V^{(3)}(\vec{r}) = -\frac{e^2}{\epsilon\,|r|}, \tag{1.22}$$

is the unscreened Coulomb potential between an electron and a hole in a medium with a dielectric constant $\epsilon$. Eigenfunctions of Eq. (1.21) describing bound states, whose eigenenergies are negative ($E^{(3)}_\nu < 0$), are labeled by $\nu = (n, l, m)$ [$n = 1, 2, 3, \cdots; l = 0, 1, 2, \cdots, n-1$; $|m| = 0, 1, 2, \cdots, l$] as

$$\psi^{(3)}_\nu(\vec{r}) = \psi^{(3)}_{nlm}(\vec{r}) = N_{nl}\, F_{nl}\left(\frac{2r}{n a_{\rm B}^{(3)}}\right) Y_{lm}(\theta, \varphi), \tag{1.23}$$

where $\vec{r} = (r, \theta, \varphi)$, $N_{nl}$ is the normalization constant, $a_{\rm B}^{(3)} \equiv \epsilon/\mu e^2$ is the Bohr radius of a 3D exciton, $Y_{lm}(\theta, \varphi)$ is the spherical harmonics, and $F_{nl}(\xi) = \xi^l e^{-\xi/2} L^{2l+1}_{n+l}(\xi)$ with the associated Laguerre polynomials: $L^s_k(\xi) = (d^s/d\xi^s)[e^\xi (d^k/d\xi^k)(\xi^k e^{-\xi})]$. Each bound state is $n^2$-fold degenerated except for the nondegenerate ground state ($n = 1$) and its energy eigenvalue is independent of $l$ and $m$, *i.e.*,

$$E^{(3)}_\nu = E^{(3)}_n = -\frac{E_R^*}{n^2}, \tag{1.24}$$

where $E_R^* = \mu e^4/2\epsilon^2$ is the effective Rydberg energy.

On the other hand, there are scattering (unbound) states with a continuous energy spectrum ($E^{(3)}_\nu \equiv |\vec{k}|^2/2\mu > 0$). These states contribute to the interband optical transition. An explicit form of the wavefunction describing the continuum states is

$$\psi^{(3)}_{klm}(\vec{r}) = N^{(3)}_k \frac{(2ikr)^l}{(2l+1)!} e^{-ikr} F\left(1+l+i|\lambda|, 2(l+1); 2ikr\right) Y_{lm}(\theta,\varphi), \tag{1.25}$$

where $F(a,c;z)$ is the confluent hypergeometric function, $\lambda \equiv -i/(a^{(3)}_{\rm B} k)$, and $N^{(3)}_k$ is the normalization constant.

The OPA strength due to the $d = 3$ exciton bound states is proportional to

$$W^{(3)}_{\rm OPA}(\omega) \propto |\langle c|\vec{\varepsilon}\cdot\vec{p}|v\rangle|^2 \sum_{\nu} \left|\psi^{(3)}_{\nu}(\vec{r}=0)\right|^2 \delta(\omega - E^{(3)}_{\rm gap} - E^{(3)}_{\nu}). \tag{1.26}$$

For $s$-series ($l = m = 0$) excitons, $|\psi^{(3)}_{n00}(0)|^2 = \{\pi[a^{(3)}_{\rm B}]^3 n^3\}^{-1}$. The $1s$ exciton ($n = 1$) induces the strongest OPA. The interband absorption intensity for $\omega > E^{(3)}_{\rm gap}$ is, on the other hand, given as $W^{(3)}_{\rm OPA}(\omega) \propto |\langle c|\vec{\varepsilon}\cdot\vec{p}|v\rangle|^2 (\omega - E^{(3)}_{\rm gap})^{1/2}\Theta(\omega - E^{(3)}_{\rm gap})$ when no excitonic effect is considered. The factor $(\omega - E^{(3)}_{\rm gap})^{1/2}$ results from the joint density of states (JDOS) in $d = 3$ [Eq. (1.1)]. The scattering states of $d = 3$ excitons, Eq. (1.25), modify the interband absorption; the OPA strength *increases* by the factor of [13]

$$S^{(3)}(\omega) = \frac{\pi\alpha_3 e^{\pi\alpha_3}}{\sinh(\pi\alpha_3)}, \tag{1.27}$$

where $\alpha_3 = \alpha_3(\omega) = \{2\mu[a^{(3)}_{\rm B}]^2(\omega - E^{(3)}_{\rm gap})\}^{-1/2}$, as shown in Fig. 1.2. This factor $S^{(d)}(\omega)$ is called the "Sommerfeld factor." The absorption spectrum near the band edge is schematically drawn in Fig. 1.3. For $d = 3$, the excitonic effects are weak but absorption is enhanced at the edge reflecting $S^{(3)}(\omega = E^{(3)}_{\rm gap}) = \infty$.

#### *1.3.1.2. The two-dimensional case*

In $d = 2$ systems, particle motions are restricted along the confinement direction (assumed to be $z$ axis) and each particle forms subbands in a confinement potential (recall $L_{\perp z} < a^{(3)}_{\rm B}$). The $e$-$h$ relative motion of the $d = 2$ exciton is possible only in the $(x, y)$ plane. The $e$-$h$ Coulomb potential may become[6]

$$V^{(2)}(x,y) = -\frac{e^2}{\epsilon}\frac{1}{\sqrt{x^2+y^2}}. \tag{1.28}$$

Exact analytic results [14] are given only in the case of ideal two dimension with a zero (infinitesimally small) thickness ($L_{\perp z} \to 0$). For real two-dimensional materials, which have a finite thickness ($L_{\perp z} > 0$), the excitons exhibit a dimensional-crossover

[6]Although the particles are confined in a $d \leq 2$ space, the electromagnetic field undertaking the Coulomb interaction is assumed to be isotropic in a $d = 3$ space. When the dielectric constant differs between inside and outside of the material, dimensionality of the Coulomb interaction should also be considered, which is called the "dielectric confinement effect" [16, 17] and will be discussed in Sec. 1.4.

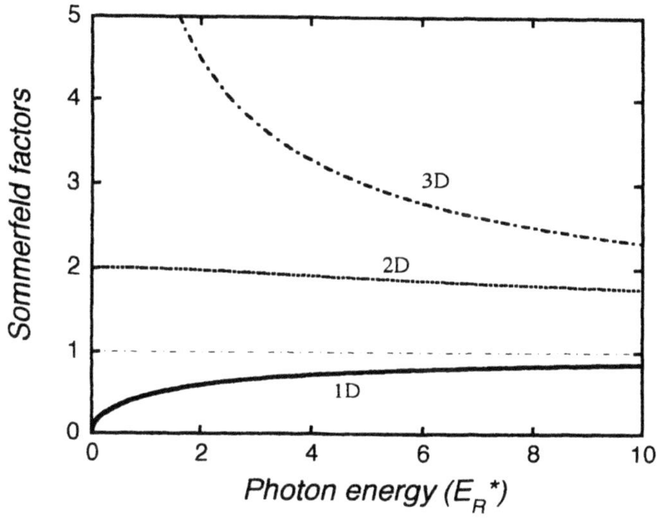


Figure 1.2. Sommerfeld factor for the one-photon-allowed direct transition in the $d = 1$ (Q1D) exciton system is plotted by a solid curve for $x_0/a_B^{(3)} = 0.5$ as a function of the scaled photon energy $(\omega - E_{gap}^{(1)})/E_R^*$. Those for the $d = 3$ (3D) and $d = 2$ (2D) cases are also shown for comparison by dash-dot and broken curves, respectively, where the photon energies are scaled by $(\omega - E_{gap}^{(3)})/E_R^*$ and $(\omega - E_{gap}^{(2)})/E_R^*$.

character between $d = 2$ and $d = 3$. Such excitons have been examined only by the variational approximations [15]. Here we concentrate on the ideal $d = 2$ case.

The energy eigenvalues $(E_\nu^{(2)} < 0)$ and wavefunctions of the bound states, characterized by two quantum numbers $n$ and $m$ $[\nu \equiv (n, m)]$, are

$$E_\nu^{(2)} = E_n^{(2)} = -\frac{E_R^*}{(n - 1/2)^2}, \tag{1.29}$$

$$\psi_\nu^{(2)}(r, \theta) = \psi_{nm}^{(2)}(r, \theta) = N_{nm}^{(2)} \, e^{im\theta - \rho/2} \rho^{|m|} L_{n+|m|-1}^{2|m|}(\rho), \tag{1.30}$$

with $n = 1, 2, 3, \cdots$, $|m| = 0, 1, 2, \cdots, n - 1$, $\rho \equiv 2r/[(n - \frac{1}{2})a_B^{(3)}]$, and $r^2 = x^2 + y^2$. From this, we see that the binding energy of the lowest-energy (1$s$) exciton in $d = 2$ is four times larger than that in $d = 3$, *i.e.*, $E_1^{(2)} = -4E_R^* = 4E_1^{(3)}$. This means that the exciton binding becomes strong quadruply by reducing the spatial dimension by one (3D $\rightarrow$ 2D). Accordingly, the effective Bohr radius of the $d = 2$ exciton is half of $a_B^{(3)}$, *i.e.*, $a_B^{(2)} = \frac{1}{2}a_B^{(3)}$. Wavefunctions describing scattering states with energy

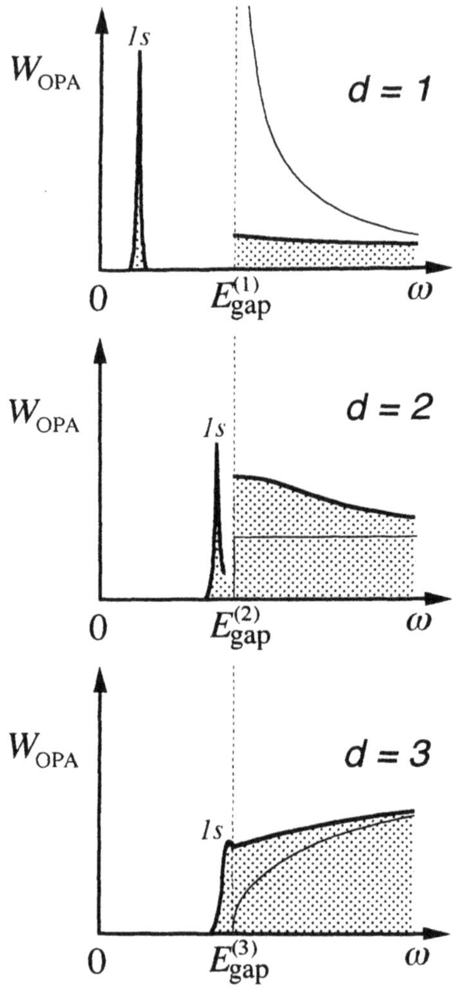


Figure 1.3. Excitonic absorption spectra for $d = 3, 2, 1$ systems are schematically plotted by thick solid curves. Below the band-gap energy, an absorption peak due to only the $1s$ exciton bound state is shown. The thin solid curves represent the joint density of states (JDOS), corresponding to the absorption spectra without excitonic effects.

$E^{(2)}_{\vec{k}^{(2)}} = |\vec{k}^{(2)}|^2/2\mu$ ($\vec{k}^{(2)}$ is the 2D wavevector), on the other hand, is given as

$$\psi^{(2)}_{km}(r,\theta,\phi) = N^{(2)}_k \frac{(2ikr)^{|m|}}{(2|m|)!} e^{-ikr} F\left(|m| + \frac{1}{2} + i|\lambda|, 2|m|+1; 2ikr\right) e^{im\phi}, \quad (1.31)$$

where $F(a,c;z)$ is the confluent hypergeometric function, $\lambda = -i/(a^{(3)}_{\rm B} k)$, and $N^{(2)}_k$ is the normalization constant.

In $d = 2$ the OPA coefficient is proportional to

$$W^{(2)}_{\rm OPA}(\omega) \propto |\langle c|\vec{\varepsilon}\cdot\vec{p}|v\rangle|^2 \sum_{\alpha\beta} |\langle\phi_\alpha|\phi_\beta\rangle|^2 \sum_{\nu} \left|\psi^{(2)}_{\nu;\alpha\beta}(r=0)\right|^2 \delta(\omega - E^{(2)}_{\rm gap} - E^{(2)}_{\nu}). \quad (1.32)$$

The subband overlap integral, $\sum_{\alpha\beta} |\langle\phi_\alpha|\phi_\beta\rangle|^2$, leads to the subband selection rule: $\alpha - \beta = \text{even} \simeq 0$. For $d = 2$ exciton states with $s$ symmetry ($m = 0$), OPA is allowed and $|\psi^{(2)}_{n,m=0;\alpha\beta}(r=0)|^2 = \{\pi[a^{(3)}_{\rm B}]^2(n-\frac{1}{2})^2\}^{-1}$. The oscillator strength of the $1s$ exciton state is $8a^{(3)}_{\rm B}/L_{\perp z}$ times as large as that in $d = 3$ par a single quantum-well layer. Here recall that the JDOS in $d = 2$ is independent of energy [Eq. (1.2)]. Consequently, in the absence of the excitonic effect, the interband absorption spectrum may become step-like, *i.e.*, $W^{(2)}_{\rm OPA}(\omega) \propto |\langle c|\vec{\varepsilon}\cdot\vec{p}|v\rangle|^2 \Theta(\omega - E^{(2)}_{\rm gap})$ with the step function $\Theta(x)$. However, this spectrum is modified by the scattering states of the $d = 2$ excitons [Eq. (1.31)]; the Sommerfeld factor in $d = 2$ systems is

$$S^{(2)}(\omega) = \frac{e^{\pi\alpha_2}}{\cosh(\pi\alpha_2)}, \quad (1.33)$$

where $\alpha_2 = \alpha_2(\omega) = \{2\mu[a^{(3)}_{\rm B}]^2(\omega - E^{(2)}_{\rm gap})\}^{-1/2}$, which is plotted in Fig. 1.2. The OPA absorption rate at the band edge ($\omega = E^{(2)}_{\rm gap}$) is twice as large as that without the excitonic effect, *i.e.*, $S^{(2)}(\omega = E^{(2)}_{\rm gap}) = 2$. The Sommerfeld factors both in $d = 3$ and $d = 2$ are also called the Coulomb enhancement factors because of $S^{(3)}(\omega) > 1$ and $S^{(2)}(\omega) > 1$. The OPA spectrum of the $d = 2$ system is illustrated in Fig. 1.3.

#### 1.3.1.3. *The one-dimensional case*

In one-dimensional ($d = 1$) systems, both the center-of-mass motion and the *e*-*h* relative motion are possible only along the $x$ axis. Bound states of the $d = 1$ hydrogenic system were studied by Loudon and Elliott [18] who assumed the $d = 1$ Coulomb potential to be

$$V^{(1)}(x) = -\frac{e^2}{\epsilon\,|x|}, \quad (1.34)$$

which is called the "bare" Coulomb potential, and they clarified pathological features of the $d = 1$ exciton, *e.g.*, (i) the divergence of the binding energy of the lowest state, (ii) the $\delta$-function-like wavefunction of the lowest state, and (iii) the violation of the nondegeneracy theorem for other bound states. Divergence of the physical quantities results from the singularity around the origin ($|x| = 0$) in $V^{(1)}(x)$.[7] In order to avoid the divergence and to obtain analytical solutions,[8] we employ a regularized potential [17]:

$$\tilde{V}^{(1)}(x; x_0) = -\frac{e^2}{\epsilon} \frac{1}{|x| + x_0}, \tag{1.35}$$

with a cutoff $x_0 \geq 0$. A physical meaning of this cutoff will be discussed afterwards. Then we have wavefunctions of the bound states, *i.e.*,

$$\psi_\nu^{(1)}(x) = N_\nu^{(1)} \tilde{x} e^{-\tilde{x}/2} \Gamma(1+\nu)[F(1-\nu, 2; \tilde{x}) - G(1-\nu, 2; \tilde{x})], \tag{1.36}$$

where $\tilde{x} = 2(|x| + x_0)/[\nu a_{\rm B}^{(3)}]$, $N_\nu^{(1)}$ is the normalization constant, $\Gamma(z)$ is the gamma function, $F(\alpha, \gamma; \tilde{x})$ and $G(\alpha, \gamma; \tilde{x})$ are fundamental solutions of the confluent hypergeometric equation. The energy eigenvalue $E_\nu^{(1)} \equiv -E_R^*/\nu^2$ is determined by $\nu$ which satisfies additional constraints:

$$\psi_\nu^{(1)}(\tilde{x})\Big|_{\tilde{x}=2x_0/\nu a_{\rm B}^{(3)}} = 0 \qquad \text{(even parity state)}, \tag{1.37}$$

$$\frac{d\psi_\nu^{(1)}(\tilde{x})}{d\tilde{x}}\Big|_{\tilde{x}=2x_0/\nu a_{\rm B}^{(3)}} = 0 \qquad \text{(odd parity state)}. \tag{1.38}$$

Figure 1.4 shows the energy of the first three bound states (corresponding to $1s$, $2p$ and $2s$) as a function of the cutoff $x_0$. As $x_0 \to 0$, the energy of the lowest state become negatively infinite, indicating the divergence of the binding energy. This is a peculiar feature of the $d = 1$ exciton. When $x_0 = 0$, other bound states become doubly degenerated for odd and even parities at finite energies $-E_R^*(n-1)^{-2}$ with $n = 2, 3, 4, \cdots$.[9]

These features suggest that the attractive force between an electron and a hole in the $d = 1$ system is more effective than in the $d = 3$ and $d = 2$ systems. Intuitively speaking, in the $d = 3$ and $d = 2$ systems, a particle can move around the origin of the Coulomb potential without directly touching the origin. On the other hand, in the $d = 1$ system, a particle is always moving through the origin because the spatial direction of motion is restricted to one dimension. Thus the particle motion in the $d = 1$ system is affected by the infinitely deep potential at the origin more strongly than in the $d = 3$ and $d = 2$ systems. In mathematical terms this situation can be

[7] Remember that although both $V^{(3)}(\vec{r})$ and $V^{(2)}(x, y)$ also have similar singularities at their origins, there is no divergence difficulties in the exciton problems in $d = 3$ and $d = 2$.

[8] Applying the WKB approximation and Bohr-Sommerfeld quantization rule to the $d = 1$ Coulomb problem leads to erroneous results. For example, the binding energy of the lowest exciton state becomes finite and is four (sixteen) times that in the $d = 2$ ($d = 3$) case. This is specious but incorrect.

[9] As shown in Fig. 1.4, only when $x_0/a_{\rm B}^{(3)} < 0.1$, the $n = 1$ ($1s$) exciton binding energy in $d = 1$, $|E_{n=1}^{(1)}|$, becomes larger than that in ideal $d = 2$, $|E_1^{(2)}| = 4E_R^*$. Therefore, exciton characteristics peculiar to $d = 1$ can be observed only in the regime of $x_0 < 0.1a_{\rm B}^{(3)}$. This also implies that the $d = 1$ treatment of the exciton relative motion is valid only for the small cutoff.

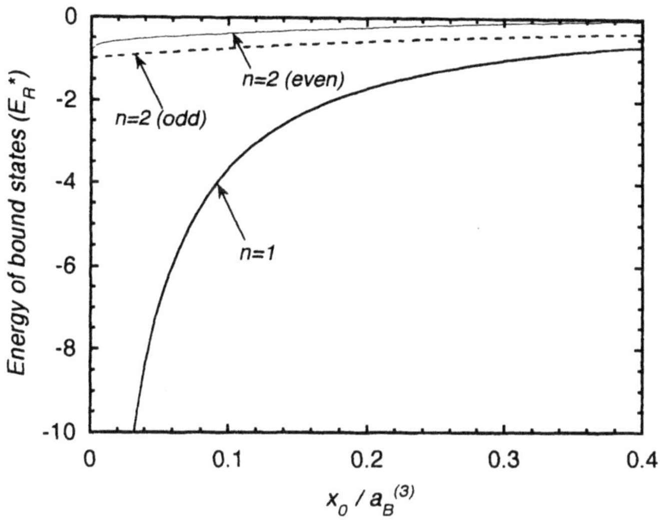


Figure 1.4. Eigenenergy of the lowest ($n = 1$) exciton state (corresponding to the 1$s$ exciton) in $d = 1$ is plotted by a thick solid curve as a function of the cutoff $x_0/a_B^{(3)}$ in the Coulomb interaction. The energy is measured from the bottom of the conduction band. The odd- (2$p$) and even-parity (2$s$) states of the $n = 2$ exciton are shown by dashed and thin solid curves, respectively. In the limit of $x_0 \to 0$, the energy of the $n = 1$ state diverges and the energies of the pair of $n = 2$ states are degenerated at $E^{(1)}_{2,\text{even}} = E^{(1)}_{2,\text{odd}} = -E^*_R$. (After [17]).

understood as follows. The spatial phase volume around the origin is given as $4\pi r^2 dr$, $2\pi r dr$ and $2dr$ for $d = 3$, 2 and 1 systems, respectively, where spherical, circular and linear shell regions from $r$ to $r + dr$ are considered. Then the $1/r$ singularity of the Coulomb potential is removed upon integration around the origin for the $d = 3$ and 2 cases, whereas a logarithmic singularity remains for the $d = 1$ case. Thus the energy and wavefunction of the exciton state in the $d = 1$ system are sensitively dependent on the potential profile near the origin.

A scattering state of the $d = 1$ exciton has an continuous eigenvalue $E^{(1)}_k = |\vec{k}|^2/2\mu > 0$ and doubly degenerated ("gerade" and "ungerade" parities) even when $x_0 \neq 0$. The normalized wavefunctions of the scattering state are exactly calculated by Ogawa and Takagahara [17] as

$$\psi^{(1)g}_k(x) = \left(\frac{e^{\pi\alpha}}{2\pi}\right)^{1/2} \frac{D^{(2)}_0 W^{(1)}(\tilde{x}) - D^{(1)}_0 W^{(2)}(\tilde{x})}{\left(|D^{(1)}_0|^2 + |D^{(2)}_0|^2\right)^{1/2}}, \tag{1.39}$$

$$\psi_k^{(1)u}(x) = \text{sgn}\,(x)\left(\frac{e^{\pi\alpha}}{2\pi}\right)^{1/2} \frac{W_0^{(2)}W^{(1)}(\tilde{x}) - W_0^{(1)}W^{(2)}(\tilde{x})}{\left(|W_0^{(1)}|^2 + |W_0^{(2)}|^2\right)^{1/2}}, \tag{1.40}$$

where $\psi_k^{(1)g}(x)$ and $\psi_k^{(1)u}(x)$ are wavefunctions with gerade (even) and ungerade (odd) parities, respectively. Here $\alpha \equiv \{2\mu[a_{\rm B}^{(3)}]^2(E_k^{(1)} - E_{\rm gap}^{(1)})\}^{-1/2}$, $W_0^{(j)} \equiv W^{(j)}(\tilde{x} = 2ikx_0)$, $D_0^{(j)} \equiv dW^{(j)}(\tilde{x})/d\tilde{x}|_{\tilde{x}=2ikx_0}$, and

$$W^{(j)}(x) \equiv \tilde{x}e^{-\tilde{x}/2}\Gamma(1 \pm i\alpha)[F(1 + i\alpha, 2; \tilde{x}) \pm G(1 + i\alpha, 2; \tilde{x})], \tag{1.41}$$

for $j = 1$ (2) corresponding to the + (−) sign of the right hand side. One of these ($\psi_k^{(1)g}$) is optically allowed.

Now we shall comment on the cutoff $x_0$. When an electron and a hole are confined perfectly in a rectangular wire whose lateral size is specified by $L_{\perp y}$ and $L_{\perp z}$, the lowest subband envelope functions have the form of $\phi_{\alpha_y}(y_e) = \sqrt{2/L_{\perp y}}\cos(\pi y_e/L_{\perp y})$ for $|y_e| \leq \frac{1}{2}L_{\perp y}$. Here, the subband in the conduction (valence) band along a confinement direction ($x_\varsigma = y$ or $z$) is labeled by $\alpha_{x_\varsigma}$ ($\beta_{x_\varsigma}$). Then an effective 1D potential is calculated by

$$U^{(1)}(x_e - x_h) \equiv \int dy_e dy_h dz_e dz_h\, V^{(3)}(\vec{r}_e, \vec{r}_h)\, w(y_e, y_h, z_e, z_h), \tag{1.42}$$

where $V^{(3)}$ is given in Eq. (1.22) and $w$ is a weight function reflecting the subband structures: $w(y_e, y_h, z_e, z_h) = \left|\phi_{\alpha_y}(y_e)\phi_{\beta_y}(y_h)\phi_{\alpha_z}(z_e)\phi_{\beta_z}(z_h)\right|^2$. Numerical results [17, 19] of this integration is nicely fitted by $\tilde{V}^{(1)}(x; x_0)$. Then, the cutoff $x_0$ is found to be proportional to the lateral size of the wire (*e.g.*, $x_0 \simeq 0.18L_{\perp y} = 0.18L_{\perp z}$ for a square cross-section). Validity of the $d = 1$ treatment is discussed here.[10] From Fig. 1.4, $x_0/a_{\rm B}^{(3)} < 0.1$ is roughly required for justifying the $d = 1$ exciton model. This means that the wire cross section should be small in order to make the exciton relative motion be $d = 1$. That is, only for a square wire with $L_\perp/a_{\rm B}^{(3)} < \frac{0.1}{0.18} \sim 0.56$, the $d = 1$ exciton binding energy is larger than that in ideal $d = 2$ and then the $d = 1$ treatment is valid.

The OPA coefficient in $d = 1$ is proportional to

$$W_{\rm OPA}^{(1)}(\omega) \propto |\langle c|\vec{\varepsilon}\cdot\vec{p}|v\rangle|^2 \sum_{i=y,z}\sum_{\alpha_i,\beta_i} |\langle\phi_{\alpha_i}|\phi_{\beta_i}\rangle|^2 \times \sum_{\nu}\left|\psi_{\nu;\{\alpha,\beta\}}^{(1)}(x=0)\right|^2 \delta(\omega - E_{\rm gap}^{(1)} - E_\nu^{(1)}). \tag{1.43}$$

Here we assume that $\phi_{\{\alpha\}}$ and $\phi_{\{\beta\}}$ take decoupled forms such as $\phi_{\{\alpha\}} \simeq \prod_{i=y,z}\phi_{\alpha_i}(i)$. From this, the subband selection rules become $\alpha_y - \beta_y = \text{even} \simeq 0$ and $\alpha_z - \beta_z =$

[10] A rough condition, $x_0/a_{\rm B}^{(3)} < 0.1$, comes from the comparison with the *ideal* $d = 2$ case. For more rigorous discussion, the binding energy should be compared with the Q2D system with a finite thickness because a nonzero cutoff $x_0 \neq 0$ means the finite wire cross section. When $x_0/a_{\rm B}^{(3)} > 0.35$, however, $|E_{n=1}^{(1)}|$ is smaller than $|E_{1s}^{(3)}| = E_R^*$. The $d = 1$ model is obviously invalid in this region.

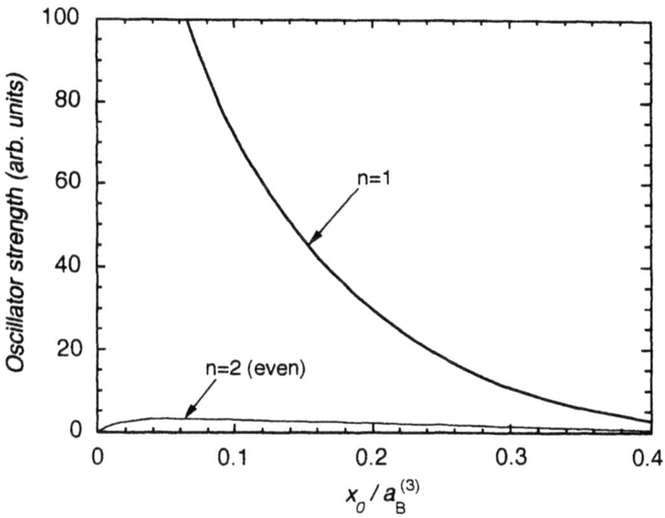


Figure 1.5. The oscillator strength of the lowest ($n = 1$) exciton ($1s$) state and the $n = 2$ even-parity exciton ($2s$) state in $d = 1$ is plotted by thick and thin solid curves, respectively, as a function of the cutoff $x_0/a_{\rm B}^{(3)}$. Strong concentration of the oscillator strength on the $n = 1$ exciton state is clearly seen. The oscillator strength of other even-parity exciton states is very small and completely vanishes in the limit of $x_0 \to 0$. The oscillator strength of all the odd-parity exciton states is always zero and irrespective of the cutoff. (After [17]).

even $\simeq 0$. The one-photon transition is allowed only for the bound states with an even parity, $|\psi^{(1)}_{\nu;\{\alpha,\beta\}}(x=0)| \neq 0$. The transition to the lowest exciton ($1s$ state) is strongest, which is the same as in $d = 3$ and $d = 2$. However, dependence of the transition rate on the cutoff $x_0$ is peculiar to one dimension; as $x_0 \to 0$, *i.e.*, the cutoff Coulomb potential $\tilde{V}^{(1)}(x; x_0)$ approaches the bare Coulomb potential $V^{(1)}(x)$, $|\psi^{(1)}_{1s;\{\alpha,\beta\}}(x=0)|^2$ diverges, which is in striking contrast to the $d = 3$ and $d = 2$ cases. The oscillator strength ($f^{(1)} \propto |\psi^{(1)}_{\nu;\{\alpha,\beta\}}(x=0)|^2$) of the first two even-parity exciton states is plotted in Fig. 1.5 as a function of the cutoff $x_0$ to illustrate the anomalously strong concentration of the oscillator strength on the lowest-energy bound state. The oscillator strength of other bound states with an even parity vanishes completely in the limit of $x_0 \to 0$. Thus, in $d = 1$ the OPA of only the lowest exciton state is extremely (divergingly) strong in comparison with the other bound states.

The interplay between effects of the $d = 1$ excitons and of the divergence of the JDOS in $d = 1$ is also of interest in the interband absorption spectrum. When no $e$-$h$ attraction is taken into account, the interband absorption spectrum may show an asymmetric peak just above the band edge due to the JDOS, *i.e.*, $W^{(1)}_{\rm OPA}(\omega) \propto$

$(\omega - E_{\text{gap}}^{(1)})^{-1/2}\Theta(\omega - E_{\text{gap}}^{(1)})$ [see Eq. (1.3)]. However, this inverse-power-law peak at the band edge does not appear in the optical spectrum when the $d = 1$ excitonic effects take into account. Ogawa and Takagahara [17] found that the $d = 1$ exciton *decreases* the interband OPA strength by the factor of

$$S_{x_0}^{(1)}(\omega) = \frac{e^{\pi\alpha_1}}{8}\frac{|D_0^{(2)}W_0^{(1)} - D_0^{(1)}W_0^{(2)}|^2}{|D_0^{(1)}|^2 + |D_0^{(2)}|^2} < 1, \tag{1.44}$$

which is plotted for $x_0/a_{\text{B}}^{(3)} = 0.5$ in Fig. 1.2. Here $\alpha_1 = \alpha_1(\omega) \equiv \{2\mu[a_{\text{B}}^{(3)}]^2(\omega - E_{\text{gap}}^{(1)})\}^{-1/2}$. The Sommerfeld factor in $d = 1$ is less than unity, which means that the $e$-$h$ Coulomb attraction suppresses the OPA of an allowed interband transition. At $\omega = E_{\text{gap}}^{(1)}$, *i.e.*, $\alpha_1 \to +\infty$, the Sommerfeld factor vanishes, $S_{x_0}^{(1)}(\omega = E_{\text{gap}}^{(1)}) = 0$, which cancels the divergence due to the JDOS. More surprisingly, in the limit of $x_0 \to 0$, *no* interband absorption takes place, *i.e.*, $S_{x_0=0}^{(1)}(\omega) = 0$ for $\forall\omega \geq E_{\text{gap}}^{(1)}$.

These anomalous results can be understood qualitatively by considering the absorption due to the lowest ($1s$) exciton state below $E_{\text{gap}}^{(1)}$. As pointed above, the oscillator strength of the $1s$ exciton state becomes very large and almost the entire oscillator strength concentrates on this bound state [17, 20, 21]. Then as a consequence of the $f$-sum rule, the oscillator strength of the interband transition becomes very weak. This is a remarkable feature of the purely $d = 1$ system in striking contrast to the $d = 3$ and $d = 2$ systems, where the $f$-sum rule is seemingly violated in the region near the band edge.[11] The OPA spectrum of the $d = 1$ system with a finite $x_0$ is plotted schematically in Fig. 1.3.

The above characteristics in 1D Wannier excitons are independent of an effective range of the $e$-$h$ interaction, that is, irrespective of the effective range of the $e$-$h$ potential. Actually, if we take the short-range $e$-$h$ interaction in the form of [17]

$$V_{\text{short}}^{(1)}(x) = -V_0 \operatorname{sech}^2\left(\frac{x}{\zeta}\right), \tag{1.45}$$

[11] In the $d = 3$ and $d = 2$ systems, excitonic effects not only induce absorption lines (due to the exciton bound states) below the band edge but also enhance the interband absorption ($S^{(3)} > 1$ and $S^{(2)} > 1$). In the $d = 1$ system, on the other hand, oscillator strength for the interband transition above the band edge approaches zero asymptotically when the $1s$-exciton absorption becomes stronger as $x_0 \to 0$. Therefore, the $f$-sum rule is seemingly invalid (valid) in the $d = 3$ and 2 ($d = 1$) systems. Unfortunately, the reason why this difference arises between the $d = 3, 2$ and $d = 1$ systems is not yet clarified. One possible reason lies in the effective mass approximation and the choice of the total wavefunction describing OPA-allowed excited states extended over the system. The total wavefunction of excited states is written as a superposition of the single-electron ground state and the excited states. Taking into account the $e$-$h$ correlation, however, the single-electron band picture is no longer valid. In the effective mass and the envelope function approximations, the exciton wavefunction and the band wavefunction are assumed to be decoupled, and the $e$-$h$ correlation affects only the exciton part not the band structure. This treatment is justified only near the band edge. Far from the band edge region, these approximations become crucially bad and the calculated absorption rates are incorrect. Therefore the $f$-sum rule, which requires a summation over *all* the transitions, seems to be violated in the present calculations. In this sense, the $f$-sum rule may be violated also in the $d = 1$ system.

which is specified by the strength $V_0$ (depth of the potential) and the effective interaction length $\zeta$ (width of the potential), the envelope function for unbound states of energy $E_k^{(1)} = k^2/2\mu > 0$ are written as a linear combination of two basis:

$$f_k(\xi) = (1-\xi^2)^{-ik\zeta/2} F\left(-ik\zeta - s, -ik\zeta + s + 1, 1 - ik\zeta; \frac{1-\xi}{2}\right), \tag{1.46}$$

$$g_k(\xi) = (1-\xi^2)^{-ik\zeta/2} \left(\frac{1-\xi}{2}\right)^{ik\zeta} F\left(-s, s+1, 1+ik\zeta; \frac{1-\xi}{2}\right), \tag{1.47}$$

where $\xi = \tanh(x/\zeta)$, $F(a,b,c;z)$ is the hypergeometric function, and

$$s \equiv \frac{1}{2}\left[-1 + \sqrt{1 + 4(V_0/E_R^*)(\zeta/a_B^{(3)})^2}\right]. \tag{1.48}$$

Also in this case, the Sommerfeld factor becomes

$$S_{\text{short}}^{(1)}(\omega) = \frac{1}{2}\frac{|g_k'(0)f_k(0) - f_k'(0)g_k(0)|^2}{|f_k'(0)|^2 + |g_k'(0)|^2} < 1, \tag{1.49}$$

which is also less than unity. Here a prime stands for the derivative of Eqs. (1.46) and (1.47) with respect to $\xi$. A major difference from the long-range case appears in the behavior at the band edge. When the parameter $s$ is an even integer, $s = 2, 4, 6, \ldots$, the absorption intensity near the band edge diverges to infinity, while it becomes zero for other cases, that is, the Sommerfeld factor approaches a finite nonzero constant for $s = 2, 4, 6, \cdots$, while it vanishes at the band edge for $s \neq 2, 4, 6, \cdots$;

$$S_{\text{short}}^{(1)}(\omega \to E_{\text{gap}}^{(1)}) \longrightarrow \begin{cases} \text{positive constants} & s = 2,4,6,\cdots \\ 0 & s \neq 2,4,6,\cdots \end{cases}. \tag{1.50}$$

In the case of the short-range potential, the number of the bound states become *finite* in contrast to the long-range case. The number is given as $[s]$, where $[\ ]$ is Gauss's symbol. Therefore, there is no quasicontinuum region in the absorption spectra just below the band edge. That is, there exists an "energy gap" between the band edge and the highest-energy bound state unless $s$ is an integer. Thus the absorption line is discontinuous near the band edge, resulting in the zero absorption at the edge. When $s = 2, 4, 6, \cdots$, however, the corresponding highest bound state ($n = 2, 4, 6, \cdots$), which is direct allowed, is degenerate to the band edge. Then the absorption spectra show the divergence there.

#### *1.3.1.4. The zero-dimensional case*

In fine particle systems [22], there are two limiting situations [23] according to the ratio of the characteristic length $R$ representing the system size to the effective Bohr radius $a_B^{(3)}$ of the exciton in the bulk material. According to the numerical calculation [23], crossover between two regimes occurs around $2 < R/a_B^{(3)} < 4$.

*(a) Exciton confinement regime:* $R/a_{\rm B}^{(3)} \gg 1$. — In this regime, the character of the exciton as a quasiparticle is conserved and the relative motion of the exciton is less restricted by the boundary condition. On the other hand, the center-of-mass motion suffers from the quantum confinement and its kinetic energy increases slightly. Taking into account the spatial extent of the *e*-*h* relative motion, total energy of the lowest exciton state is given by

$$E_{\rm total}^{(0)} \sim -E_R^* + \frac{\pi^2}{2(m_e^* + m_h^*)[R - \eta(\sigma)a_{\rm B}^{(3)}]^2}, \tag{1.51}$$

where $\sigma = m_h^*/m_e^*$ and $\eta(\sigma) \approx 1$ is an increasing function of $\sigma$ of order of unity. The term $\eta(\sigma)a_{\rm B}^{(3)}$ corresponds to the dead layer: The center of mass of the exciton cannot reach the surface because it requires a strong deformation in the *e*-*h* relative motion. If the exciton Bohr radius is sufficiently small (*e.g.*, Frenkel excitons), this dead-layer correction is unnecessary.

*(b) Individual particle confinement regime:* $R/a_{\rm B}^{(3)} \ll 1$. — In this limit, an electron and a hole occupy primarily an individual lowest eigenstate in a confinement potential with relatively little spatial correlation. The wavefunction of the *e*-*h* relative motion may be squeezed due to the strong confinement. Because the kinetic energy of each particle becomes very large in comparison with the Coulomb energy, no bound state (exciton) is formed. An asymptotic wavefunction of the ground state in a sphere structure is given by

$$\psi^{(0)}(\vec{r}_e, \vec{r}_h) \sim N^{(0)}\left(1 - \alpha|\vec{r}_e - \vec{r}_h|\right) j_0\left(\frac{\pi r_e}{R}\right) j_0\left(\frac{\pi r_h}{R}\right), \tag{1.52}$$

where $N^{(0)}$ is the normalization constant, $j_0(x)$ is a spherical Bessel function of the 0th order and $\alpha = 0.498/a_{\rm B}^{(3)}$. Its total energy is

$$E_{\rm total}^{(0)} \sim \frac{1}{2\mu}\frac{\pi^2}{R^2} - 1.786\frac{e^2}{\epsilon R} - 0.248E_R^*. \tag{1.53}$$

The third term stands for the remnant of the exciton effect.

In the exciton confinement regime $[R/a_{\rm B}^{(3)} \gg 1]$, there are two types of excited states of the exciton; one is the states where the *e*-*h* relative motion is in the 1*s* (lowest) state and the center-of-mass motion is excited (characterized by a quantum number $N$) in a confinement potential. The other is that the *e*-*h* relative motion is excited to the 2*s*, 3*s*, $\cdots$ states. In the $d = 0$ spherical systems with a finite size, all the *s*-like states is one-photon allowed in principle because of the breakdown of the translational invariance. For example, the 1*s* exciton with the $N$th excited state of the center-of-mass motion has the oscillator strength per unit volume as

$$f_{1s;N}^{(0)} = f_{1s}^{(3)}\left[1 - \eta(\sigma)\frac{a_{\rm B}^{(3)}}{R}\right]^3 \frac{6}{\pi^2 N^2}, \tag{1.54}$$

where $N = 1, 2, 3, \cdots$ and $f_{1s}^{(3)}$ is the oscillator strength per unit volume of the $1s$ exciton in bulk material [23]. Quantization of the center-of-mass motion leads to a small blue shift of the exciton's total energies [the second term of the right-hand-side of Eq. (1.51)]. In the individual particle confinement regime ($R/a_{\mathrm{B}}^{(3)} \ll 1$), an electron (a hole) forms subbands in all the confinement directions. Then, the OPA spectra consist of discrete lines, each corresponding to a pair of subbands which satisfy the selection rules: $\alpha_i - \beta_i = \text{even} \simeq 0$ with $i = x, y, z$ in a cubic geometry.

### *1.3.2. Excitonic Effects on Two-Photon Absorption Processes in d Dimension*

In the OPA process, as mentioned above, the polarization dependence appears only in the interband matrix element, $|\langle c|\vec{\varepsilon}\cdot\vec{p}|v\rangle|$. Therefore, OPA anisotropy for the polarization directions reflects only that of the Bloch (band) electrons not of the excitons [24]. On the other hand, the two-photon absorption (TPA) process reflects directly the dimensionality of excitons, *i.e.*, squeezing of the wavefunction of the exciton's relative motion itself in a low-dimensional geometry. In this subsection, we shall introduce a unified theory of the TPA in an arbitrary dimension ($d = 3, 2, 1, 0$), developed by Shimizu, Ogawa, and Sakaki [25, 26], and discuss the polarization anisotropy of the TPA comparing with that of the OPA process.

A general expression of the TPA rate is proportional to

$$W_{\mathrm{TPA}} \propto \left| \sum_f \sum_n \left[ \frac{\langle f|\hat{V}_2|n\rangle\langle n|\hat{V}_1|i\rangle}{E_n - \omega_1} + (1 \leftrightarrow 2) \right] \right|^2 , \tag{1.55}$$

where $|i\rangle$ ($|f\rangle$) is an initial (final) state of an exciton (electron) system, $\hat{V}_j$ ($j = 1$, 2) is the Hamiltonian describing the interaction between the $j$th light beam and the exciton (electron) system, and $E_n$ is the energy of the $n$th intermediate state $|n\rangle$. We shall consider the situation [25, 27] where the photon energies $\omega_1$ and $\omega_2$ of incident light beams are both close to $\frac{1}{2}E_{\mathrm{gap}}^{(d)}$ and these two beams have a common polarization vector $\vec{\varepsilon}$. The complete summation over all the intermediate states is performed by approximating off-resonant energy denominators as a constant. Mahan [28] showed that resulting errors are indeed negligible.

#### *1.3.2.1. The three-dimensional case*

The TPA rate in $d = 3$ is proportional to

$$W_{\mathrm{TPA}}^{(3)}(\omega_1 + \omega_2; \vec{\varepsilon} \parallel \hat{x}_\xi) \propto \frac{|\langle c|\vec{\varepsilon}\cdot\vec{p}|v\rangle|^2}{[\mu E_{\mathrm{gap}}^{(3)}]^2} \sum_\nu \left| \frac{\partial}{\partial x_\xi} \psi_\nu^{(3)}(\vec{r}) \right|_{\vec{r}=0}^2 \times \delta(\omega_1 + \omega_2 - E_{\mathrm{gap}}^{(3)} - E_\nu^{(3)}), \tag{1.56}$$

where $x_\xi$ is a relative coordinate parallel to $\vec{\varepsilon}$. When $d = 3$, there is no confinement direction. Hence, if the bulk band structure is isotropic, the TPA spectra are also isotropic. The lowest exciton peak is the $2p$ exciton, whose oscillator strength is much smaller than those of the lowest excitons for $d = 1$ and 2.

### *1.3.2.2. The two-dimensional case*

When $d = 2$ the TPA spectra strongly depend on the polarization direction $\vec{\varepsilon}$. For $\vec{\varepsilon} \parallel \hat{x}$ or $\hat{y}$ (an unconfinement direction $\hat{x}_\xi$), the TPA rate is given by

$$W^{(2)}_{\mathrm{TPA}}(\omega_1 + \omega_2; \vec{\varepsilon} \parallel \hat{x}_\xi) \propto \frac{|\langle c|\vec{\varepsilon}\cdot\vec{p}|v\rangle|^2}{[\mu_\xi E^{(2)}_{\mathrm{gap}}]^2} G^{(2)}(\vec{\varepsilon} \parallel \hat{x}_\xi), \tag{1.57}$$

where $G^{(2)}$ is called an envelope-function part of the TPA rate:

$$G^{(2)}(\vec{\varepsilon} \parallel \hat{x}_\xi) = \sum_{\alpha\beta} |\langle\phi_\alpha|\phi_\beta\rangle|^2 \sum_\nu \left| \frac{\partial}{\partial x_\xi} \psi^{(2)}_{\nu;\alpha\beta}(x,y) \right|^2_{x=y=0} \delta(\omega_1 + \omega_2 - E^{(2)}_{\mathrm{gap}} - E^{(2)}_\nu). \tag{1.58}$$

Here $\mu_\xi$ is the *e-h* reduced mass for the motion along an unconfinement direction and $x_\xi$ ($= x$ or $y$) is a relative coordinate parallel to $\vec{\varepsilon}$. We then find the subband selection rule: $\alpha - \beta = \text{even} \simeq 0$. Exciton states with $m = \pm 1$ (so-called $p$ series) are TPA-allowed along an unconfinement direction. To see the most fundamental features coming from the $G^{(2)}$ part, we assume common band parameter for any $d$; typical band parameters of the $c$ and light-hole ($lh$) valence bands of GaAs/$Al_{0.45}Ga_{0.55}As$ QW are employed. Figure 1.6(a) shows the $G^{(2)}(\vec{\varepsilon} \parallel \hat{x}_\xi)$ part of the TPA spectrum.

On the other hand, when $\vec{\varepsilon} \parallel \hat{z}$ (the confinement direction), the TPA rate becomes

$$W^{(2)}_{\mathrm{TPA}}(\omega_1 + \omega_2; \vec{\varepsilon} \parallel \hat{z}) \propto \frac{|\langle c|\hat{z}\cdot\vec{p}|v\rangle|^2}{[\mu_z E^{(2)}_{\mathrm{gap}}]^2} G^{(2)}(\vec{\varepsilon} \parallel \hat{z}), \tag{1.59}$$

with

$$G^{(2)}(\vec{\varepsilon} \parallel \hat{z}) = \sum_{\alpha\beta} \left| \left\langle \phi_\alpha \middle| \frac{\partial}{\partial z} \middle| \phi_\beta \right\rangle \right|^2 \sum_\nu \left| \psi^{(2)}_{\nu;\alpha\beta}(x = y = 0) \right|^2 \delta(\omega_1 + \omega_2 - E^{(2)}_{\mathrm{gap}} - E^{(2)}_\nu), \tag{1.60}$$

where $\mu_z$ is the *e-h* reduced mass for the motion along the confinement direction ($\hat{z}$). We find the subband selection rule: $\alpha - \beta = \text{odd} \simeq \pm 1$. For $m = 0$ excitons ($s$ series), $\psi^{(2)}_{\nu;\alpha\beta}(x = y = 0)$ is nonzero and then TPA-allowed along the confinement direction, as plotted in Fig. 1.6(b). Note that the lowest exciton peak for $\vec{\varepsilon} \parallel \hat{z}$ is much stronger than that of bulk crystal, despite the decreased volume density of optically active regions.

We here comment on the physical meaning of Eqs. (1.57) and (1.59). In the first term, $|\langle c|\vec{\varepsilon}\cdot\vec{p}|v\rangle|^2$, the parity is inverted. The TPA process requires an additional inversion of the parity for the polarization direction [see Eq. (1.55)]. The residual part of the TPA rate, which contains $\phi_\alpha\phi_\beta$ and $\psi^{(2)}_{\nu;\alpha\beta}$, undertakes this additional inversion of the parity (parity recovery). For an unconfinement direction ($\vec{\varepsilon} \parallel \hat{x}_\xi$), only the $\psi^{(2)}_{\nu;\alpha\beta}$ part can invert the parity, while $\phi_\alpha\phi_\beta$ can do so only for the confinement direction ($\vec{\varepsilon} \parallel \hat{z}$). Therefore, $|\langle\phi_\alpha|\phi_\beta\rangle|^2 |(\partial/\partial x_\xi)\psi^{(2)}_{\nu;\alpha\beta}(0,0)|^2$ appears for $\vec{\varepsilon} \parallel \hat{x}_\xi$, and $|\langle\phi_\alpha|(\partial/\partial z)|\phi_\beta\rangle|^2 |\psi^{(2)}_{\nu;\alpha\beta}(0,0)|^2$ for $\vec{\varepsilon} \parallel \hat{z}$.

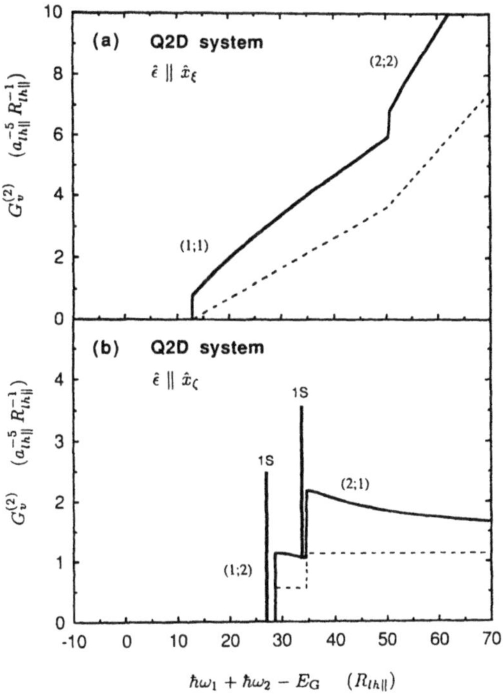


Figure 1.6. Envelope-function part $G^{(2)}(\vec{\varepsilon})$ of a Q2D structure for the polarization $\vec{\varepsilon}$ parallel to (a) an unconfinement direction ($\vec{\varepsilon} \parallel \hat{x}_\xi$) and (b) the confinement direction ($\vec{\varepsilon} \parallel \hat{x}_\zeta = \hat{z}$). In (a) $2p$ exciton peaks are invisible in this scale, and quasicontinuous spectra just below the continuum ones are not shown. Origins of the TPA are shown by the subband indices $(\alpha, \beta)$ and the exciton quantum numbers $n$ and $m$. The bulk band-gap energy is denoted as $E_G$. The dashed lines represent $G^{(2)}$ in the absence of exciton effects. (After [25]).

### 1.3.2.3. *The one-dimensional case*

The anisotropic features become more drastic when $d = 1$. When $\vec{\varepsilon} \parallel \hat{x}$ (the unconfinement direction), the TPA rate is proportional to

$$W_{\rm TPA}^{(1)}(\omega_1 + \omega_2; \vec{\varepsilon} \parallel \hat{x}) \propto \frac{|\langle c|\hat{x} \cdot \vec{p}|v\rangle|^2}{[\mu_x E_{\rm gap}^{(1)}]^2} G^{(1)}(\vec{\varepsilon} \parallel \hat{x}), \tag{1.61}$$

where

$$G^{(1)}(\vec{\varepsilon} \parallel \hat{x}) = \sum_{i=y,z} \sum_{\alpha_i,\beta_i} |\langle \phi_{\alpha_i} | \phi_{\beta_i} \rangle|^2 \sum_\nu \left| \frac{\partial}{\partial x} \psi^{(1)}_{\nu;\{\alpha,\beta\}}(x) \right|^2_{x=0} \delta(\omega_1 + \omega_2 - E_{\rm gap}^{(1)} - E_\nu^{(1)}), \tag{1.62}$$

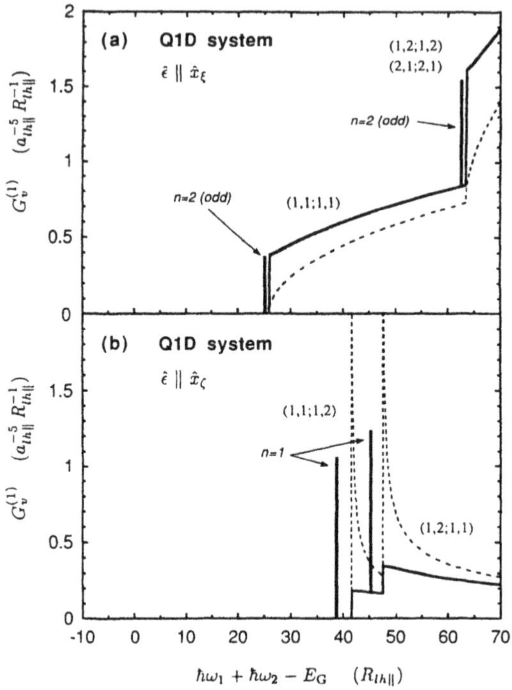


Figure 1.7. Envelope-function part $G^{(1)}(\vec{\varepsilon})$ of a Q1D structure for the polarization $\vec{\varepsilon}$ parallel to (a) the unconfinement direction ($\vec{\varepsilon} \parallel \hat{x}_\xi = \hat{x}$) and (b) a confinement direction ($\vec{\varepsilon} \parallel \hat{x}_\zeta$). In (b) discrete absorption lines due to the $n = 2$ even excitons are invisible. Origins of the TPA are shown by the subband indices ($\{\alpha\}; \{\beta\}$) and the 1D exciton quantum number $n$ and parity. In (b) $\vec{\varepsilon} \parallel \hat{z}$ is assumed. The bulk band-gap energy is denoted as $E_G$. The dashed lines represent $G^{(1)}$ in the absence of exciton effects. (After [25]).

as plotted in Fig. 1.7(a). Here we find the subband selection rule: $\alpha_i - \beta_i = \text{even} \simeq 0$ for $i = y$ or $z$. The derivative of the exciton wavefunction was given from Eq. (1.36), which is nonzero only for odd-parity states.

On the other hand, when $\vec{\varepsilon} \parallel \hat{y}$ or $\hat{z}$ (a confinement direction $\hat{x}_\zeta$), the TPA coefficient is proportional to

$$W_{\text{TPA}}^{(1)}(\omega_1 + \omega_2; \vec{\varepsilon} \parallel \hat{x}_\zeta) \propto \frac{|\langle c|\vec{\varepsilon} \cdot \vec{p}|v\rangle|^2}{[\mu_\zeta E_{\text{gap}}^{(1)}]^2} G^{(1)}(\vec{\varepsilon} \parallel \hat{x}_\zeta), \tag{1.63}$$

with

$$G^{(1)}(\vec{\varepsilon} \parallel \hat{x}_\zeta) = \sum_{\alpha_\zeta, \beta_\zeta} \left| \left\langle \phi_{\alpha_\zeta} \left| \frac{\partial}{\partial x_\zeta} \right| \phi_{\beta_\zeta} \right\rangle \right|^2 \sum_{\alpha_{\zeta'}, \beta_{\zeta'}} \left| \langle \phi_{\alpha_{\zeta'}} | \phi_{\beta_{\zeta'}} \rangle \right|^2$$

$$\times \sum_{\nu} \left| \psi^{(1)}_{\nu;\{\alpha,\beta\}}(x=0) \right|^2 \delta(\omega_1 + \omega_2 - E^{(1)}_{\text{gap}} - E^{(1)}_{\nu}), \tag{1.64}$$

where $\mu_\zeta$ is the $e$-$h$ reduced mass for the motion along the confinement direction ($\hat{x}_\zeta = \hat{y}$ or $\hat{z}$). The subband selection rules are $\alpha_\zeta - \beta_\zeta$ = odd $\simeq \pm 1$ for the polarization direction ($\hat{x}_\zeta \parallel \vec{\varepsilon}$) and $\alpha_{\zeta'} - \beta_{\zeta'}$ = even $\simeq 0$ for the other direction ($\hat{x}_{\zeta'}$). Only for even-parity excitons, $\psi^{(1)}_{\nu;\{\alpha,\beta\}}(x=0)$ is nonzero [see Eq. (1.37)], as shown in Fig. 1.7(b).

Comparing the $d = 1$ case with the $d = 2$ and $d = 3$ cases, the relative magnitude of the discrete spectra to the continuous spectra becomes larger for $d = 1$. This is due to increased oscillator strength for the lowest exciton state in the $d = 1$ system. In the regions of continuous spectra, the exciton effects enhance the TPA for $\vec{\varepsilon} \parallel \hat{x}$ (the unconfinement direction), whereas they reduce the TPA for $\vec{\varepsilon} \parallel \hat{x}_\zeta$ (a confinement direction). By contrast, the TPA is always enhanced by excitonic effects for $d \geq 2$. As a result, in the case of the TPA in $d = 1$, one can observe both the reduction (for $\vec{\varepsilon} \parallel \hat{x}$) and enhancement (for $\vec{\varepsilon} \parallel \hat{x}_\zeta$) in the same sample simply by rotating the polarization direction.

#### *1.3.2.4. The zero-dimensional case*

In the individual particle confinement regime ($R/a^{(3)}_{\text{B}} \ll 1$), an electron (also a hole) forms subbands in all the confinement directions. Then, excitonic effects are negligible as compared with strong effects of subband quantization. The TPA spectra consist of discrete lines as similar to the OPA spectra, each line corresponding to a pair of subbands which satisfies the selection rules: $\alpha_\zeta - \beta_\zeta$ = odd $\simeq \pm 1$ for the polarization direction and $\alpha_{\zeta'} - \beta_{\zeta'}$ = even $\simeq 0$ for the other directions in a cubic geometry.

#### *1.3.2.5. Origins of the TPA anisotropy*

The anisotropy of the TPA spectra results from the anisotropic subband selection rules, the squeezed wavefunction of the exciton relative motion, and the interband matrix element, which is in striking contrast to the OPA spectra whose anisotropy comes only from the interband matrix element. We shall summarize here the differences between the OPA and the TPA in terms of the polarization dependence.

*(a) The subband selection rule.* — The subband selection rule for the OPA is independent of the polarization direction, while for the TPA, different rules are derived depending on the polarization direction. Using TPA, we can observe transitions to the subband which are forbidden for the OPA process.
*(b) The parity of exciton wavefunction.* — For the TPA, excitons with either the $s$-type symmetry (allowed also for the OPA) or with the $p$-type symmetry (forbidden for the OPA) can be observed by controlling the polarization direction. If the polarization direction is parallel to the free direction (along the exciton's relative motion), excitons with odd-parity relative wavefunction (mainly $p$ symmetry) are detectable in TPA process, while excitons with even-parity relative wavefunction (mainly $s$ symmetry) can be observable for the polarization direction parallel to the confinement direction.

*(c) The anisotropic Bloch functions.* — The matrix element of the interband dipole transition,

$$\langle c|\vec{\varepsilon}\cdot\vec{p}|v\hat{k}\rangle = \frac{1}{v_0}\int_{\rm cell} d^3r\, u_c^*(\vec{r})\,\frac{\hbar}{i}\nabla u_{v\hat{k}}(\vec{r}), \tag{1.65}$$

depends generally on the polarization direction. Here $u_c$ and $u_{v\hat{k}}$ are, respectively, the cell-periodic parts of the $\vec{k}\to 0$ Bloch functions of the electron and hole in $d$ dimension. In III-V semiconductor compounds in general, $u_{v\vec{k}}$ strongly depends on the direction of $\vec{k}$, *i.e.*, $u_{v\vec{k}}\to u_{v\hat{k}}$ as $\vec{k}\to 0$. In GaAs QW systems, for example, the exciton wavefunction is a highly anisotropic superposition of the Bloch functions, so that only $u_{v\hat{k}}$ for $\hat{k}=\hat{z}$ (the confinement direction) has appeared in the above formulae. We define as usual $M^2 = |\langle c|p_i|v\hat{k}\rangle|^2$ averaged over $\hat{k}$. Considering the mixing of heavy-hole ($hh$) and light-hole ($lh$) bands, the interband matrix element is given in a QW (Q2D) by $|\langle c|\hat{p}_z|v\hat{z}\rangle|^2 = q_z^v M^2$ and $|\langle c|\hat{p}_x|v\hat{z}\rangle|^2 = q_x^v M^2$, respectively, where $q_z^{hh}=0$, $q_z^{lh}=2$, $q_x^{hh}=\frac{3}{2}$, $q_x^{lh}=\frac{1}{2}$ [27, 29]. In the quantum wire structure, on the other hand, several authors investigated the anisotropy of the Bloch states. The results depend still on the numerical scheme and the model employed [30, 31, 32, 33].

One understands now that the shape and the anisotropy of the TPA spectra depend drastically on the dimension $d$ both of the Bloch electrons and excitons. Hence, the dimensionality of excitons can be probed directly by the TPA spectroscopy.[12] Anisotropies of the OPA spectra cannot be an evidence of the exciton's quantum confinements. Finally, we comment on an application. The TPA rate is proportional to $\mathrm{Im}\,\chi^{(3)}$. In the $\omega\sim\frac{1}{2}E_{\rm gap}^{(d)}$ region, $\mathrm{Im}\,\chi^{(3)}$ is connected to $\mathrm{Re}\,\chi^{(3)}$ via the Kramers-Kronig-like relation. At the photon energy just below the TPA edge, $\mathrm{Re}\,\chi^{(3)}(\omega;\omega,-\omega,\omega)$ becomes substantially large. Because the OPA is also negligibly small in this region, this large $\mathrm{Re}\,\chi^{(3)}$ may lead to various applications and devices. Note that the nonlinear susceptibility $\chi^{(3)}$ and its numerical evaluation are discussed in detail in Chapter 2.

### *1.3.3. A Two-Photon Absorption Experiment in GaAs Quantum Wires*

We here introduce an experimental study of TPA in GaAs/AlGaAs quantum wires by Cingolani *et al.* [34]. The investigated quantum wire array has been fabricated by holographic patterning and subsequent reactive etching of a molecular-beam-epitaxy-grown MQW heterostructure. The QW structure consisted of 25 GaAs wells of width $L_{\perp z}=103$ Å and $Al_{0.34}Ga_{0.66}As$ barriers of width $L_{\rm barrier}=148$ Å. The processed QW structure resulted in a regular array of quantum wires with 280-nm periodicity and a crystalline wire width $L_{\perp y}=60\pm 5$ nm. The nonlinear absorption is studied by measuring the TPA-induced photoluminescence excitation (PLE) spectra at 10 K for different polarization directions of the exciting laser beam. The detection energy

[12] Note that we can obtain the formula for the TPA rate in the absence of excitonic effects by making the substitution $\psi_v^{(d)}(\vec{r}_\parallel)\to L^{-d/2}\exp(i\vec{k}_\parallel\cdot\vec{r}_\parallel)$ in the above equations. Here $L$ is a normalization length along an unconfinement direction.

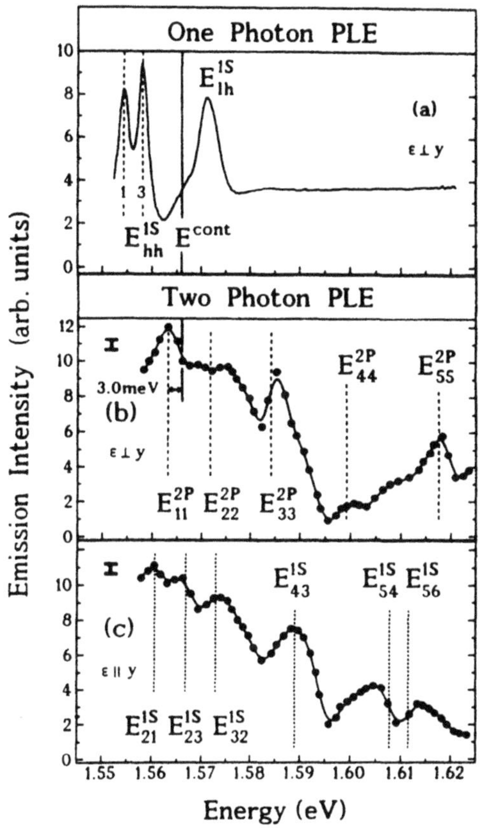


Figure 1.8. (a) Linear photoluminescence excitation (PLE) spectrum of the quantum wire array at $T = 10$ K. The detection energy is set at the low-energy tail of the fundamental quantum-wire exciton state (around 1.552 eV). $E^{\rm cont}$ indicates the energy of the exciton continuum. (b) Two-photon-absorption-induced photoluminescence excitation spectrum (TPA-PLE) of the quantum wires measured in the $\vec{\varepsilon} \parallel \hat{x}$ configuration at $T = 10$ K. The vertical lines indicate the calculated energy positions of the $2p$ exciton states associated with $\Delta n_y = 0$ (*i.e.*, $\alpha_y - \beta_y = 0$) transition. The 3.0 meV splitting indicates the estimated binding energy of the $2p$ exciton state of the $n_y = 1$ exciton in the quantum wire. The vertical error bar deriving from signal-to-noise ratio and slope-2 regression on experimental points is indicated. The horizontal error bar coincides with the solid dots. (c) The same as (b) but in the $\vec{\varepsilon} \parallel \hat{y}$ configuration. The vertical lines stand for the calculated energy positions of the $1s$ exciton states associated with $\Delta n_y = \pm 1$ (*i.e.*, $\alpha_y - \beta_y = \pm 1$) transitions. For both TPA spectra the detection energy is set at around 1.554 eV. (After [34]).

is set at the fundamental quantum-wire exciton state ($E^{(1)}_{1s;\{1,1;1,1\}} = 1.554$ eV). The quadratic behavior of the detected luminescence signal versus excitation intensity is checked as each experimental point and several measurement runs are carried out at all exciting wavelengths.

As a result of the relatively large width of the investigated quantum wires, the quantum effects induced by the lateral confinement along $y$ direction are expected to be different under different excitation conditions. In the rather wide quantum wires the lateral confinement leads to a quantization of the excitonic center-of-mass motion, which leaves the internal structure of the exciton unchanged and therefore 2D-like. Therefore the linear PLE profile of Fig. 1.8(a) still maintains some 2D character and does not exhibit pure 1D states. Depending on the wire width $L_{\perp y}$ one can find 1D excitons if $L_{\perp y} < a_{\mathrm{B}}^{(3)}$ and 2D excitons if $L_{\perp y} \gg a_{\mathrm{B}}^{(3)}$. In the sample, roughly $L_{\perp y} \simeq 4a_{\mathrm{B}}^{(3)}$; hence, $1s$ excitons behave almost like 2D excitons, while $2p$ excitons, having a larger spatial extent of the relative motion, should exhibit an almost 1D character.

In the $\vec{\varepsilon} \parallel \hat{x}$ configuration [Fig. 1.8(b)], we observe five structures whose relative energy splittings are the same as those observed for the $\Delta n_y = 0$ (*i.e.*, $\alpha_y - \beta_y = 0$) transitions in the photoluminescence (PL) spectra under a high-photoexcitation condition, *i.e.*, they directly reflect the 1D subband spacing. Furthermore, the first resonance of the TPA-PLE spectrum is about 3 meV below the onset of the continuum in Fig. 1.8(a). This confirms that indeed $2p$ excitonic states are observed. The spectral positions of the higher-energy transitions (which are indicated by dashed lines) have been obtained by adding the 1D subband spacing to the transitions labeled $E^{(1)}_{2p;\{1,1;1,1\}}$. In the $\vec{\varepsilon} \parallel \hat{y}$ configuration, on the other hand, the TPA-PLE spectrum appears to be even more structured, and exhibits several transjtions. The vertical lines of Fig. 1.8(c) represent the energy positions of the $1s$ exciton states associated with $\Delta n_y = \pm 1$ (*i.e.*, $\alpha_y - \beta_y = \pm 1$) transitions in the quantum wire. These are calculated by adding the net confinement energies of the 1D conduction and valence subbands to the $E^{(1)}_{1s;\{1,1;1,1\}}$ exciton energy. Reasonable agreement between the experimental data points and the calculated transitions demonstrates that in this configuration $1s$ states with $\Delta n_y = \pm 1$ (*i.e.*, $\alpha_y - \beta_y = \pm 1$) are allowed. However, the TPA-PLE spectrum exhibits rather broad peaks, which come from the dimensional crossover effects of excitons (see next subsection).

### 1.3.4. *Dimensional Crossover of Excitons*

Here the dimensional crossover between $d$ and $d \pm 1$ is considered. Ogawa and Shimizu [35] investigated theoretically the OPA and the TPA of Wannier excitons confined in an intermediate dimension $1 < d < 2$, where the dimensionality of the excitons is intermediate between $d = 2$ and $d = 1$. The intermediate exciton is described as a hybrid state between $d = 1$ exciton states and $d = 2$ exciton states with a quantized center-of-mass motion (QCMM). We shall take the $z$ axis along the confinement direction of the QW, and the $y$ axis along the lateral confinement direction, with the $x$ axis being along the free motion. The lateral width $L_{\perp y}$ of the wire sample is assumed to be several times $a_B^{(2)}$, where $a_B^{(1)} < a_B^{(2)} < a_B^{(3)}$ in general. In accordance with the experiment [34], the photon energy is $\omega \simeq \frac{1}{2}E_{gap}$. The cross section of the wire is assumed to be rectangular, $L_{\perp y} \times L_{\perp z}$. Since we are interested in the crossover between $d = 1$ and $d = 2$, we vary $L_{\perp y}$ as a parameter, while $L_{\perp z}$ is fixed and small. That is, considering that energy is inversely proportional to the square of the length scale, we assume that

$$L_{\perp z}^2 \ll L_{\perp y}^2 \quad \text{and} \quad a^2 \ll L_{\perp z}^2 \ll [a_B^{(2)}]^2, \tag{1.66}$$

where $a$ is a lattice constant. Taking GaAs into consideration as an example, we denote the cell-periodic part of the Bloch function $u$ as $u^{(d)}$ to stress the dimensionality of the GaAs valence band structure. The $z$-direction confinement causes the valence band mixing, and Q2D subbands are formed. The cell-periodic part $u^{(3)}$ of the Bloch functions (around the $\Gamma$ point) is altered into a Q2D form, $u^{(2)}$. We then take account of the lateral ($y$) confinement and the exciton effect. Under condition Eq. (1.66), we find $u^{(3)} \simeq u^{(2)}$ because energy separation between Q2D subbands is much larger than these perturbations. On the other hand, these perturbations cause state mixing and correlation within each pair of Q2D conduction and valance subbands. Therefore, we have a marginal situation that $u$'s are of Q2D forms while the envelope functions are of lower-dimensional forms. This allows us to treat each pair of Q2D conduction and valance subbands separately, and the overall spectrum is just the sum over the pairs.

#### *1.3.4.1. A Q2D exciton with a quantized center-of-mass motion*

We first consider a Q2D exciton with a QCMM, which we call a Q2′D exciton. Its envelope function is

$$\Psi_{nmN}^{(2')}(\vec{r}_e, \vec{r}_h; [\alpha_z; \beta_z]) \equiv G_N(Y)\, \psi_{nm}^{(2)}(x, y; [\alpha_z; \beta_z])\, \phi_{\alpha_z}(z_e)\, \phi_{\beta_z}(z_h), \tag{1.67}$$

where $\psi_{nm}^{(2)}(x, y; [\alpha_z; \beta_z])$ is given in Eq. (1.30) [14, 27] associated with the $\alpha_z$th conduction and the $\beta_z$th valence subbands. The QCMM is represented by a normalized function $G_N(Y)$ with a quantum number $N = 1, 2, \cdots$, where $Y$ is the $y$-coordinate of the center-of-mass. A subband envelope-function in the $\alpha_z$th conduction ($\beta_z$th valence) band is written as $\phi_{\alpha_z}$ ($\phi_{\beta_z}$).

We find that the OPA or TPA rate, $W^{(2')}$, of a Q2′D exciton is simply given by[13]

$$W^{(2')} = \frac{1}{L_{\perp y}} \left| \int G_N(Y) dY \right|^2 W^{(2)}\{2'\}, \tag{1.68}$$

where $W^{(2)}\{2'\}$ denotes the OPA or TPA rate obtained from $W^{(2)}$ of a Q2D exciton [27] by replacing its Q2D band parameters with the corresponding Q2′D ones. We

[13]One might wonder why a term including a derivative of $G_N$ does not appear in the TPA rate. Intuitively, this is because such a term should be proportional to the square of the integral $\int [dG_N(Y)/dY] dY$, which is zero since it just gives the difference between two boundary values of $G_N$, both of which are zero.

see that the selection rules are those for the Q2D excitons [27] and, additionally, $N = 1, 3, 5, \cdots$. More importantly, the polarization anisotropy in the $xy$ plane can only come from a possible anisotropy of Q2′D effective masses (which results in an anisotropy in the Q2D envelope-function part [25]) and/or of Q2′D Bloch functions (which results in anisotropy in the interband matrix element). However, these Q2′D band parameters under the condition (1.66) approximately equal the Q2D band parameters, which are isotropic in the $xy$ plane. Therefore, when the condition (1.66) is satisfied, the Q2′D exciton does not lead to significant $xy$ anisotropy in either the OPA or TPA spectrum.

#### *1.3.4.2. Exciton states in an intermediate dimension*

Let us consider exciton states in an intermediate dimension, which is called Q1-2D excitons. One notices that the wire structure is invariant under $x \to -x$, and under $y \to -y$. This allows us to classify Q1-2D exciton states using corresponding parities, $\sigma_x = \pm 1$ and $\sigma_y = \pm 1$. We note that the Q1D envelope function,

$$\Psi^{(1)}_{n\sigma}(\vec{r}_e, \vec{r}_h; [\alpha_y, \alpha_z; \beta_y, \beta_z]) = \psi^{(1)}_{n\sigma}(x; [\alpha_y, \alpha_z; \beta_y, \beta_z])\, \phi_{\alpha_y}(y_e)\, \phi_{\beta_y}(y_h)\, \phi_{\alpha_z}(z_e)\, \phi_{\beta_z}(z_h), \tag{1.69}$$

also has definite parities of $\sigma_x = \sigma$ and $\sigma_y = (-1)^{\alpha_y + \beta_y}$. Here, $\psi^{(1)}_{n\sigma}(x; [\alpha_y, \alpha_z; \beta_y, \beta_z])$ is, as given in Eq. (1.36), the Q1D exciton relative motion associated with the $(\alpha_y, \beta_y)$ and $(\alpha_z, \beta_z)$ subband pairs. We will simply call $\{n = 1, \sigma = +1\}$ and $\{n = 2, \sigma = -1\}$ states "$1S^{(1)}$" and "$2P^{(1)}$" states, respectively. We can also arrange the Q2′D states of Eq. (1.67) to have definite parities $(\sigma_x \sigma_y)$, by taking a linear combination of $m = +|m|$ and $m = -|m|$ states. For example, linear combinations of $\{n = 2, m = \pm 1, N = 1\}$ states yield $\{n = 2, |m| = 1, N = 1\}$ states of $(\sigma_x \sigma_y) = (+-)$ and $(-+)$, which we will call $\{2P^{(2')}_y : N = 1\}$ and $\{2P^{(2')}_x : N = 1\}$ states, respectively. A schematic energy diagram of low-lying levels of the Q1D and Q2′D states is shown in Fig. 1.9, where $(\sigma_x \sigma_y)$ of each state is indicated.

We may approximate a Q1-2D envelope function of $(\sigma_x \sigma_y)$ parities by a linear combination of the Q1D and Q2D states of the same parities, that is,

$$\Psi^{(1-2)}_{\sigma_x \sigma_y} \simeq \sum_{\text{same parity}} C^{(1)} \Psi^{(1)} + \sum_{\text{same parity}} C^{(2')} \Psi^{(2')}, \tag{1.70}$$

where the sums are taken over the states of the same parities. Here, we have assumed that the band parameters can be approximated by the Q2D band parameters. To determine the superposition coefficients $C^{(1)}$ and $C^{(2')}$, we have to solve the matrix equation obtained by substituting Eq. (1.70) into the Schrödinger equation and taking the inner-product with each of $\Psi^{(1)}$ and $\Psi^{(2')}$ [35].

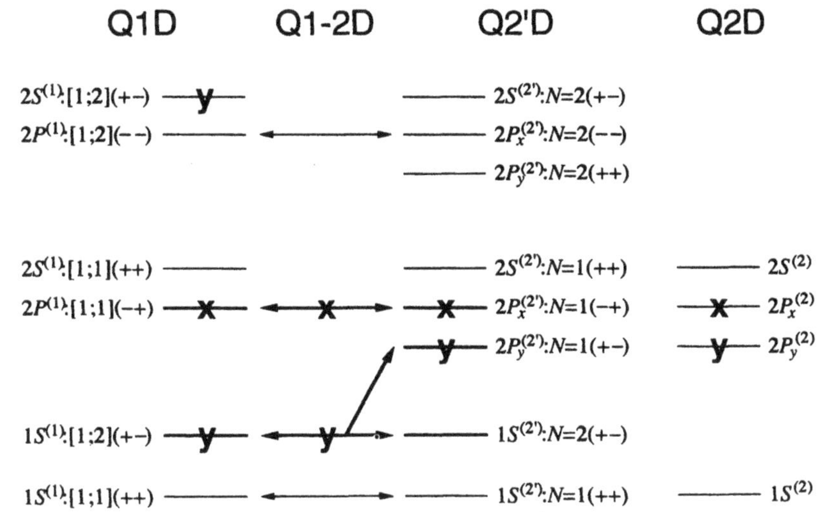


Figure 1.9. Schematic level diagram of low-lying exciton states of Q1D, Q1-2D, Q2′D, and Q2D semiconductors. The $z$-subband indices $\alpha_z = \beta_z = 1$ are assumed and suppressed. The TPA activity is indicated by the bold letters **x** and **y**, corresponding to $\vec{\varepsilon} \parallel \hat{x}$ and $\vec{\varepsilon} \parallel \hat{y}$ polarization, respectively. (After [35]).

### *1.3.4.3. TPA spectra in an intermediate dimension*

Using Eq. (1.70), the OPA and TPA rates $W^{(1-2)}$ can be evaluated and can be cast in the form

$$W^{(1-2)} \simeq \left| \sum_{\text{same parity}} C^{(1)} w^{(1)}\{2\} + \sum_{\text{same parity}} C^{(2')} w^{(2')}\{2\} \right|^2 , \tag{1.71}$$

where $|w^{(1)}\{2\}|^2$ and $|w^{(2')}\{2\}|^2$ are the OPA or TPA rate of a Q1D exciton [25] and of a Q2′D exciton in Eq. (1.68), respectively, and in both of them, the band parameters are replaced with the Q2D band parameters.

An immediate conclusion of Eq. (1.71) is that the OPA spectrum of the Q1-2D excitons, under condition (1.66), is almost isotropic in the $xy$ plane. This is because, in the case of OPA, $w^{(1)}_{\text{OPA}}\{2\}$ and $w^{(2')}_{\text{OPA}}\{2\}$ are totally isotropic. Moreover, the OPA oscillator strength varies only slowly as a function of $L_{\perp y}$. Hence the total oscillator strength becomes insensitive to the state mixing of Eq. (1.70). As a result, the OPA spectrum in the intermediate regime becomes similar to the Q2D spectrum, *i.e.*, we can observe neither significant anisotropy nor sharp dependence on $L_{\perp y}$. Therefore, the OPA spectroscopy is a poor probe of the dimensionality in this regime. The situation is very different for the TPA spectrum, which varies drastically, in both anisotropy and oscillator strength, as a function of $L_{\perp y}$. The strong dependence of the $xy$ anisotropy on $L_{\perp y}$ appears because, in the case of TPA, $w_{\rm TPA}^{(1)}\{2\}$ becomes strongly anisotropic whereas $w_{\rm TPA}^{(2')}\{2\}$ is isotropic. Therefore, the TPA spectrum becomes very anisotropic for small $L_{\perp y}$ (for which $C^{(1)}$'s become large), and almost isotropic for large $L_{\perp y}$ (where $C^{(2')}$'s become dominant).

*(a) In the Q1D regime.* — The TPA edge spectrum for $\vec{\varepsilon} \parallel \hat{x}$ is composed of $2P^{(1)}$ excitons of $[\alpha_y;\beta_y] = [1;1]$ subband pairs, whereas that for $\vec{\varepsilon} \parallel \hat{y}$ is composed of $1S^{(1)}$ excitons of $[1;2]$ subband pairs. In Fig. 1.9 these states are indicated by the bold letters **x** and **y**, which denote that the state contributes to the TPA spectrum for $\vec{\varepsilon} \parallel \hat{x}$ and for $\vec{\varepsilon} \parallel \hat{y}$, respectively.
*(b) In the Q2′D regime.* — On the other hand, the TPA edge spectrum for $\vec{\varepsilon} \parallel \hat{x}$ is composed of $2P_x^{(2')}$ excitons of $N = 1$ QCMM subband, whereas that for $\vec{\varepsilon} \parallel \hat{y}$ is composed of $2P_y^{(2')}$ excitons of the same QCMM subband. These states, as well as the corresponding Q2D states for reference, are indicated in Fig. 1.9 by **x** or **y**.
*(c) In the Q1-2D regime.* — We note that a hybridization of Eq. (1.70) occurs between states with similar energies. Hence, the Q1-2D excitons can be obtained by the superposition indicated by the arrows in Fig. 1.9. For example, the Q1-2D exciton of the lowest energy of $(\sigma_x\sigma_y) = (-+)$ parities would be a linear combination of the $\{2P_x^{(2')} : N = 1\}$ Q2′D exciton and the $\{2P^{(1)} : [1;1]\}$ Q1D exciton. Since both components have **x** activity, with oscillator strengths of similar magnitude, this state exhibits a TPA peak only for $\vec{\varepsilon} \parallel \hat{x}$, whose strength varies slowly with $L_{\perp y}$. On the other hand, the lowest exciton of $(+-)$ parities would be a linear combination of the $\{1S^{(1)} : [1;2]\}$ exciton (with **y** activity), the $\{1S^{(2')} : N = 2\}$ exciton (no activity), and the $\{2P_y^{(2')} : N = 1\}$ exciton (**y** activity). In contrast to the $(-+)$ state, this $(+-)$ state exhibits a TPA peak only for $\vec{\varepsilon} \parallel \hat{y}$. As a result, the total spectrum exhibits a strong $xy$ anisotropy, in agreement with the experiment [34].[14]

Although the above discussions concern quantum wires, similar crossover effects will also occur in QWs and quantum boxes when they are not small enough. Moreover, imperfections in their structures can induce a similar effect because it causes state mixing [24]. These effects will distort the pure Q*d*D TPA spectrum in [36], depending on the nature of a sample, while the OPA spectrum will be almost unchanged.

### 1.3.5. *Geometrical Effects on the Exciton Center-of-Mass Motion*

A main chain of conjugated polymers usually shows some globally deformed structures deviating from straight lines [37]. Tokihiro and Hanamura [38] investigated the problem of how much the optical responses of these systems are influenced by their geometrical structures. Here we shall introduce briefly their argument on

[14]Note that our argument relies completely on the excitonic effects. In fact, if we had ignored the *e-h* Coulomb attraction, the length scale $a_{\rm B}^{(d)}$ would not have appeared, and no qualitative change would have resulted when $L_{\perp y}$ varied.

the Wannier excitons. The Schrödinger equation for a Wannier exciton is written in terms of the center-of-mass coordinate $X$ and the relative coordinate $x$ as $\hat{H}(X,x)\Psi^{(1)}(X,x) = E^{(1)}_{\text{total}}\Psi^{(1)}(X,x)$. Here

$$\hat{H}(X,x) = -\frac{1}{2M}\frac{\partial^2}{\partial X^2} - \frac{1}{8\mu}[\kappa(X)]^2 - \frac{1}{2\mu}\frac{\partial^2}{\partial x^2} + V^{(1)}(X,x), \tag{1.72}$$

where $M = m_e^* + m_h^*$ and the wire is assumed to be rather smoothly curved. Here $\kappa(X)$ is the curvature at arc $X$ and the term proportional to $[\kappa(X)]^2$ arises from the global deformation of the wire. When the mean $e$-$h$ separation $l$ ($l \sim 2a_{\text{B}}^{(1)}$) is much smaller than $1/\kappa(X)$, the Coulomb potential $V^{(1)}(X,x)$ can be expanded in terms of $x$ as $V^{(1)}(X,x) \sim -(e^2/\epsilon)\{|x|^{-1} + [\kappa(X)]^2|x|/4! + \cdots\}$. Note that the stronger curvature gives the deeper potential to the exciton. The higher-order terms contain the derivative of $\kappa(X)$ and torsion $\tau(X)$ with respect to $X$. When the spatial extent $l$ is small enough to satisfy $l\dot{\kappa} \ll \kappa$ and $l\tau \ll \kappa$, the exciton wavefunction can be decoupled as $\Psi^{(1)}(X,x) \simeq G^{(1)}(X)\psi^{(1)}(x)$. Then the wavefunction $G^{(1)}(X)$ for the center-of-mass coordinate obeys

$$\left[-\frac{1}{2M}\frac{d^2}{dX^2} - V_0[\Delta\kappa(X)]^2\right] G^{(1)}(X) \simeq \left(E^{(1)}_{\text{total}} - E^{(1)}_{\nu}\right) G^{(1)}(X), \tag{1.73}$$

where $V_0 \equiv (1/8\mu)(2a_{\text{B}}^{(1)}/3a_{\text{B}}^{(3)} + 1)$, $[\Delta\kappa(X)]^2$ is the deviation of $[\kappa(X)]^2$ from the mean value $\langle\kappa^2\rangle$, and $E^{(1)}_{\nu}$ is the energy of the relative motion, as given in Sec. 1.3.1.3. Equation (1.73) implies that the center-of-mass motion of the Wannier exciton feels attractive potentials in the regions with large curvature (large $[\Delta\kappa]^2$) on the chain. Therefore we expect certain "trapped states" to appear below the free exciton states.

### 1.3.6. *Exciton Confinement on a Spherical Shell*

Kanemitsu *et al.* [39] examined extensively the PL properties of oxidized Si nanoballs (nanometer-sized spheres). This material consists of a spherical crystalline Si core and an amorphous $SiO_2$ surface layer. To explain their experimental results, they proposed a model where the oxidized Si nanoballs consist of three regions, and excitons are confined in a middle, spherical shell region (an interfacial layer between the crystalline Si core and the amorphous $SiO_2$ surface layer). Slow decay dynamics of PL were also observed [40] in this material, which indicates the hopping-limited recombination of carriers in the Q2D interfacial region. Details will be discussed in Chapter 5.

The Wannier exciton on a spherical shell has been considered theoretically by Kayanuma and Saito [41], who found that the motional state changes as a function of $R/a_{\text{B}}^{(3)}$ (where $R$ is the radius of the sphere). In the limit of $R/a_{\text{B}}^{(3)} \ll 1$, the Coulomb interaction becomes relatively unimportant. Then the system shows the free rotation of an independent pair. The leading term of eigenenergy of the states with totally $s$-symmetry is given by

$$E_n^{(\text{shell})} \sim \begin{cases} -2a_{\text{B}}^{(3)}/R & n = 1 \\ n(n-1)\left[a_{\text{B}}^{(3)}/R\right]^2 & n = 2,3,4,\cdots \end{cases}. \tag{1.74}$$

The kinetic energy of rotation vanishes for the lowest ($n = 1$) state. The oscillator strength per unit area for the $n$th state is proportional to

$$f_n^{(\text{shell})} \propto \frac{2n-1}{4\pi R^2}. \tag{1.75}$$

In the opposite limit $R/a_B^{(3)} \gg 1$, on the other hand, an electron and a hole do not feel a curvature of the sphere and form a tightly bound two-dimensional Wannier exciton. Intermediate regime ($R/a_B^{(3)} \sim 1$) has been studied only numerically [41].

### 1.3.7. *Excitonic Molecules in One Dimension*

When two electrons in the valence band are excited up to the conduction band, the system has four excited particles, two electrons and two holes, which can form an excitonic molecule (EM, often called biexciton). In 3D, the molecular binding energy of the EM has been evaluated by the variational methods [42]. In 2D, Kleinmann [43] investigated the properties of the EM. Here we confine ourselves to the EM in a 1D semiconductor, where the geometrical confinement becomes more distinct than in higher dimensions. Unlike the exciton binding energy, the EM binding energy does not blow up as the wire width decreases to zero. First, we shall review a conventional Heitler-London approach to an EM by a continuum model. Next the shortcomings of such treatments are discussed and then alternative 1D lattice model of an EM will be introduced.

#### 1.3.7.1. *Heitler-London scheme for 1D excitonic molecule*

We shall overview the procedure of evaluating the EM binding energy by the Heitler-London approximation [19]. A continuum model start with the 1D effective Hamiltonian,

$$\begin{aligned}\hat{H}_{\text{EM}}^{(1)} &= -\frac{\partial^2}{\partial \xi_1^2} - \frac{\partial^2}{\partial \xi_2^2} + \frac{2\mu}{m_h^*}\left(\frac{\partial}{\partial \xi_1} - \frac{\partial}{\partial \xi_2}\right)\frac{\partial}{\partial x_{hh}} - \frac{2\mu}{m_h^*}\frac{\partial^2}{\partial x_{hh}^2} - \frac{\mu}{2M}\frac{\partial^2}{\partial X^2} \\ &\quad +2U_{\text{EM}}^{(1)}(\xi_1, \xi_2, x_{hh}),\end{aligned} \tag{1.76}$$

where the coordinates are defined as $\xi_1 \equiv x_{e1} - x_{h1}$, $\xi_2 \equiv x_{e2} - x_{h2}$, $x_{hh} \equiv x_{h1} - x_{h2}$, and $X \equiv (2M)^{-1}[m_e^*(x_{e1} + x_{e2}) + m_h^*(x_{h1} + x_{h2})]$, and the potential $U_{\text{EM}}^{(1)}$ is given as

$$\begin{aligned}U_{\text{EM}}^{(1)}(\xi_1, \xi_2, x_{hh}) &= \tilde{V}^{(1)}(\xi_1 - \xi_2 + x_{hh}; x_0) + \tilde{V}^{(1)}(x_{hh}; x_0) \\ &\quad -\tilde{V}^{(1)}(\xi_1; x_0) - \tilde{V}^{(1)}(\xi_2; x_0) \\ &\quad -\tilde{V}^{(1)}(\xi_1 + x_{hh}; x_0) - \tilde{V}^{(1)}(\xi_2 - x_{hh}; x_0).\end{aligned} \tag{1.77}$$

Here $\tilde{V}^{(1)}(x; x_0)$ is the regularized 1D Coulomb potential [Eq. (1.35)]. To evaluate the ground-state EM binding energy, we look for a solution of the form

$$\psi_{\text{EM}}^{(1)}(\xi_1, \xi_2, x_{hh}) = \phi_{hh}^{(1)}(|x_{hh}|)\Phi_{\text{corr}}^{(1)}(\xi_1, \xi_2, x_{hh}), \tag{1.78}$$

which consists of the hole-hole wavefunction $\phi_{hh}^{(1)}(|x_{hh}|)$ for the relative motion of the two holes and $\Phi_{\rm corr}^{(1)}$ describing the exciton correlations, $\Phi_{\rm corr}^{(1)}(\xi_1, \xi_2, x_{hh}) = N(x_{hh})\, \phi_{\rm corr}^{(1)}(\xi_1, \xi_2, x_{hh})$, with the normalization function $N(x_{hh})$. Employing the Heitler-London approximation, we set

$$\phi_{\rm corr}^{(1)}(\xi_1, \xi_2, x_{hh}) = \psi_\nu^{(1)}(\xi_1)\psi_\nu^{(1)}(\xi_2) + \psi_\nu^{(1)}(\xi_1 + x_{hh})\psi_\nu^{(1)}(\xi_2 - x_{hh}), \tag{1.79}$$

where $\psi_\nu^{(1)}$ is the relative wavefunction of an exciton in 1D [Eq. (1.36)]. Within this approximation, no polarization of one exciton by the other is accounted for. In 3D this polarization leads to a significant enhancement of the EM binding energy. In 1D, on the other hand, problem due to the complicated nature of the exciton wavefunctions an analogous scheme is computationally much more involved.

The 1D EM binding energy, $E_{\rm EM}^{(1)}$, is obtained by solving the Schrödinger equation for the relative hole-hole wavefunction, that is,

$$\left[-\frac{2\mu}{m_h^*}\frac{\partial^2}{\partial x_{hh}^2} + V_{\rm hole}^{(1)}(x_{hh})\right]\psi_{hh}^{(1)} = -E_{\rm EM}^{(1)}\psi_{hh}^{(1)}, \tag{1.80}$$

with the effective hole potential:

$$V_{\rm hole}^{(1)}(x_{hh}) = -2E^{(1)} + \int_{-\infty}^{\infty} d\xi_1 \int_{-\infty}^{\infty} d\xi_2\, \Phi_{\rm corr}^{(1)}(\xi_1, \xi_2, x_{hh})\hat{H}_{\rm EM}^{(1)}\Phi_{\rm corr}^{(1)}(\xi_1, \xi_2, x_{hh}). \tag{1.81}$$

Here $E^{(1)}$ is the ground-state excitonic energy in 1D. The actual value of $E_{\rm EM}^{(1)}$ is relatively large [19], more than a factor of 5 larger than the bulk EM binding energy, $E_{\rm EM}^{(3)} \simeq 0.04E_R^*$, although this calculation does not include the excitonic polarization and deformation effects.

### 1.3.7.2. *Four-fermion problem for an excitonic molecule on a 1D lattice*

Conventionally as in the previous section, an EM is regarded as a bound state of two Wannier excitons. More rigorously, however, we should solve a four-fermion problem taking into full account the correlation effect on the length scale of a lattice constant. In other words, a usual bosonic picture of the EM cannot capture the binding mechanism quantitatively. We need to look into the way in which two electrons and two holes are reshuffled. In addition, in lower dimensions, the electronic correlation effects lead to the crossover between the Frenkel-type EM and the Wannier-type EM. In order to treat consistently this crossover, both a lattice model and the long-range interaction are required for the description of an EM. Ishida, Aoki, and Ogawa [44] solved the four-fermion model of an EM by exact diagonalization on a finite 1D lattice. The EM hamiltonian which they employed is

$$\begin{aligned}\hat{\mathcal{H}}_{\rm EM}^{(1)} &= -t_e\sum_{i,\sigma}(\hat{a}_{i+1,\sigma}^\dagger\hat{a}_{i\sigma} + \text{h.c.})\\ &\quad -t_h\sum_{i,\sigma}(\hat{d}_{i+1,\sigma}^\dagger\hat{d}_{i\sigma} + \text{h.c.}) + E_0\sum_{i,\sigma}\hat{d}_{i\sigma}^\dagger\hat{d}_{i\sigma}\end{aligned}$$

$$- \sum_{i,j,\sigma,\sigma'} U_{ij}\, \hat{a}^\dagger_{i\sigma}\hat{a}_{i\sigma}\hat{d}^\dagger_{j\sigma'}\hat{d}_{j\sigma'}$$
$$+ \sum_{i,j,\sigma,\sigma'} V_{ij} \left(\hat{a}^\dagger_{i\sigma}\hat{a}_{i\sigma}\hat{a}^\dagger_{j\sigma'}\hat{a}_{j\sigma'} + \hat{d}^\dagger_{i\sigma}\hat{d}_{i\sigma}\hat{d}^\dagger_{j\sigma'}\hat{d}_{i\sigma'}\right)$$
$$- V_{\text{dipole}} \sum_{i,\sigma} \left(\hat{a}^\dagger_{i+1,\sigma}\hat{d}^\dagger_{i+1,-\sigma}\hat{d}_{i,-\sigma}\hat{a}_{i\sigma} + \text{h.c.}\right), \tag{1.82}$$

where $\hat{a}^\dagger_{i\sigma}$ ($\hat{d}^\dagger_{i\sigma}$) creates an electron in the conduction band (a hole in the valence band) with spin $\sigma$ at site $i$, $t_e$ ($t_h$) is the transfer energy of electrons (holes), and $E_0$ is the energy difference between the two orbitals from which conduction and valence bands arise. The attractive $e$-$h$ interaction between the sites $i$ and $j$ is denoted by $U_{ij}$, while $V_{ij}$ represents the repulsive electron-electron ($e$-$e$) and hole-hole ($h$-$h$) interactions. The last term describes the transfer of an $e$-$h$ pair from site to site, which results from the electromagnetic dipole-dipole coupling between atoms. Thus, $V_{\text{dipole}}$, which induces a center-of-mass motion of the exciton, increases the bandwidth of the exciton. At large distances $U_{ij}$ should behave like $1/|i-j|$ (the Pople type), since screening effect is negligible for a system with two $e$-$h$ pairs. To take care of the localized nature of Wannier functions, we have in addition considered the intra-atomic energy when an electron and a hole come to the same unit cell, which we denote by $U_{ii} \equiv U_0$ and $V_{ii} \equiv V_0$. Thus we end up with

$$U_{ij} = \begin{cases} U_0 & (i=j) \\ U_1/|i-j| & (i \neq j) \end{cases}, \tag{1.83}$$

$$V_{ij} = \begin{cases} V_0 & (i=j) \\ V_1/|i-j| & (i \neq j) \end{cases}, \tag{1.84}$$

where physically $U_0 > U_1 \geq 0, V_0 > V_1 \geq 0$, and we assume $V_1 = U_1$ for simplicity.

The numerical diagonalization of the Hamiltonian Eq. (1.82) is carried out for a finite 1D system of length $N = 81$ with periodic boundary condition. In Fig. 1.10, $E_{\text{EM}}$ (normalized by the 1D exciton binding energy $E_{\text{B}}$) is plotted as a function of the transfer energy $t \equiv t_e = t_h$. One finds that the Heitler-London scheme gives only about half the exact result for the whole range of $t$. This discrepancy indicates that the correlation effect is in fact of significance. The Heitler-London approximation is expected to give reasonable results only when the $e$-$h$ relative motion is spatially extended. The Wannier excitons in higher dimensions correspond to this case. We stress, however, that in 1D systems the correlated motion of electrons and holes over the length scale of a lattice constant is essential even in the weak-coupling case. Moreover, continuum models for an EM holds only in the weak-coupling regime, where the hole-hole correlation is peaked at a finite distance larger than a lattice constant, while it decays faster for the strong-coupling regime where an EM is more compact in space and its size is of an atomic order. In 1D, the coupling among particles becomes essentially stronger in comparison with that in higher dimensions. Therefore, the lattice model is suitable for description of the elementary excitations in 1D such as a 1D EM in the whole range of the $e$-$h$ coupling constant. Other physical quantities of the 1D EM are given in [44].

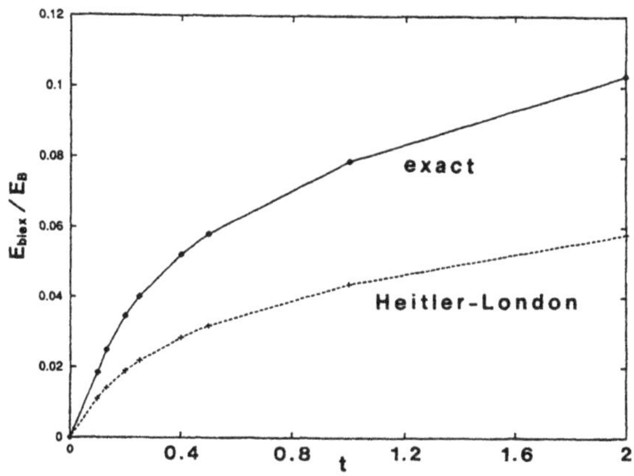


Figure 1.10. The 1D EM binding energy $(E_{\text{biex}})$ normalized by the 1D exciton binding energy $(E_{\text{B}})$ is plotted as a function of the transfer energy $t \equiv t_e = t_h$. Here $V_1 = U_1 = 0.75$, $V_{\text{dipole}} = 0$, and $V_0/U_0 = 1$. A result from the exact numerical diagonalization (solid line) is much larger than that with the Heitler-London approximation (dotted line). (After [44]).

## 1.4. DIMENSIONALITY IN THE DIELECTRIC CONFINEMENT EFFECTS

In determining the effective Coulomb potential in $d \leq 2$ dimension, we need to include the dielectric confinement effect arising from the difference in the dielectric constant between the material and the surroundings (including vacuum). This is a confinement effect on the Coulomb interaction, called the "dielectric confinement effect" or the image charge effect. If the surrounding materials have a smaller dielectric constant and a larger energy gap than the relevant material, the *e*-*h* Coulomb attraction in the relevant material works very effectively through the surroundings with reduced screening. As a result, the exciton binding energy is expected to increase much more than in a normal low-dimensional system without this effect. This effect was firstly predicted by Keldysh [45] and Rytova [46] in 2D systems. A variational calculation of the exciton quantities in a single QW was carried out by Hanamura *et al.* [16, 47]. Ishihara *et al.* [48] employed this idea for the interpretation of experiments in PbI-based compounds (Chapter 6). In Q1D semiconductors, Bányai *et al.* [19] and Ogawa *et al.* [17, 49] examined this effect theoretically.

### 1.4.1. *Quasi-Two-Dimensional Systems*

First, we shall review results by Keldysh [50]. Assume a layered system with thickness $L_{\perp z}$ ($|z| \leq \frac{1}{2}L_{\perp z}$) and a dielectric constant $\epsilon$ and a surrounding material with a (smaller) dielectric constants $\epsilon_1$ and $\epsilon_2$ ($\epsilon > \epsilon_1, \epsilon_2$). The interaction of two point charges (an electron and a hole) both inside a semiconductor layer is

$$V^{(2)}_{\text{dielectric}}(\vec{\rho}; z_e, z_h) = -\frac{e^2}{\epsilon}\int d^2k \, \frac{e^{i\vec{k}\cdot\vec{\rho}}}{k\sinh(kL_{\perp z}+\eta_1+\eta_2)} \times \cosh\left[k\left(\frac{L_{\perp z}}{2}-z_e\right)+\eta_2\right] \times \cosh\left[k\left(\frac{L_{\perp z}}{2}+z_h\right)+\eta_1\right], \tag{1.85}$$

where $k = |\vec{k}| = \sqrt{k_x^2+k_y^2}$, $\eta_i = \frac{1}{2}\ln[(\epsilon+\epsilon_i)/(\epsilon-\epsilon_i)]$, $\vec{\rho} = (x_e - x_h, y_e - y_h)$ is the 2D vector connecting projections of the positions of both particles on the layer plane, and $z_e$ ($z_h$) is a coordinate of the electron (the hole) in the direction normal to the layer ($z = 0$ is the middle plane of the layer). There exist three distinct ranges of $e$-$h$ distance where the general formula Eq. (1.85) can be simplified essentially.

(i) The case of very small distance: The $e$-$h$ distance is small compared to the layer thickness, *i.e.*, $r^2 = |\vec{r}|^2 \equiv |\vec{\rho}|^2+(z_e-z_h)^2 \ll (L_{\perp z})^2$. Then the interaction becomes the same as that in the bulk semiconductor [Eq. (1.22)]: $V^{(2)}_{\text{dielectric}} \to V^{(3)}(\vec{r}) = -e^2/(\epsilon r)$.
(ii) The case of very large distance: $\rho \gg (\epsilon/\epsilon_i)L_{\perp z}$. The interaction is also Coulomb-like but with the (averaged) dielectric constant of the surrounding medium:

$$V^{(2)}_{\text{dielectric}} \to V^{(2)}_{\text{strong}}(\vec{\rho}) = -\frac{2e^2}{(\epsilon_1+\epsilon_2)\,\rho}. \tag{1.86}$$

(iii) If the ratio of the dielectric constants is large enough, there exists also a broad range of intermediate distances, $L_{\perp z} \ll \rho \ll [2e^2/(\epsilon_1+\epsilon_2)]L_{\perp z}$. Here the dependence of the interaction potential on the interparticle distance is very weak, describing the transition from the "weak" Coulomb interaction $V^{(3)}(\vec{r})$ to the "strong" one $V^{(2)}_{\text{strong}}(\vec{\rho})$, *i.e.*,

$$V^{(2)}_{\text{dielectric}} \to V^{(2)}_{\text{intermediate}}(\vec{\rho}) = -\frac{2e^2}{\epsilon L_{\perp z}}\left\{\ln\left[\frac{2\epsilon L_{\perp z}}{(\epsilon_1+\epsilon_2)\rho}\right]-\gamma\right\}, \tag{1.87}$$

where $\gamma = 0.577$ is Euler's constant.

If the layer thickness satisfies $a_{\text{B}}^{(3)} \ll L_{\perp z} \ll a_{\text{B}}^{(3)}(\epsilon_1+\epsilon_2)^2/4\epsilon^2$, the effective exciton radius $\tilde{a}_{\text{B}}^{(3)}$ and the exciton binding energy $\tilde{E}^{(3)}_{n=1}$ taking into account the dielectric confinement effects are

$$\tilde{a}_{\text{B}}^{(3)} = \frac{1}{2}\left(a_{\text{B}}^{(3)}L_{\perp z}\right)^{1/2}, \tag{1.88}$$

$$|\tilde{E}_1^{(3)}| \simeq \frac{e^2}{\epsilon L_{\perp z}}\ln\left[\frac{4\epsilon^2 L_{\perp z}}{a_{\text{B}}^{(3)}(\epsilon_1+\epsilon_2)^2}\right]. \tag{1.89}$$

In the case of very thin layer and not too large ratio of the dielectric constants, $L_{\perp z} < a_{\text{B}}^{(3)}(\epsilon_1+\epsilon_2)^2/4\epsilon^2$, on the other hand, the exciton effective radius is larger than

$(\epsilon/\epsilon_i)L_{\perp z}$. Then we have

$$\tilde{a}_{\rm B}^{(2)} = \frac{\epsilon_1+\epsilon_2}{4\epsilon} a_{\rm B}^{(3)}, \tag{1.90}$$

$$|\tilde{E}_1^{(2)}| = \left(\frac{2\epsilon}{\epsilon_1+\epsilon_2}\right)^2 |E_1^{(3)}|. \tag{1.91}$$

These are limiting values which can be achieved in that way for a given pair of semiconductor and insulators. For numerical calculations, we write here a formula which is valid in the whole range of the potential and the layer thickness; the Coulomb attraction between an electron at $\vec{r}_e$ and a hole at $\vec{r}_h$ is

$$V_{\rm dielectric}^{(2)}(\vec{r}_e,\vec{r}_h) = -\frac{e^2}{\epsilon}\sum_{n=-\infty}^{\infty}\left(\frac{\epsilon-\epsilon_1}{\epsilon+\epsilon_1}\right)^{|n|}\left[(x_e-x_h)^2+(y_e-y_h)^2+z_n^2\right]^{-1/2}, \tag{1.92}$$

where $z_n = z_e - nL_{\perp z} - (-1)^n z_h$ and $\epsilon_1 = \epsilon_2$ is assumed. Using this Coulomb interaction, exciton properties were evaluated numerically. Details are given in literatures [16, 47].

### 1.4.2. *Quasi-One-Dimensional Systems*

Bányai and coworkers [19] studied the 1D effective Coulomb potential in a cylindrical wire structure with radius $R$ in a cylindrical coordinates $(\rho,\theta,x)$. The cladding medium outside the wire has a dielectric constant $\epsilon_1$ smaller than that within the wire $\epsilon$. They found an effective Coulomb potential:

$$V_{\rm dielectric}^{(1)\bigcirc}(x_e-x_h) = \int_0^R \rho_e d\rho_e \int_0^R \rho_h d\rho_h \int_0^{2\pi} d\theta_e \int_0^{2\pi} d\theta_h\, U_{\rm dielectric}^{\bigcirc}(\vec{r}_e,\vec{r}_h)|\phi_0(\rho_e)|^2|\phi_0(\rho_h)|^2, \tag{1.93}$$

where $\phi_0(\rho)$ is the envelope function of the ground-state subband, which is given as $\phi_0(\rho) = J_0(\alpha_0\rho/R)/[\sqrt{\pi}RJ_1(\alpha_0)]$, where $\alpha_0$ is the first zero of the Bessel function $J_0$. Here $U_{\rm dielectric}^{\bigcirc}(\vec{r}_e,\vec{r}_h)$ means the electrostatic potential inside the wire:

$$\begin{aligned} U_{\rm dielectric}^{\bigcirc}(\vec{r}_e,\vec{r}_h) &= -\frac{e^2}{\epsilon|\vec{r}_e-\vec{r}_h|} \\ &\quad -\frac{2e^2}{\pi}\left(\frac{1}{\epsilon_1}-\frac{1}{\epsilon}\right)\sum_{m=-\infty}^{\infty}\int_0^{\infty} dk\, \cos[k(x_e-x_h)]e^{im(\theta_e-\theta_h)} \\ &\quad \times C_m(kR)I_m(k\rho_e)I_m(k\rho_h), \end{aligned} \tag{1.94}$$

where

$$C_m(kR) = \frac{K_m(kR)K_m'(kR)}{I_m(kR)K_m'(kR)-(\epsilon/\epsilon_1)I_m'(kR)K_m(kR)}, \tag{1.95}$$

with the modified Bessel functions, $I_m$ and $K_m$. This potential, $V_{\rm dielectric}^{(1)\bigcirc}$ apart from a front factor $e^2/\epsilon R$, is a universal function of $|x_e-x_h|/R$, *i.e.*, it depends only on

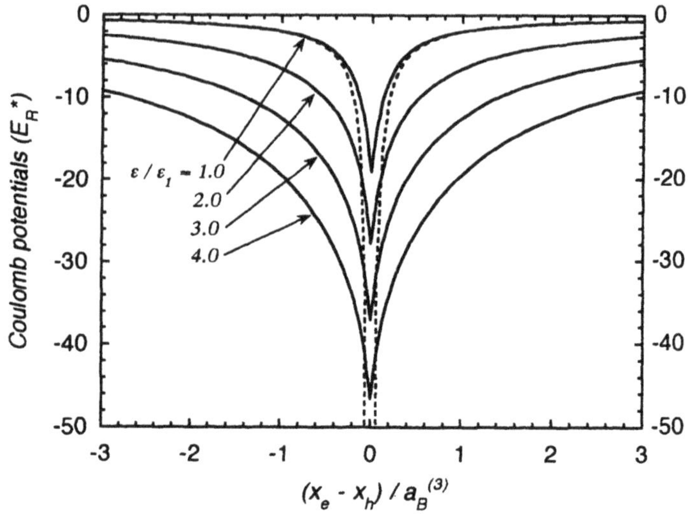


Figure 1.11. The effective 1D Coulomb potential, $V^{(1)\square}_{\text{dielectric}}$, including the dielectric confinement effect is plotted for a rectangular wire structure having the lateral size of $L_{\perp y} = L_{\perp z} = \frac{1}{2}a_B^{(3)}$ and for $\epsilon/\epsilon_1 = 1.0$, 2.0, 3.0, and 4.0 from top to bottom. The dotted curve represents the bare Coulomb potential without the dielectric confinement effect. (After [17]).

the ratio of the $e$-$h$ separation and the wire radius and not on each individually. An analytic approximation to this potential $V^{(1)\bigcirc}_{\text{dielectric}}(x_e - x_h)$ is the regularized Coulomb potential, $\tilde{V}^{(1)}(x; x_0)$ in Eq. (1.35), with $x_0 = x_0(\epsilon, \epsilon_1, L_{\perp y}, L_{\perp z})$ which is given only numerically.

In a rectangular wire structure ($|y| \le \frac{1}{2}L_{\perp y}$, $|z| \le \frac{1}{2}L_{\perp z}$), on the other hand, the $e$-$h$ interaction including the dielectric confinement effect is derived as [17]

$$U^{\square}_{\text{dielectric}}(\vec{r}_e, \vec{r}_h) = -\frac{e^2}{\epsilon} \sum_{m=-\infty}^{\infty} \sum_{n=-\infty}^{\infty} q^{|m|+|n|} \left[(x_e - x_h)^2 + y_m^2 + z_n^2\right]^{-1/2}, \tag{1.96}$$

where $q \equiv (\epsilon - \epsilon_1)/(\epsilon + \epsilon_1)$, $y_m \equiv y_e - mL_{\perp y} - (-1)^m y_h$, and $z_n \equiv z_e - nL_{\perp z} - (-1)^n z_h$. After averaging over $y_e$, $z_e$, $y_h$ and $z_h$ as in Eq. (1.42), we have

$$V^{(1)\square}_{\text{dielectric}}(x_e - x_h) = -\frac{4E_R^*}{\bar{L}_{\perp y}\bar{L}_{\perp z}} \int_0^\infty dk_y \int_0^\infty dk_z \, \frac{F(k_y)F(k_z)}{|G(k_y, k_z)|} e^{-\pi\bar{x}|G(k_y,k_z)|}, \tag{1.97}$$

where $x = |x_e - x_h|$, the overbar ( $^-$ ) means the scaling by $a_B^{(3)}$, and

$$F(k) = \frac{1 + 2q(1-q^2)\cos(\pi k) - q^4}{1 - 2q^2\cos(2\pi k) + q^4} f(k), \tag{1.98}$$

$$G^2(k_y, k_z) = \left(\frac{k_y}{\bar{L}_{\perp y}}\right)^2 + \left(\frac{k_z}{\bar{L}_{\perp z}}\right)^2. \tag{1.99}$$

Here $f(k)$ is given using $j_0(x) = \sin x/x$ as

$$\begin{aligned} f(k) = {} & j_0^2\Big(\frac{\pi}{2}k\Big) + j_0\Big(\frac{\pi}{2}(k-2)\Big)j_0\Big(\frac{\pi}{2}k\Big) + j_0\Big(\frac{\pi}{2}(k+2)\Big)j_0\Big(\frac{\pi}{2}k\Big) \\ & + \frac{1}{4}j_0^2\Big(\frac{\pi}{2}(k-2)\Big) + \frac{1}{4}j_0^2\Big(\frac{\pi}{2}(k+2)\Big) \\ & + \frac{1}{2}j_0\Big(\frac{\pi}{2}(k-2)\Big)j_0\Big(\frac{\pi}{2}(k+2)\Big). \end{aligned} \tag{1.100}$$

Figure 1.11 shows effective 1D Coulomb potentials including the dielectric confinement effect. In this case also, the potential can be well fitted by Eq. (1.35). As the ratio $\epsilon/\epsilon_1$ increases, the attractive potential becomes more long-ranged and is enhanced for all regions of $x$ over the bare Coulomb potential $-(e^2/\epsilon)|x|^{-1}$. Therefore, the peculiar features of the $d = 1$ excitons would be more pronounced by the dielectric confinement effect.

## 1.5. DIMENSIONALITY IN THE FERMI-SURFACE EFFECTS

Fermi-surface effects in metals or conductors have been one of the central topics in condensed-matter physics [51]. Low-lying excitations near the Fermi level determine ground-state properties of the electronic systems. Electronic many-body effects in lower dimensions are of particular interest because the interactions among elementary excitations and quantum fluctuations are enhanced due to the reduced dimensionality. In this section, we concentrate on the Fermi-surface effects in $d$D interacting Fermion systems.

### 1.5.1. *The Peierls Instability*

We shall consider the $d$D free electronic system, whose eigenenergy is $\varepsilon_{\vec{k}} = |\vec{k}|^2/2m^*$ with the $d$D wavevector $\vec{k}$. Consider a periodic potential with wave number $\vec{q}$,

$$V(\vec{r}) = V_{\vec{q}}\left(e^{i\vec{q}\cdot\vec{r}} + e^{-i\vec{q}\cdot\vec{r}}\right), \tag{1.101}$$

which is applied to this $d$D electronic system. The Fourier component of the electron number density $n(\vec{r})$ of this system is given as $n_{\vec{q}} = -2\chi^0_{\vec{q}}V_{\vec{q}}$, where the susceptibility $\chi_{\vec{q}}$ is evaluated by the perturbation method:

$$\chi_{\vec{q}} = \sum_{\vec{k}} \frac{f(\varepsilon_{\vec{k}}) - f(\varepsilon_{\vec{k}+\vec{q}})}{\varepsilon_{\vec{k}+\vec{q}} - \varepsilon_{\vec{k}}}. \tag{1.102}$$

Here $f(\varepsilon) = [\exp\{(\varepsilon-\mu)/T\}+1]^{-1}$ with the chemical potential $\mu$. At $T = 0$, in particular, we have $\chi^{(d)}_{\vec{q}} = dn_e/(8\varepsilon_F)F^{(d)}(\bar{q})$ for $d = 1, 2, 3$. Here $n_e$ is the average

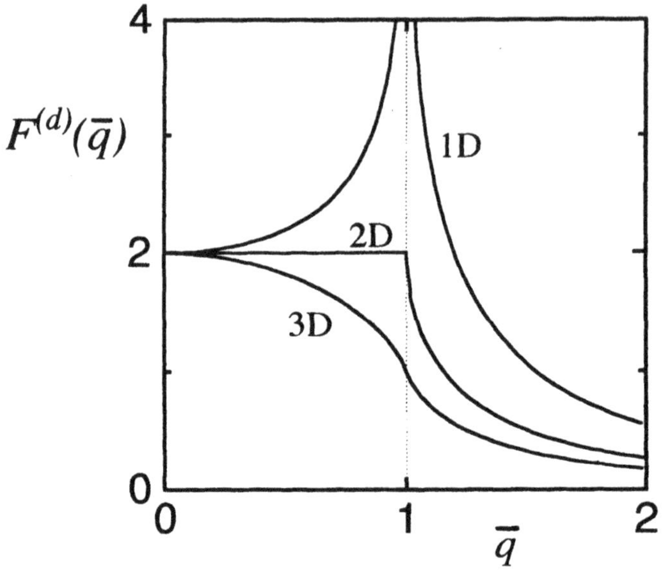


Figure 1.12. Susceptibility of the electron density fluctuation under a periodic perturbation is schematically drawn as a function of $\bar{q} \equiv |\vec{q}|/2k_F$.

number density of electrons and $F^{(d)}(\bar{q})$ is given as

$$F^{(3)}(\bar{q}) = 1 + \frac{1-\bar{q}^2}{2\bar{q}} \ln \left| \frac{1+\bar{q}}{1-\bar{q}} \right|, \tag{1.103}$$

$$F^{(2)}(\bar{q}) = \begin{cases} 2 & (\bar{q} < 1) \\ 2(1 - \sqrt{\bar{q}^2 - 1}/\bar{q}) & (\bar{q} > 1), \end{cases} \tag{1.104}$$

$$F^{(1)}(\bar{q}) = \frac{1}{\bar{q}} \ln \left| \frac{1+\bar{q}}{1-\bar{q}} \right|, \tag{1.105}$$

where $\bar{q} \equiv |\vec{q}|/|\vec{Q}|$. Here $\vec{Q}$ is called "the nesting vector." In ideal 1D cases, $Q = 2k_F$. These functions are drawn in Fig. 1.12. At $\bar{q} = 1$, *i.e.*, $\vec{q} = \pm\vec{Q}$, the summand of Eq. (1.102) diverges due to the presence of the sharp Fermi surface, hence the susceptibility $\chi_{\vec{q}}^{(d)}$ becomes singular at $\bar{q} = 1$. In the 3D and 2D cases, however, the contribution of this divergence to the $\vec{k}$-integral is relatively small, $\chi_{\vec{q}}^{(3)}$ and $\chi_{\vec{q}}^{(2)}$ show no divergence. On the other hand, in the 1D case, $\chi_{\vec{q}}^{(1)}$ itself diverges at $\bar{q} = 1$.

At finite temperature, the singularity of $\chi_{\vec{q}}^{(1)}$ at $\bar{q} = 1$ becomes absent mathematically since the discontinuous jump at $\bar{q} = 1$ in the Fermi distribution function disappears. But $\chi_{\bar{q}}^{(1)}$ still shows a sharp peak at $\bar{q} = 1$ if the temperature is low enough ($T \ll \varepsilon_{\rm F}$). Its peak value is evaluated as

$$\chi_{\bar{q}=1}^{(1)} \simeq \frac{m^*}{\pi Q} \ln \left( \frac{4\gamma \varepsilon_{\rm F}}{\pi T} \right), \tag{1.106}$$

where $\gamma$ is the Euler constant. In 1D systems, an infinitesimally small periodic perturbation with a wave number $Q = 2k_{\rm F}$ makes the 1D Fermi surface unstable, which is called the "Peierls instability." This is peculiar to the 1D electronic systems and leads to the Peierls gap in the band. The nesting of the Fermi surface play essential roles in formation of the charge-density wave (CDW) and the spin-density wave (SDW).

### *1.5.2. The Tomonaga-Luttinger Liquid*

Here we review the interacting 1D electron liquid. An interacting electron system in 1D is in general described not as a Fermi liquid [52] but as a Tomonaga-Luttinger (TL) liquid [53, 54, 55, 56] provided that the system has a gapless excitation.[15] This was shown in weak-coupling regime by using abelian bosonization methods. The validity of the TL liquid picture has been confirmed recently also in the strong-coupling regime by conformal field theory techniques and the Bethe ansatz methods [57, 58]. In the lower dimensions, the quantum fluctuations become crucial in general and can never be neglected in discussing the low-energy electronic properties. In this sense, mean-field treatments are less effective in the study of low-dimensional systems. The TL model is appropriate for taking into account the quantum fluctuations.

The 1D interacting fermion system has the same static properties as those of the conventional Fermi liquid. We shall now exploit characteristics which show that a 1D interacting fermion system is not the Fermi liquid. The Fermi surface of 1D electron systems consists of two Fermi points, $k_{\rm F}$ and $-k_{\rm F}$, around which we linearize the energy dispersion of an free electron as

$$\varepsilon(k) = \sum_{j=1,2} \varepsilon_j(k) = v_{\rm F}(k - k_{\rm F}) - v_{\rm F}(k + k_{\rm F}), \tag{1.107}$$

where $v_{\rm F}$ is the Fermi velocity. We have two branches ($j = 1, 2$) of the energy dispersion. Although this linearization is justified only near the Fermi points, we use the dispersion relation Eq. (1.107) for all momentum $k$. The negative-energy states ($k < k_{\rm F}$ for branch $j = 1$ and $k > -k_{\rm F}$ for $j = 2$) are all occupied at zero temperature if there is no interaction between electrons. The error arising from this

[15] The term the "Luttinger liquid" was first used by Haldane, who noticed that a wide class of 1D many-body systems has a low-energy excitation spectrum similar to the Luttinger model spectrum. The Luttinger model is closely related to the Tomonaga model. A difference between these two models lies only in the high-energy cutoff; the Luttinger model deals with Dirac fermions whereas in the Tomonaga model the energy dispersion is cut off by a band width. In so far as the low-energy properties that we are interested in are concerned, both models are essentially equivalent.

approximation should be small when we are dealing with low-energy processes only. The kinetic energy part of the Hamiltonian becomes

$$\hat{\mathcal{H}}_{\mathrm{kin}} = v_{\mathrm{F}} \sum_{k,\sigma} (k - k_{\mathrm{F}})\, \hat{a}^{\dagger}_{1k\sigma} \hat{a}_{1k\sigma} - v_{\mathrm{F}} \sum_{k,\sigma} (k + k_{\mathrm{F}})\, \hat{a}^{\dagger}_{2k\sigma} \hat{a}_{2k\sigma}, \tag{1.108}$$

where $\hat{a}^{\dagger}_{jk\sigma}$ ($\hat{a}_{jk\sigma}$) is a creation (annihilation) operator of an electron of branch $j$ with momentum $k$ and spin $\sigma = \pm 1 = \uparrow, \downarrow$.

The Coulomb interactions in 1D can be parameterized at low temperature. We shall consider here only the forward scatterings (with small momentum transfer), the main processes of scatterings in 1D, which are represented by $g_2$ and $g_4$ according to the $g$-ology notation [59]. The scattering between parallel-spin ($\|$) particles and that between antiparallel-spin ($\perp$) ones are in general different in magnitude. Then we parametrize them separately as $g_{i\|}$ and $g_{i\perp}$, respectively, for $i = 2, 4$. The backward, umklapp, and spin-flip scatterings are neglected for simplicity. Then the interaction part of the Hamiltonian is

$$\begin{aligned}
\hat{\mathcal{H}}^{\mathrm{forward}}_{\mathrm{int}} &= \frac{1}{L} \sum_{k,k'} \sum_{\sigma,\sigma'} \sum_{p} \left( g_{2\|} \delta_{\sigma,\sigma'} + g_{2\perp} \delta_{\sigma,-\sigma'} \right) \hat{a}^{\dagger}_{1k\sigma} \hat{a}_{1,k-p,\sigma} \hat{a}^{\dagger}_{2k'\sigma'} \hat{a}_{2,k'+q,\sigma'} \\
&+ \frac{1}{L} \sum_{k,k'} \sum_{\sigma,\sigma'} \sum_{p>0} \left( g_{4\|} \delta_{\sigma,\sigma'} + g_{4\perp} \delta_{\sigma,-\sigma'} \right) \\
&\times \left( \hat{a}^{\dagger}_{1k\sigma} \hat{a}_{1,k-p,\sigma} \hat{a}^{\dagger}_{1k'\sigma'} \hat{a}_{1,k'+q,\sigma'} + \hat{a}^{\dagger}_{2k\sigma} \hat{a}_{2,k+p,\sigma} \hat{a}^{\dagger}_{2k'\sigma'} \hat{a}_{2,k'-q,\sigma'} \right), 
\end{aligned} \tag{1.109}$$

where $L$ is the length of the system. The total Hamiltonian, $\hat{\mathcal{H}} = \hat{\mathcal{H}}_{\mathrm{kin}} + \hat{\mathcal{H}}^{\mathrm{forward}}_{\mathrm{int}}$, can be expressed in terms of the density operator as

$$\begin{aligned}
\hat{\mathcal{H}} &= \frac{\pi}{L} \sum_{\sigma,\sigma'} \left( g_{2\|} \delta_{\sigma,\sigma'} + g_{2\perp} \delta_{\sigma,-\sigma'} \right) \sum_{p>0} \left[ \hat{\rho}_{1\sigma}(p) \hat{\rho}_{2\sigma'}(-p) + \hat{\rho}_{1\sigma}(-p) \hat{\rho}_{2\sigma'}(p) \right] \\
&+ \frac{\pi}{L} \sum_{\sigma,\sigma'} \left( g_{4\|} \delta_{\sigma,\sigma'} + g_{4\perp} \delta_{\sigma,-\sigma'} \right) \sum_{p>0} \left[ \hat{\rho}_{1\sigma}(p) \hat{\rho}_{1\sigma'}(-p) + \hat{\rho}_{2\sigma}(-p) \hat{\rho}_{2\sigma'}(p) \right] \\
&+ \frac{2\pi}{L} v_{\mathrm{F}} \sum_{\sigma} \sum_{p>0} \left[ \hat{\rho}_{1\sigma}(p) \hat{\rho}_{1\sigma}(-p) + \hat{\rho}_{2\sigma}(-p) \hat{\rho}_{2\sigma}(p) \right],
\end{aligned} \tag{1.110}$$

where $\hat{\rho}_{j\sigma}(p) \equiv \sum_k \hat{a}^{\dagger}_{j,k+p,\sigma} \hat{a}_{jk\sigma}$ is the spin-dependent density operator, which obeys boson commutation relations. Equation (1.110) can be diagonalized by using the Bogolyubov transformation, yielding

$$\hat{\mathcal{H}} = v_{\mathrm{charge}} \sum_{p} |p|\, \hat{\alpha}^{\dagger}_{p} \hat{\alpha}_{p} + v_{\mathrm{spin}} \sum_{p} |p|\, \hat{\xi}^{\dagger}_{p} \hat{\xi}_{p} + \mathrm{const.}, \tag{1.111}$$

where $\hat{\alpha}_p$ and $\hat{\xi}_p$ are new boson operators describing normal modes of the charge and spin density fluctuations, respectively, defined by

$$\hat{\alpha}_p \cosh \lambda_{\mathrm{charge}} - \hat{\alpha}^{\dagger}_{-p} \sinh \lambda_{\mathrm{charge}} = \begin{cases} (\pi/pL)^{1/2} \sum_\sigma \hat{\rho}_{1\sigma}(-p) & p > 0 \\ (-\pi/pL)^{1/2} \sum_\sigma \hat{\rho}_{2\sigma}(-p) & p < 0 \end{cases}, \tag{1.112}$$

$$\hat{\xi}_p \cosh \lambda_{\mathrm{spin}} - \hat{\xi}^{\dagger}_{-p} \sinh \lambda_{\mathrm{spin}} = \begin{cases} (\pi/pL)^{1/2} \sum_\sigma \sigma \hat{\rho}_{1\sigma}(-p) & p > 0 \\ (-\pi/pL)^{1/2} \sum_\sigma \sigma \hat{\rho}_{2\sigma}(-p) & p < 0 \end{cases}, \tag{1.113}$$

where $\tanh(2\lambda_{\text{charge}}) = (g_{2\parallel} + g_{2\perp})/(2v_F + g_{4\parallel} + g_{4\perp})$ and $\tanh(2\lambda_{\text{spin}}) = (g_{2\parallel} - g_{2\perp})/(2v_F + g_{4\parallel} - g_{4\perp})$. Both normal modes have linear gapless dispersion relations. The charge and spin collective modes have different velocities:

$$v_{\text{charge}} = \left[\left(v_F + \frac{g_{4\parallel}}{2} + \frac{g_{4\perp}}{2}\right)^2 - \left(\frac{g_{2\parallel}}{2} + \frac{g_{2\perp}}{2}\right)^2\right]^{1/2}, \tag{1.114}$$

$$v_{\text{spin}} = \left[\left(v_F + \frac{g_{4\parallel}}{2} - \frac{g_{4\perp}}{2}\right)^2 - \left(\frac{g_{2\parallel}}{2} - \frac{g_{2\perp}}{2}\right)^2\right]^{1/2}. \tag{1.115}$$

This is a peculiar feature of the electron gas in 1D, that is, the "separation of charge and spin degrees of freedom," which occurs both in the continuum limit and the strong coupling limit. In these limits, the Hamiltonian separates into a sum of independent charge and spin parts [Eq. (1.111)], and consequently the wavefunctions and space-time correlation functions factorize into products.

Now we define the phase fields:

$$\hat{\theta}_\pm(x) \equiv i\eta_\rho^{\pm 1/2} \sum_{k>0} \sqrt{\frac{\pi}{Lk}} e^{-\alpha k/2} \left[e^{-ikx}(\hat{\alpha}_k^\dagger \pm \hat{\alpha}_{-k}) - e^{ikx}(\hat{\alpha}_k \pm \hat{\alpha}_{-k}^\dagger)\right], \tag{1.116}$$

$$\hat{\phi}_\pm(x) \equiv i\eta_\sigma^{\pm 1/2} \sum_{k>0} \sqrt{\frac{\pi}{Lk}} e^{-\alpha k/2} \left[e^{-ikx}(\hat{\xi}_k^\dagger \pm \hat{\xi}_{-k}) - e^{ikx}(\hat{\xi}_k \pm \hat{\xi}_{-k}^\dagger)\right], \tag{1.117}$$

where $\alpha$ is a positive infinitesimal; later it will be taken as a finite number of the order of a lattice spacing. The phase field $\hat{\theta}_+(x)$ can be regarded as the phase of the CDW, whereas the phase $\hat{\theta}_-(x)$ corresponds to the Josephson phase, *i.e.*, the phase of a Cooper pair. The phase fields $\hat{\phi}_+(x)$ and $\hat{\phi}_-(x)$, on the other hand, are related to the spin degrees of freedom. In the above definitions, we use

$$\eta_\rho = \left[\frac{v_F + \frac{1}{2}(g_{4\parallel} + g_{4\perp}) - \frac{1}{2}(g_{2\parallel} + g_{2\perp})}{v_F + \frac{1}{2}(g_{4\parallel} + g_{4\perp}) + \frac{1}{2}(g_{2\parallel} + g_{2\perp})}\right]^{1/2}, \tag{1.118}$$

$$\eta_\sigma = \left[\frac{v_F + \frac{1}{2}(g_{4\parallel} - g_{4\perp}) - \frac{1}{2}(g_{2\parallel} - g_{2\perp})}{v_F + \frac{1}{2}(g_{4\parallel} - g_{4\perp}) + \frac{1}{2}(g_{2\parallel} - g_{2\perp})}\right]^{1/2}. \tag{1.119}$$

Various correlation functions decay ***algebraically*** at zero temperature, and their exponents are related to $\eta_\rho$ and $\eta_\sigma$. In this sense, $\eta_\rho$ and $\eta_\sigma$ are the most important parameters characterizing the TL system. In particular, for noninteracting systems, $\eta_\rho = \eta_\sigma = 1$. One advantage of the bosonization method comes from the fact that electron field operators can be written in terms of the phase fields, or equivalently, boson creation and annihilation operators, *e.g.*,

$$\hat{\psi}_{1\sigma}(x) = (2\pi\alpha)^{-1/2} e^{ik_F x} e^{\frac{i}{2}\left\{\hat{\theta}_+(x) + \hat{\theta}_-(x) + \sigma[\hat{\phi}_+(x) + \hat{\phi}_-(x)]\right\} + i\hat{\varphi}_{1\sigma}}, \tag{1.120}$$

$$\hat{\psi}_{2\sigma}(x) = (2\pi\alpha)^{-1/2} e^{-ik_F x} e^{-\frac{i}{2}\left\{\hat{\theta}_+(x) - \hat{\theta}_-(x) + \sigma[\hat{\phi}_+(x) - \hat{\phi}_-(x)]\right\} + i\hat{\varphi}_{2\sigma}}, \tag{1.121}$$

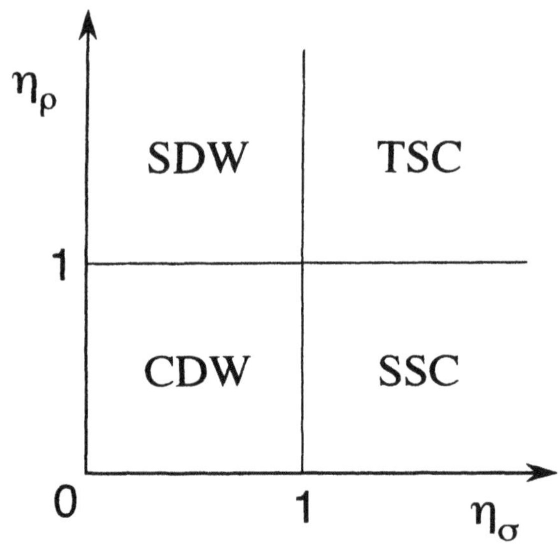


Figure 1.13. The phase diagram of the ground state of the spin-dependent TL model at zero temperature.

for $j = 1$ and 2, respectively, where $\hat{\varphi}_{j\sigma}$ is necessary for ensuring the anticommutation relation of $\hat{\psi}_{j\sigma}$ with different $j$ and $\sigma$.[16] Using this phase fields, the Hamiltonian is rewritten as

$$\hat{\mathcal{H}} = v_{\text{charge}} \int dx \left[ \pi\eta_\rho \hat{P}_+^2 + \frac{(\nabla\hat{\theta}_+)^2}{4\pi\eta_\rho} \right] + v_{\text{spin}} \int dx \left[ \pi\eta_\sigma \hat{\Pi}_+^2 + \frac{(\nabla\hat{\phi}_+)^2}{4\pi\eta_\sigma} \right], \tag{1.122}$$

where $\hat{P}_+(x) \equiv -(2\pi)^{-1}d\hat{\theta}_-(x)/dx$ and $\hat{\Pi}_+(x) \equiv -(2\pi)^{-1}d\hat{\phi}_-(x)/dx$. We can readily calculate various correlation functions. At zero temperature correlation functions for charge-density wave (CDW), spin-density wave (SDW), spin-singlet superconductivity (SSC), and triplet superconductivity (TSC) are evaluated and fall off algebraically for large $x$, which means that the system is just on the critical point at zero temperature. Figure 1.13 shows the phase diagram of the ground state of the TL liquid at zero temperature, which we infer from the most long-lived correlation functions of several order parameters.

The TL liquid has the power-law dependence of the momentum distribution, $n_k$,

[16]Explicitly $\hat{\varphi}_{j\sigma}$'s are given by $\hat{\varphi}_{1\uparrow} = 0$, $\hat{\varphi}_{1\downarrow} = \pi \int dx\, \hat{\psi}_{1\uparrow}^\dagger(x)\hat{\psi}_{1\uparrow}(x)$, $\hat{\varphi}_{2\uparrow} = \pi \sum_\sigma \int dx\, \hat{\psi}_{1\sigma}^\dagger(x)\hat{\psi}_{1\sigma}(x)$, and $\hat{\varphi}_{2\downarrow} = \hat{\varphi}_{2\uparrow} + \pi \int dx\, \hat{\psi}_{2\uparrow}^\dagger(x)\hat{\psi}_{2\uparrow}(x)$. These play only a minor role of giving a factor $\pm 1$ to $\hat{\psi}_{j\sigma}$.

around $k_F$. One finds for the momentum distribution function in the vicinity of $k_F$:

$$n_k \sim n_{k_F} - C\,\mathrm{sgn}\,(k-k_F)|k-k_F|^b, \tag{1.123}$$

with $C$ is a model-dependent constant and

$$b = \frac{1}{4}\left(\eta_\rho + \eta_\rho^{-1} - 2\right). \tag{1.124}$$

Note that for any $\eta_\rho \neq 1$, *i.e.*, for any nonvanishing interaction, the momentum distribution function has power-law singularities at the Fermi level, with vanishing single particle density states at $\varepsilon_F$. This behavior is quite different from a standard Fermi liquid, which have a finite density of states and a step-like singularity in $n_k$. The absence of a step at $k_F$ in the momentum distribution function implies the absence of a quasiparticle pole in the one-particle Green function.

### 1.5.3. *Fermi-Edge Singularity in Optical Processes*

One of the most striking optical phenomena due to existence of the Fermi-surface effects is the Fermi-edge singularity (FES) [60, 61, 62]; the optical absorption [emission] spectrum $W(\omega)$ in the vicinity of the absorption [emission] edge $E_F$ behaves as $W_{\rm OPA}(\omega) \sim (\omega - E_F)^\beta \Theta(\omega - E_F)$ $[W_{\rm PL}(\omega) \sim (E_F - \omega)^\beta \Theta(E_F - \omega)]$, which is a power-law anomaly. Here $\beta$ is called the FES exponent, which characterizes the shape of optical edge spectra; for $\beta < 0$, optical spectra exhibit a power-law divergence at the Fermi edge $\omega \sim E_F$, while it is a convergent one for $\beta > 0$. We here investigate the FES in one-dimensional (1D) systems with one or two Fermi surface(s). The FES in 2D have been studied in [63, 64, 65, 66]. In examining the FES in 1D, spatial transfer of an optical hole in a valence band yields a peculiar feature in 1D [67]; contrary to the higher dimensional cases, where the hole transfer destroys the FES [68, 69, 70], the FES in 1D systems persists against the hole delocalization.

According to Eq. (1.18), the important function to study for optical processes is

$$D(\omega) = 2\pi \sum_n |\langle \Psi_n|\hat{V}|\Psi_0\rangle|^2 \delta(E_n - E_0 - \omega), \tag{1.125}$$

where $\hat{V}$ is a one-photon operator, $\hat{V} = \sum_i \hat{v}_i$ in absorption and emission $\hat{v} = \exp(\pm i\vec{k}\cdot\vec{r})\vec{\varepsilon}\cdot\vec{p}$, and in optical scattering $\hat{v} = \exp(-i\vec{q}\cdot\vec{r})$. In absorption and emission the exponent can be neglected (the dipole approximation) unless we are high up in the X-ray or $\gamma$-ray region where $k$ becomes large. There is a simple relation between matrix elements of $\hat{V}_p = \sum_i \vec{\varepsilon}\cdot\vec{p}_i$ and $\hat{V}_r = \sum_i \vec{\varepsilon}\cdot\vec{r}_i$: $\langle \Psi_2|\hat{V}_p|\Psi_1\rangle = im(E_2 - E_1)\langle \Psi_2|V_r|\Psi_1\rangle$, and thus $D_p(\omega) = (m\omega)^2 D_r(\omega)$. The second-quantized form of $\hat{V}_p$ is $\hat{V}_p = \int d\vec{x}\,\hat{\psi}^\dagger(\vec{x})\,\vec{\varepsilon}\cdot\vec{p}\,\hat{\psi}(\vec{x}) = \sum_{kl} \vec{\varepsilon}\cdot\vec{p}_{kl}\,\hat{a}_k^\dagger \hat{a}_l$. The element $\langle \Psi_n|\hat{a}_k^\dagger \hat{a}_l|\Psi_0\rangle$ is zero unless $l$ is an occupied state and $k$ an unoccupied state in $|\Psi_0\rangle$. Further $E_n - E_0 = \epsilon_k - \epsilon_l$ and we have

$$D_p(\omega) = 2\pi \sum_k^{\rm unocc} \sum_l^{\rm occ} |\vec{\varepsilon}\cdot\vec{p}_{kl}|^2 \delta(\epsilon_k - \epsilon_l - \omega), \tag{1.126}$$

where $\epsilon_k$ ($\epsilon_l$) is the energy of the unoccupied (occupied) one-particle states. The general expression for $D(\omega)$ can be written by using $2\pi\delta(\omega) = \int \exp(i\omega t)dt$ as

$$\begin{aligned} D(\omega) &= \sum_n \langle \Psi_0|\hat{V}^\dagger|\Psi_n\rangle\langle\Psi_n|\hat{V}|\Psi_0\rangle \int_{-\infty}^{\infty} dt\, \exp[i(E_0+\omega-E_n)t] \\ &= \int_{-\infty}^{\infty} dt\, e^{i\omega t} \sum_n \langle \Psi_0|e^{i\hat{\mathcal{H}}t}\hat{V}^\dagger e^{-i\hat{\mathcal{H}}t}|\Psi_n\rangle\langle\Psi_n|\hat{V}|\Psi_0\rangle \\ &= \int_{-\infty}^{\infty} dt\, \langle\Psi_0|\hat{V}^\dagger(t)\hat{V}(0)|\Psi_0\rangle. \end{aligned} \tag{1.127}$$

In the integrand, $\langle\Psi_0|\hat{V}^\dagger(t)\hat{V}(0)|\Psi_0\rangle$ is just a correlation function. The relation in Eq. (1.127) is called a Van Hove expression. We define an excitation state $|\Phi(0)\rangle = \hat{V}|\Psi_0\rangle$ and its "overlap integral," $C(t) \equiv \langle\Phi(0)|\Phi(t)\rangle$. Clearly $C(t)$ is identical to the above correlation function (apart from a trivial factor), and $D(\omega)$ is essentially the Fourier transform of $C(t)$:

$$D(\omega) = \int_{-\infty}^{\infty} dt\, C(t) \exp[i(\omega+E_0)t] = 2\mathrm{Re}\int_0^{\infty} dt\, C(t)\exp[i(\omega+E_0)t]. \tag{1.128}$$

Here we use the relation $C^*(t) = C(-t)$.

### *1.5.3.1. FES in the single-band Tomonaga-Luttinger liquids*

Consider a 1D conductor with a Fermi sea in a conduction band with no spin splitting: $v_F \equiv v_F^\uparrow = v_F^\downarrow$, which is described as the single-band TL liquid. A single hole (created by an optical absorption process) is assumed to exist in the valence band, whose dispersion is $\varepsilon_{\mathrm{hole}}(k)$. The interaction matrix elements are assumed to be spin-independent for simplicity. Then we have three interaction parameters in the forward-scattering model: $g_2^{\mathrm{eh}}$, $g_2^{\mathrm{e}} \equiv g_{2\parallel} = g_{2\perp}$, and $g_4^{\mathrm{e}} \equiv g_{4\parallel} = g_{4\perp}$. Here we neglect the backward scattering, $g_1$, [71, 72, 73, 74] and the umklapp scattering, $g_3$.

Employing the TL bosonization [75], the system is described by the following 1D Hamiltonian:

$$\begin{aligned} \hat{\mathcal{H}}_{\mathrm{1TL}} &= \sum_k v_{\mathrm{charge}}|k|\, \hat{\alpha}_k^\dagger\hat{\alpha}_k + \sum_k v_{\mathrm{spin}}|k|\, \hat{\xi}_k^\dagger\hat{\xi}_k + \sum_{k,\sigma}[E_0 - \varepsilon_{\mathrm{hole}}(k)]\, \hat{d}_{k\sigma}^\dagger\hat{d}_{k\sigma} \\ &\quad - \frac{g_2^{\mathrm{eh}}\sqrt{\eta_\rho}}{2}\left(\frac{\pi}{L}\right)^{1/2} \sum_k\sum_{n\sigma} \hat{d}_{n\sigma}^\dagger\hat{d}_{n\sigma} e^{ikan}\sqrt{|k|}\,(\hat{\alpha}_k + \hat{\alpha}_{-k}^\dagger), \end{aligned} \tag{1.129}$$

where $\eta_\rho$ was defined by Eq. (1.118), $\hat{d}_{n\sigma}$ is an operator of the valence hole at site $n$, and $\hat{\alpha}_k$ and $\hat{\alpha}_k^\dagger$ ($\hat{\xi}_k$ and $\hat{\xi}_k^\dagger$) are operators of the charge (spin) of the TL boson in the conduction band, whose velocity is given by Eqs. (1.114) and (1.115): $v_{\mathrm{charge}} = [(v_F + g_4^{\mathrm{e}})^2 - (g_2^{\mathrm{e}})^2]^{1/2}$ and $v_{\mathrm{spin}} = v_F$. To treat the motion of a hole in the valence band, Ogawa, Furusaki, and Nagaosa [76] employ intermediate-coupling theory and introduce the following unitary transform:

$$\exp\left[\sum_k\sum_{n\sigma} f_k\, \hat{d}_{n\sigma}^\dagger\hat{d}_{n\sigma}\left(e^{ikan}\hat{\alpha}_k - e^{-ikan}\hat{\alpha}_k^\dagger\right)\right]. \tag{1.130}$$

In the case of a single valence hole, the variational parameter $f_k$ obeys

$$f_k = -\frac{g_2^{\rm eh}\sqrt{\eta_\rho}}{2}\left(\frac{\pi}{L}\right)^{1/2}\frac{e^{-\alpha|k|/2}\sqrt{|k|}}{v_{\rm charge}\,k - 2\mathcal{T}e^{-S}(1-\cos ak)}. \tag{1.131}$$

Here $\mathcal{T}$ is the transfer integral of the valence hole and $S \equiv \sum_k |f_k|^2(1-\cos ak)$, which given by a solution of the integral equation:

$$S = \frac{(g_2^{\rm eh})^2\eta_\rho}{4}\int_0^\infty dk\,\frac{k\,(1-\cos ak)e^{-\alpha k}}{[v_{\rm charge}k - 2\mathcal{T}e^{-S}(1-\cos ak)]^2}. \tag{1.132}$$

The effective transfer integral $\mathcal{T}^* \equiv \mathcal{T}e^{-S}$ of the hole is evaluated. Near $k \simeq 0$ the integrand becomes $k/2(v_{\rm charge})^2 \simeq 0$ without any divergence. Thus $S$ is finite and the hole in 1D is not localized ($\mathcal{T}^* \neq 0$) even when the low-energy excitations of the 1D Fermi sea are taken into account [77, 78]. Then, we need to examine the FES for a finite hole-mass.

Now we turn to the FES [79] for an arbitrary $\mathcal{T}^*$ at zero temperature. The optical spectrum is given by Eq. (1.128), where

$$C(t) \equiv \sum_{k,k'}\sum_{\sigma}\left\langle e^{i\hat{\mathcal{H}}_{\rm 1TL}t}\hat{d}_{-k,\sigma}\hat{a}_{k\sigma}e^{-i\hat{\mathcal{H}}_{\rm 1TL}t}\hat{a}^\dagger_{k'\sigma}\hat{d}^\dagger_{-k',\sigma}\right\rangle. \tag{1.133}$$

Using the standard boson algebra and Eq. (1.130), $C(\tau)$ is evaluated in the continuum approximation. The FES exponent $\beta_{\rm 1TL}$ for this single-band TL system is determined by the asymptotic form of $C(t) \sim t^{-\beta_{\rm 1TL}-1}$ for large $t$. The FES exponent consists of two parts: $\beta_{\rm 1TL} = \beta^{\rm ex}_{\rm 1TL} + \beta^{\rm oc}_{\rm 1TL}$. These two terms correspond to the "many-body excitonic (ex) correlation" part and the "orthogonality catastrophe (oc)" part, respectively. The latter part, $\beta^{\rm oc}_{\rm 1TL}$, is also an exponent of the hole Green function. Generally speaking, the excitonic correlation enhances the optical transition, while the orthogonality of the Fermi sea prohibits the transition [80, 81]. These components are given as [76]

$$\beta^{\rm ex}_{\rm 1TL} = \mathcal{F}(\bar{g}_2^{\rm eh}, \bar{g}_2^{\rm e}) - \frac{1}{2}, \tag{1.134}$$

$$\beta^{\rm oc}_{\rm 1TL} = \mathcal{G}(\bar{g}_2^{\rm eh}, \bar{g}_2^{\rm e}), \tag{1.135}$$

where

$$\mathcal{F}(x,y) \equiv \frac{1}{2}(1-x)(1-y^2)^{-1/2}, \tag{1.136}$$

$$\mathcal{G}(x,y) \equiv \frac{1}{4}x^2(1-y)^{-1/2}(1+y)^{-3/2}, \tag{1.137}$$

and $\bar{g} \equiv g/(v_{\rm F} + g_4^{\rm e})$. Both components are independent of the hole transfer $\mathcal{T}^*$. Thus in 1D, the hole recoil results in no effect on the FES. The numerical calculation [82] of the finite-size Hubbard model also supports this property.

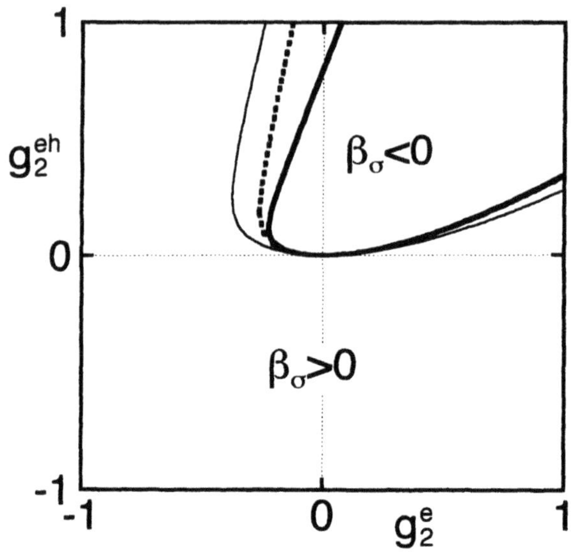


Figure 1.14. Phase diagram for the sign of the FES exponent $\beta_{1\mathrm{TL}}$ is drawn in the ($g_2^{\mathrm{e}} \equiv g_{2\|}^{\mathrm{e}} = g_{2\perp}^{\mathrm{e}}, g_2^{\mathrm{eh}} \equiv g_{2\|}^{\mathrm{eh}} = g_{2\perp}^{\mathrm{eh}}$) plane. A thin solid line represents the boundary between the divergent ($\beta_{1\mathrm{TL}} < 0$) and convergent ($\beta_{1\mathrm{TL}} > 0$) edge spectra for $v_F^{\uparrow} = v_{\mathrm{F}}^{\downarrow}$. The thick solid (broken) line corresponds to the boundary for the sign of $\beta_{\downarrow}$ ($\beta_{\uparrow}$) when $v_{\mathrm{F}}^{\uparrow}/v_{\mathrm{F}}^{\downarrow} = \frac{1}{2}$.

The region of the divergent edge ($\beta_{1\mathrm{TL}} < 0$) is shown in the ($g_2^{\mathrm{e}}$, $g_2^{\mathrm{eh}}$) plane (Fig. 1.14). The divergent edge requires the attractive interaction between electrons and a hole ($g_2^{\mathrm{eh}} > 0$). Weak repulsive interaction among the conduction electrons ($g_2^{\mathrm{e}} > 0$) weakens the orthogonality of the 1D Fermi sea, and consequently makes the FES peak sharper. However, the edge spectrum becomes convergent in the case of stronger repulsive correlation.

Here we linearize the renormalized dispersion of the valence band as $\varepsilon_{\mathrm{hole}}(k)e^{-S} \approx \mp v_{\mathrm{hole}}^{*}(k \mp k_{\mathrm{F}})$ near $k = \pm k_{\mathrm{F}}$. Energy at which the FES occurs, $E_{\mathrm{F}}$, is evaluated as

$$E_{\mathrm{F}} = E_0 - \varepsilon_{\mathrm{hole}}(k_{\mathrm{F}})e^{-S} + \frac{2}{\pi L}\eta_\rho \sum_{k>0} \left(v_{\mathrm{charge}} h_k^2 - \pi g_2^{\mathrm{eh}} h_k\right), \tag{1.138}$$

where $h_k \equiv \frac{1}{2}\pi g_2^{\mathrm{eh}}[v_{\mathrm{charge}} - 2\mathcal{T}^{*}(1-\cos ak)/|k|]^{-1}$. We have the following two situations:

(i) When $v_{\mathrm{hole}}^{*} < v_{\mathrm{charge}}$,

$$W_{\mathrm{OPA}}(\omega) \sim \begin{cases} 0 & \omega < E_{\mathrm{F}} \\ (\omega - E_{\mathrm{F}})^{\beta_{1\mathrm{TL}}} & \omega > E_{\mathrm{F}}. \end{cases} \tag{1.139}$$

In this case, $E_F$ is the energy of the absorption edge.
(ii) When $v^*_{\rm hole} > v_{\rm charge}$, on the other hand,

$$W_{\rm OPA}(\omega) \sim \begin{cases} (E_F - \omega)^{\beta_{\rm 1TL}} & \omega < E_F \\ (\omega - E_F)^{\beta_{\rm 1TL}} & \omega > E_F. \end{cases} \tag{1.140}$$

In this case, an "indirect" transition takes place at $\omega \sim E_{\rm ind} \equiv E_F - (v^*_{\rm hole} - v_{\rm charge})k_F < E_F$ from the top of the valence band ($k = 0$) to the Fermi point in the conduction band ($k = \pm k_F$), together with an excitation of a TL boson with momentum $\mp k_F$ in the conduction band [83]. Then the FES at $\omega \sim E_F$ will be observed in a continuum spectrum. This is a feature peculiar to the 1D system.

### 1.5.3.2. *An experimental observation of the FES in Q1D*

An experimental observation of the FES in Q1D systems, doped GaAs quantum wires, is reported by Calleja and co-workers [67]. The sample has a 250 Å thick QW and two monolayers of Si in the top $Al_{0.32}Ga_{0.68}As$ barrier to create a Q2D electron gas with electron density and mobility of $n_{2D} = 3.2 \times 10^{11}{\rm cm}^{-2}$ and $\mu_{2D} = 1.1 \times 10^6 {\rm cm}^2/{\rm Vs}$. The 1D pattern consists of lines 1000 Å wide with 2000 Å period. The quantum wires are formed by depletion of the QW in the openings between the lines, using low-energy ion bombardment.

The strong, exciton-like peaks observed in the PLE spectra of Q1D sample are interpreted as the FES of the 1D electron gas. The sharp peaks observed disappear at temperatures around 30 K. Their temperature dependence is in agreement with this interpretation because the singularities disappear at temperature comparable to the Fermi energy. The 1D FES shows two main differences with respect to the 2D case. (i) The singularity is sharper and stronger in 1D and (ii) it is also clearly observed in the emission spectra. These differences can be understood if one considers that the PLE threshold for a 2D sample is an indirect transition from the top of the valence band to the Fermi level, together with a zero-energy excitation of the Fermi sea. For high electron concentrations the energy difference between this indirect transition and the direct one at the Fermi wavevector $k_F$ is larger than the binding energy of the Mahan exciton [61] and no FES appears. At low electron densities, the FES is weak but observable, even in the emission spectra, for the temperature range where there is a noticeable population of holes at $k_F$ and yet the Fermi distribution is not smeared out. In 1D systems, on the other hand, indirect transitions cannot take place [corresponding to the case of Eq. (1.139)] and the PLE onset is a vertical transition at $k_F$. This means that the FES will be present in 1D systems at any electron concentration and for finite hole mass. This experiment shows that the hole recoil plays no role in the many-body reaction of the Fermi sea during the absorption of light in 1D systems, in agreement with theoretical results.

### *1.5.3.3. FES in the spin-split Tomonaga-Luttinger liquids*

Thus far, all the previous theories of the FES have assumed both that the two-body scatterings are spin-independent and that there is no spin splitting in the bands. Very recently, however, a spin-split character of the conduction band was predicted (even at zero magnetic field) in III–V zincblende semiconductor quantum wires [84]. In this subsection, therefore, we shall discuss the FES exponent for 1D conductors with spin-dependent electronic correlations and with spin-split bands [85]. The interaction is then described by six parameters: $g^{\rm e}_{2\parallel}$, $g^{\rm e}_{2\perp}$, $g^{\rm e}_{4\parallel}$, $g^{\rm e}_{4\perp}$, $g^{\rm eh}_{2\parallel}$, and $g^{\rm eh}_{2\perp}$. Hereafter we use the abbreviated notations, $g^{\nu}_{i+} \equiv \frac{1}{2}(g^{\nu}_{i\parallel} + g^{\nu}_{i\perp})$ and $g^{\nu}_{i-} \equiv \frac{1}{2}(g^{\nu}_{i\parallel} - g^{\nu}_{i\perp})$. A valence hole can be safely assumed to be localized since the hole recoil does not affect the FES exponent in 1D systems, as discussed previously. When the conduction band is split into two spin subbands, the band dispersion near Fermi points is linearized as

$$\varepsilon(k) \approx \sum_{j=1,2}\sum_{\sigma=\uparrow,\downarrow} \varepsilon^{\sigma}_{j}(k) = \sum_{\sigma=\uparrow,\downarrow} v^{\sigma}_{\rm F}(k - k^{\sigma}_{\rm F}) - v^{\sigma}_{\rm F}(k + k^{\sigma}_{\rm F}), \tag{1.141}$$

as shown in Fig. 1.15. Here $v^{\sigma}_{\rm F}$ is the spin-dependent bare Fermi velocity of the conduction electrons with spin $\sigma = \uparrow, \downarrow = \pm 1$. We note that the time-reversal symmetry is broken in this system, corresponding to the case of applying an external magnetic field [86].

The TL liquid with the spin splitting is described by the bosonized Hamiltonian: $\hat{\mathcal{H}} = \frac{1}{2}\sum_{p>0} p\, \boldsymbol{b}^{\dagger}_{p}\mathbf{H}\boldsymbol{b}_{p}$, where $\boldsymbol{b}^{\dagger}_{p} \equiv (2\pi/Lp)^{1/2}\ [\hat{\rho}_{1\uparrow}(p), \hat{\rho}_{2\uparrow}(p), \hat{\rho}_{1\downarrow}(p), \hat{\rho}_{2\downarrow}(p)]$ is a boson operator. Here

$$\mathbf{H} = \begin{bmatrix} 2v^{\uparrow}_{\rm F} + g^{\rm e}_{4\parallel} & g^{\rm e}_{2\parallel} & g^{\rm e}_{4\perp} & g^{\rm e}_{2\perp} \\ g^{\rm e}_{2\parallel} & 2v^{\uparrow}_{\rm F} + g^{\rm e}_{4\parallel} & g^{\rm e}_{2\perp} & g^{\rm e}_{4\perp} \\ g^{\rm e}_{4\perp} & g^{\rm e}_{2\perp} & 2v^{\downarrow}_{\rm F} + g^{\rm e}_{4\parallel} & g^{\rm e}_{2\parallel} \\ g^{\rm e}_{2\perp} & g^{\rm e}_{4\perp} & g^{\rm e}_{2\parallel} & 2v^{\downarrow}_{\rm F} + g^{\rm e}_{4\parallel} \end{bmatrix}. \tag{1.142}$$

This can be diagonalized by employing a unitary transformation [87]:

$$\boldsymbol{b}_{p} = \mathbf{U}\boldsymbol{c}_{p} \equiv \begin{bmatrix} x^{\uparrow}_{a} & y^{\uparrow}_{a} & x^{\uparrow}_{b} & y^{\uparrow}_{b} \\ y^{\uparrow}_{a} & x^{\uparrow}_{a} & y^{\uparrow}_{b} & x^{\uparrow}_{b} \\ x^{\downarrow}_{a} & y^{\downarrow}_{a} & x^{\downarrow}_{b} & y^{\downarrow}_{b} \\ y^{\downarrow}_{a} & x^{\downarrow}_{a} & y^{\downarrow}_{b} & x^{\downarrow}_{b} \end{bmatrix} \begin{bmatrix} \hat{B}_{ap} \\ \hat{A}^{\dagger}_{ap} \\ \hat{B}_{bp} \\ \hat{A}^{\dagger}_{bp} \end{bmatrix}, \tag{1.143}$$

where $\hat{A}_{\mu p}$ and $\hat{B}_{\mu p}$ for $\mu = a, b$ are new boson operators. The matrix elements of $\mathbf{U}$, $x^{\uparrow}_{\mu}$, $y^{\uparrow}_{\mu}$, $x^{\downarrow}_{\mu}$, and $y^{\downarrow}_{\mu}$, are given as $x^{\uparrow}_{a} = f^{\uparrow\downarrow}_{2a} f^{\downarrow\downarrow}_{2a}\Delta_a$, $x^{\downarrow}_{a} = f^{\downarrow\uparrow}_{2a} f^{\uparrow\downarrow}_{2a}\Delta_a$, $y^{\uparrow}_{a} = f_{4a} f^{\downarrow\downarrow}_{2a}\Delta_a$, and $y^{\downarrow}_{a} = f_{4a} f^{\uparrow\downarrow}_{2a}\Delta_a$, where

$$\Delta_a \equiv \left[\frac{N_b(f^{\uparrow\downarrow}_{2b})^2 - M_b(f^{\downarrow\downarrow}_{2b})^2}{M_a N_b (f^{\uparrow\downarrow}_{2b})^2 (f^{\downarrow\downarrow}_{2a})^2 - M_b N_a (f^{\uparrow\downarrow}_{2a})^2 (f^{\downarrow\downarrow}_{2b})^2}\right]^{1/2}, \tag{1.144}$$

$$f^{\uparrow\downarrow}_{2\mu} \equiv g^{\rm e}_{2\perp}[(g^{\rm e}_{2\parallel})^2 + (g^{\rm e}_{4\perp})^2 - (g^{\rm e}_{2\perp})^2 + V^{\uparrow}_{\mu+}V^{\downarrow}_{\mu-}] - g^{\rm e}_{2\parallel}g^{\rm e}_{4\perp}(V^{\uparrow}_{\mu+} + V^{\downarrow}_{\mu-}), \tag{1.145}$$

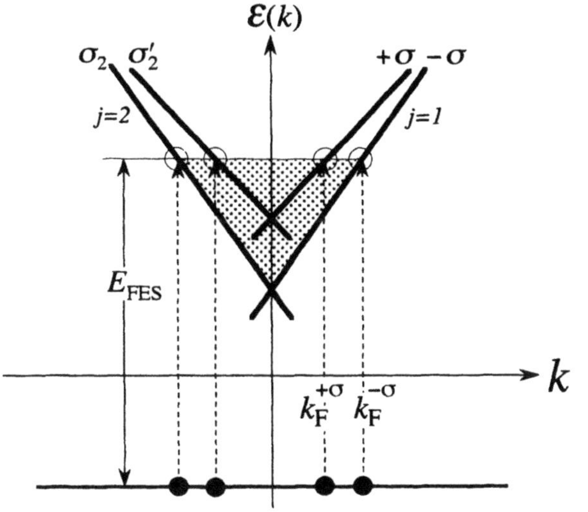


Figure 1.15. The single-electron band structure of the TL liquid with split spin subbands is schematically drawn. The case of $\sigma_2 = +\sigma$ and $\sigma_2' = -\sigma$ ($\sigma_2 = -\sigma$ and $\sigma_2' = +\sigma$) corresponds to the spin-split system with (without) the time-reversal symmetry. The valence hole is assumed to be localized and the dispersion of the conduction bands is linearized near the Fermi points (denoted as open circles). Two types of optical absorption processes (denoted as vertical dashed lines with an upward arrow) can be chosen selectively by using the circularly polarization of incident light.

$$f_{2\mu}^{\downarrow\downarrow} \equiv g_{2\parallel}^{\rm e}[(g_{2\perp}^{\rm e})^2 + (g_{4\perp}^{\rm e})^2 - (g_{2\parallel}^{\rm e})^2 + V_{\mu+}^{\downarrow}V_{\mu-}^{\downarrow}] - g_{2\perp}^{\rm e}g_{4\perp}^{\rm e}(V_{\mu+}^{\downarrow} + V_{\mu-}^{\downarrow}), \quad (1.146)$$

$$f_{2\mu}^{\downarrow\uparrow} \equiv g_{2\perp}^{\rm e}[(g_{2\parallel}^{\rm e})^2 + (g_{4\perp}^{\rm e})^2 - (g_{2\perp}^{\rm e})^2 + V_{\mu+}^{\downarrow}V_{\mu-}^{\uparrow}] - g_{2\parallel}^{\rm e}g_{4\perp}^{\rm e}(V_{\mu+}^{\downarrow} + V_{\mu-}^{\uparrow}), \quad (1.147)$$

$$f_{4\mu} \equiv g_{4\perp}^{\rm e}[(g_{2\parallel}^{\rm e})^2 + (g_{2\perp}^{\rm e})^2 - (g_{4\perp}^{\rm e})^2 + V_{\mu-}^{\uparrow}V_{\mu-}^{\downarrow}] - g_{2\parallel}^{\rm e}g_{2\perp}^{\rm e}(V_{\mu-}^{\uparrow} + V_{\mu-}^{\downarrow}), \quad (1.148)$$

with $M_\mu \equiv (f_{2\mu}^{\uparrow\downarrow})^2 - (f_{4\mu})^2$, $N_\mu \equiv (f_{2\mu}^{\downarrow\uparrow})^2 - (f_{4\mu})^2$, $V_{\mu+}^{\sigma} \equiv 2v_{\rm F}^{\sigma} + g_{4\parallel}^{\rm e} + 2v_\mu$, and $V_{\mu-}^{\sigma} \equiv 2v_{\rm F}^{\sigma} + g_{4\parallel}^{\rm e} - 2v_\mu$. Other elements, $x_b^{\uparrow}$, $y_b^{\uparrow}$, $x_b^{\downarrow}$, and $y_b^{\downarrow}$, are given by replacing the index $a$ ($b$) in the above equations with $b$ ($a$).

After the diagonalization procedure, we find that $\hat{A}_{ap}$ and $\hat{B}_{ap}$ ($\hat{A}_{bp}$ and $\hat{B}_{bp}$) describe the new normal modes of collective excitation, labeled $a$ ($b$), which is a hybrid collective mode of the charge-density ($\hat{\rho}_\uparrow + \hat{\rho}_\downarrow$) and spin-density ($\hat{\rho}_\uparrow - \hat{\rho}_\downarrow$) excitations with momentum $p$. The velocity of the $\mu$ mode is given as $v_\mu \equiv [\Gamma \pm \Lambda]^{1/2}$, where $\mu = a$ ($b$) corresponds to the + (−) sign, and

$$\Gamma = \frac{1}{2}\left[(v_{\rm F}^{\uparrow})^2 + (v_{\rm F}^{\downarrow})^2\right] + \frac{1}{4}\left[(g_{4\parallel}^{\rm e})^2 + (g_{4\perp}^{\rm e})^2 - (g_{2\parallel}^{\rm e})^2 - (g_{2\perp}^{\rm e})^2\right] + g_{4\parallel}^{\rm e}v_{\rm F}^{+}, \quad (1.149)$$

$$\Lambda = \left[ \left\{ g_{4\perp}^{e} v_F^+ - \frac{1}{2} \left( g_{2\parallel}^{e} g_{2\perp}^{e} - g_{4\parallel}^{e} g_{4\perp}^{e} \right) \right\}^2 + (v_F^-)^2 \left\{ (2v_F^+ + g_{4\parallel}^{e})^2 - (g_{2\perp}^{e})^2 \right\} \right]^{1/2}, \quad (1.150)$$

with $v_F^+ \equiv \frac{1}{2}(v_F^\uparrow + v_F^\downarrow)$ and $v_F^- \equiv \frac{1}{2}(v_F^\uparrow - v_F^\downarrow)$.

Along a similar manner in the previous subsection, the FES exponent is evaluated as $\beta_\sigma = \beta_\sigma^{ex} + \beta_\sigma^{oc}$, where

$$\beta_\sigma^{ex} = \sum_{\mu=a,b} \left[ (x_\mu^\sigma)^2 + (y_\mu^\sigma)^2 - (x_\mu^\sigma - y_\mu^\sigma) \frac{G_\mu^\sigma}{v_\mu} \right] - 1, \quad (1.151)$$

$$\beta_\sigma^{oc} = \frac{1}{2} \sum_{\mu=a,b} \left( \frac{G_\mu^\sigma}{v_\mu} \right)^2, \quad (1.152)$$

with $G_\mu^\sigma = \frac{1}{2}\sum_{\sigma'=\uparrow,\downarrow} \left( g_{2\parallel}^{eh}\delta_{\sigma,\sigma'} + g_{2\perp}^{eh}\delta_{\sigma,-\sigma'} \right) \left( x_\mu^{\sigma'} + y_\mu^{\sigma'} \right)$. Note that $\beta_\uparrow \neq \beta_\downarrow$ in general. Hence the FES exponent depends on the spin orientation of an excited electron $\sigma$ (or equivalently that of an excited hole $-\sigma$), that is, $\beta_\sigma$ depends on the rotation direction of circularly polarized incident light. In the case of $v_F^\uparrow = v_F^\downarrow \equiv v_F$ but $k_F^\uparrow \neq k_F^\downarrow$, the FES exponent becomes independent of the hole spin, *i.e.*, $\beta_\uparrow = \beta_\downarrow \equiv \beta = \beta^{ex} + \beta^{oc}$, where

$$\beta^{ex} = \mathcal{F}(\bar{g}_{2+}^{eh}, \bar{g}_{2+}^{e}) + \mathcal{F}(\bar{g}_{2-}^{eh}, \bar{g}_{2-}^{e}) - 1, \quad (1.153)$$

$$\beta^{oc} = \mathcal{G}(\bar{g}_{2+}^{eh}, \bar{g}_{2+}^{e}) + \mathcal{G}(\bar{g}_{2-}^{eh}, \bar{g}_{2-}^{e}). \quad (1.154)$$

Here $\mathcal{F}(x,y)$ and $\mathcal{G}(x,y)$ have been defined in Eqs. (1.136) and (1.137), and $\bar{g}_{2\pm}^{\gamma} \equiv g_{2\pm}^{\gamma}/(v_F + g_{4\pm}^{e})$. Thus when $v_F^\uparrow = v_F^\downarrow$, the FES exponent is independent of the rotation direction of circularly polarized light even for $g_{i\parallel}^{\gamma} \neq g_{i\perp}^{\gamma}$.

The phase diagrams for the sign of $\beta_\sigma$ are drawn in Fig. 1.14 on the ($g_2^e$, $g_2^{eh}$) plane and Fig. 1.16 on the ($g_{2\parallel}^{eh}$, $g_{2\perp}^{eh}$) plane. The region where $\beta_\sigma < 0$ exists only in the right half-plane ($g_{2\parallel}^{eh} > 0$) of Fig. 1.16. This means that the spin-singlet "exciton" contribution ($g_{2\parallel}^{eh} > 0$) plays a main role in yielding the power-law divergence. When $v_F^\uparrow \neq v_F^\downarrow$, there are two striking features; (i) we find the region where $\beta_\uparrow < 0$ but $\beta_\downarrow > 0$. In such a region, the divergent or convergent edge spectrum can be observed by simply changing the rotation direction of circularly polarized light. (ii) The exponent can be negative even when the *e-h* interaction is repulsive ($g_{2\parallel}^{eh} < 0$). In fact, there exists in Fig. 1.16 the region where $\beta_\uparrow < 0$ even for $g_{2\parallel}^{eh} < 0$ and $g_{2\perp}^{eh} < 0$. This is in striking contrast to the results of previous higher-dimensional theories [60, 76]. One finds that $\beta_\sigma < 0$ in the case of $g_{2\parallel}^{eh} < 0$ requires $v_F^\sigma < v_F^{-\sigma}$, hence existence of this region in the phase diagram is peculiar to the spin-split system with $v_F^\uparrow \neq v_F^\downarrow$.

There exists another spin-split band system *with* the time-reversal symmetry (see Fig. 1.15), where the band dispersion is

$$\tilde{\varepsilon}(k) \approx \sum_{j=1,2} \sum_{\sigma=\uparrow,\downarrow} \tilde{\varepsilon}_j^\sigma(k) = \sum_{\sigma=\uparrow,\downarrow} v_F^\sigma(k - k_F^\sigma) - v_F^{-\sigma}(k + k_F^\sigma). \quad (1.155)$$

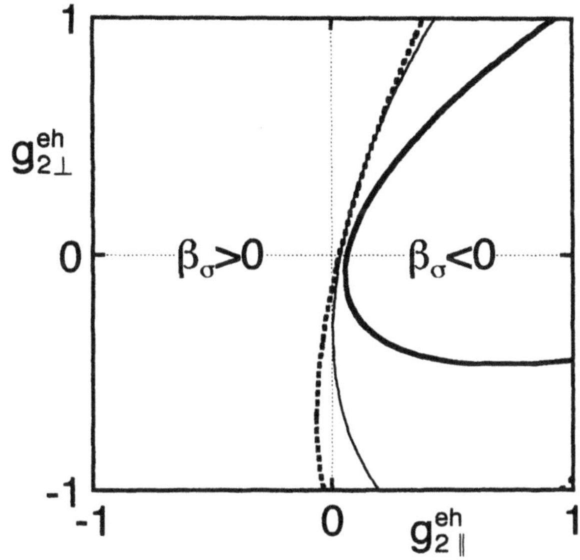


Figure 1.16. The phase diagram for the sign of the FES exponent in the plane of $g^{\rm eh}_{2\parallel}$ and $g^{\rm eh}_{2\perp}$, showing the boundary between the divergent edge ($\beta_\sigma < 0$) and the convergent one ($\beta_\sigma > 0$). Here $g^{\rm e}_{2\parallel} = g^{\rm e}_{4\parallel} = 0.1$ and $g^{\rm e}_{2\perp} = g^{\rm e}_{4\perp} = 0.4$. The thick solid (broken) curve represents $\beta_\downarrow = 0$ ($\beta_\uparrow = 0$) for $v^\uparrow_{\rm F} = 0.5$ and $v^\downarrow_{\rm F} = 1.0$. The thin solid curve represents $\beta_\uparrow = \beta_\downarrow = 0$ for $v^\uparrow_{\rm F} = v^\downarrow_{\rm F} = 1.0$.

Such a spin-split character is expected in III–V zincblende semiconductor wires [84]. The FES exponent in this system, denoted $\tilde\beta^{\rm ex} + \tilde\beta^{\rm oc}$, is independent of the hole spin $-\sigma$ in contrast to the case of broken time-reversal symmetry [Eqs. (1.151) and (1.152)]. Thus the edge spectra do not depend on the rotating direction of circularly polarized light. The explicit form of the exponent is

$$\tilde\beta^{\rm ex} = \min_{\sigma=\uparrow,\downarrow} \sum_{\mu=a,b} \left[ (\tilde x^\sigma_\mu)^2 + (\tilde y^\sigma_\mu)^2 - \tilde x^\sigma_\mu \frac{\tilde G^{-\sigma}_\mu}{\tilde v_\mu} + \tilde y^\sigma_\mu \frac{\tilde G^\sigma_\mu}{\tilde v_\mu} \right] - 1, \tag{1.156}$$

$$\tilde\beta^{\rm oc} = \frac{1}{4} \sum_{\sigma=\uparrow,\downarrow} \sum_{\mu=a,b} \left( \frac{\tilde G^\sigma_\mu}{\tilde v_\mu} \right)^2, \tag{1.157}$$

where $\tilde G^\sigma_\mu = \frac{1}{2} \sum_{\sigma'=\uparrow,\downarrow} \left( g^{\rm eh}_{2\parallel} \delta_{\sigma,\sigma'} + g^{\rm eh}_{2\perp} \delta_{\sigma,-\sigma'} \right) \left( \tilde x^{-\sigma'}_\mu + \tilde y^{\sigma'}_\mu \right)$. Here $\tilde x^\sigma_\mu$, $\tilde y^\sigma_\mu$, and $\tilde v_\mu$ are given respectively by $x^\sigma_\mu$, $y^\sigma_\mu$, and $v_\mu$ with interchanging $g^{\rm e}_{2\parallel} \rightleftharpoons g^{\rm e}_{2\perp}$.

We mention here the FES exponent in the spin-split chiral Luttinger liquid, whose dispersion is defined as $\varepsilon_{\mathrm{chiral}}(k) \approx \sum_{\sigma=\uparrow,\downarrow} \varepsilon^{\sigma}_{\mathrm{chiral}}(k)$ with $\varepsilon^{\sigma}_{\mathrm{chiral}}(k) = v_{\mathrm{F}}(k - k^{\sigma}_{\mathrm{F}})$. We write down here the FES exponent $\beta_{\mathrm{chiral}} = \beta^{\mathrm{ex}}_{\mathrm{chiral}} + \beta^{\mathrm{oc}}_{\mathrm{chiral}}$:

$$\beta^{\mathrm{ex}}_{\mathrm{chiral}} = -\frac{1}{2}\left(\bar{g}^{\mathrm{eh}}_{2+} + \bar{g}^{\mathrm{eh}}_{2-}\right), \tag{1.158}$$

$$\beta^{\mathrm{oc}}_{\mathrm{chiral}} = \frac{1}{8}\left[(\bar{g}^{\mathrm{eh}}_{2+})^2 + (\bar{g}^{\mathrm{eh}}_{2-})^2\right]. \tag{1.159}$$

Note that the $g^{\mathrm{e}}_2$ process is absent in the chiral Luttinger liquid. Comparing these with Eqs. (1.153) and (1.154) for $g^{\mathrm{e}}_{2\|} = g^{\mathrm{e}}_{2\perp} = 0$, one finds $\beta^{\mathrm{ex}}_{\mathrm{chiral}} = \beta^{\mathrm{ex}}$ and $\beta^{\mathrm{oc}}_{\mathrm{chiral}} = \frac{1}{2}\beta^{\mathrm{oc}}$. Since the orthogonality part, $\beta^{\mathrm{oc}}_{\mathrm{chiral}} > 0$, is smaller, the edge spectra tend to diverge more strongly even in the case of $g^{\mathrm{eh}}_{2\|} < 0$, $g^{\mathrm{eh}}_{2\perp} < 0$ and $v^{\uparrow}_{\mathrm{F}} = v^{\downarrow}_{\mathrm{F}}$, in comparison with the spin-split TL liquid.

We comment on the effects of the *e-h* backward scattering, denoted as $g^{\mathrm{eh}}_{1\|}$ and $g^{\mathrm{eh}}_{1\perp}$, on the FES exponent. In the case of $g^{\mathrm{eh}}_{1\|} = g^{\mathrm{eh}}_{1\perp} \equiv g^{\mathrm{eh}}_1 > 0$, the system is decoupled into two semi-infinite wires below a characteristic energy scale, $\varepsilon_1$ [73]. Then the exponent contains additionally a universal constant, independent of $g^{\mathrm{eh}}_1$ and $g^{\mathrm{eh}}_2$. Therefore, our results may be valid only when $|\omega - E_{\mathrm{F}}| > \varepsilon_1$. However, the main features of the edge spectra are determined solely by the forward-scattering correlations, *i.e.*, by Eqs. (1.151) and (1.152). Furthermore, interface roughness, impurity scatterings, and the Peierls transition may introduce the mass gap, $\varepsilon_{\mathrm{mass}}$, to all the collective excitations. Then the FES is cut off and can be seen only when $|\omega - E_{\mathrm{F}}| > \varepsilon_{\mathrm{mass}}$.

#### *1.5.3.4. FES in the two-band Tomonaga-Luttinger liquids*

With irradiation of an intense laser pulse to undoped semiconducting wires, we can explore the many-body effects in a high-density *e-h* system in 1D, which is treated as a two-band TL liquid [88, 89]. Densities of electrons (in the conduction band) and holes (in the valence band) are assumed to be equal to each other, both of which are high enough so that the Fermi points are well-defined in both bands. We further assume that this system is in quasi-thermal equilibrium within a radiative lifetime (so-called ***rigid*** Fermi-sea picture). Although the Fermi wavevectors $k_{\mathrm{F}}$'s are generally different for the two bands [90, 91, 92], we here concentrate on the 1D high-density *e-h* system with the same Fermi wavevector $k_{\mathrm{F}}$ for both the conduction and valence bands. Note that the valence holes are mobile as well as the conduction electrons.

The bosonization method [59, 93, 94, 95, 96, 97] is used to treat the collective excitations of the degenerate electrons and holes in 1D. To this end, we linearize the dispersions near the Fermi points as

$$\varepsilon_{\mathrm{e}}(k) = \sum_{j=1,2} \varepsilon^{\mathrm{e}}_j(k) = v^{\mathrm{e}}_{\mathrm{F}}(k - k_{\mathrm{F}}) - v^{\mathrm{e}}_{\mathrm{F}}(k + k_{\mathrm{F}}), \tag{1.160}$$

$$\varepsilon_{\mathrm{h}}(k) = \sum_{j=1,2} \varepsilon^{\mathrm{h}}_j(k) = v^{\mathrm{h}}_{\mathrm{F}}(k + k_{\mathrm{F}}) - v^{\mathrm{h}}_{\mathrm{F}}(k - k_{\mathrm{F}}), \tag{1.161}$$

where $v^{\mathrm{e}}_{\mathrm{F}}$ ($v^{\mathrm{h}}_{\mathrm{F}}$) is the Fermi velocity of the conduction (valence) band. The interaction matrix element is assumed to be spin-independent and parameterized to the six interaction matrix elements: $g_i^{\mathrm{e}}$, $g_i^{\mathrm{h}}$, and $g_i^{\mathrm{eh}}$ for $i = 1, 2$. We study the forward-scattering model, which includes only $g_2$'s ($g_2^{\mathrm{e}}$, $g_2^{\mathrm{h}}$, and $g_2^{\mathrm{eh}}$) and is solved exactly, and next treat the backward scatterings as a perturbation. There exist other forward scattering processes denoted as $g_4$'s, which contribute only to the correction of the Fermi velocities as $v_{\mathrm{F}}^{\nu} \to v_{\mathrm{F}}^{\nu} + g_4^{\nu}$ with $\nu =$ e or h.

After bosonization, using the phase variables, we obtain a phase Hamiltonian of the two-band TL (2TL) liquid [98, 99]: $\hat{\mathcal{H}}_{\mathrm{2TL}} \equiv \hat{\mathcal{H}}_{\mathrm{forward}}^{\mathrm{charge}} + \hat{\mathcal{H}}_{\mathrm{forward}}^{\mathrm{spin}} + \hat{\mathcal{H}}_{\mathrm{back}}$, where

$$\begin{aligned}\hat{\mathcal{H}}_{\mathrm{forward}}^{\mathrm{charge}} &= \int dx \left\{ A_{\mathrm{e}} \hat{P}_{\mathrm{e}}^2(x) + C_{\mathrm{e}} [\nabla \hat{\theta}_{\mathrm{e}}(x)]^2 + A_{\mathrm{h}} \hat{P}_{\mathrm{h}}^2(x) + C_{\mathrm{h}} [\nabla \hat{\theta}_{\mathrm{h}}(x)]^2 \right. \\ &\quad \left. + 2C_{\mathrm{eh}} \nabla \hat{\theta}_{\mathrm{e}}(x) \nabla \hat{\theta}_{\mathrm{h}}(x) \right\}, \qquad (1.162) \\ \hat{\mathcal{H}}_{\mathrm{forward}}^{\mathrm{spin}} &= \int dx \left\{ B_{\mathrm{e}} \hat{\Pi}_{\mathrm{e}}^2(x) + D_{\mathrm{e}} [\nabla \hat{\phi}_{\mathrm{e}}(x)]^2 + B_{\mathrm{h}} \hat{\Pi}_{\mathrm{h}}^2(x) + D_{\mathrm{h}} [\nabla \hat{\phi}_{\mathrm{h}}(x)]^2 \right\}, \qquad (1.163) \\ \hat{\mathcal{H}}_{\mathrm{back}} &= \frac{4g_1^{\mathrm{e}}}{(2\pi\alpha)^2} \int dx \cos\left[\sqrt{2}\hat{\phi}_{\mathrm{e}}(x)\right] + \frac{4g_1^{\mathrm{h}}}{(2\pi\alpha)^2} \int dx \cos\left[\sqrt{2}\hat{\phi}_{\mathrm{h}}(x)\right] \\ &\quad + \frac{16 g_1^{\mathrm{eh}}}{(2\pi\alpha)^2} \int dx \cos\left[\frac{\hat{\theta}_{\mathrm{e}}(x) + \hat{\theta}_{\mathrm{h}}(x)}{\sqrt{2}}\right] \cos\frac{\hat{\phi}_{\mathrm{e}}(x)}{\sqrt{2}} \cos\frac{\hat{\phi}_{\mathrm{h}}(x)}{\sqrt{2}}. \qquad (1.164)\end{aligned}$$

Here $\hat{\theta}_{\mathrm{e}}(x)$ $[\hat{\theta}_{\mathrm{h}}(x)]$ is the charge phase of the conduction [valence] electrons [see Eqs. (1.116) and (1.117)], whose conjugate momentum is $\hat{P}_{\mathrm{e}}(x)$ $[\hat{P}_{\mathrm{h}}(x)]$, while $\hat{\phi}_{\mathrm{e}}(x)$ $[\hat{\phi}_{\mathrm{h}}(x)]$ is the spin phase, whose conjugate momentum is $\hat{\Pi}_{\mathrm{e}}(x)$ $[\hat{\Pi}_{\mathrm{h}}(x)]$. Coefficients are given by

$$\begin{aligned} A_{\nu} &\equiv 2\pi(v_{\mathrm{F}}^{\nu} + g_4^{\nu} - g_2^{\nu}), \qquad (1.165) \\ B_{\nu} &\equiv 2\pi(v_{\mathrm{F}}^{\nu} + g_4^{\nu}), \qquad (1.166) \\ C_{\nu} &\equiv (v_{\mathrm{F}}^{\nu} + g_4^{\nu} + g_2^{\nu})/(8\pi), \qquad (1.167) \\ D_{\nu} &\equiv (v_{\mathrm{F}}^{\nu} + g_4^{\nu})/(8\pi), \qquad (1.168) \\ C_{\mathrm{eh}} &\equiv g_2^{\mathrm{eh}}/(4\pi), \qquad (1.169)\end{aligned}$$

where $\nu =$ e or h. The charge part of the forward-scattering Hamiltonian, $H_{\mathrm{forward}}^{\mathrm{charge}}$, is diagonalized by the transformation: $(\hat{\theta}_{\mathrm{e}}, \hat{\theta}_{\mathrm{h}}) \to (\hat{\theta}_1, \hat{\theta}_2)$ via

$$\begin{bmatrix} \hat{\theta}_{\mathrm{e}} \\ \hat{\theta}_{\mathrm{h}} \end{bmatrix} = \begin{bmatrix} \sqrt{A_{\mathrm{e}}} \cos\Theta & -\sqrt{A_{\mathrm{e}}} \sin\Theta \\ \sqrt{A_{\mathrm{h}}} \sin\Theta & \sqrt{A_{\mathrm{h}}} \cos\Theta \end{bmatrix} \begin{bmatrix} \hat{\theta}_1 \\ \hat{\theta}_2 \end{bmatrix}, \qquad (1.170)$$

with $\Theta$ being given by $\tan 2\Theta \equiv 2C_{\mathrm{eh}}\sqrt{A_{\mathrm{e}}A_{\mathrm{h}}}/\left(A_{\mathrm{e}}C_{\mathrm{e}} - A_{\mathrm{h}}C_{\mathrm{h}}\right)$. Consequently, the diagonalized forward-scattering Hamiltonian becomes

$$\hat{\mathcal{H}}_{\mathrm{forward}} = \int dx \sum_{i=1,2} \left\{ \hat{P}_i^2(x) + [v_{\mathrm{charge}}^{(i)} \nabla \hat{\theta}_i(x)]^2 \right\} + \hat{\mathcal{H}}_{\mathrm{forward}}^{\mathrm{spin}}, \qquad (1.171)$$

where $\hat{P}_i$ is the conjugate momentum for $\hat{\theta}_i$ ($\hat{\mathcal{H}}_{\mathrm{forward}}^{\mathrm{spin}}$ is already diagonalized). We find that the motions of four phases, $\hat{\theta}_1$, $\hat{\theta}_2$, $\hat{\phi}_{\mathrm{e}}$, and $\hat{\phi}_{\mathrm{h}}$, are described by massless acoustic

Table 1.2. Definition of order parameters in the two-band TL liquid. Here $\hat{\psi}^{\nu}_{j\sigma}(x)$ is the field operator for an electron ($\nu = \mathrm{e}$) or a hole ($\nu = \mathrm{h}$) with momentum around $(-1)^{j-1}k_{\mathrm{F}}$ and spin $\sigma$. Two blanches of a band are labeled as $i, j = 1$ or 2.

| Type of order | Order parameter |
|---|---|
| $\nu$-CDW | $\hat{O}^{\nu}_{\mathrm{CDW}}(x) = \sum_{i\neq j}\sum_{\sigma}\hat{\psi}^{\nu\dagger}_{i\sigma}(x)\,\hat{\psi}^{\nu}_{j\sigma}(x)$ |
| $\nu$-SDW | $\hat{O}^{\nu}_{\mathrm{SDW}}(x) = \sum_{i\neq j}\sum_{\sigma}\sigma\,\hat{\psi}^{\nu\dagger}_{i\sigma}(x)\,\hat{\psi}^{\nu}_{j\sigma}(x)$ |
| $\nu$-SSC | $\hat{O}^{\nu}_{\mathrm{SSC}}(x) = \sum_{i\neq j}\sum_{\sigma}\hat{\psi}^{\nu}_{i\sigma}(x)\,\hat{\psi}^{\nu}_{j,-\sigma}(x)$ |
| $\nu$-TSC | $\hat{O}^{\nu}_{\mathrm{TSC}}(x) = \sum_{i\neq j}\sum_{\sigma}\hat{\psi}^{\nu}_{i\sigma}(x)\,\hat{\psi}^{\nu}_{j\sigma}(x)$ |
| cv-SSC | $\hat{O}^{\mathrm{eh}}_{\mathrm{SSC}}(x) = \sum_{i\neq j}\sum_{\sigma}\hat{\psi}^{\mathrm{e}}_{i\sigma}(x)\,\hat{\psi}^{\mathrm{h}}_{j,-\sigma}(x)$ |
| cv-TSC | $\hat{O}^{\mathrm{eh}}_{\mathrm{TSC}}(x) = \sum_{i\neq j}\sum_{\sigma}\hat{\psi}^{\mathrm{e}}_{i\sigma}(x)\,\hat{\psi}^{\mathrm{h}}_{j\sigma}(x)$ |
| cv-SSC($2k_{\mathrm{F}}$) | $\hat{O}^{\mathrm{eh}}_{\mathrm{SSC2k_F}}(x) = \sum_{i=j}\sum_{\sigma}\hat{\psi}^{\mathrm{e}}_{i\sigma}(x)\,\hat{\psi}^{\mathrm{h}}_{j,-\sigma}(x)$ |
| cv-TSC($2k_{\mathrm{F}}$) | $\hat{O}^{\mathrm{eh}}_{\mathrm{TSC2k_F}}(x) = \sum_{i=j}\sum_{\sigma}\hat{\psi}^{\mathrm{e}}_{i\sigma}(x)\,\hat{\psi}^{\mathrm{h}}_{j\sigma}(x)$ |
| S-EBC | $\hat{O}_{\mathrm{SEBC}}(x) = \sum_{i=j}\sum_{\sigma}\hat{\psi}^{\mathrm{e}\dagger}_{i\sigma}(x)\,\hat{\psi}^{\mathrm{h}}_{j\sigma}(x)$ |
| T-EBC | $\hat{O}_{\mathrm{TEBC}}(x) = \sum_{i=j}\sum_{\sigma}\hat{\psi}^{\mathrm{e}\dagger}_{i\sigma}(x)\,\hat{\psi}^{\mathrm{h}}_{j,-\sigma}(x)$ |
| S-EDW | $\hat{O}_{\mathrm{SEDW}}(x) = \sum_{i\neq j}\sum_{\sigma}\hat{\psi}^{\mathrm{e}\dagger}_{i\sigma}(x)\,\hat{\psi}^{\mathrm{h}}_{j\sigma}(x)$ |
| T-EDW | $\hat{O}_{\mathrm{TEDW}}(x) = \sum_{i\neq j}\sum_{\sigma}\hat{\psi}^{\mathrm{e}\dagger}_{i\sigma}(x)\,\hat{\psi}^{\mathrm{h}}_{j,-\sigma}(x)$ |

modes, which have linear dispersions in their excitation spectra. The velocities of the transformed charge phases, $\hat{\theta}_1$ and $\hat{\theta}_2$, are given by

$$v^{(i)}_{\mathrm{charge}} = \left\{2(A_{\mathrm{e}}C_{\mathrm{e}} + A_{\mathrm{h}}C_{\mathrm{h}}) \pm 2\Big[(A_{\mathrm{e}}C_{\mathrm{e}} - A_{\mathrm{h}}C_{\mathrm{h}})^2 + 4C_{\mathrm{eh}}^2 A_{\mathrm{e}}A_{\mathrm{h}}\Big]^{1/2}\right\}^{1/2}, \tag{1.172}$$

respectively, where $i = 1$ (2) corresponds to the $+$ ($-$) sign of the right-hand side. On the other hand, velocities of the spin phases, $\hat{\phi}_{\mathrm{e}}$ and $\hat{\phi}_{\mathrm{h}}$, are $v_{\mathrm{F}}^{\mathrm{e}}$ and $v_{\mathrm{F}}^{\mathrm{h}}$, respectively: $v^{\nu}_{\mathrm{spin}} = v^{\nu}_{\mathrm{F}}$.

Now we shall discuss the optical absorption spectrum. The absorption spectrum, $W_{\mathrm{OPA}}(\omega)$, is related to the correlation function, $\langle \hat{P}(t)\hat{P}^{\dagger}(0)\rangle$, where the polarization $\hat{P}$ (namely the dipole operator) is given by

$$\hat{P}(t) = |M|\int dx \sum_{\sigma}\hat{\psi}^{\mathrm{e}\dagger}_{\sigma}(x,t)\hat{\psi}^{\mathrm{h}}_{\sigma}(x,t)\,e^{iE_0 t} + \text{h.c.}, \tag{1.173}$$

where $\hat{\psi}^{\nu}_{\sigma}(x,t) = \exp[i\hat{\mathcal{H}}_{\mathrm{forward}}t]\hat{\psi}^{\nu}_{\sigma}(x)\exp[-i\hat{\mathcal{H}}_{\mathrm{forward}}t]$ is the Heisenberg representation of the field operator, $M$ is an interband matrix element (assumed to be constant), and $E_0$ is the energy of the absorption and emission edge in the single-electron picture. The correlation function of the dipole operator, $\langle \hat{P}(t)\hat{P}^{\dagger}(0)\rangle$, contains the correlation, $\langle \hat{O}_{\mathrm{SEBC}}(x,\tau)\hat{O}^{\dagger}_{\mathrm{SEBC}}(0,0)\rangle$, where $\hat{O}_{\mathrm{SEBC}}$ is the order parameter describing the spin-singlet exciton Bose condensation (S-EBC), that is (see Table 1.2), $\hat{O}_{\mathrm{SEBC}}(x) \equiv$

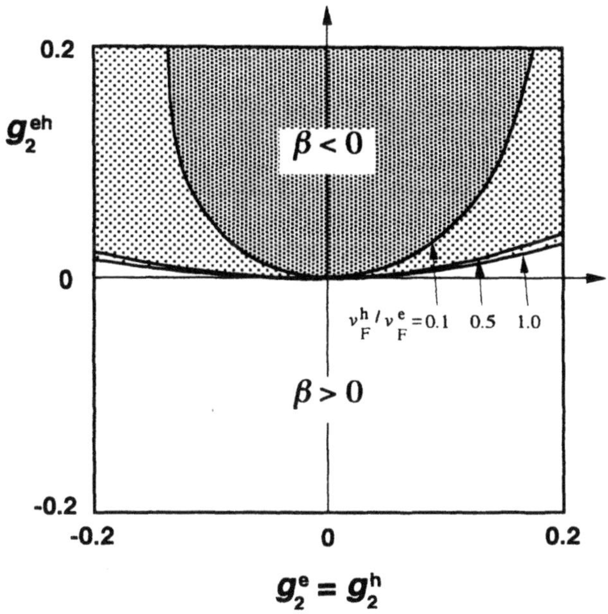


Figure 1.17. In a shaded region, the divergent edge anomaly ($\beta_{2\text{TL}} < 0$) in optical spectra is observable in the two-band TL liquid for several values of $v_F^h/v_F^e = 0.1$, 0.5, and 1.0. Irrespective of the ratio $v_F^h/v_F^e$, the region where $\beta_{2\text{TL}} < 0$ exists only in the upper half-plane [$\bar{g}_2^{eh} \equiv g_2^{eh}/(v_F^e + g_4^e) > 0$], which corresponds to the attractive interaction between electrons and holes. (After [99]).

$\sum_{j=1,2} \sum_\sigma \hat{\psi}_{j\sigma}^{e\dagger}(x)\hat{\psi}_{j\sigma}^{h}(x)$, and $\hat{O}(x,\tau) \equiv \exp[\hat{\mathcal{H}}_{\text{forward}}\tau]\hat{O}(x)\exp[-\hat{\mathcal{H}}_{\text{forward}}\tau]$ with an imaginary time $\tau$. For large $x$ and $\tau$, the correlation function behaves like

$$\left\langle \hat{O}_{\text{SEBC}}(x,\tau)\hat{O}_{\text{SEBC}}^{\dagger}(0,0)\right\rangle \sim [\max(x,\tau)]^{-\eta_{\text{SEBC}}}. \tag{1.174}$$

This means that the optical spectrum shows the power-law singularity like $W_{\text{OPA}}(\omega) \sim |\omega - E_F|^{\beta_{2\text{TL}}}$ in the vicinity of $E_F$, where $E_F$ is the edge energy and $\beta_{2\text{TL}}$ is the FES exponent of this two-band TL system. This FES exponent is given as $\beta_{2\text{TL}} = \eta_{\text{SEBC}} - 2$. We stress that the FES exponent, $\beta_{2\text{TL}}$, is determined by the correlation exponent, $\eta_{\text{SEBC}}$, of the order parameter describing the S-EBC. We successfully found an analytical expression of $\beta_{2\text{TL}}$ as $\beta_{2\text{TL}} = \beta_{2\text{TL}}^{\text{ex}} + \beta_{2\text{TL}}^{\text{c-oc}} + \beta_{2\text{TL}}^{\text{v-oc}}$, where

$$\beta_{2\text{TL}}^{\text{ex}} = -\frac{\pi[v_{\text{charge}}^{(1)} - v_{\text{charge}}^{(2)}]}{2\sqrt{A_e A_h}}\left[1 + \frac{A_e A_h}{4\pi^2 v_{\text{charge}}^{(1)} v_{\text{charge}}^{(2)}}\right]\sin 2\Theta, \tag{1.175}$$

$$\beta_{2\text{TL}}^{\text{c-oc}} = \frac{A_h s_1}{8\pi v_{\text{charge}}^{(1)} v_{\text{charge}}^{(2)}} + \frac{\pi s_2}{2A_h} - \frac{1}{2}, \tag{1.176}$$

$$\beta_{2\mathrm{TL}}^{\mathrm{v-oc}} = \frac{A_{\mathrm{e}} s_2}{8\pi v_{\mathrm{charge}}^{(1)} v_{\mathrm{charge}}^{(2)}} + \frac{\pi s_1}{2A_{\mathrm{e}}} - \frac{1}{2}. \tag{1.177}$$

Here $s_i \equiv \frac{1}{2}[v_{\mathrm{charge}}^{(1)} + v_{\mathrm{charge}}^{(2)} \pm (v_{\mathrm{charge}}^{(1)} - v_{\mathrm{charge}}^{(2)}) \cos 2\Theta]$ with $i = 1$ (2) corresponding to the + (−) sign. The region where $\beta_{2\mathrm{TL}} < 0$ is shaded in the phase diagram of Fig. 1.17.[17] Roughly speaking, the divergent edge spectrum can be observed when the *e*-*h* interaction is attractive ($g_2^{\mathrm{eh}} > 0$) and rather weak *e*-*e* and *h*-*h* correlations, which is a usual situation. When the excitons lie completely at the condensation state at zero temperature, an optical spectrum is expected to show a $\delta$-function-like peak if we can ignore the quantum fluctuations due to the low dimensionality. Actually, however, quantum fluctuation can never be neglected in 1D systems and it tends to destroy the exciton condensation. Consequently, $\delta$-function-like peak spectrum vanishes and is replaced by the power-law peak appears, which is really the FES and is a remnant of the EBC. In this sense, the FES singularity results from a fluctuating condensed state of 1D excitons. The FES exponent depends on the ratio of the Fermi velocities of the bands, $|v_{\mathrm{F}}^{\mathrm{h}}/v_{\mathrm{F}}^{\mathrm{e}}|$. Unlike the FES in bulk normal metals, the FES in this system persists against the hole motion with momentum $\pm k_{\mathrm{F}}$.

Now we shall turn to discussion on the competition among various Fermi-surface instabilities [98, 100]. Possible order parameters, $\hat{O}(x)$'s, are listed in Table 1.2. All the correlation functions, $\langle \hat{O}(x,\tau)\, \hat{O}^{\dagger}(0,0)\rangle$, behave like $[\max(x,\tau)]^{-\eta}$ for large $x$ and $\tau$. The exponent $\eta$ for each order parameter is given in Table 1.3. The order with the smallest $\eta$ predominates over the others. According to this criterion, the phase diagram of the quasi-thermal-equilibrium state is given in Fig. 1.18 for $v_{\mathrm{F}}^{\mathrm{h}}/v_{\mathrm{F}}^{\mathrm{e}} = 0.1$ at zero temperature. We see that the phase diagram is divided into four domains.

*(a) EBC phase.* — When the repulsive interaction between conduction electrons and valence ones is strong (positive and large $g_2^{\mathrm{eh}}$), *i.e.*, the strong attraction between electrons and holes, the EBC at $k = 0$ predominates. Spin-singlet (S-EBC) and triplet exciton Bose condensation (T-EBC) are degenerate in our spin-independent forward-scattering model. The exciton density wave (EDW) at $k = \pm 2k_{\mathrm{F}}$ cannot prevail in the whole plane [101].

*(b) Density-wave phase.* — When the repulsion between the electrons within each band is strong (positive and large $g_2^{\mathrm{e}}$ or $g_2^{\mathrm{h}}$), the CDW and the SDW predominate, which are degenerate in the forward-scattering model. The valence-band electrons with heavier mass have the stronger tendency toward the ordering. When the velocity ratio $|v_{\mathrm{F}}^{\mathrm{h}}/v_{\mathrm{F}}^{\mathrm{e}}|$ is decreased from unity, the CDW and SDW region extends and invades the EBC region.

*(c) Ordinary SC phase.* — The attractive interactions, *i.e.*, negative $g_2$'s, yield conventional SC. The negative $g_2^{\mathrm{h}}$ results in intraband pairing with zero total momentum $k = 0$. The valence-band electrons with heavier mass have the stronger tendency toward the ordering.

[17]This is exactly the criterion for the generalized susceptibility for EBC to diverge in the low-temperature limit.

Table 1.3. Exponents of correlation functions of the order parameters for the forward scattering model. Here $V_1 \equiv v^{(1)}_{\rm charge}\cos^2\Theta + v^{(2)}_{\rm charge}\sin^2\Theta$, $V_2 \equiv v^{(1)}_{\rm charge}\sin^2\Theta + v^{(2)}_{\rm charge}\cos^2\Theta$, $\bar{v}^{(i)}_{\rm charge} \equiv \frac{1}{2}v^{(i)}_{\rm charge}(A_{\rm e}A_{\rm h})^{-1/2}$ for $i = 1,2$, $\Delta\bar{v}_{\rm charge} \equiv \bar{v}^{(1)}_{\rm charge} - \bar{v}^{(2)}_{\rm charge}$, and $\sum\eta \equiv \eta^{\rm e}_{\rm CDW} + \eta^{\rm h}_{\rm CDW} + \eta^{\rm e}_{\rm SSC} + \eta^{\rm h}_{\rm SSC}$.

| Type of order | Exponent of the correlation function |
|---|---|
| c-CDW | $\eta^{\rm e}_{\rm CDW} = 1 + A_{\rm e}V_2/(2\pi v^{(1)}_{\rm charge}v^{(2)}_{\rm charge})$ |
| c-SDW | $\eta^{\rm e}_{\rm SDW} = \eta^{\rm e}_{\rm CDW}$ |
| v-CDW | $\eta^{\rm h}_{\rm CDW} = 1 + A_{\rm h}V_1/(2\pi v^{(1)}_{\rm charge}v^{(2)}_{\rm charge})$ |
| v-SDW | $\eta^{\rm h}_{\rm SDW} = \eta^{\rm h}_{\rm CDW}$ |
| c-SSC | $\eta^{\rm e}_{\rm SSC} = 1 + 2\pi V_1/A_{\rm e}$ |
| c-TSC | $\eta^{\rm e}_{\rm TSC} = \eta^{\rm e}_{\rm SSC}$ |
| v-SSC | $\eta^{\rm h}_{\rm SSC} = 1 + 2\pi V_2/A_{\rm h}$ |
| v-TSC | $\eta^{\rm h}_{\rm TSC} = \eta^{\rm h}_{\rm SSC}$ |
| cv-SSC | $\eta^{\rm eh}_{\rm SSC} = \frac{1}{4}\sum\eta + \pi\,\Delta\bar{v}_{\rm charge}[1 - (16\pi^2\bar{v}^{(1)}_{\rm charge}\bar{v}^{(2)}_{\rm charge})^{-1}]\sin 2\Theta$ |
| cv-TSC | $\eta^{\rm eh}_{\rm TSC} = \eta^{\rm eh}_{\rm SSC}$ |
| cv-SSC($2k_{\rm F}$) | $\eta^{\rm eh}_{\rm SSC2k_F} = \frac{1}{4}\sum\eta + \pi\,\Delta\bar{v}_{\rm charge}[1 + (16\pi^2\bar{v}^{(1)}_{\rm charge}\bar{v}^{(2)}_{\rm charge})^{-1}]\sin 2\Theta$ |
| cv-TSC($2k_{\rm F}$) | $\eta^{\rm eh}_{\rm TSC2k_F} = \eta^{\rm eh}_{\rm SSC2k_F}$ |
| S-EBC | $\eta_{\rm SEBC} = \frac{1}{4}\sum\eta - \pi\,\Delta\bar{v}_{\rm charge}[1 + (16\pi^2\bar{v}^{(1)}_{\rm charge}\bar{v}^{(2)}_{\rm charge})^{-1}]\sin 2\Theta$ |
| T-EBC | $\eta_{\rm TEBC} = \eta_{\rm SEBC}$ |
| S-EDW | $\eta_{\rm SEDW} = \frac{1}{4}\sum\eta - \pi\,\Delta\bar{v}_{\rm charge}[1 - (16\pi^2\bar{v}^{(1)}_{\rm charge}\bar{v}^{(2)}_{\rm charge})^{-1}]\sin 2\Theta$ |
| T-EDW | $\eta_{\rm TEDW} = \eta_{\rm SEDW}$ |

*(d) Unconventional SC phase.* — On the other hand, the negative $g_2^{\rm eh}$ causes *interband* Cooper pairing at $k = \pm 2k_{\rm F}$ [102]. This unconventional superconductivity is a peculiar feature of the 1D system. Singlet (SSC) and triplet (TSC) SC's are degenerate in both the above cases.

We shall here take into account the backward scattering, $\hat{\mathcal{H}}_{\rm back}$ in Eq. (1.164), which is treated by the self-consistent harmonic approximation (SCHA) [98, 99]. The intraband backward scatterings, $g_1^{\rm e}$ and $g_1^{\rm h}$, are "marginal" and therefore we can neglect them. The interband $g_1^{\rm eh}$ interaction, on the other hand, becomes relevant when

$$\eta_1^{\rm eh} \equiv \frac{1}{2}\left(\eta^{\rm e}_{\rm CDW} + \eta^{\rm h}_{\rm CDW}\right) - \frac{\sin 2\Theta}{8\pi}\,\frac{\bar{v}^{(1)}_{\rm charge} - \bar{v}^{(2)}_{\rm charge}}{\bar{v}^{(1)}_{\rm charge}\bar{v}^{(2)}_{\rm charge}} < 2, \tag{1.178}$$

with $\bar{v}^{(i)}_{\rm charge} \equiv v^{(i)}_{\rm charge}/2\sqrt{A_{\rm e}A_{\rm h}}$ for $i = 1$ and 2. In such a case, a mass gap $\Delta\varepsilon \propto (g_1^{\rm eh})^{1/(2-\eta_1^{\rm eh})}$ is introduced[18] to the three collective modes, $\hat{\theta}_{\rm e} + \hat{\theta}_{\rm h}$, $\hat{\phi}_{\rm e}$, and $\hat{\phi}_{\rm h}$. Whenever $g_1^{\rm eh}$ is relevant, CDW is found to predominate. Then the FES is cutoff at

[18]The current operator is proportional to $\partial(v_{\rm F}^{\rm e}\hat{\theta}_{\rm e} + v_{\rm F}^{\rm h}\hat{\theta}_{\rm h})/\partial x$, which is also massive and the system

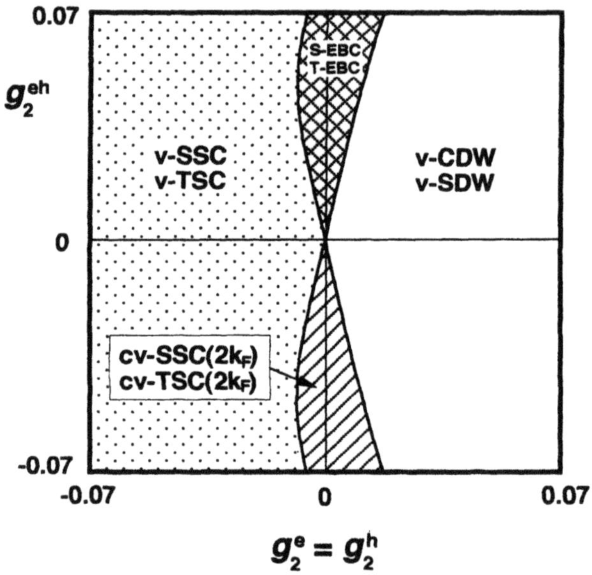


Figure 1.18. Phase diagram of the quasi-thermal-equilibrium state at zero temperature of the two-band TL liquid is drawn in the plane of $g_2^{\rm e} = g_2^{\rm h}$ and $g_2^{\rm eh}$ for $v_{\rm F}^{\rm h}/v_{\rm F}^{\rm e} = 0.1$. "S-EBC" ("T-EBC") stands for the spin-singlet (triplet) exciton Bose condensation, "v-CDW" ("v-SDW") the charge (spin) density wave in the valence band, "v-SSC" ("v-TSC") the spin-singlet (triplet) superconductivity in the valence band, and "cv-SSC($2k_{\rm F}$)" ["cv-TSC($2k_{\rm F}$)"] is the spin-singlet [triplet] interband cooper pairing with $\pm 2k_{\rm F}$ center-of-mass momentum.

$|\omega - E_{\rm F}| \sim \Delta\varepsilon$.

We also clarify the role of randomness in this system. Assuming that the spatial change of the random potential $u(x)$ is slow compared with the lattice constant, the intraband scattering is a major effect, which may be written as

$$\begin{aligned}\hat{\mathcal{H}}_{\rm random} &= \int dx\, u(x)\Big[\hat{\psi}_\sigma^{\rm e\dagger}(x)\hat{\psi}_\sigma^{\rm e}(x) + \hat{\psi}_\sigma^{\rm h\dagger}(x)\hat{\psi}_\sigma^{\rm h}(x)\Big] \\ &= \frac{2}{\pi\alpha}\int dx\, u(x)\bigg[\cos\Big(\frac{\hat{\theta}_{\rm e}}{\sqrt{2}} - 2k_{\rm F}x\Big)\cos\frac{\hat{\phi}_{\rm e}}{\sqrt{2}} \\ &\quad + \cos\Big(\frac{\hat{\theta}_{\rm h}}{\sqrt{2}} - 2k_{\rm F}x\Big)\cos\frac{\hat{\phi}_{\rm h}}{\sqrt{2}}\bigg]. \qquad (1.179)\end{aligned}$$

becomes an insulator only when $|v_{\rm F}^{\rm e}| = |v_{\rm F}^{\rm h}|$, while it decomposes into massive $(\hat{\theta}_{\rm e} + \hat{\theta}_{\rm h})$ and massless $(-\sqrt{A_{\rm h}/A_{\rm e}}\,\hat{\theta}_{\rm e} + \sqrt{A_{\rm e}/A_{\rm h}}\,\hat{\theta}_{\rm h})$ modes in the general case.

Here we take only the component with large momentum transfer. In terms of the phase variables, the Anderson localization is viewed as a pinning of the phases [103]. Applying the SCHA, the random potential becomes relevant when

$$\eta_{\mathrm{CDW}} \equiv \min\left[\eta_{\mathrm{CDW}}^{\mathrm{e}}, \eta_{\mathrm{CDW}}^{\mathrm{h}}\right] < 3. \tag{1.180}$$

This region covers the whole range of Fig. 1.18 [99]. The localization length $\xi$ is proportional to $\langle u^2\rangle^{-1/[2(3-\eta_{\mathrm{CDW}})]}$, and the gap $\Delta\varepsilon'$ ($\propto \xi^{-1}$) is introduced to all the four modes. Then the system becomes insulating [104, 105], but finite conductivity can be observed only when the sample size is less than $\xi$ or the frequency is higher than the gap. The FES in optical spectra can be seen only for $|\omega - E_{\mathrm{F}}| > \Delta\varepsilon'$.

## 1.6. DIMENSIONALITY IN THE SELF TRAPPING

In a perfect rigid lattice both excitons and electrons are in delocalized free states, which are described by the Bloch states. When the lattice is deformable, an exciton or an electron forms a polaron state as long as the exciton (or electron)-lattice interaction is weak. In such a case, the crystal is filled with the Bloch states of polarons, which are not localized. If the exciton (or electron)-lattice coupling is strong enough, on the other hand, a localized state of them with a spontaneous local lattice distortion becomes lower in energy than the free states. This phenomenon is called "self trapping" [106, 107, 108]. In this section, we focus our attention to the dimensionality of the self trapping process, according to Toyozawa [109] and Sumi [110].

As is well known, localization of an electron due to randomness and disorders depends strongly on the electron's dimensionality [111]. Localization in 3D differs qualitatively from that in lower dimensions [112]; in lower (2D and 1D) dimensions, electrons are trapped by an infinitesimally weak lattice disorder and all the eigenstates become localized. In 3D crystals, on the other hand, free states can exist as long as the disorder is not strong. These situations hold also for the case of a sole lattice defect in a crystal. In this case, a bound state of electrons appears in 1D even when the attractive potential for the electrons is infinitesimally weak, while an electron can be bounded in 3D only when the strength of the potential exceeds a critical value. The 2D system is in intermediate between localized and extended characteristics, yielding the "weak localization." These differences of the localization process in 3D and in lower dimensional systems might be reflected in the feature of the self trapping in $d$D.[19]

### 1.6.1. *Overview of the Dimensionality in the Self Trapping*

First we comment on the difference between electrons in ionic crystals and excitons concerning the formation of the self-trapped state. Since an exciton is electrically neutral, its interaction with lattices is of a short-ranged type even in ionic crystals.

[19]The term of "low dimensions" is concerned only with a degree of freedom of exciton's or electron's motion on a 3D lattice. A lattice with dimensions lower than three is unstable thermodynamically.

In this case, the self-trapped state is stabilized suddenly as a strongly-localized state below the continuum of the free states when the coupling constant exceeds a critical value. This situation can be visualized in the adiabatic-potential diagram written as a function of the amplitude of a local lattice distortion. The interaction of electrons with lattices is, on the other hand, of an electrostatic long-ranged type in ionic crystals, obeying the Fröhlich interaction [113], although it is still of a short-ranged type in homopolar crystals, obeying the deformation-potential interaction [114]. Then an electron in ionic crystals is always associated with a nonvanishing lattice distortion around it.

We will first study the local stabilities of the free and self-trapped states against lattice distortions [115, 116], by introducing the electron-lattice coupling and the intersite transfer as a perturbation. The kinetic energy of the lattice will be neglected in this adiabatic argument. The self trapping has usually been investigated in the molecular-crystal model where a free exciton in a perfect rigid lattice is regarded as having only the translational freedom with the internal freedom frozen, while in a deformable lattice it is regarded as having a short-ranged interaction with dispersionless Einstein phonons excitable at each lattice site. The Einstein phonons can be described by an amplitude of lattice distortion $Q_n$ and its conjugate momentum $P_n$ distinguished by an integer $n$ specifying each unit cell. The distortion energy associated with $Q_n$ is taken as $\frac{1}{2}Q_n^2$ at each unit cell $n$.

Starting with the Bloch eigenstates $|\vec{K}\rangle = N^{-1/2}\sum_n \exp(i\vec{K}\cdot\vec{R}_n)|n\rangle$ with energy $E(\vec{K})$, where $N$ is the total number of unit cells containing in the crystal, $|n\rangle$ is the Wannier state localized at the $n$th unit cell, let us calculate the energy change of the lowest state (assumed to be $\vec{K}=0$) to second order in the on-site electron-lattice interaction:

$$H_{e-l} = -\sqrt{2S}\sum_n |n\rangle Q_n \langle n|, \tag{1.181}$$

where $S$ represents the energy of lattice relaxation. Introducing the Fourier transform: $Q_n = N^{-1/2}\sum_{\vec{K}}\exp(i\vec{K}\cdot\vec{R}_n)Q_{\vec{K}}$, one has $\delta_2 E(0) = -N^{-1}\sum_{\vec{K}} c^2 Q^\dagger_{\vec{K}}Q_{\vec{K}}[E(\vec{K})-E(0)]^{-1}$. Adding this to $\frac{1}{2}\sum_{\vec{K}} Q^\dagger_{\vec{K}}Q_{\vec{K}}$, one finds the first candidate modes $\vec{K}$ to become unstable are those with the lowest excitation energy $E(\vec{K})-E(0)$. With the cyclic boundary condition for the $d$D cube with $N^{1/d}$ sites for each edge, we obtain the condition for instability:

$$\frac{2d}{\pi^2}gN^{2/d-1} > 1. \tag{1.182}$$

Here a dimensionless coupling constant of the exciton-lattice interaction is defined by $g \equiv S/B$, where $B$ is the halfwidth of the exciton band. By letting $N \to \infty$, one finds that the free state is always unstable against lattice distortion for $d=1$ but is always stable (at least locally in the $Q_n$ space) for $d=3$. In the marginal case of $d=2$, the instability sets in when $g$ exceeds $\pi^2/4$ (this value depends on the boundary shape of the macroscopic crystal). This argument for short-range coupling is extended to the general case of the force range index $\gamma$ defined by $K^{-\gamma}$ ($K \to 0$) for the Fourier transform of the electron-lattice interaction. The local stability criterion for the free state is then given by the sign of the stability index $s$:

$$s \equiv d - 2(\gamma + 1) \begin{cases} < 0 & \text{self trapping} \\ = 0 & \text{marginal} \\ > 0 & \text{no self trapping} \end{cases} \tag{1.183}$$

Long-range electron-lattice coupling, such as the Frölich and piezoelectric interactions ($\gamma = 1$) in a 3D lattice belongs to the self-trapping case, being similar to short-range coupling $\gamma = 0$ in a 1D lattice. To investigate the dimensionality dependence of the self trapping, therefore, it is sufficient to consider only the short-ranged interaction.

We shall consider the opposite limit of a localized state $|n\rangle$ with lattice distortion $Q_n = c$, and introduced the nearest neighbor transfer $\mathcal{T}$ as a perturbation. The second-order energy turns out to be $\delta_2 E = -2d\mathcal{T}^2/c^2 = -S/4dg^2$. The binding energy of this optimally localized state, the self-trapped state, can thus be calculated in the form of a strong coupling expansion, $|E_S| = S\left[1 + (4dg^2)^{-1} + O\left(4dg^2\right)^{-2}\right]$, where the electron-induced lattice distortions at other sites ($n' \neq n$) contribute to higher-order terms. The self-trapped state with the binding energy $|E_S|$ becomes more stable than the free state with energy $(-B)$ when

$$g > g_c \equiv 1 - \frac{1}{4d} + O\left(\frac{1}{4d}\right)^2, \tag{1.184}$$

for $d = 2, 3$. Such a comparison of the free and self-trapped states is meaningless for $d = 1$ since the free state is locally unstable.

### 1.6.2. *A Unified Treatment of the Self Trapping in d Dimension*

In the short-ranged interaction in 3D, the self-trapped state appears as a strongly-localized state, whose spatial extension has shrunk to the order of a lattice constant. In the short-ranged interaction in 1D, on the other hand, excitons or electrons are always associated with a finite lattice distortion even for a small coupling constant, as in the long-ranged interaction in 3D. This means that the self-trapped state obtained for a small coupling constant has a spatial extention much larger than the order of the lattice constant. Therefore, whole aspect in the dimensionality dependence of the self trapping should be investigated by taking into account the discreteness of the lattice structure correctly. Sumi and Sumi [110] found an approximation in which the binding energy can be obtained correctly in the both limits of a strongly-localized and a strongly-extended lattice distortion. In the following only excitons are mentioned, but obtained results can be applied also to electrons as long as they have the short-ranged interaction with lattices.

A lattice distortion extending spatially over many unit cells can be described by a linear combination of many $Q$'s as $Q = \sum_n b_n Q_n$ with $b_n > 0$, where the coefficient $b_n$, which is real and normalized ($\sum_n b_n^2 = 1$), is usually taken as $b_n \propto \exp(-|n|^2/l^2)$ with a characteristic spatial extension of the lattice distortion $l$. We shall introduce the number of $N_b \equiv \left(\sum_n b_n\right)^2$ with $N_b \geq 1$. This $N_b$ can be regarded as an average number of unit cells covered by the lattice distortion $Q$. There are two types of the self-trapped states in low dimensions; one has a large spatial extension ($N_b \gg 1$) and the other is a strongly-localized one ($N_b \sim 1$). Equation (1.181) is rewritten as $H_{e-l}(Q) = -\sqrt{2S}Q\sum_n b_n|n\rangle\langle n|$. The adiabatic potential in $d$D is given by the sum of the bound-state energy $E^{(d)}(Q)$ and the distortion energy $\frac{1}{2}Q^2$ as $P^{(d)}(Q;N_b) \equiv \frac{1}{2}Q^2 + E^{(d)}(Q)$, which is also a function of $N_b$. Since $E^{(d)}(Q) \to 0$ as $N_b \to \infty$ irrespective of the dimensionality, we get $P^{(d)}(Q;N_b) \to \frac{1}{2}Q^2$ as $N_b \to \infty$.

### *1.6.2.1. Self trapping in one dimension*

We first note that the bound state can be obtained even for an infinitesimally small amplitude $Q$ of the lattice distortion in 1D. The adiabatic potential is evaluated in 1D for $N_b \gg 1$ as

$$P^{(1)}(Q;N_b \gg 1) \approx \begin{cases} \frac{1}{2}(1-2gN_b)Q^2 & \text{for (a)} \\ \sqrt{\frac{S}{2}}Q\left[\frac{Q}{\sqrt{2S}} - g^{-1}N_b^{-3/2} + (1-2gN_b)g^{-1}N_b^{-3/2}\right] & \text{for (b)} \end{cases}, \tag{1.185}$$

where the cases (a) and (b) correspond to $Q/\sqrt{2S} \ll g^{-1}N_b^{-3/2}$ and $Q/\sqrt{2S} \gg g^{-1}N_b^{-3/2}$, respectively. We find that $P^{(1)}(Q;N_b)$ is always positive as long as $gN_b < \frac{1}{2}$. When $gN_b > \frac{1}{2}$, however, there appears a region of $P^{(1)}(Q;N_b) < 0$ at least for small $Q$'s in the case of $Q/\sqrt{2S} \ll g^{-1}N_b^{-3/2}$. Therefore $gN_b = \frac{1}{2}$ gives a phase boundary of the self trapping. Since there exists an appropriate $N_b$ for any $g$ in this phase boundary however small $g$ is, the self trapping takes place even for the weak exciton-lattice interaction in 1D. This is the most peculiar feature in the self trapping in 1D. In the opposite limit of $N_b = 1$ where $b_n = \delta_{n,0}$, $E^{(1)}(Q) = B - \sqrt{B^2 + 2SQ^2}$. Then the adiabatic potential can be expressed as

$$P^{(1)}(Q;N_b = 1) = \frac{1}{2}Q^2\left[1 - \frac{4g}{1+\sqrt{1+2g^2Q^2/S}}\right], \tag{1.186}$$

for $N_b = 1$. In the entire region of $N_b \geq 1$, the equation of the phase boundary of the self trapping in 1D is $gN_b = \frac{1}{2}$ approximately.

There exists no potential barrier for moving from the free to the self-trapped states when the self-trapped states become stable for $g > (2N_b)^{-1}$. Since $(2N_b)^{-1} \to 0$ as $N_b \to \infty$, however, the free states are always unstable without potential barrier for the self trapping for any $g$. When $g < \frac{1}{2}$, the self trapping takes place with a lattice distortion with spatial extension of $N_b > (2g)^{-1}$. When $g > \frac{1}{2}$, on the other hand, the self trapping occurs with any lattice distortion of various spatial extensions.

Figure 1.19 shows the phase boundaries of the self trapping. The self trapping takes place in the hatched region edged by the phase boundary. The value of $N_b$ starts from $\infty$ at $g = 0$ and decreases with increasing $g$. When $g$ increases to about 0.75 where $N_b$ decreases to about 2.9, the lowest self-trapped state jumps suddenly to that of $N_b = 1$, where the lattice distortion in the self-trapped state is localized only

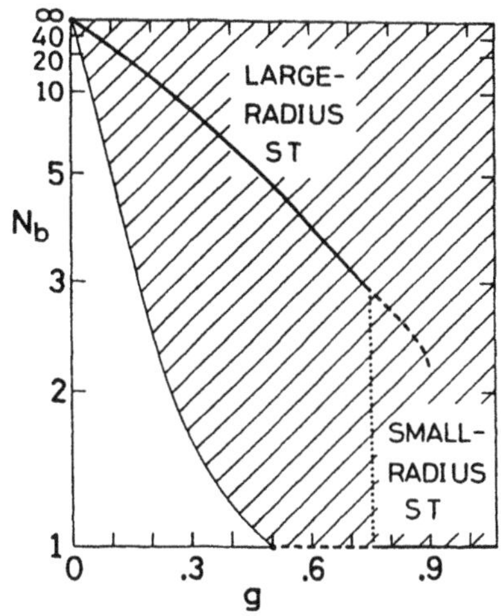


Figure 1.19. Phase boundary of the self trapping in 1D in a plane of $g$ and $N_b$. The $g$ dependence of $N_b$ of the lowest self-trapped state is shown by a solid line. (After [110]).

at a single unit cell. After $g$ exceeds about 0.75, the lowest self-trapped state remains always $N_b = 1$. For $g < 0.75$, therefore, large-radius self-trapped states are realized. Even after $g$ exceeds about 0.75, however, the large-radius self-trapped state survives as a metastable state whose location is shown by a dashed line in Fig. 1.19.

### *1.6.2.2. Self trapping in two dimension*

The adiabatic potential $P^{(2)}(Q; N_b)$ is essentially a function of the single scaled variable $x \equiv -N_b E^{(2)}(Q)/B > 0$ alone. Actually, $P^{(2)}(Q; N_b)$ is given by the form of

$$P^{(2)}(Q; N_b \gg 1) = \frac{B}{N_b} \frac{[g^{-1} - u^{(2)}(x)]x}{u^{(2)}(x)}, \tag{1.187}$$

with

$$u^{(2)}(x) = 4x \left[ \int_0^\infty dt \, \frac{e^{-t}}{\pi t + x} \right]^2 . \tag{1.188}$$

Function $u^{(2)}(x)$ has a peak at $x = x_c \approx 0.83$. The peak value is denoted by $u^{(2)}(x_c) \equiv 1/g_{c2} \approx 1/1.74$. When $g < g_{c2}$, Eq. (1.187) is positive irrespective of the value of $x$, and the self trapping does not take place. When $g > g_{c2}$, on the other hand, Eq. (1.187) becomes negative in a region of $x$ around $x_c$, and the self trapping takes place. When $g$ reaches $g_{c2}$ from below, the self trapping becomes possible simultaneously with various lattice distortions. For $N_b = 1$, putting $b_n = \delta_{n,0}$, we

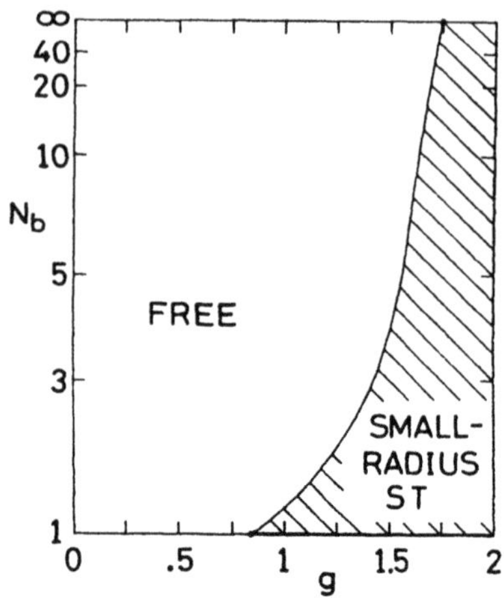


Figure 1.20. Phase boundary of the self trapping in 2D in a plane of $g$ and $N_b$. Location of the lowest self-trapped state is shown by a solid line. It is always located on the abscissa for $N_b = 1$. (After [110]).

can determine the critical value $g_{c1}$ of $g$ where the minimum point of $P^{(2)}(Q; N_b)$ for $N_b = 1$ first reaches zero at $Q \neq 0$ as $g_{c1} = 0.84$.

We can get the phase boundary of the self trapping connecting $g = g_{c1}$ at $N_b = 1$ and $g = g_{c2}$ at $N_b = \infty$. It is shown in Fig. 1.20, where the self trapping takes place in the hatched region. In 2D, a strongly-localized small-radius self-trapped state appears suddenly when $g$ exceeds $g_{c1}$, and it persists afterward, although the self trapping with an infinitely extended lattice distortion becomes possible when $g$ exceeds $g_{c2} > g_{c1}$.

### *1.6.2.3. Self trapping in three dimension*

There exists in 3D the critical amplitude of lattice distortions, given by

$$Q_c = \frac{\pi\sqrt{2S}}{6gN_b^{1/6}} \quad \text{for} \quad N_b \gg 1, \tag{1.189}$$

necessary to bind an exciton, the situation different from those in 1D and 2D. The adiabatic potential can be written in a form of

$$P^{(3)}(Q; N_b \gg 1) = \frac{2\pi B}{3N_b^{2/3}} \frac{\left[N_b^{1/3}g^{-1} - u^{(3)}(y)\right] y}{u^{(3)}(y)}, \tag{1.190}$$

with $y = -3N_b^{2/3}E/2\pi B > 0$ and

$$u^{(3)}(y) = \frac{24}{\pi} y \left[1 - \sqrt{\pi y}\, e^y \operatorname{erfc}(\sqrt{y})\right]^2, \tag{1.191}$$

where $\operatorname{erfc}(z)$ is the complementary error function. Function $u^{(3)}(y)$ has a peak at $y$ of the order of unity. The peak is given by 0.459 at $y = 0.675$. Then Eq. (1.190) gives the equation of the phase boundary of the self trapping for $N_b \gg 1$ as $N_b = (0.459)^3 g^3$, where the self trapping takes place with a lattice distortion of $N_b < (0.459)^3 g^3$. Since $(0.459)^3 g^3$ is finite, the self trapping does not take place with an infinitely-extended lattice distortion of the limit of $N_b \to \infty$ however large the coupling constant $g$ is as long as $g$ is finite. This is just opposite to those in the 1D and 2D cases.

Also in the opposite limit of $N_b = 1$, there exists a critical lower bound $Q_c$ of the amplitude of a lattice distortion for appearance of a bound state. It is given by

$$Q_c = 0.33 \frac{\sqrt{2S}}{g} \quad \text{for} \quad N_b = 1. \tag{1.192}$$

This situation is just the same as that for $N_b \gg 1$. The adiabatic potential gets a minimum point only when the coupling constant $g$ is sufficiently large. The critical value of $g$ where the minimum energy of the adiabatic potential first reaches zero is $g_c = 0.90$ for $N_b = 1$.

For other fixed values of $N_b$, the critical amplitude $Q_c$ of a lattice distortion for appearance of a bound state is obtained at a value intermediate between Eq. (1.192) for $N_b = 1$ and Eq. (1.189) for $N_b = \infty$. The phase boundary of the self trapping in 3D, which starts from $g_c = 0.90$ at $N_b = 1$, is shown in Fig. 1.21. The lowest point of the adiabatic potential with respect to both $Q$ and $N_b$ is always located at $N_b = 1$ for $g > g_c$, similarly in 2D. In 3D, therefore, a strongly-localized small-radius self-trapped state appears suddenly when $g$ exceeds $g_c$. In 3D, the free states cannot become unstable toward the self trapping even for large $g$. Thus, the free states are always stable against the self trapping however large $g$ is. The self trapping in 3D is different from that in 2D only in this point.

To summarize, in 1D tendency of localization is predominant in exciton states since the free states are always unstable with the self-trapped states always stable irrespective of the magnitude of the coupling constant $g$. In 3D, on the other hand, opposite tendency of delocalization is predominant since the free states do not become unstable even for large $g$. In 2D, these two types of tendency cancel each other. It occurs, then, that a small change in the 2D systems reverses the situation competitively. For example, when the exciton-lattice interaction is inter-site type, the critical coupling constant $g_{c2}$ for $N_b \gg 1$ becomes smaller than $g_{c1}$ for $N_b = 1$. Thus the situation concerning localization is just reversed by a change in the type of exciton-lattice interaction (*i.e.*, its force range) in 2D. Such a large change is never brought about in other dimensions.

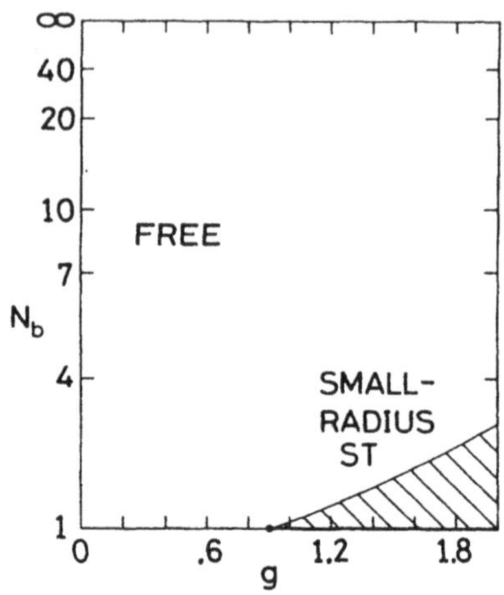


Figure 1.21. Phase boundary of the self trapping in 3D in a plane of $g$ and $N_b$. Location of the lowest self-trapped state is shown by a solid line. (After [110]).

## 1.7. DIMENSIONALITY IN THE PHOTOINDUCED STRUCTURE CHANGES

The electron-lattice coupling can also lead to the macroscopic change of structure of the system, which is a result of relaxation after optical pumping. In this section, we introduce theories for photoinduced structure changes [117], which have been observed experimentally as photopolymerization and photoisomerization of diacetylene crystals and TTF-CA [118].

We investigate simple linear chain model which consists of the localized electronic two levels $|e_n\rangle$ and $|g_n\rangle$ at every site $n$ with the Franck-Condon energy $E_{\mathrm{FC}}$. The excited state $|e_n\rangle$ is coupled with the interaction mode $Q_n$ with the lattice relaxation energy $S$. The interaction modes at sites $n$ and $n'$ are coupled with each other with the coupling constant $K_{nn'}$. Because the relaxation of the interaction mode $Q_n$ is very rapid within the time of order picosecond, it is reasonable to assume that $Q_n$ is always in equilibrium with the given electronic state $|e_n\rangle$ and $|g_n\rangle$. Then we start with the Hamiltonian of the linear chain:

$$H = \sum_n |e_n\rangle \left(E_{\mathrm{FC}} - \sqrt{2S}Q_n\right) \langle e_n| + \frac{1}{2}\sum_n Q_n^2 - \frac{1}{2}\sum_{n,n'} K_{nn'}Q_nQ_{n'}, \qquad (1.193)$$

and $|g_n\rangle\langle g_n| + |e_n\rangle\langle e_n| = 1$. Here the diagonal element $K_{nn}$ is zero and $\sqrt{2S}Q_n$ is the Stokes shift in the excited electronic state at the $n$th site. The summation over $n$ is taken over the sites of the excited molecules. The last two terms of Eq. (1.193) work in both the electronic ground and excited states. We took the restoring force constant of the lattice vibration at each molecule as a unit. This linear chain treatment can be easily extended to other dimensions, *i.e.*, 0D, 2D, and 3D.

### 1.7.1. *Nucleation Picture*

Due to the theory of the nucleation process [119], the initial metastable state decays to the absolutely stable state with the creation of a critical droplet. The nucleation rate is determined by the creation energy of the critical droplet, and is quite different between in one- and in higher-dimensional systems. Since the nucleation process dominates in the linear chain model, we shall investigate the stability of a cluster of excited molecules with size $m$. In this model, there is no "phase transition" in the thermodynamical sense because of the one dimensionality. Only the interaction between two neighboring excited sites is taken into account here.

According to Hanamura and Nagaosa [120], let us consider that $n_1$, $n_2$, $\cdots$, $n_m$th molecules are optically excited and others are in the ground state. For this system we rewrite Eq. (1.193) in such a way as the lattice coordinate $Q_k = Q_k^{(m)} + \Delta^{(m)}(k)$ fluctuates around the stationary displacement $\Delta^{(m)}(k)$:

$$H_g^{(m)} = \frac{1}{2}\sum_k \omega_k^2 Q_k^{(m)} Q_{-k}^{(m)} + mE_{\rm FC} - \frac{1}{2}\sum_k \omega_k^2 \Delta^{(m)}(k)\Delta^{(m)}(-k) \tag{1.194}$$

$$= \frac{1}{2}q_0^2 + E_g^{(m)}, \tag{1.195}$$

where

$$\Delta^{(m)}(k) \equiv \sqrt{\frac{2S}{N}}\frac{1}{\omega_k^2}\sum_{n=1}^{m} e^{ikn}, \tag{1.196}$$

$$q_0 \equiv \frac{\bar{\omega}}{\sqrt{N}}\sum_k \frac{q_k}{\omega_k}, \tag{1.197}$$

with $\omega_k^2 = 1 - 2\sum_{n\neq 0} K_{n0}e^{ikn}$ and $\bar{\omega}^{-2} = N^{-1}\sum_k \omega_k^{-2}$. Here we note that the 0th site is in the ground state, and $q_0$ is the interaction mode at the 0th site. $E_g^{(m)}$ contains the energy of $m$ excited molecules and the lattice vibration energy besides the interaction mode $q_0$. The last term of Eq. (1.194) denotes stabilization energy due to the lattice distortion. The absolute value of this stabilization energy increases by stronger than the first power of $m$. Therefore under some conditions, this energy overcomes the Franck-Condon energy of the second term of Eq. (1.194), and the distorted cluster becomes more stable. When the 0th site is in the excited state, on the other hand,

$$\begin{aligned} H_e^{(m+1)} &= H_g^{(m)} + E_{\rm FC} - \sqrt{2S}Q_0 \\ &= \frac{1}{2}q_0^2 + E_g^{(m)} + E_{\rm FC} - \frac{\sqrt{2S}}{\bar{\omega}}q_0 - \delta_g^{(m)} \\ &= \frac{1}{2}\left(q_0 - \frac{\sqrt{2S}}{\bar{\omega}}\right)^2 + E_g^{(m)} + \frac{S}{\bar{\omega}^2} + E_{\rm FC} - \delta_e^{(m+1)}, \end{aligned} \tag{1.198}$$

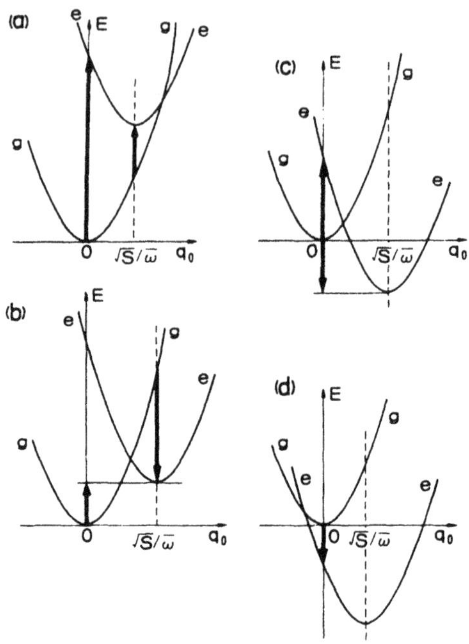


Figure 1.22. Energy dispersions of the 0th molecule: (a) The excited cluster is unstable, (b) metastable, (c) expands thermally, and (d) expands spontaneously. (After [120]).

where

$$\delta_e^{(m+1)} = \delta_g^{(m)} + \frac{2S}{\bar{\omega}^2} = \frac{2S}{\sqrt{1-(4K)^2}} \frac{1 - \left(\frac{1-\sqrt{1-(4K)^2}}{4K}\right)^{m+1}}{1 - \frac{1-\sqrt{1-(4K)^2}}{4K}}, \tag{1.199}$$

$$\delta_g^{(m)} = \frac{2S}{N} \sum_k \sum_{n=1}^{m} \frac{e^{ikn}}{\omega_k^2} = \frac{1-\sqrt{1-(4K)^2}}{4K} \delta_e^{(m)}. \tag{1.200}$$

Here we took into account the nearest neighbor coupling (*i.e.*, $\omega_k^2 = 1 - 4K\cos k$). The superscripts $(m)$ and $(m+1)$ denote the numbers of excited sites and the subscript $g$ or $e$ of $H$ means whether the 0th molecule is in the electronic ground ($g$) or excited ($e$) state. We shall discuss the stability of a cluster of $m$ excited molecules.

*(a) The case of* $E_{\rm FC} - \delta_e^{(m+1)} > 0$. — The energy dispersion is shown in Fig. 1.22(a). In this case, the minimum energy of the excited state is above the corresponding displaced ground state. Then the excited state $|e\rangle$ at the 0th and $(m+1)$th sites is unstable against radiative as well as nonradiative decay into $|g\rangle$. All the excited sites

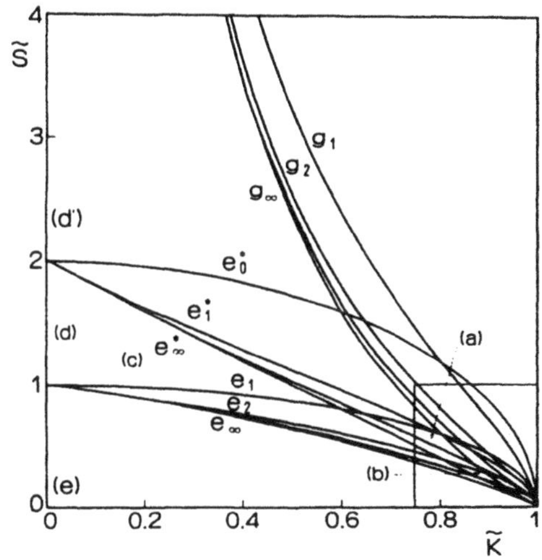


Figure 1.23. The phase diagram over $\tilde{S} \equiv 2S/E_{\rm FC}$ and $\tilde{K} \equiv 4K$. $\mathrm{e}_m$, $\mathrm{e}^*_m$, and $\mathrm{g}_m$ lines satisfy $E_{\rm FC} = \delta_e(m)$, $E_{\rm FC} = \frac{1}{2}[\delta_e(m+1) + \delta_g(m)]$, and $E_{\rm FC} = \delta_g(m)$, respectively. (After [120]).

relax into the ground state.

*(b) The case of* $\delta_e^{(m+1)} \geq E_{\rm FC} > \frac{1}{2}[\delta_e^{(m+1)} + \delta_g^{(m)}]$. — The excited minimum energy becomes lower than the ground state energy at the displaced coordinate $q_0 = \sqrt{2S}/\bar{\omega}$. Simultaneously, the energy-minimum ground state is stable relative to that of the excited state. Then the electronic excited and ground states at the 0th site are metastable and absolutely stable, respectively [see Fig. 1.22(b)]. The cluster of the $m+1$ excited molecules survives at low temperature $T \ll \Delta E_e$ (the potential barrier in the excited state).

*(c) The case of* $\frac{1}{2}[\delta_e^{(m+1)} + \delta_g^{(m)}] \geq E_{\rm FC} > \delta_g^{(m)}$. — The horizontal relative stability of the ground and excited energy-minimum states at the 0th and $(m+1)$th sites is exchanged as shown in Fig. 1.22(c). So the excited cluster extends thermally (without optical pumping) beyond the size $m$ at the lattice temperature larger than the potential barrier $\Delta E_g$ in the ground state.

*(d) The case of* $\delta_g^{(m)} \geq E_{\rm FC}$. — The ground state of the molecule neighboring from both ends of the excited molecules is unstable and changes spontaneously into the excited state. Therefore the cluster size increases without any additional pumping and irrespective of lattice temperature.

The several kinds of photoinduced structure changes in this nucleation picture are characterized by two parameters, $S/E_{\rm FC}$ and $K$. We consider the following situation as an example:

$$E_{\rm FC} = \delta_e(m_{c1} + 1), \tag{1.201}$$

$$E_{\rm FC} = \frac{1}{2}\left[\delta_e(m_{c2} + 1) + \delta_g(m_{c2})\right], \tag{1.202}$$

$$E_{\rm FC} = \delta_g(m_{c3}). \tag{1.203}$$

(i) Such a small excited cluster as $m < m_{c1}$ disappears promptly. (ii) An excited cluster with $m$ satisfying $m_{c1} < m < m_{c2}$ is stable at low temperature and decays thermally at high temperature $T \geq \Delta E_e$. (iii) An excited cluster with $m_{c2} < m < m_{c3}$ is stable at low temperature and expands thermally at high temperature $T \geq \Delta E_g$. (iv) Such a large excited cluster as $m > m_{c3}$ expands promptly to the system size with no pumping and independent of the temperature. Thus we can understand photoinduced structure changes basing upon the phase diagram of Fig. 1.23. For $4K \approx 1$ but rather small $2S/E_{\rm FC}$, we have very colorful phenomena of expansion, division, and coagulation of photoexcited molecule clusters, which depend very sensitively on the parameters, and excitation degree through three critical sizes of excited clusters as well as temperature. The spontaneous amplification is brought about once the $m$ excited molecule cluster is formed above $g_m$ line. The probability to excite the neighboring more than $m$ molecules depends on the excitation power. The photoinduced structure change in one linear chain does not interfere with other chains so that we can write the information with grading by changing the excitation power [121]. This point is characteristic to the nucleation model.

### 1.7.2. *Mean-Field Picture*

When the 3D long-range interactions among the excited clusters are taken into account, the critical droplet becomes large and the nucleation rate is very small. In this case, we can neglect the nucleation process within the time scale we are interested in, and the phase-transition-like threshold behavior can be expected [122]. Because the fluctuation due to the nucleation process is small, we can employ the mean-field approximation, and the problem is reduced to that of the single site. After integrating out $Q_n$'s, the model is reduced to the Ising model, where spin-up and spin-down states correspond to $|e_n\rangle$ and $|g_n\rangle$, respectively. The transition rates between these two electronic states are composed of two contributions: optical and thermal transitions. The optical transitions occur vertically with respect to the interaction mode $Q_n$ (Franck-Condon principle), while the thermal transition rates are governed by the potential barrier height between the two states.

Introducing the spin $\sigma_n = \pm\frac{1}{2}$, according to Nagaosa and Ogawa [122], where $+$ $(-)$ sign corresponds to the $e$ $(g)$ state, the partition function $Z$ of the Hamiltonian (1.193) is

$$Z = {\rm Tr}_{\{\sigma\}} \int \mathcal{D}Q \, \exp\left[-\frac{H(\{\sigma\},\{Q\})}{T}\right], \tag{1.204}$$

where

$$H(\{\sigma\},\{Q\}) = \sum_n \left(E_{\rm FC} - \sqrt{2S}Q_n\right)\left(\sigma_n + \frac{1}{2}\right) + \frac{1}{2}\sum_n Q_n^2 - \frac{1}{2}\sum_{n,n'} K_{nn'}Q_nQ_{n'}, \quad (1.205)$$

and $\mathcal{D}Q = \prod_n dQ_n$. The trace is over the spin configuration. Here we assume that $Q_n$ is always distributed around the bottom of one of the parabolas with distribution functions, $P$'s, which are given as

$$P_g(Q_n) = (2\pi T)^{-1/2}\exp\left[-\frac{(Q_n - \Delta_n)^2}{2T}\right], \quad (1.206)$$

$$P_e(Q_n) = (2\pi T)^{-1/2}\exp\left[-\frac{(Q_n - \Delta_n - \sqrt{2S})^2}{2T}\right], \quad (1.207)$$

with $\Delta_n = \sum_{n'\neq n} K_{nn'}Q_{n'}$ and temperature $T$. Then the integration with respect to $Q_n$'s are easily performed to give $Z \propto {\rm Tr}_{\{\sigma\}}\exp\left[-H_{\rm eff}(\{\sigma\})/T\right]$, where the effective Hamiltonian is

$$H_{\rm eff}(\{\sigma\}) = -h\sum_n \sigma_n - \frac{1}{2}\sum_{n\neq n'} J_{nn'}\sigma_n\sigma_{n'}. \quad (1.208)$$

This is nothing but the Ising model, and the "field" $h$ and the "exchange" $J_{nn'}$ now correspond to

$$h = -E_{\rm FC} + \frac{S}{1-K}, \quad (1.209)$$

$$J_{nn'} = \frac{2S}{N}\sum_{\vec{k}} \frac{\cos[\vec{k}\cdot(\vec{R}_n - \vec{R}_{n'})]}{\omega_{\vec{k}}^2}, \quad (1.210)$$

where $K = \sum_{n'} K_{nn'}$ and $\vec{R}_n$ is the lattice vector of the $n$th site, and $\omega_{\vec{k}}^2 = 1 - \sum_{n'} K_{nn'}\exp[i\vec{k}\cdot(\vec{R}_n - \vec{R}_{n'})]$. The summation with respect to the wave vector $\vec{k}$ in Eq. (1.210) runs over the first Brillouin zone, *i.e.*, $-\pi \leq k_x, k_y, k_z \leq \pi$.

Applying the mean-field approximation to this Ising model,[20] *i.e.*, $J_{nn'}$ is replaced by an infinite ranged constant $N^{-1}\sum_{n'\neq n} J_{nn'} \equiv J/N$, we find the followings. (i) The transition temperature $T_c$ in the case of $h = 0$ is $T_c = \frac{J}{4} \equiv \frac{SK}{2(1-K)}$. (ii) Below this temperature, the free energy $F(M)$ has double minima as a function of the magnetization $M = \langle\sigma_n\rangle \equiv n_e - \frac{1}{2}$. Applying the "magnetic field" $h$, the relative height of the two minima of the free energy $F(M) - hM$ changes, and one of the minima vanishes. The switching phenomena are expected only when the double minima exist. The boundary of the switching and nonswitching regions is given by

$$f_+\left(\frac{T}{J}\right) < \frac{1}{2} - \frac{h}{J} < f_-\left(\frac{T}{J}\right), \quad (1.211)$$

[20]This mean-field approximation can be obtained also by replacing $K_{nn'}$ by $K/N$ in the original Hamiltonian.

where

$$f_{\pm}(x) \equiv \frac{2x}{1 \pm \sqrt{1-4x}} + x \ln \left[ \frac{1 - 2x \pm \sqrt{1-4x}}{2x} \right]. \tag{1.212}$$

This condition is reduced to $|h| < \frac{1}{2}J$ at $T = 0$.

Now we make comments on the effects of the fluctuations which enter when the interaction $K_{nn'}$ is short ranged. We assume two kinds of coupling constants, $J_{\parallel}$ and $J_{\perp}$, for the exchange integral $J_{nn'}$ in Eq. (1.210) parallel and perpendicular to the chain, respectively. Applying the nucleation theory [119] to this anisotropic 3D system, the critical droplet size is

$$L_{\parallel} \sim \frac{J_{\parallel}^{2/3} J_{\perp}^{1/3}}{|h|} \quad \text{along the chain,} \tag{1.213}$$

$$L_{\perp} \sim \frac{J_{\parallel}^{1/6} J_{\perp}^{5/6}}{|h|} \quad \text{perpendicular to the chain.} \tag{1.214}$$

The quasi-one-dimensionality reflects in $J_{\perp} \ll J_{\parallel}$, *i.e.*, $L_{\perp}/L_{\parallel} \ll 1$. When $L_{\perp} < 1$, *i.e.*, $J_{\parallel}^{1/6} J_{\perp}^{5/6} < |h|$, the nucleation process occurs in each chain separately. This case has been discussed in Sec. 1.7.1. Therefore, in this section, we restrict ourselves to the case $L_{\perp} > 1$.

We now discuss the dynamics of the photoinduced structure changes after an optical pumping by applying the mean-field approximation to the kinetic Ising model. The influence from the surrounding sites on the transition probabilities at $n$th site is replaced by the average value. Therefore $\Delta_n$ is replaced by its average value $\Delta = K\langle Q \rangle$ and $\langle Q \rangle$ is the average value of the displacement $Q_n$'s over all the sites, which is $\langle Q \rangle = n_e \sqrt{2S}/(1-K) = \Delta/K$. Therefore the state is specified only by the ratio of the $e$ sites, $n_e \equiv N_e/N$, in this mean-field picture.

The time evolution of the system is described by the rate equation for $n_e(t)$:

$$\frac{d}{dt} n_e(t) = I(t) + (P^{\text{th}}_{g \to e} + P^{\text{opt}}_{g \to e})[1 - n_e(t)] - (P^{\text{th}}_{e \to g} + P^{\text{opt}}_{e \to g}) n_e(t), \tag{1.215}$$

where $I(t)$ is the optical pumping, which is positive when the $e$ state is created while negative when the $g$ state is created. Here $P^{\text{th}}_{e \to g}$ ($P^{\text{th}}_{g \to e}$) and $P^{\text{opt}}_{e \to g}$ ($P^{\text{opt}}_{g \to e}$) are the transition probabilities from $e$ state ($g$ state) to $g$ state ($e$ state) via the thermal and the optical processes, respectively [122]. We confine ourselves to the low-temperature case $T/S \ll 1$, where we can neglect the finite temperature effect on the optical transition. At $T = 0$, only the spontaneous optical emission survives. Whether the spontaneous emission is possible or not depends on the relative relation of the two parabolas in the energy diagram. The system is optically stable against the spontaneous emission when the following condition is satisfied:

$$\sqrt{2S}\Delta = \sqrt{2S} K \langle Q \rangle < E_{\text{FC}} < 2S + \sqrt{2S}\Delta = 2S + \sqrt{2S} K \langle Q \rangle, \tag{1.216}$$

which is also written as $n_{\text{L}} < n_e < n_{\text{U}}$, where $n_{\text{L}} \equiv E_{\text{FC}}(1-K)(1 - 2S/E_{\text{FC}})/(2KS)$ and $n_{\text{U}} \equiv E_{\text{FC}}(1-K)/(2KS)$. The ratio of $e$ state $n_e$ lies between 0 and 1, and

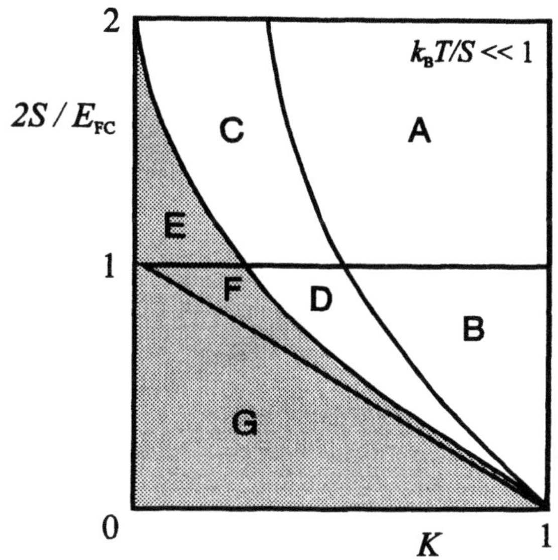


Figure 1.24. Characteristics of the photoinduced structure changes at zero temperature on the parameter space $(K, 2S/E_{FC} \equiv \tilde{S})$. In regions A–D, the system yields switching, but the regions E–G are nonswitching.

the relative relations of $n_L$, $n_U$, 0, and 1 results in the qualitative difference in the dynamics of the structure change. Figure 1.24 summarized the results. The relations $0 < n_U$ and $n_L < n_U$ always hold, and there are three boundaries which correspond to $n_U = 1$, $n_L = 0$, and $n_L = 1$. Together with the condition of $|h| < \frac{1}{2}J$ for switching at $T = 0$ [Eq. (1.211)], the plane is divided into seven regions. Here we note that the mean-field approximation works only when $|h| \ll J$, *i.e.*, near the diagonal line, $S/E_{FC} = 1 - K$. Only in the region A, $n_e = 1$ state is absolutely stable. The spontaneous emission in the relaxation process (from $n_e = 0$ to $n_e = 1$) occurs in the regions A and B, while no spontaneous emission takes place in the region C. In the region D, it occurs only in the process back to the initial $n_e = 0$ state below a critical pumping strength.

We take the region D as an example and discuss a threshold of the optical pumping. Starting from the minimum point $n_e = 0$, we apply the following optical pumping:

$$I(t) = \begin{cases} I_{\text{pump}} & \text{for } 0 \le t \le t_{\text{pump}} \\ 0 & \text{otherwise.} \end{cases} \tag{1.217}$$

The product of $|I_{\text{pump}}|$ and $t_{\text{pump}}$ is the number of the $e$-state sites created by the pumping divided by the total number of sites. When we neglect the saturation of absorption, there is a critical value of the duration time $t_{\text{pump}}$ above which the system

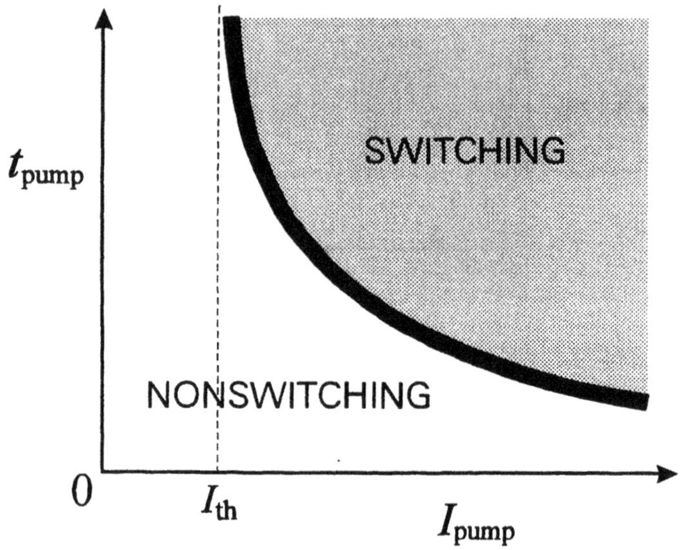


Figure 1.25. The threshold behavior of the switching in a plane of the intensity $I_{pump}$ and the duration time $t_{pump}$ of a pumping pulse. In the hatched region, the system can switch to the other state.

is switched from $n_i = 0$ to $n_f = 1$. As a function of $I_{pump}$, this critical time $t_{cr}$ is inversely proportional to $I_{pump}$ for large $I_{pump}$ and diverges as $I_{pump} \rightarrow I_{th}$, as shown in Fig. 1.25. An asymptotic form of $t_{cr}$ near the divergent point $I_{th}$ is given as $t_{cr} \sim (I_{pump} - I_{th})^{-1/2}$, where

$$I_{th} = \tau_t^{-1} \frac{1-K}{K} \frac{1-2S/E_{FC}}{2S/E_{FC}}, \tag{1.218}$$

with the attempt frequency $\tau_t^{-1}$ of the order of the Debye frequency. The extension to other regions is straightforward. Finite temperature effects should be referred to [122].

## 1.8. CONCLUDING REMARKS

We have overviewed several relations between the dimensionality of materials and their electronic and optical properties from theoretical viewpoints. In particular, we stress universal (material-independent) features of low-dimensional systems, not limited to a special material. Previously, "natural" low-dimensional materials, *e.g.*, organic semiconducting chains, organic synthetic metals, or layered compounds, are typical low dimensional materials. While recently, "artificial" low-dimensional systems, *e.g.*, inorganic semiconductor compounds with controlled geometry and carrier doping, can be obtained due to rapid progress of crystal-growth and nanometer-scale fabrication techniques. Here we remark the differences between them.

*Natural (mainly organic) low-dimensional materials.* — The low dimensionality is an inherent property of these materials, which are composed of low-dimensional molecular structures. Constituting elements of these materials are often soft and anisotropic molecules. Therefore, (quasi)particles moves on a soft lattice with a large fluctuation, which results from low-dimensional structural instability. Conjugate $\pi$-electron polymers, delocalized $\sigma$-electron polymers, halogen-bridged mixed valence complex, and charge-transfer complex are typical examples. Such systems exhibit peculiar phenomena resulting from a strong electron (or exciton)-lattice interaction, which phenomena cannot be expected only in a single-particle band picture. For example, the Peierls transition (metal-insulator transition) has been elucidated in Sec. 1.5.1, which is a spontaneous lattice distortion coupled to the density wave formation. An electron or a hole in an excited state induces a local lattice distortion to relax its total energy, and then solitons, polarons, self-trapped excitons (see Sec. 1.6), or charge-transfer excitons can be formed which are nonlinear elementary excitations peculiar to the 1D electron-lattice systems (see also Sec. 1.7).

*Artificial (mainly inorganic) low-dimensional materials.* — The low dimensionality is introduced "by hand" into bulk semiconductor compounds with the use of the microfabrication technology. Constituting elements of these materials are often hard-core atoms, then (quasi)particles moves in a rigid and distortionless perfect crystal. Band characteristics and fundamental optical properties can be controlled by changing the ratio of components of compounds and the carrier doping density. Several state-of-the-art technologies are available particularly in the III–V and IV groups. In these materials, single-electron band structures are well known. In turn, the band picture can describe well their characteristics. A particle-lattice coupling is often weak in such materials.

Silicon in the IV group is intermediate between the above two categories. Recent development of chemical synthesis of Si polymers also enables us to control the dimensionality of the material. For example, the following Si-based materials are now available: octasilacubans (Si molecules) ($d \simeq 0$), hydrogenated Si clusters ($0 < d < 3$), polysilane chains ($d \simeq 1$), Si quantum wires ($1 < d < 3$), branched polysilane chains and ladder polysilanes ($1 < d < 2$), network Si polymers (planar polysilanes) ($d \simeq 2$), a bulk Si crystal and an amorphous Si ($d = 3$). They offer a new field for the study of the dimensionality and the dimensional crossover effects in their electronic and optical properties. The readers should go to Chapter 5 for more detailed information.

The readers may often notice the importance of the dimensionality of materials in optical responses of meso- and microscopic systems. Geometrical confinement of excitons and the dielectric confinement effects as well as the single-electron band structures are a key for designing novel materials. Moreover, many-body effects, *e*-

$e$ correlations, the Fermi-surface effects also play an important role in the optical responses (*e.g.*, the Fermi-edge singularity). The lattice distortion also induces a variety of phenomena, *e.g.*, the self trapping and the photoinduced structure changes. The exciton (electron)-lattice interaction as well as electronic correlations becomes of crucial importance in the study of the low-dimensional materials.

Since the volume of this chapter is limited, we have to omit several interesting phenomena due to the low dimensionality. For example, the CDW and SDW are peculiar to low-dimensional conductors, which yield novel collective excitations to show nonlinear conductivity. The phase of the CDW or SDW satisfies the sine-Gordon equation, whose solution has been named solitons. A disordering of bound sequences in polymers can be treated as a domain wall, *i.e.*, a topological soliton [123]. Solitons, polarons, and charge-transfer exciton "strings" [124] in organic chains are peculiar excited states, which can be probed in optical experiments. Superconductivity in organic 1D systems is also owing to the dimensionality [125]. These are all omitted from this chapter.

Although the theories do not always include the actual details, the obtained results are universal to the optical features of low-dimensional systems. Our new findings will also be of significance in the interpretation of experimental results. In addition, this chapter will offer a guiding principle in the materials design of new low-dimensional materials. In the future, the following problems will be important in the study of the optical responses of low-dimensional materials.

*Dimensional crossover.* — In the study on universal features of low-dimensional systems, theoreticians usually assume an ***ideal*** dimension. However, the dimensional crossover between $d$ and $d \pm 1$ is also important, which plays a crucial role in an optical anisotropy in the TPA process, as discussed in Sec. 1.3.4. Interlayer [126] and interchain couplings should be payed more attention. Actually, a quantum-dot lattice structure of semiconductor, a semiconductor wire array, and a multilayer semiconductor in a magnetic field are recently studied extensively in terms of their transport and nonlinear optical properties.

*Dimensionality of the photon field.* — When we study the dimensionality of materials, the confinement of photon fields should be simultaneously considered. This radiation confinement is closely related to the cavity QED (quantum electrodynamics) for controlling the spontaneous emission. Similar to the control of the electron Bloch states by crystal periodicity, the photonic band structure can also be artificially controlled in dielectric structures. Moreover, photon localization problems are also concerned with the dimensionality of the photon field.

*Quantum fluctuations in material and photon fields.* — In low-dimensional systems, their quantum fluctuations play crucial roles and then cannot be neglected. When the radiation field is quantized, quantum fluctuation of the photons is an interesting quantity as well as that of electrons, excitons, and lattices. However, the relation between fluctuations of photons and of materials has not been understood in detail. The transfer dynamics is also an open problem.

## REFERENCES


[1] *Highly Conducting One-Dimensional Solids*, ed. by J. T. Devreese, R. P. Evrard, and V. E. van Doren (Plenum, New York, 1979).

[2] H. Haug and S. W. Koch, *Quantum Theory of the Optical and Electronic Properties of Semiconductors*, 3rd Edition (World Scientific, Singapore, 1994).

[3] R. E. Peierls, *Quantum Theory of Solids* (Oxford Press, London, 1955).

[4] M. Pope and C. E. Swenberg, *Electronic Processes in Organic Crystals* (Oxford Univ. Press, New York, 1982).

[5] A. Graja, *Low-Dimensional Organic Conductors*, (World Scientific, Singapore, 1992).

[6] T. Ohno, K. Shiraishi, and T. Ogawa, *Phys. Rev. Lett.* **69**, 2400 (1992).

[7] G. D. Sanders and Y.-C. Chang, *Phys. Rev. B* **45**, 9202 (1992); M. S. Hybertsen, in *Light Emission from Silicon*, ed. by S. S. Iyer, L. T. Canham, and R. T. Collins (Material Research Society, Pittsburgh, 1992), p. 179; M. S. Hybertsen and M. Needels, *Phys. Rev. B* **48**, 4608 (1993); A. J. Read *et al.*, *Phys. Rev. Lett.* **69**, 1232 (1992); F. Buda, J. Kohanoff, and M. Parrinello, *Phys. Rev. Lett.* **69**, 1272 (1992).

[8] L. T. Canham, *Appl. Phys. Lett.* **57**, 1046 (1990); A. G. Cullis and L. T. Canham, *Nature* **353**, 335 (1991).

[9] T. Takagahara and K. Takeda, *Phys. Rev. B* **46**, 15578 (1992).

[10] J. Ihm, A. Zunger, and M. L. Cohen, *J. Phys. C* **12**, 4409 (1979); L. Kleinman and D. M. Bylander, *Phys. Rev. Lett.* **48**, 1425 (1982).

[11] T. Ogawa, in *Light Emission from Novel Silicon Materials*, ed. by Y. Kanemitsu, M. Kondo, and K. Takeda (Phys. Soc. Jpn., Tokyo, 1994), p. 120.

[12] M. Kumagai and T. Ohno, *Soild State Commun.* **83**, 837 (1992).

[13] R. J. Elliott, *Phys. Rev.* **108**, 1384 (1957).

[14] M. Shinada and S. Sugano, *J. Phys. Soc. Jpn.* **21**, 1936 (1966).

[15] Y. Shinozuka and M. Matsuura, *Phys. Rev. B* **28**, 4878 (1983); **29**, 3717(E) (1984).

[16] M. Kumagai and T. Takagahara, *Phys. Rev. B* **40**, 12359 (1989).

[17] T. Ogawa and T. Takagahara, *Phys. Rev. B* **43**, 14325 (1991); **44**, 8138 (1991); *Surf. Sci.* **263**, 506 (1992).

[18] R. Loudon, *Am. J. Phys.* **27**, 649 (1959); R. J. Elliott and R. Loudon, *J. Phys. Chem. Solids* **8**, 382 (1959); **15**, 196 (1960).

[19] L. Bányai, I. Galbraith, C. Ell, and H. Haug, *Phys. Rev. B* **36**, 6099 (1987); H. Ando, H. Oohashi, and H. Kanbe, *J. Appl. Phys.* **70**, 7024 (1991).

[20] S. Abe, *J. Phys. Soc. Jpn.* **58**, 62 (1989).

[21] K. Ishida, H. Aoki, and T. Chikyu, *Phys. Rev. B* **47**, 7594 (1993).

[22] L. Bányai and S. W. Koch, *Semiconductor Quantum Dots* (World Scientific, Singapore, 1993).

[23] Y. Kayanuma, *Phys. Rev. B* **38**, 9797 (1988); Al. L. Efros and A. L. Efros, *Fiz. Tekh. Poluprovodn.* **16**, 1209 (1982) [*Sov. Phys.-Semicond.* **16**, 2555 (1986)].

[24] G. E. W. Bauer and H. Sakaki, *Surf. Sci.* **267**, 442 (1992).

[25] A. Shimizu, T. Ogawa, and H. Sakaki, *Phys. Rev. B* **45**, 11338 (1992); *Surf. Sci.* **263**, 512 (1992).

[26] H. N. Spector, *Phys. Rev. B* **35**, 5876 (1987).

[27] A. Shimizu, *Phys. Rev. B* **40**, 1403 (1989).

[28] G. D. Mahan, *Phys. Rev.* **170**, 825 (1968).

[29] M. Yamanishi and I. Suemune, *Jpn. J. Appl. Phys.* **23**, L35 (1984).

[30] H. Ando, S. Nojima, and H. Kanbe, *J. Appl. Phys.* **74**, 6383 (1993).

[31] P. C. Sercel and K. J. Vahala, *Phys. Rev. B* **42**, 3690 (1990); **44**, 5681 (1991).

[32] I. Suemune and L. A. Coldren, *IEEE J. Quantum Electron.* **QE-24**, 1778 (1988).

[33] U. Bockelmann and G. Bastard, *Phys. Rev. B* **45**, 1688 (1992).

[34] R. Cingolani, M. Lepore, R. Tommasi, I. M. Catalano, H. Lage, D. Heitmann, K. Ploog, A. Shimizu, H. Sakaki, and T. Ogawa, *Phys. Rev. Lett.* **69**, 1276 (1992); *Superlattices and Microstructures* **13**, 71 (1993).

[35] T. Ogawa and A. Shimizu, *Phys. Rev. B* **48**, 4910 (1993).

[36] Experimental work on the TPA of QWs for $\omega \simeq \frac{1}{2}E^{(2)}_{\mathbf{gap}}$ includes K. Tai *et al.*, *Phys. Rev. Lett.* **62**, 1784 (1989); M. Nithisoontorn *et al.*, **62**, 3078 (1989); K. Fujii *et al.*, **65**, 1808 (1990); I. M. Catalano *et al.*, *Solid State Commun.* **71**, 217 (1989); *Phys. Rev. B* **41**, 12937 (1990).

[37] T. Hasegawa *et al.*, *Phys. Rev. Lett.* **69**, 668 (1992).

[38] T. Tokihiro and E. Hanamura, *Phys. Rev. Lett.* **71**, 1423 (1993).

[39] Y. Kanemitsu, T. Ogawa, K. Shiraishi, and K. Takeda, *Phys. Rev. B* **48**, 4883 (1993); *J. Lumin.* **60&61**, 337 (1994).

[40] Y. Kanemitsu, *Phys. Rev. B* **48**, 12357 (1993).

[41] Y. Kayanuma and N. Saito, *Solid State Commun.* **84**, 771 (1992).

[42] W. F. Brinkman, T. M. Rice, and B. Bell, *Phys. Rev. B* **8**, 1570 (1973); O. Akimoto and E. Hanamura, *J. Phys. Soc. Jpn.* **33**, 1537 (1972).

[43] D. A. Kleinman, *Phys. Rev. B* **28**, 871 (1983).

[44] K. Ishida, H. Aoki, and T. Ogawa, *Phys. Rev. B* (1995).

[45] L. V. Keldysh, *Pis'ma Zh. Eksp. Teor. Fiz.* **29**, 716 (1979) [*JETP Lett.* **29**, 658 (1979)].

[46] N. S. Rytova, *Vestnik Moskovsk. Univ.* **3**, 30 (1967).

[47] E. Hanamura, N. Nagaosa, M. Kumagai, and T. Takagahara, *Mater. Sci. Eng.* **B1**, 255 (1988).

[48] T. Ishihara, J. Takahashi, and T. Goto, *Solid State Commun.* **69**, 933 (1989); X. Hong, T.Ishihara, and A. V. Nurmikko, *Phys. Rev. B* **45**, 6961 (1992).

[49] T. Ogawa and H. Tsuda, (unpublished).

[50] L. V. Keldysh, *Superlattices and Microstructures* **4**, 637 (1988).

[51] *Fermi Surface Effects*, ed. by J. Kondo and A. Yoshimori (Springer-Verlag, Berlin, 1988).

[52] D. Pines and P. Noziéres, *The Theory of Quantum Liquids, Volume I: Normal Fermi Liquids* (Benjamin, New York, 1966).

[53] S. Tomonaga, *Prog. Theor. Phys.* **5**, 544 (1950).

[54] J. M. Luttinger, *J. Math. Phys.* **4**, 1154 (1963).

[55] F. D. M. Haldane, *Phys. Rev. Lett.*, **45**, 1358 (1980).

[56] H. J. Schulz, *Phys. Rev. Lett.* **64**, 2831 (1990); **71**, 1864 (1993).

[57] *Correlated Electron Systems*, ed. V. J. Emery (World Scientific, Singapore, 1993).

[58] *Correlation Effects in Low-Dimensional Electron Systems*, ed. by A. Okiji and N. Kawakami (Springer-Verlag, Berlin, 1994).

[59] V. J. Emery, in *Highly Conducting One-Dimensional Solids*, ed. by J. T. Devreese, R. P. Evrard, and V. E. van Doren (Plenum, New York, 1979).

[60] K. Ohtaka and Y. Tanabe, *Rev. Mod. Phys.* **62**, 929 (1990).

[61] G. D. Mahan, *Phys. Rev.* **153**, 882 (1967); **163**, 612 (1967).

[62] P. Noziéres and C. T. de Dominicis, *Phys. Rev.* **178**, 1097 (1969).

[63] M. S. Skolnick, J. M. Rorison, K. J. Nash, D. J. Mowbray, P. R. Tapster, S. J. Bass, and A. D. Pitt, *Phys. Rev. Lett.* **58**, 2130 (1987).

[64] A. E. Ruckenstein and S. Schmitt-Rink, *Phys. Rev. B* **35**, 7551 (1987).

[65] S. Schmitt-Rink, C. Ell, and H. Haug, *Phys. Rev. B* **33**, 1183 (1986).

[66] J. F. Mueller, *Phys. Rev. B* **42**, 11189 (1990).

[67] J. M. Calleja, A. R. Goñi, B. S. Dennis, J. S. Weiner, A. Pinczuk, S. Schmitt-Rink, L. N. Pfeiffer, K. W. West, J. F. Müller, and A. E. Ruckenstein, *Solid State Commun.* **79**, 911 (1991).

[68] E. Müller-Hartmann, T. V. Ramakrishnan, and G. Toulouse, *Phys. Rev. B* **3**, 1102 (1971).

[69] T. Kopp, A. E. Ruckenstein, and S. Schmitt-Rink, *Phys. Rev. B* **42**, 6850 (1990).

[70] T. Uenoyama and L. J. Sham, *Phys. Rev. Lett.* **65**, 1048 (1990).

[71] A. Luther and V. J. Emery, *Phys. Rev. Lett.* **33**, 589 (1974).

[72] A. O. Gogolin, *Phys. Rev. Lett.* **71**, 2995 (1993).

[73] N. V. Prokof'ev, *Phys. Rev. B* **49**, 2148 (1994).

[74] C. L. Kane and M. P. A. Fisher, *Phys. Rev. Lett.* **68**, 1220 (1992); *Phys. Rev. B* **49**, 2253 (1994).

[75] K. D. Schotte and U. Schotte, *Phys. Rev.* **182**, 479 (1969).

[76] T. Ogawa, A. Furusaki, and N. Nagaosa, *Phys. Rev. Lett.* **68**, 3638 (1992).

[77] A. Schmid, *Phys. Rev. Lett.* **51**, 1506 (1983).

[78] F. Guinea, V. Hakim, and A. Muramatsu, *Phys. Rev. Lett.* **54**, 263 (1985).

[79] D. K. K. Lee and Y. Chen, *Phys. Rev. Lett.* **69**, 1399 (1992).

[80] P. W. Anderson, *Phys. Rev. Lett.* **18**, 1049 (1967).

[81] K. Yamada and K. Yoshida, *Prog. Theor. Phys.* **60**, 353 (1978).

[82] J. F. Mueller, A. E. Ruckenstein, and S. Schmitt-Rink, *Phys. Rev. B* **45**, 8902 (1992).

[83] T. Ogawa, A. Furusaki, and N. Nagaosa, *Jpn. J. Appl. Phys. Suppl.* **32**, 76 (1993).

[84] G. Fasol and H. Sakaki, *Phys. Rev. Lett.* **70**, 3643 (1993).

[85] T. Ogawa, *Jpn. J. Appl. Phys. Suppl.* **34**, 146 (1995).

[86] J. M. Calleja, A. R. Goñi, A. Pinczuk, B. S. Dennis, J. S. Weiner, L. N. Pfeiffer, and K. W. West, *Phys. Rev. B* **51**, 4285 (1995); F. J. Rodriguez and C. Tejedor, *Phys. Rev. B* **49**, 16781 (1994).

[87] T. Ogawa and H. Otani, *J. Phys. Soc. Jpn.* **64**, (1995); *Surf. Sci.* (in print).

[88] C. Tanguy and M. Combescot, *Phys. Rev. Lett.* **68**, 1935 (1992).

[89] J.-P. Foing, D. Hulin, M. Joffre, M. K. Jackson, J.-L. Ouder, C. Tanguy, and M. Combescot, *Phys. Rev. Lett.* **68**, 110 (1992).

[90] C. M. Varma and A. Zawadowski, *Phys. Rev. B* **32**, 7399 (1985).

[91] K. A. Muttalib and V. J. Emery, *Phys. Rev. Lett.* **57**, 1370 (1986).

[92] K. Penc and J. Solyom, *Phys. Rev. B* **41**, 704 (1990).

[93] J. Solyom, *Adv. Phys.* **28**, 201 (1979).

[94] G. D. Mahan, *Many-Particle Physics*, Second Edition (Plenum, New York, 1990).

[95] H. Fukuyama and H. Takayama, in *Dynamical Properties of Quasi-One-Dimensional Conductors*, p. 41, ed. by P. Monceau (Reidel, 1984).

[96] Y. Suzumura, *Prog. Theor. Phys.* **61**, 1 (1979).

[97] Yu. A. Bychkov, L. P. Gor'kov, and I. E. Dzyaloshinskii, *Sov. Phys. JETP* **23**, 489 (1966).

[98] N. Nagaosa and T. Ogawa, *Solid State Commun.* **88**, 295 (1993).

[99] T. Ogawa, *phys. stat. sol. (b)* **188**, 83 (1995).

[100] K. Kuroki and H. Aoki, *Phys. Rev. Lett.* **69**, 3820 (1992); *Phys. Rev. B* **48**, 7598 (1993); *Phys. Rev. Lett.* **72**, 2947 (1994).

[101] A. L. Ivanov and H. Haug, *Phys. Rev. Lett.* **71**, 3182 (1993).

[102] Y. Takada and M. Kohmoto, *Phys. Rev. B* **41**, 8872 (1990).

[103] Y. Suzumura and H. Fukuyama, *J. Phys. Soc. Jpn.* **52**, 2870 (1983).

[104] T. Kimura, K. Kuroki, H. Aoki, and M. Eto, *Phys. Rev. B* **49**, 16852 (1994).

[105] N. F. Mott, *Metal-Insulator Transitions*, Second Edition (Taylor & Francis, London, 1990).

[106] E. I. Rashba, *Opt. Spektrosk.* **2**, 88 (1957); **3**, 568 (1957).

[107] Y. Toyozawa, *Technical Report of ISSP* **A119**, (1964).

[108] H. Sumi and Y. Toyozawa, *J. Phys. Soc. Jpn.* **31**, 342 (1972).

[109] M. Ueta, H. Kanzaki, K. Kobayashi, Y. Toyozawa, and E. Hanamura, *Excitonic Processes in Solids* (Springer-Verlag, Berlin, 1986), Chapter 4.

[110] H. Sumi and A. Sumi, *J. Phys. Soc. Jpn.* **63**, 637 (1994).

[111] P. W. Anderson, *Phys. Rev.* **109**, 1492 (1958).

[112] E. Abrahams, P. W. Anderson, D. C. Licciardello, and T. V. Ramakrishnan, *Phys. Rev. Lett.* **42**, 673 (1979).

[113] H. Fröhlich, *Proc. R. Soc. A* **160**, 230 (1937); H. Fröhlich, H. Pelzer, and S. Zienau, *Philos. Mag.* **41**, 221 (1950).

[114] J. Bardeen and W. Schockley, *Phys. Rev.* **80**, 72 (1950).

[115] D. Emin and T. Holstein, *Phys. Rev. Lett.* **36**, 323 (1976).

[116] Y. Toyozawa and Y. Shinozuka, *J. Phys. Soc. Jpn.* **48**, 472 (1980); Y. Shinozuka and Y. Toyozawa, *J. Phys. Soc. Jpn.* **46**, 505 (1979).

[117] Y. Toyozawa, *Solid State Commun.* **84**, 255 (1992).

[118] S. Koshihara, Y. Tokura, K. Takeda, K. Koda, and A. Kobayashi, *J. Chem. Phys.* **92**, 7581 (1990); S. Koshihara, Y. Tokura, K. Takeda, and K. Koda, *Phys. Rev. Lett.* **68**, 1148 (1992); S. Koshihara, Y. Tokura, T. Mitani, G. Saito, and T. Koda, *Phys. Rev. B* **42**, 6853 (1990).

[119] J. D. Gunton and M. Droz, *Introduction to the Theory of Metastable and Unstable States*, Vol. 183 of Lecture Notes in Physics (Springer-Verlag, Berlin, 1983).

[120] E. Hanamura and N. Nagaosa, *J. Phys. Soc. Jpn.* **56**, 2080 (1987).

[121] T. Nagai, N. Nagaosa, and E. Hanamura, *J. Lumin.* **38**, 314 (1987).

[122] N. Nagaosa and T. Ogawa, *Phys. Rev. B* **39**, 4472 (1989).

[123] W. P. Su, J. R. Schrieffer, and A. J. Heeger, *Phys. Rev. B* **22**, 2099 (1980); H. Takayama, Y. R. Lin-Liu, and K. Maki, *Phys. Rev. B* **21**, 2388 (1980).

[124] M. Kuwata-Gonokami, N. Peyghambarian, K. Meissner, B. Fluegel, Y. Sato, K. Ema, R. Shimano, S. Mazumdar, F. Guo, T. Tokihiro, H. Ezaki, and E. Hanamura, *Nature* **367**, 47 (1994).

[125] W. A. Little, *Phys. Rev.* **134**, 1416 (1964).

[126] T. Fukuzawa, S.S. Kano, T.K. Gustafson, and T. Ogawa, *Surf. Sci.* **228**, 482 (1990).

# *CHAPTER 2*

# *Ab initio* CALCULATION OF NONLINEAR OPTICAL SUSCEPTIBILITY

Takashi Nakayama
*Department of Physics, Chiba University*

## 2.1. INTRODUCTION

### 2.1.1. *Purpose of This Chapter*

One of recent remarkable trends in condensed matter physics is a vigorous interaction among physics, material science, and technology. Especially in the nonlinear optics field, rapidly-advancing nonlinear optical techniques have often been applied to a variety of newly-synthesized materials to clarify their fundamental electronic and optical properties. On the other hand, because the development of new advanced nonlinear optical materials is of crucial importance in the technical areas such as optelectronics, intense efforts have been made to fabricate various nonlinear optical materials, including semiconductors, ionic compounds, liquid crystals, organic polymers, glasses, and their superstructures. In these views, the microscopic understanding of nonlinear optical processes in materials is extremely important both in clarifying optical properties of new materials and in searching for other new optelectronic materials.

Nonlinear responses of materials to the applied electromagnetic fields are completely described in terms of the nonlinear optical susceptibilities, $\chi^{(n)}$. Although the microscopic formulation of $\chi^{(n)}$ had been carried out in 1960s, a predictive ***ab initio*** calculation of nonlinear optical properties of materials has been of long-standing difficulty mainly due to the lack of large computational facilities and accurate information on electronic structures. However, owing to developments in the computer science and the ***ab initio*** (first-principles) methods in the electronic-structure calculations, an exact evaluation of complex microscopic $\chi^{(n)}$ has become possible very recently. To explain in great detail the method of such ***ab initio*** nonlinear-optical-susceptibility calculation and to demonstrate its usefulness with comparing to experiments are main purpose of this chapter.

Since the demonstration of the first working laser by Maiman in 1960, we have been possessing sufficiently intense and coherent light sources, thus the study of nonlinear optical properties being possible. Let us review the history of such study by considering the two-photon absorption (TPA) as an example. TPA properties are fully described by the third-order susceptibility, $\chi^{(3)}$. Figure 2.1 shows the experimentally-measured and theoretically-calculated values of TPA coefficient, $\beta$, of GaAs for years beginning with 1966 [1]. It is seen that the earlier experimental values lie scattered over three orders of magnitude. This is because the determination of TPA magnitude was difficult due to its sensitivity to experimental variables [2], such as laser-pulse duration, peak power, spatial and temperature coherence of the lasers, presence of free carriers, etc. Especially, wide-pulse and multimode lasers have often induced large TPA magnitude through the reexcitation of photogenerated carriers. However, the recent developments of laser equipment and the superior experimental methods such as "Z-scan" and "four-beam coupling" have eliminated these disadvantages [11], thus enabling the precise TPA measurement since around 1990.

From the theoretical viewpoint, numerous model-band calculations of TPA spectra have been performed [2]. Typical one is to employ the bond-charge model where the polarization is expressed as a sum of bond polarizabilities [12]. Another one is

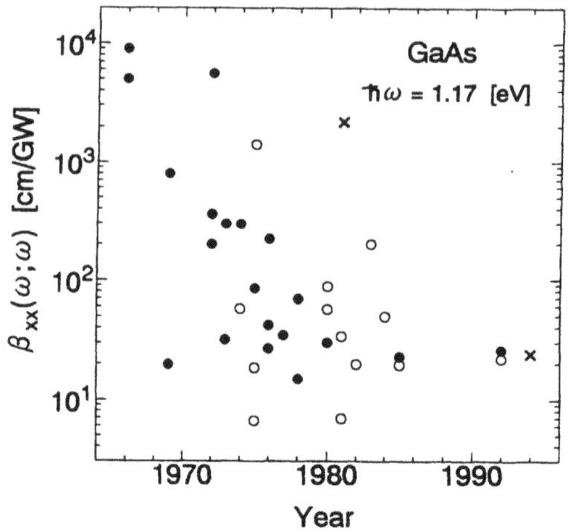


Figure 2.1. Semilogarithmic plot of measured and calculated values of two-photon-absorption coefficient for GaAs as a function of year beginning with 1966. Filled circles correspond to experiments, while open circles and crosses to calculations respectively based on the model and full-band-structure calculations. Experimental data are taken from [1, 2] except the recent ones [3], while calculated values are from [1, 2, 4, 5, 6, 7, 8, 9, 10].

to adopt the two-band model where the polarization is caused through the inter- and intra-band transitions among the single valence and single conduction bands [8, 13]. These are empirical approaches, thus the results strongly depending on band parameters adopted. Therefore, although model-band calculations were sometimes useful to qualitatively study the characteristics of TPA, the calculated values scatters widely, as shown in Fig. 2.1, as well as the measured ones. The recent works based on the Kane and Luttinger-Kohn band-structure models [9, 14] for various zinc-blende semiconductors are noteworthy because these calculations were successful to give the TPA magnitude comparable to recent accurate experiments. However, these calculations are limited to the TPA around the center of the Brillouin zone of direct band-gap materials with zinc-blende structure, where band parameters are *well known*. In order to evaluate the TPA spectra of materials having transitions over the whole Brillouin zone and having various crystal structures, including new materials of low-dimensional structures, the developments of the *ab initio* calculation coupled with the full-band calculation are indispensable.

At the present stage, the *ab initio* nonlinear-optical-susceptibility calculations coupled with full-band calculations are very few. The representative works are (a) Moss, Sipe, van Driel, and Ghahramani [15, 16, 17] calculated the second and third-harmonic-generation (SHG and THG) spectra for a number of cubic semiconductors and superlattices, using the empirical tight-binding model and the *semi-ab-initio* linear-combination-of-atomic-orbitals method. However, due to the empirical nature in their electronic-structure calculations, the degree of agreement with the measured data was less than desirable. (b) Levine and Allan [18, 19] have calculated the zero-frequency-limit SHG magnitude of Si, Ge, AlP, AlAs, GaP, and GaAs, using the *ab initio* pseudopotential method in the local density approximation (LDA) with a self-energy correction in the form of a scissors operator including local-field effects, and obtained values in good agreement with experiments. (c) Huang and Ching [20, 21] have calculated the SHG and THG spectra of various cubic semiconductors, using the orthogonalized linear-combination-of-atomic-orbitals method in the LDA with a scissors operator but without the corresponding self-energy correction. (d) Murayama and Nakayama [10] have calculated the TPA spectra for various semiconductors and superlattices using the *ab initio* pseudopotential method in the LDA with a self-energy correction in the form of a scissors operator neglecting local-field effects, and obtained values in good agreement with experiments.

In these views, the application of the *ab initio* methods is still limited to bulk-semiconductors and some semiconductor-superlattice systems. However, it should be noted that these methods can apply to materials having arbitrary components and structures, particularly new materials of any dimensionalities reviewed in this book, since the methods are fully "*ab initio*" in a sense that they need no input parameters in principles (see Fig. 2.2). In fact, in the recent *ab initio* electronic-structure calculations, once the material-component atoms are given, the ground-state properties of materials such as the crystal structure, cohesive energy, charge density, structural and magnetic phase transitions, and phonon spectra, can be evaluated by using the density-functional-theory and quantum-Monte-Carlo approaches [22, 23]. While the excited-state properties such as the electron and hole spectra can be evaluated by the many-body Green's function approach [24], and other elementary excitation spectra such as exciton are calculated based on the conventional theoretical techniques using the input parameters derived from such *ab initio* calculations. In this way, one can evaluate much elemental electronic data in materials without any experimental information. The linear and nonlinear (optical) responses correspond to include additional external-field freedom into consideration, and the *ab initio* nonlinear-optical-property calculation reviewed in this chapter is based on thus-obtained material data. In this sense it is fully *ab initio*. (The above-mentioned procedure corresponds to the research of material properties, whose direction is denoted by the upward arrows in Fig. 2.2. While by feeding back such procedure (down-arrows) one can also perform the material design.)

The plan for this chapter is as follows. In the rest of this section, the derivation of the microscopic formula of nonlinear optical susceptibilities are outlined, then various nonlinear optical phenomena and their relations to the nonlinear optical susceptibilities are presented. In Sec. 2.2, we describe the method of the *ab initio* nonlinear susceptibility calculation in great detail. To demonstrate the usefulness of the *ab initio* method, the calculated results with respect to the two-photon-absorption in various bulk semiconductors are in detail presented and compared with the recent experiments in Sec. 2.3. The applications of the *ab initio* method to other nonlinear

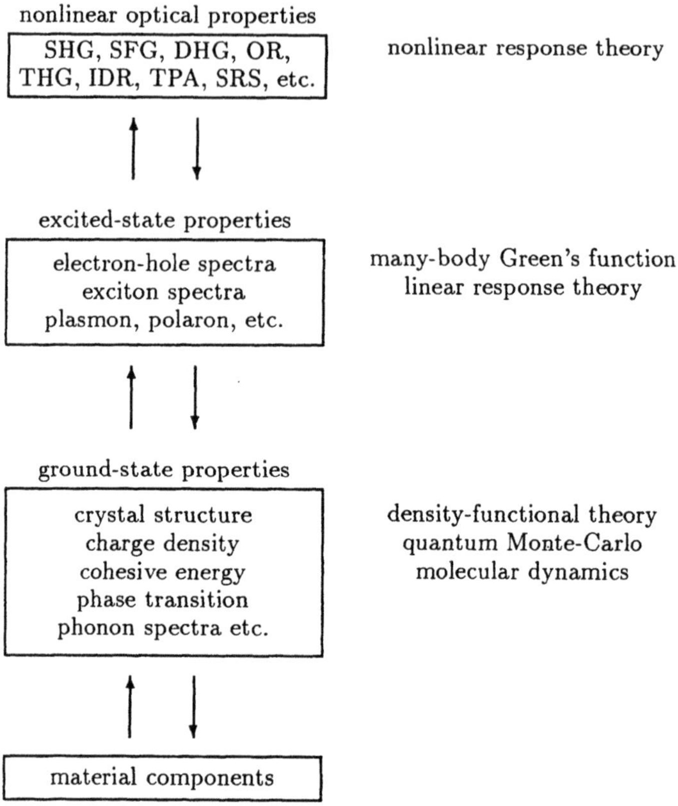


**Figure 2.2.** ***Ab initio*** **approach to nonlinear optical properties.**

spectra (harmonic-generation) in bulk semiconductors and low-dimensional superlattices are presented in Sec. 2.4.2. Section 2.5 is devoted to prospects and problems in the future ***ab initio*** calculations.

### *2.1.2. Microscopic Nonlinear Optical Susceptibility*

Now, we derive the microscopic expression of the nonlinear optical susceptibilities, or more exactly the nonlinear optical susceptibility *function*, $\tilde{\chi}^{(n)}$. The nonlinear optical susceptibility functions are the coefficients of the induced polarization with respect to the products of applied electromagnetic fields, and are frequently defined in the wavenumber (momentum) and frequency (energy) spaces. For example, the third-order susceptibility function is defined by

$$\begin{aligned}\tilde{\mathrm{P}}_a^{(3)}(\mathbf{q}_a,\omega_a) &= \sum_{bcd}\int\frac{d\mathbf{q}_b}{(2\pi)^3}\int\frac{d\mathbf{q}_c}{(2\pi)^3}\int\frac{d\mathbf{q}_d}{(2\pi)^3}\int\frac{d\omega_c}{(2\pi)}\int\frac{d\omega_b}{(2\pi)}\int\frac{d\omega_d}{(2\pi)}\\ &\quad \tilde{\chi}^{(3)}_{abcd}(\mathbf{q}_a,\omega_a;\mathbf{q}_b,\omega_b,\mathbf{q}_c,\omega_c,\mathbf{q}_d,\omega_d)\\ &\quad\quad\times\tilde{\mathrm{E}}_b(\mathbf{q}_b,\omega_b)\tilde{\mathrm{E}}_c(\mathbf{q}_c,\omega_c)\tilde{\mathrm{E}}_d(\mathbf{q}_d,\omega_d),\end{aligned} \tag{2.1}$$

where $\tilde{\mathrm{P}}_a^{(3)}(\mathbf{q}_a,\omega_a)$ is the Fourier-transformed function of the third-order polarization, $\tilde{\mathrm{P}}_a^{(3)}(\mathbf{x}_a,t_a)$, having the polarization-, wavenumber- and frequency-variables, $(a,\mathbf{q}_a,\omega_a)$, while $\tilde{\mathrm{E}}_b(\mathbf{q}_b,\omega_b)$'s are those of applied electromagnetic fields. Note that this definition is the most general one displaying the full dependence on the polarizations, wavenumber-vectors and frequencies of field quantities.

The microscopic derivation starts with Hamiltonian of electrons interacting with applied electromagnetic fields such as

$$\begin{aligned}\hat{H}(t) &= \int d\mathbf{x}\hat{\psi}^\dagger(\mathbf{x},t)\frac{1}{2m_0}(\hat{\mathbf{p}}-\frac{e}{c}\tilde{\mathbf{A}}(\mathbf{x},t))^2\hat{\psi}(\mathbf{x},t)+\int d\mathbf{x}\hat{\psi}^\dagger(\mathbf{x},t)e\tilde{\phi}(\mathbf{x},t)\hat{\psi}(\mathbf{x},t)\\ &\quad +\hat{H}_{int},\end{aligned} \tag{2.2}$$

where $m_0$ and $e$ are respectively the electron mass and charge ($e>0$), $c$ the speed of light, $\hat{\mathbf{p}}$ the momentum operator, $\hbar/i\cdot\nabla$. $\tilde{\mathbf{A}}(\mathbf{x},t)$ and $\tilde{\phi}(\mathbf{x},t)$ are respectively the vector and scalar potentials originating from the applied electromagnetic fields. $\hat{H}_{int}$ is an operator representing the interaction among electrons and ions. $\hat{\psi}^\dagger(\mathbf{x},t)$ and $\hat{\psi}(\mathbf{x},t)$ are the ordinary field operators of electrons in a Heisenberg representation. From now on, the radiation gauge, $\tilde{\phi}=0$, is adopted. Moreover the long-wavelength limit, $\nabla\cdot\tilde{\mathbf{A}}=0$, is assumed. This choice of gauge is allowed when the electromagnetic fields considered have wavelength much larger than the crystal unit cell, or equivalently when the linear and higher-order terms to the wavenumber, $\mathbf{q}$, is neglected in deriving the $\tilde{\chi}^{(n)}$ formula.

Most responses of system to the applied electromagnetic fields are completely characterized by the current density due to the motion of electrons. The operator representing such current density in the presence of electromagnetic fields is [25]

$$\begin{aligned}\hat{\mathrm{J}}_a(\mathbf{x},t) &= -c\frac{\delta\hat{H}(t)}{\delta\tilde{\mathrm{A}}_a(\mathbf{x},t)}\\ &= \hat{\mathrm{j}}_a(\mathbf{x},t)-\frac{e}{m_0c}\hat{\rho}(\mathbf{x},t)\tilde{\mathrm{A}}_a(\mathbf{x},t),\end{aligned} \tag{2.3}$$

where

$$\hat{\mathrm{j}}_a(\mathbf{x},t) = \frac{e}{m_0}\hat{\psi}^\dagger(\mathbf{x},t)\hat{\mathrm{p}}_a\hat{\psi}(\mathbf{x},t), \tag{2.4}$$

$$\hat{\rho}(\mathbf{x},t) = e\hat{\psi}^\dagger(\mathbf{x},t)\hat{\psi}(\mathbf{x},t). \tag{2.5}$$

$\hat{\mathrm{j}}_a(\mathbf{x},t)$ and $\hat{\rho}(\mathbf{x},t)$ are the operators respectively corresponding to the current and charge densities in the absence of electromagnetic fields. The current density is obtained by the expectation value, $\langle\hat{\mathrm{J}}_a(\mathbf{x},t)\rangle$, with respect to the electronic state of system, which is an unperturbed state normally being an equilibrium and/or ground state. When such current density is expanded in a power series of the external vector potential $\tilde{\mathbf{A}}$ as

$$\begin{aligned}\langle\hat{\mathrm{J}}_a(\mathbf{x},t)\rangle &= -\frac{e}{m_0c}\langle\hat{\rho}(\mathbf{x},t)\rangle\tilde{\mathrm{A}}_a(\mathbf{x},t)\\ &\quad+\langle\hat{\mathrm{j}}_a(\mathbf{x},t)\rangle^{(1)}+\langle\hat{\mathrm{j}}_a(\mathbf{x},t)\rangle^{(2)}+\langle\hat{\mathrm{j}}_a(\mathbf{x},t)\rangle^{(3)}+\cdots,\end{aligned}\tag{2.6}$$

the $n$th-order current density, $\langle\hat{\mathrm{j}}_a(\mathbf{x},t)\rangle^{(n)}$, is obtained according to the general response theory [26] as the $(n+1)$-th-order correlation function of current density operators. For example,

$$\begin{aligned}\langle\hat{\mathrm{j}}_a(\mathbf{x}_a,t_a)\rangle^{(3)} &= \left(-\frac{1}{c}\right)^3\left(\frac{1}{i\hbar}\right)^3\sum_{bcd}\int d\mathbf{x}_b\int d\mathbf{x}_c\int d\mathbf{x}_d\int_{-\infty}^{t_a}dt_b\int_{-\infty}^{t_b}dt_c\int_{-\infty}^{t_c}dt_d\\ &\quad\hat{\mathcal{P}}^{bcd}\mathrm{Tr}\{\hat{\rho}_0[[[\hat{\mathrm{j}}^{\mathrm{I}}_a(\mathbf{x}_a,t_a),\hat{\mathrm{j}}^{\mathrm{I}}_b(\mathbf{x}_b,t_b)],\hat{\mathrm{j}}^{\mathrm{I}}_c(\mathbf{x}_c,t_c)],\hat{\mathrm{j}}^{\mathrm{I}}_d(\mathbf{x}_d,t_d)]\}\\ &\quad\times\tilde{\mathrm{A}}_b(\mathbf{x}_b,t_b)\tilde{\mathrm{A}}_c(\mathbf{x}_c,t_c)\tilde{\mathrm{A}}_d(\mathbf{x}_d,t_d),\end{aligned}\tag{2.7}$$

where $\hat{\rho}_0$ is a density-matrix operator in the absence of applied electromagnetic fields, which is given by $\hat{\rho}_0=\exp[-\beta\hat{H}(\tilde{\mathbf{A}}=\tilde{\phi}=0)]$ with the effective inverse temperature, $\beta$. $\hat{\mathrm{j}}^{\mathrm{I}}_a$ is the current-density operator in an interaction representation. $\hat{\mathcal{P}}^{bcd}$ is the intrinsic permutation operator on the quantities characterizing the applied electromagnetic fields, which are the polarization and the space- and time-coordinates of fields in Eq. (2.7). This permutation originates from the fact that the order of applied electromagnetic fields in Eq. (2.7) is not uniquely distinguished in the nonlinear responses.

It should be noted here that the anomalous current proportional to $\tilde{\mathbf{A}}$, the second term in Eq. (2.3), appears in the linear response as explicitly shown in Eq. (2.6). However, it does not appear in nonlinear responses as shown in Eq. (2.7) because it is much smaller than the normal current of $\hat{\mathrm{j}}_a$ in the long wavelength limit [27].

The current density and the vector potential are respectively related to the polarization and the electromagnetic field, such as $\langle\hat{\mathrm{J}}_a(\mathbf{x}_a,t_a)\rangle=-\partial\tilde{\mathrm{P}}_a(\mathbf{x}_a,t_a)/\partial t_a$ and $\tilde{\mathrm{E}}_a(\mathbf{x}_a,t_a)=1/c\cdot\partial\tilde{\mathrm{A}}_a(\mathbf{x}_a,t_a)/\partial t_a$. Thus, by Fourier-transforming all the quantities in Eq. (2.7) into the wavenumber and frequency spaces and by comparing the results with Eq. (2.1), the microscopic expression of $\tilde{\chi}^{(n)}$ is obtained. For the third-order susceptibility function,

$$\begin{aligned}&\tilde{\chi}^{(3)}_{abcd}(\mathbf{q}_a,\omega_a;\mathbf{q}_b,\omega_b,\mathbf{q}_c,\omega_c,\mathbf{q}_d,\omega_d)\\ &=\frac{1}{|\hat{\mathcal{P}}^{bcd}|}\left(\frac{1}{\hbar}\right)^3\left(\frac{e}{m_0}\right)^4\frac{i^4}{\omega_a\omega_b\omega_c\omega_d}2\pi\delta(\omega_a-\omega_b-\omega_c-\omega_d)\sum_{klmn}e^{-\beta\hbar\omega_n}\\ &\hat{\mathcal{P}}^{bcd}\left\{\frac{\langle n|\hat{\mathrm{P}}^a(-\mathbf{q}_a)|k\rangle\langle k|\hat{\mathrm{P}}^b(\mathbf{q}_b)|l\rangle\langle l|\hat{\mathrm{P}}^c(\mathbf{q}_c)|m\rangle\langle m|\hat{\mathrm{P}}^d(\mathbf{q}_d)|n\rangle}{(\omega_b+\omega_c+\omega_d-\omega_{kn}+i\delta)(\omega_c+\omega_d-\omega_{ln}+i\delta)(\omega_d-\omega_{mn}+i\delta)}\right.\end{aligned}$$

$$
\begin{aligned}
&- \frac{\langle n|\hat{\mathrm{P}}^b(\mathbf{q}_b)|k\rangle\langle k|\hat{\mathrm{P}}^a(-\mathbf{q}_a)|l\rangle\langle l|\hat{\mathrm{P}}^c(\mathbf{q}_c)|m\rangle\langle m|\hat{\mathrm{P}}^d(\mathbf{q}_d)|n\rangle}{(\omega_b+\omega_c+\omega_d+\omega_{kl}+i\delta)(\omega_c+\omega_d-\omega_{ln}+i\delta)(\omega_d-\omega_{mn}+i\delta)} \\
&- \frac{\langle n|\hat{\mathrm{P}}^c(\mathbf{q}_c)|k\rangle\langle k|\hat{\mathrm{P}}^a(-\mathbf{q}_a)|l\rangle\langle l|\hat{\mathrm{P}}^b(\mathbf{q}_b)|m\rangle\langle m|\hat{\mathrm{P}}^d(\mathbf{q}_d)|n\rangle}{(\omega_b+\omega_c+\omega_d+\omega_{kl}+i\delta)(\omega_c+\omega_d+\omega_{km}+i\delta)(\omega_d-\omega_{mn}+i\delta)} \\
&+ \frac{\langle n|\hat{\mathrm{P}}^c(\mathbf{q}_c)|k\rangle\langle k|\hat{\mathrm{P}}^b(\mathbf{q}_b)|l\rangle\langle l|\hat{\mathrm{P}}^a(-\mathbf{q}_a)|m\rangle\langle m|\hat{\mathrm{P}}^d(\mathbf{q}_d)|n\rangle}{(\omega_b+\omega_c+\omega_d+\omega_{lm}+i\delta)(\omega_c+\omega_d+\omega_{km}+i\delta)(\omega_d-\omega_{mn}+i\delta)} \\
&- \frac{\langle n|\hat{\mathrm{P}}^d(\mathbf{q}_d)|k\rangle\langle k|\hat{\mathrm{P}}^a(-\mathbf{q}_a)|l\rangle\langle l|\hat{\mathrm{P}}^b(\mathbf{q}_b)|m\rangle\langle m|\hat{\mathrm{P}}^c(\mathbf{q}_c)|n\rangle}{(\omega_b+\omega_c+\omega_d+\omega_{kl}+i\delta)(\omega_c+\omega_d+\omega_{km}+i\delta)(\omega_d+\omega_{kn}+i\delta)} \\
&+ \frac{\langle n|\hat{\mathrm{P}}^d(\mathbf{q}_d)|k\rangle\langle k|\hat{\mathrm{P}}^b(\mathbf{q}_b)|l\rangle\langle l|\hat{\mathrm{P}}^a(-\mathbf{q}_a)|m\rangle\langle m|\hat{\mathrm{P}}^c(\mathbf{q}_c)|n\rangle}{(\omega_b+\omega_c+\omega_d+\omega_{lm}+i\delta)(\omega_c+\omega_d+\omega_{km}+i\delta)(\omega_d+\omega_{kn}+i\delta)} \\
&+ \frac{\langle n|\hat{\mathrm{P}}^d(\mathbf{q}_d)|k\rangle\langle k|\hat{\mathrm{P}}^c(\mathbf{q}_c)|l\rangle\langle l|\hat{\mathrm{P}}^a(-\mathbf{q}_a)|m\rangle\langle m|\hat{\mathrm{P}}^b(\mathbf{q}_b)|n\rangle}{(\omega_b+\omega_c+\omega_d+\omega_{lm}+i\delta)(\omega_c+\omega_d+\omega_{ln}+i\delta)(\omega_d+\omega_{kn}+i\delta)} \\
&\left. - \frac{\langle n|\hat{\mathrm{P}}^d(\mathbf{q}_d)|k\rangle\langle k|\hat{\mathrm{P}}^c(\mathbf{q}_c)|l\rangle\langle l|\hat{\mathrm{P}}^b(\mathbf{q}_b)|m\rangle\langle m|\hat{\mathrm{P}}^a(-\mathbf{q}_a)|n\rangle}{(\omega_b+\omega_c+\omega_d+\omega_{mn}+i\delta)(\omega_c+\omega_d+\omega_{ln}+i\delta)(\omega_d+\omega_{kn}+i\delta)} \right\}, \quad (2.8)
\end{aligned}
$$

where we use the definition of the Fourier transformation for the function $\mathrm{f}(\mathbf{x},t)$ such as

$$\mathrm{f}(\mathbf{q},\omega) \;=\; \int d\mathbf{x}\int dt\ \mathrm{f}(\mathbf{x},t)e^{-i\mathbf{q}\mathbf{x}+i\omega t}. \tag{2.9}$$

In deriving Eq. (2.8), we have used the completeness relation, $\sum_k |k\rangle\langle k| = \hat{1}$, for the many-body electronic eigenstates, $|k\rangle$. $\hbar\omega_{kn} = \hbar\omega_k - \hbar\omega_n$ is the energy difference between $|k\rangle$ and $|n\rangle$. $\hat{\mathcal{P}}^{bcd}$ is an intrinsic permutation operator on applied electromagnetic fields, thus interchanging $(b, \mathbf{q}_b, \omega_b)$, $(c, \mathbf{q}_c, \omega_c)$ and $(d, \mathbf{q}_d, \omega_d)$, while $|\hat{\mathcal{P}}^{bcd}|$ is the number of such permutations. $\hat{\mathrm{P}}^a(-\mathbf{q}_a)$'s are momentum operators for many-body states defined by

$$\hat{\mathrm{P}}^a(-\mathbf{q}_a) \equiv \int d\mathbf{x}\,\hat{\psi}^\dagger(\mathbf{x})\hat{\mathrm{p}}_a e^{-i\mathbf{q}_a\mathbf{x}}\hat{\psi}(\mathbf{x}). \tag{2.10}$$

$\delta$'s in the denominator of Eq. (2.8) are damping factors representing the inverse of electronic-state life time, $\Gamma_{kn}$, and/or those of the applied electromagnetic fields, $\gamma_b$. Namely, they respectively emerge in the denominator of Eq. (2.8) such as $\omega_{kn} \to \omega_{kn} - i\Gamma_{kn}$ and $\omega_b \to \omega_b + i\gamma_b$, where the negative and positive signs are from the result of causality.

The expression Eq. (2.8) has eight terms, each of which comes from the third-order commutator in Eq. (2.7). When the damping factors are additive, *e.g.*, $\Gamma_{kn} = \Gamma_{kl} + \Gamma_{ln}$, these terms are collected together and the compressed expression with the full permutation operator is available . This is especially true in the infinitesimal-damping-limit case, $\Gamma_{kn} \to 0$, or for the real part of $\tilde{\chi}^{(n)}$. For $\tilde{\chi}^{(3)}$,

$\tilde{\chi}^{(3)}_{abcd}(\mathbf{q}_a, \omega_a; \mathbf{q}_b, \omega_b, \mathbf{q}_c, \omega_c, \mathbf{q}_d, \omega_d)$

$$
\begin{aligned}
= & \frac{1}{|\hat{\mathcal{P}}^{bcd}|}\left(\frac{1}{\hbar}\right)^3\left(\frac{e}{m_0}\right)^4\frac{(-i)^2}{\omega_a\omega_b\omega_c\omega_d}2\pi\delta(\omega_a-\omega_b-\omega_c-\omega_d)\sum_{klmn}\mathrm{e}^{-\beta\hbar\omega_n} \\
& \hat{\mathcal{P}}^{abcd}\frac{\langle n|\hat{\mathrm{P}}^a(-\mathbf{q}_a)|k\rangle\langle k|\hat{\mathrm{P}}^b(\mathbf{q}_b)|l\rangle\langle l|\hat{\mathrm{P}}^c(\mathbf{q}_c)|m\rangle\langle m|\hat{\mathrm{P}}^d(\mathbf{q}_d)|n\rangle}{(\omega_{kn}-\omega_b-\omega_c-\omega_d-i\delta)(\omega_{ln}-\omega_c-\omega_d-i\delta)(\omega_{mn}-\omega_d-i\delta)},
\end{aligned}
\tag{2.11}
$$

where $\hat{\mathcal{P}}^{abcd}$ is the full permutation operator interchanging $(a,-\mathbf{q}_a,-\omega_a)$, $(b,\mathbf{q}_b,\omega_b)$, $(c,\mathbf{q}_c,\omega_c)$ and $(d,\mathbf{q}_d,\omega_d)$. The similar expressions are obtained for any $\tilde{\chi}^{(n)}$, such as

$$
\begin{aligned}
& \tilde{\chi}^{(1)}_{ab}(\mathbf{q}_a,\omega_a;\mathbf{q}_b,\omega_b) \\
= & \frac{1}{\hbar}\left(\frac{e}{m_0}\right)^2\frac{1}{\omega_a\omega_b}2\pi\delta(\omega_a-\omega_b)\sum_{mn}\mathrm{e}^{-\beta\hbar\omega_n} \\
& \hat{\mathcal{P}}^{ab}\frac{\langle n|\hat{\mathrm{P}}^a(-\mathbf{q}_a)|m\rangle\langle m|\hat{\mathrm{P}}^b(\mathbf{q}_b)|n\rangle}{\omega_{mn}-\omega_b-i\delta},
\end{aligned}
\tag{2.12}
$$

$$
\begin{aligned}
& \tilde{\chi}^{(2)}_{abc}(\mathbf{q}_a,\omega_a;\mathbf{q}_b,\omega_b,\mathbf{q}_c,\omega_c) \\
= & \frac{1}{|\hat{\mathcal{P}}^{bc}|}\left(\frac{1}{\hbar}\right)^2\left(\frac{e}{m_0}\right)^3\frac{-i}{\omega_a\omega_b\omega_c}2\pi\delta(\omega_a-\omega_b-\omega_c)\sum_{lmn}\mathrm{e}^{-\beta\hbar\omega_n} \\
& \hat{\mathcal{P}}^{abc}\frac{\langle n|\hat{\mathrm{P}}^a(-\mathbf{q}_a)|m\rangle\langle m|\hat{\mathrm{P}}^b(\mathbf{q}_b)|l\rangle\langle l|\hat{\mathrm{P}}^c(\mathbf{q}_c)|n\rangle}{(\omega_{ln}-\omega_b-\omega_c-i\delta)(\omega_{mn}-\omega_c-i\delta)}.
\end{aligned}
\tag{2.13}
$$

It should be noted that these expressions apply to any nonlinear optical phenomena and to arbitrary systems, as long as the exact many-body states are used.

So far, we have derived the microscopic formula of nonlinear susceptibility *functions*, $\tilde{\chi}^{(n)}$. Apparently these formula possess the delta functions

$$(2\pi)^3\Delta(\mathbf{q}_a-\mathbf{q}_b-\mathbf{q}_c\cdots), \tag{2.14}$$

$$2\pi\quad\delta(\omega_a-\omega_b-\omega_c\cdots), \tag{2.15}$$

where $(\mathbf{q}_a,\omega_a)$ is the wavenumber and frequency of the induced polarization, while $(\mathbf{q}_b,\omega_b)$, $(\mathbf{q}_c,\omega_c)\cdots$ etc. are those of the applied electromagnetic fields. The latter delta function is explicitly included in Eqs. (2.11) to (2.13), while the former one is implicitly included in a numerator of these equations as follows. When the periodic system is considered, $\langle n|\hat{\mathrm{P}}^a(-\mathbf{q}_a)|k\rangle$ has non-zero value for wavenumbers satisfying $\mathbf{q}_n=\mathbf{q}_k-\mathbf{q}_a$ (mod $\{\mathbf{G}_i\}_{i=1,3}$), where $\mathbf{G}_i$'s are the reciprocal lattice vectors, and $\mathbf{q}_n$ and $\mathbf{q}_k$ are the Bloch wavenumbers of states, $|n\rangle$ and $|k\rangle$. For example, applying this property to the numerator of $\tilde{\chi}^{(3)}$ in Eq. (2.11),

$$
\begin{aligned}
\mathbf{q}_n &= \mathbf{q}_k-\mathbf{q}_a \quad (\mathrm{mod}\,\{\mathbf{G}_i\}), \\
\mathbf{q}_k &= \mathbf{q}_l+\mathbf{q}_b \quad (\mathrm{mod}\,\{\mathbf{G}_i\}), \\
\mathbf{q}_l &= \mathbf{q}_m+\mathbf{q}_c \quad (\mathrm{mod}\,\{\mathbf{G}_i\}), \\
\mathbf{q}_m &= \mathbf{q}_n+\mathbf{q}_d \quad (\mathrm{mod}\,\{\mathbf{G}_i\}),
\end{aligned}
\tag{2.16}
$$

thus the following equation being satisfied in $\tilde{\chi}^{(3)}$,

$$0 \overset{\mathrm{mod}\,\{\mathbf{G}_i\}}{=} \mathbf{q}_a-\mathbf{q}_b-\mathbf{q}_c-\mathbf{q}_d. \tag{2.17}$$

In this way, $\Delta$ in Eq. (2.15) is the three-dimensional delta function having the period of reciprocal lattice vectors. When the system is non-periodic along the direction $i$, thus having the infinitely large unit cell along this direction, $\mathbf{G}_i \to 0$ and there appears to be no longer such delta function with respect to this direction.

In most literatures [28, 29], the nonlinear optical susceptibility (component), $\chi^{(n)}$, is sometimes defined by excluding the above-mentioned delta functions from $\tilde{\chi}^{(n)}$, *i.e.*,

$$
\begin{aligned}
&\tilde{\chi}^{(n)}(\mathbf{q}_a, \omega_a; \mathbf{q}_b, \omega_b, \mathbf{q}_c, \omega_c, \cdots) = \\
&\quad \chi^{(n)}(\mathbf{q}_a \overset{\mathrm{mod}\,\{\mathbf{G}_i\}}{=} \mathbf{q}_b + \mathbf{q}_c + \cdots, \omega_a = \omega_b + \omega_c + \cdots; \mathbf{q}_b, \omega_b, \cdots) \\
&\quad \times (2\pi)^3 \Delta(\mathbf{q}_a - \mathbf{q}_b - \mathbf{q}_c - \cdots) 2\pi\delta(\omega_a - \omega_b - \omega_c - \cdots).
\end{aligned}
\tag{2.18}
$$

When the polarization and electromagnetic fields have discrete number of components such as

$$
\begin{aligned}
\tilde{\mathrm{P}}_a(\mathbf{x}_a, t_a) &= \sum_{(\mathbf{q}_1, \omega_1)} \mathrm{P}_a(\mathbf{q}_1, \omega_1) \mathrm{e}^{i\mathbf{q}_1 \mathbf{x}_a - i\omega_1 t_a}, \\
\tilde{\mathrm{E}}_b(\mathbf{x}_b, t_b) &= \sum_{(\mathbf{q}_2, \omega_2)} \mathrm{E}_b(\mathbf{q}_2, \omega_2) \mathrm{e}^{i\mathbf{q}_2 \mathbf{x}_b - i\omega_2 t_b},
\end{aligned}
\tag{2.19}
$$

their Fourier transformations are given by

$$
\begin{aligned}
\tilde{\mathrm{P}}_a(\mathbf{q}_a, \omega_a) &= \sum_{(\mathbf{q}_1, \omega_1)} \mathrm{P}_a(\mathbf{q}_1, \omega_1) \cdot (2\pi)^3 \delta(\mathbf{q}_a - \mathbf{q}_1) \cdot 2\pi\delta(\omega_a - \omega_1), \\
\tilde{\mathrm{E}}_b(\mathbf{q}_b, \omega_b) &= \sum_{(\mathbf{q}_2, \omega_2)} \mathrm{E}_b(\mathbf{q}_2, \omega_2) \cdot (2\pi)^3 \delta(\mathbf{q}_b - \mathbf{q}_2) \cdot 2\pi\delta(\omega_b - \omega_2).
\end{aligned}
\tag{2.20}
$$

Note here that $\tilde{\mathrm{P}}_a(\mathbf{q}_a.\omega_a)$ and $\tilde{\mathrm{E}}_b(\mathbf{q}_b, \omega_b)$ are *functions*, while $\mathrm{P}_a(\mathbf{q}_1, \omega_1)$ and $\mathrm{E}_b(\mathbf{q}_2, \omega_2)$ are *components*. Inserting Eq. (2.20) into (2.1) yields the familiar definition of $\chi^{(3)}$,

$$
\begin{aligned}
&\mathrm{P}_a(\mathbf{q}_1, \omega_2) = \sum_{(b, \mathbf{q}_2, \omega_2)} \sum_{(c, \mathbf{q}_3, \omega_3)} \sum_{(d, \mathbf{q}_4, \omega_4)} \\
&\chi^{(3)}_{abcd}(\mathbf{q}_1 \overset{\mathrm{mod}\,\{\mathbf{G}_i\}}{=} \mathbf{q}_2 + \mathbf{q}_3 + \mathbf{q}_4, \omega_1 = \omega_2 + \omega_3 + \omega_4; \mathbf{q}_2, \omega_2, \mathbf{q}_3, \omega_3, \mathbf{q}_4, \omega_4) \\
&\times \mathrm{E}_b(\mathbf{q}_2, \omega_2) \mathrm{E}_c(\mathbf{q}_3, \omega_3) \mathrm{E}_d(\mathbf{q}_4, \omega_4).
\end{aligned}
\tag{2.21}
$$

### 2.1.3. *Nonlinear Optical Phenomena*

#### 2.1.3.1. *Conservative nature in $\chi^{(n)}$*

In this subsection, we present qualitative explanation how a variety of nonlinear optical phenomena are described by using nonlinear susceptibilities. The polarization of materials plays a key role in such descriptions because a time-varying polarization can act as the *source* of new components of the electromagnetic fields. This can be easily understood by seeing the wave equation for the propagation of light through a polarizable materials

$$
\mathbf{q}_1^2 \mathrm{E}_a(\mathbf{q}_a, \omega_1) - \frac{\omega_1^2}{c^2} \mathrm{E}_a(\mathbf{q}_a, \omega_1) = 4\pi \frac{\omega_1^2}{c^2} \mathrm{P}_a(\mathbf{q}_a, \omega_1),
\tag{2.22}
$$

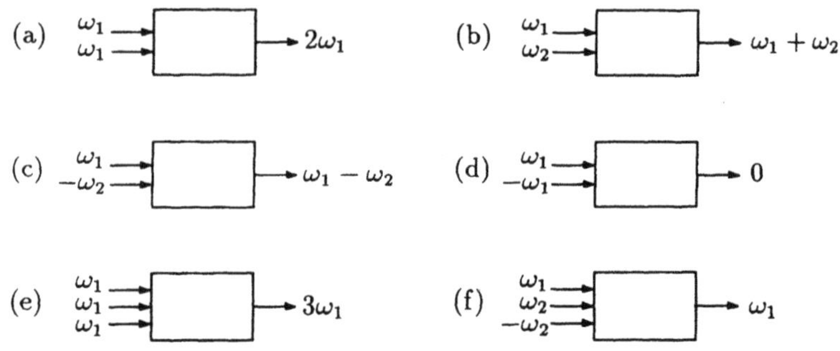


Figure 2.3. Various second and third-order nonlinear optical processes. (a) SHG, (b) SFG, (c) DHG, (d) OR, (e) THG, and (f) IDR, TPA, and SRS. With respect to abbreviations, see the text.

where we adopt the wavenumber- and frequency-component expression. The right-handed term comes from $-4\pi/c^2 \cdot \partial^2\tilde{\mathbf{P}}(\mathbf{x},t)/\partial t^2$. This equation is an inhomogeneous wave equation in which the polarization, $\mathrm{P}_a(\mathbf{q}_1,\omega_1)$, drives the electric field $\mathrm{E}_a(\mathbf{q}_1,\omega_1)$.

Then what kinds of *sources*, *i.e.*, $\mathrm{P}_a(\mathbf{q}_1,\omega_1)$, are produced in nonlinear optical processes ? We have already noted that the general formula of $\tilde{\chi}^{(n)}$ has the delta functions shown in Eqs. (2.15) and (2.15). These functions appear physically due to the wavenumber and frequency conservation in nonlinear optical processes. As a result, for example, when the incident applied electromagnetic fields have the wavelength components such as $\mathbf{q}_2, \mathbf{q}_3 \cdots$ etc., only the electromagnetic field having the wavelength $\mathbf{q}_1 \equiv \mathbf{q}_2 + \mathbf{q}_3 + \cdots (\mathrm{mod}\,\{\mathbf{G}_i\})$ is produced and propagates in materials. This feature is sometimes called the ***phase matching*** condition of electromagnetic fields [28, 29]. Similarly, the tensor nature in $\chi^{(n)}$ represents that the polarization directions are changed in nonlinear optical processes. This appears physically due to the *conservation* of polarization directions due to the crystal symmetry in materials. It should be noted here that these conservative properties are not always satisfied for electromagnetic fields having any polarizations, wavenumbers and frequencies. This is because the polarization, wavenumber and frequency of the electromagnetic field in material are not the independent variables, as seen in the dispersion relation $\omega_2(\mathbf{q}_2)$.

Using these conservative properties, we can convert the polarization direction, wavenumber, and frequency in incident electromagnetic fields to different ones in output electromagnetic fields. In what follows, we demonstrate one of these features, *i.e.*, the frequency conservation, neglecting the polarization and wavenumber dependence (see Fig. 2.3). When the second-order optical process described by $\chi^{(2)}$ occurs with the incident electromagnetic fields having the frequency components, $\omega_2$ and $\omega_3$, the polarizations $\mathbf{P}^{(2)}$ having the following frequency components are independently induced [28];

$$
\begin{aligned}
\mathbf{P}^{(2)}(2\omega_2), \mathbf{P}^{(2)}(2\omega_3) \quad &; \quad \text{SHG}, \\
\mathbf{P}^{(2)}(\omega_2 + \omega_3) \quad &; \quad \text{SFG}, \\
\mathbf{P}^{(2)}(\omega_2 - \omega_3) \quad &; \quad \text{DHG}, \\
\mathbf{P}^{(2)}(0 = \omega_2 - \omega_2 = \omega_3 - \omega_3) \quad &; \quad \text{OR}.
\end{aligned} \tag{2.23}
$$

Here we have labeled each expression by the name of the physical process that it describes, such as second-harmonic-generation (SHG), sum-frequency-generation (SFG), difference-frequency-generation (DFG), and optical rectification (OR). Note that, because the reality of the electromagnetic fields requires the coexistence of the frequencies, $-\omega_2$ and $\omega_2$, the latter two processes are possible because of the frequency sums such as $\omega_2 + (-\omega_3)$. On the other hand, although there is also a response at the negative of each of nonzero frequencies given above, *i.e.*, at $-2\omega_2, -2\omega_3, -(\omega_2 + \omega_3)$ and $-(\omega_2 - \omega_3)$, we need not take explicit account of such frequency components, again due to the reality of the polarization $\mathbf{P}(\mathbf{x}, t)$.

Similarly, when the third-order optical processes described by $\chi^{(3)}$ occur with the incident electromagnetic fields with frequencies, $\omega_2, \omega_3$ and $\omega_4$, the induced polarization $\mathbf{P}^{(3)}$ has 22 different frequency components [28]:

$$
\begin{aligned}
\omega_2, \omega_3, \omega_4 \quad &; \quad \text{IDR, TPA, SRS}, \\
3\omega_2, 3\omega_3, 3\omega_4 \quad &; \quad \text{THG}, \\
(\omega_2 + \omega_3 + \omega_4), (\omega_2 + \omega_3 - \omega_4), & \\
(\omega_2 - \omega_3 + \omega_4), (-\omega_2 + \omega_3 + \omega_4), & \\
(2\omega_2 \pm \omega_3), (2\omega_2 \pm \omega_4), (2\omega_3 \pm \omega_2), & \\
(2\omega_3 \pm \omega_4), (2\omega_4 \pm \omega_2), (2\omega_4 \pm \omega_3). &
\end{aligned} \tag{2.24}
$$

Because some processes out of these occur with less than three incident electromagnetic-field components, thus being easily observed in experiments, they are specially labeled by the names, such as third-harmonic-generation (THG), intensity-dependent-refraction (IDR), two-photon-absorption (TPA), and stimulated-Raman-scattering (SRS). For example, for the incident electromagnetic fields having only one frequency component the THG occurs through the frequency-sum such as $3\omega_2 = \omega_2 + \omega_2 + \omega_2$, while the TPA is possible for the incident electromagnetic fields having one and/or two frequency components because of the frequency-sums such as $\omega_2 = \omega_2 + \omega_2 - \omega_2$ and/or $\omega_2 = \omega_2 + \omega_3 - \omega_3$.

### *2.1.3.2. Complex nature in $\chi^{(n)}$*

Next, we consider other features of nonlinear optical susceptibilities. As an example, the third-order nonlinear optical processes named as IDR and TPA are considered in detail for materials having the cubic symmetry. The cubic symmetry allows four nonzero tensor elements such as $\chi^{(3)}_{aaaa}, \chi^{(3)}_{abba}, \chi^{(3)}_{abab}$ and $\chi^{(3)}_{aabb}$, where $a$ and $b$ indicate the crystallographic axes, $x$, $y$ and $z$, while the IDR/TPA process demands $\chi^{(3)}_{abba} = \chi^{(3)}_{abab}$ (see Sec. 2.3.3.1). Assuming that the incident electromagnetic fields have at most two components, $(a, \mathbf{q}_1, \omega_1)$ and $(b, \mathbf{q}_2, \omega_2)$, and that they are linearly polarized, which is the reason why the $\chi^{(3)}_{aabb}$ element does not appear in the following expression, the induced polarization is written by

$$\mathrm{P}_a(\mathbf{q}_1, \omega_1) = (\chi^{(1)}_{eff} + \chi^{(3)}_{eff})\mathrm{E}_a(\mathbf{q}_1, \omega_1), \tag{2.25}$$

where we have introduced the effective susceptibilities,

$$\begin{aligned} \chi^{(1)}_{eff} &\equiv \chi^{(1)}_{aa}(\mathbf{q}_1\omega_1; \mathbf{q}_1\omega_1), \\ \chi^{(3)}_{eff} &\equiv 3\chi^{(3)}_{aaaa}(\mathbf{q}_1\omega_1; -\mathbf{q}_1 - \omega_1, \mathbf{q}_1\omega_1, \mathbf{q}_1\omega_1)|\mathrm{E}_a(\mathbf{q}_1, \omega_1)|^2 \\ &\quad +6\chi^{(3)}_{abba}(\mathbf{q}_1\omega_1; -\mathbf{q}_2 - \omega_2, \mathbf{q}_2\omega_2, \mathbf{q}_1\omega_1)|\mathrm{E}_b(\mathbf{q}_2, \omega_2)|^2 \\ &\quad \times(1 - \delta_{a,b}\delta_{\mathbf{q}_1,\mathbf{q}_2}\delta_{\omega_1,\omega_2}). \end{aligned} \tag{2.26}$$

Here we are also assuming that the $\chi^{(2)}$ components does not exist, which is true for materials with the inversion symmetry. When the induced electromagnetic field caused by the induced polarization is propagating along the $c$ direction perpendicular to the $a$ and $b$ directions, $\mathbf{q}_1^2$ is equal to $\mathrm{q}_{c,1}^2$. By substituting Eq. (2.25) into the wave equation (2.22), we obtain

$$\begin{aligned} \mathrm{q}_{c,1} &= \frac{\omega_1}{c}\{1 + 4\pi(\chi^{(1)}_{eff} + \chi^{(3)}_{eff})\}^{1/2} \\ &\cong \frac{\omega_1}{c}(1 + 4\pi\chi^{(1)}_{eff})^{1/2}(1 + \frac{2\pi\chi^{(3)}_{eff}}{1 + 4\pi\chi^{(1)}}). \end{aligned} \tag{2.27}$$

Let us consider the case, $\mathrm{Im}\chi^{(1)}_{eff} = \mathrm{Im}\chi^{(3)}_{eff} = 0$. As shown below, this condition indicates that there are no linear and two-photon absorption in materials, which is true for the incident electromagnetic fields with energy less than the half of the band-gap energy of a material. In this case, the refractive index, $n$, is given by

$$n_a(\omega_1) = n^{(1)}_a(\omega_1) + \frac{2\pi\chi^{(3)}_{eff}}{n^{(1)}_a(\omega_1)}, \tag{2.28}$$

where $n^{(1)}$ denotes the usual linear refractive index,

$$n^{(1)}_a(\omega_1) = (1 + 4\pi\chi^{(1)}_{eff})^{1/2}. \tag{2.29}$$

Because $\mathrm{q}_{c,1}$ in Eq. (2.27)is a real quantity in the present case, the electromagnetic field, $\mathrm{E}_a(\mathrm{q}_{c,1}, \omega_1)$, propagates in materials without losing its intensity, *i.e.*, its energy. However, its velocity is modulated from $c$ to $c/n_a(\omega_1)$, especially depending on the incident-field intensities, $|\mathrm{E}_a(\mathrm{q}_{c,1}, \omega_1)|^2$ and $|\mathrm{E}_b(\mathbf{q}_2, \omega_2)|^2$, owing to the second term in the right-handed side of Eq. (2.26). Thus, $n_a(\omega_1)$ is sometimes called the

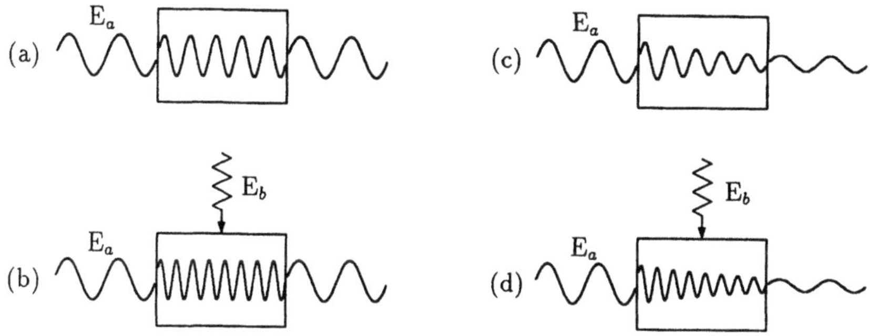


Figure 2.4. Schematic picture of electromagnetic field in IDR and TPA phenomena caused by $\chi^{(3)}$. (a) linear refraction, (b) nonlinear refraction (IDR), (c) linear absorption (OPA), and (d) nonlinear absorption (TPA). In the cases (b) and (d), the refraction (light velocity in material) and absorption (light-amplitude decay) are respectively changed from (a) and (c) cases due to the presence of another field, $E_b$.

intensity-dependent refractive (IDR) index. Since this velocity renormalization of the electromagnetic field occurs without energy dissipation, it is easily understood that the real part of $\chi^{(3)}$ is caused by the *virtual* (nonresonant) interaction between the electromagnetic fields and electronic states in materials. The optical processes without such energy dissipation is called the *parametric* processes.

Next, consider the case where $\mathrm{Im}\chi_{eff}^{(1)} = 0$ but $\mathrm{Im}\chi_{eff}^{(3)} \neq 0$. Because of the presence of the imaginary part in $q_{c,1}$ in this case, the electromagnetic field, $E_a(q_{c,1}, \omega_1)$, propagates in materials with losing its intensity. Writing such intensity by $I(x_c)$ as a function of the coordinate, $x_c$, along the $c$ direction, the decaying rate of the electromagnetic field is described using the absorption coefficient, $\alpha_a(\omega_1)$, such as

$$dI(x_c) = - \alpha_a(\omega_1)\, I(x_c)\, dx_c. \tag{2.30}$$

Using the familiar relation for the time-averaged intensity,

$$I(x_c) = \frac{c}{2\pi} n_a^{(1)}(\omega_1) |E_a(q_{c,1}, \omega_1) e^{iq_{c,1} \cdot x_c}|^2, \tag{2.31}$$

the absorption coefficient is

$$\begin{aligned} \alpha_a(\omega_1) &= 2\mathrm{Im}q_{c,1} \\ &= \frac{4\pi\omega_1}{c n_a^{(1)}(\omega_1)} \mathrm{Im}\chi_{eff}^{(3)}. \end{aligned} \tag{2.32}$$

From this equation, one note that the absorption of the field, $E_a(\mathbf{q}_1, \omega_1)$, is proportional to both the intensity of the field itself, $|E_a(\mathbf{q}_1, \omega_1)|^2$, and the intensity of another field, $|\mathrm{E}_b(\mathbf{q}_2, \omega_2)|^2$, as long as the imaginary parts of $\chi^{(3)}_{aaaa}$ and $\chi^{(3)}_{abba}$ have nonvanishing values. As seen in the general expression of $\chi^{(3)}$ derived in Sec. 2.1.2, the imaginary part of $\chi^{(3)}$ is caused by the two-photon-energy ($\omega_1 + \omega_1$ or $\omega_1 + \omega_2$) resonance to the energy difference between the electronic eigenstates in materials. Since this resonance is followed by the electromagnetic-field-energy dissipation into materials, it is easily understood that the imaginary part of $\chi^{(3)}$ is caused by the ***real*** electronic processes in materials. The optical processes with such energy dissipation is called the ***nonparametric*** processes.

Finally, it is noted that the parametric (virtual, nonresonant) and the nonparametric (real, resonant) processes are tightly related to each other. This can be easily recognized by recalling that both processes originate from the single denominator in Eqs. (2.11) to (2.13) using the identity [Eq. (2.47)]. This relation is often called the Kramers-Krönig causality dispersion relation both in linear and nonlinear responses [30, 31].

## 2.2. METHOD OF *ab initio* CALCULATION

To calculate the microscopic formula of $\chi^{(n)}$ derived in Sec. 2.1.2, we have first to know the electronic eigenstates in materials. In most semiconductors, the primary contribution to $\chi^{(n)}$ comes from the one-particle band-state excitation among various electronic excitations in materials. Thus, in this section, we first explain the recent status in the ***ab initio*** band-structure calculations. Then, the $\chi^{(n)}$ formula in the band-state representation is derived, followed by the numerical details on the ***ab initio*** evaluation of momentum matrix elements and nonlinear optical susceptibilities.

### 2.2.1. *Ab initio Band Structure Calculations*

With respect to the recent ***ab initio*** electronic structure calculation of solids such as semiconductors, insulators and metals, there are two methods that are widely accepted. One is the calculational method of ground-state electronic properties of solids, such as the crystal structure, lattice vibration, electron density, and cohesive energy, based on the density functional theory [22, 32], while the other is the calculational method of quasiparticle excitation spectra of solids based on the many-body Green's function scheme [24, 33]. Since the quasiparticle states (normally the electron and hole band-states) are the one-particle-like excited states among various many-body excited states, the latter method is suit for the susceptibility calculation. According to this method, the energy and wavefunction of quasiparticle state are obtained by solving the following equation;

$$(-\frac{\hbar^2}{2m_0}\nabla^2 + \hat{V}_{ext} + \hat{V}_{ee})\varphi_\mu(\mathbf{x}) = \varepsilon_\mu \varphi_\mu(\mathbf{x}), \tag{2.33}$$

where $m_0$ is the free-electron mass and $\mu$ the state index. The potential operators, $\hat{V}_{ext}$ and $\hat{V}_{ee}$, respectively caused by the electron-nucleus and electron-electron interactions are given by

$$\hat{V}_{ext}\varphi_\mu(\mathbf{x}) = -\sum_i \frac{Z_i e^2}{|\mathbf{x}-\mathbf{X}_i|}\varphi_\mu(\mathbf{x}), \quad (2.34)$$

$$\hat{V}_{ee}\varphi_\mu(\mathbf{x}) = \int d\mathbf{x}' \frac{\rho(\mathbf{x}')e^2}{|\mathbf{x}-\mathbf{x}'|}\varphi_\mu(\mathbf{x}) + \int d\mathbf{x}'\Sigma(\mathbf{x},\mathbf{x}',\varepsilon_\mu)\varphi_\mu(\mathbf{x}'), \quad (2.35)$$

where $\mathbf{X}_i$ and $Z_i e$ are respectively the coordinate and charge of $i$-th nucleus, and $\rho(\mathbf{x})$ the electron density. The first term in $\hat{V}_{ee}$ is the Hartree potential, while $\Sigma$ in the second term is called the self-energy operator that is nonlocal and depends on the quasiparticle energy, $\varepsilon_\mu$, to be solved.

Even when the simplest approximation (GW approximation) is adopted [24], the electronic-structure calculation in the Green's function method is hard, *i.e.*, time-consuming and uneconomic, in the present stage, due to the multi-variable ($\mathbf{x}, \mathbf{x}'$, and $\varepsilon_\mu$) dependence in $\Sigma$. This is especially true in the case of nonlinear-susceptibility calculation where a great number of band states must be calculated. Because of such practical reasons, most electronic-structure calculations have been performed in the density functional theory, especially in the local density approximation (LDA) [34, 35], where we needs much less computing time than in the case of the Green's function approach. In the LDA calculation, the energies and wavefunctions of the band-states are obtained by solving the Kohn-Sham variational equation, which has the similar form to Eq. (2.33) except that the electron-electron potential is given by

$$\hat{V}_{ee}\varphi_\mu(\mathbf{x}) = \int d\mathbf{x}' \frac{\rho(\mathbf{x}')e^2}{|\mathbf{x}-\mathbf{x}'|}\varphi_\mu(\mathbf{x}) + \frac{\delta E_{xc}[\rho(\mathbf{x})]}{\delta\rho(\mathbf{x})}\varphi_\mu(\mathbf{x}). \quad (2.36)$$

Here $E_{xc}[\rho(\mathbf{x})]$ is the exchange-correlation energy that is a functional of the electron density, $\rho(\mathbf{x})$. Fortunately, in most weak-correlated electron systems such as semiconductors and metals, there is a systematic difference in the calculated state energy, $\varepsilon_\mu$, and wavefunction, $\varphi_\mu(\mathbf{x})$, between the Green's function approach and the LDA approach. This is partially due to the fact that the potential operators, $\hat{V}_{ee}$, have comparable magnitude between two approaches because the Hartree potential is dominant in such $\hat{V}_{ee}$'s. For example, in most semiconductors, the valence- and conduction-band quasiparticle energies are respectively obtained with an error of about 0.1 eV by more or less uniform down- and up-shifts from the corresponding LDA-state energies [36]. Moreover, the LDA-calculated wavefunction is 99% similar to that by the Green's function approach [24].

The details of these approaches and their calculational procedures are seen in the literatures [22, 24]. The important points in these *ab initio* band-structure calculations are that the electronic structures of a number of materials are calculated by the same method without any fitting parameters, and that the results are in agreement with experiments with an error of $\sim 0.1$ eV.

### 2.2.2. Band-State Expression of $\tilde{\chi}^{(n)}$

As an example, the band-state expression of $\tilde{\chi}^{(3)}$ is derived here. When the many-body states are approximated to single Slater determinants of one-particle Bloch states, $\varphi_{n\mathbf{k}}(\mathbf{x}) = \langle \mathbf{x}|n\mathbf{k}\rangle$, where $n$ is the band index and $\mathbf{k}$ the Bloch wavenumber, the field operator of electrons can be expanded in terms of such Bloch states as

$$\hat{\psi}(\mathbf{x}) = \sum_{n\mathbf{k}} \hat{C}_{n\mathbf{k}}\varphi_{n\mathbf{k}}(\mathbf{x}). \tag{2.37}$$

Substituting this into Eq. (2.11), we obtain

$$\begin{aligned}
&\tilde{\chi}^{(3)}_{abcd}(\mathbf{q}_a,\omega_a;\mathbf{q}_b,\omega_b,\mathbf{q}_c,\omega_c,\mathbf{q}_d,\omega_d)\\
&= \frac{1}{|\hat{p}^{bcd}|}\left(\frac{1}{\hbar}\right)^3\left(\frac{e}{m_0}\right)^4\frac{-1}{\omega_a\omega_b\omega_c\omega_d}2\pi\delta(\omega_a-\omega_b-\omega_c-\omega_d)\sum_{klmn} e^{-\beta\hbar\omega_n}\\
&\quad \sum_1\sum_2\sum_3\sum_4\sum_5\sum_6\sum_7\sum_8 \langle n|\hat{C}_8^\dagger\hat{C}_7|k\rangle\langle k|\hat{C}_6^\dagger\hat{C}_5|l\rangle\langle l|\hat{C}_4^\dagger\hat{C}_3|m\rangle\langle m|\hat{C}_2^\dagger\hat{C}_1|n\rangle\\
&\quad \hat{p}^{abcd}\frac{\mathrm{P}^a(8,7)\mathrm{P}^b(6,5)\mathrm{P}^c(4,3)\mathrm{P}^d(2,1)}{(\omega_{kn}-\omega_b-\omega_c-\omega_d-i\delta)(\omega_{ln}-\omega_c-\omega_d-i\delta)(\omega_{mn}-\omega_d-i\delta)},
\end{aligned} \tag{2.38}$$

where

$$\sum_1 \equiv \sum_{n_1}\int\frac{d\mathbf{k}_1}{(2\pi)^3}, \tag{2.39}$$

$$\begin{aligned}
\mathrm{P}^a(8,7) &\equiv \langle n_8\mathbf{k}_8|e^{-i\mathbf{q}_a\mathbf{x}}\,\hat{\mathrm{p}}^a|n_7\mathbf{k}_7\rangle\\
&= \int d\mathbf{x}\,\varphi^*_{n_8\mathbf{k}_8}(\mathbf{x})\mathrm{p}_a e^{-i\mathbf{q}_a\mathbf{x}}\varphi_{n_7\mathbf{k}_7}(\mathbf{x}),
\end{aligned} \tag{2.40}$$

and we have adopted the compressed notation, $1 \equiv (n_1,\mathbf{k}_1)$, to display the dependence on the Bloch band-state, $|n_1\mathbf{k}_1\rangle$, and its energy, $\omega(n_1\mathbf{k}_1)$. When the low-temperature limit ($\beta \to 0$) is assumed, the initial state, $|n\rangle$, becomes the Fermi-vacuum state, $|g\rangle$. As a result, the energy differences between the excited and ground states are changed such as

$$\begin{aligned}
\omega_{mn} &\to \omega_{mg} \equiv -\omega(n_1\mathbf{k}_1)+\omega(n_2\mathbf{k}_2),\\
\omega_{ln} &\to \omega_{lg} \equiv -\omega(n_1\mathbf{k}_1)+\omega(n_2\mathbf{k}_2)-\omega(n_3\mathbf{k}_3)+\omega(n_4\mathbf{k}_4),\\
\omega_{kn} &\to \omega_{kg} \equiv -\omega(n_1\mathbf{k}_1)+\omega(n_2\mathbf{k}_2)-\omega(n_3\mathbf{k}_3)+\omega(n_4\mathbf{k}_4)-\omega(n_5\mathbf{k}_5)+\omega(n_6\mathbf{k}_6).
\end{aligned} \tag{2.41}$$

In this case, the denominator in Eq. (2.38) does not depend on the many-body states, $|k\rangle, |l\rangle$, and $|m\rangle$. Thus, using the completeness relations, $\sum_k |k\rangle\langle k| = \hat{1}$ etc.,

$$\tilde{\chi}^{(3)}_{abcd}(\mathbf{q}_a,\omega_a;\mathbf{q}_b,\omega_b,\mathbf{q}_c,\omega_c,\mathbf{q}_d,\omega_d)$$

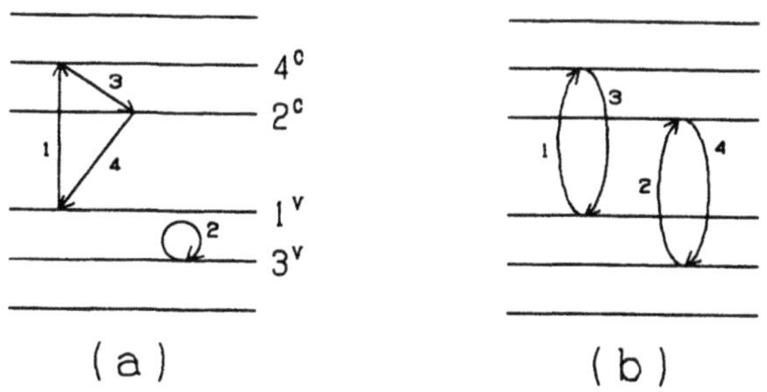


Figure 2.5. Schematic picture of third-order single-particle processes which are canceled out in the final expression of $\tilde{\chi}^{(3)}$. $1^v$ and $3^v$ are valence bands, while $2^c$ and $4^c$ conduction bands. The numbers associated with arrows indicate the order of transitions.

$$= \frac{1}{|\hat{\mathcal{P}}^{bcd}|}\left(\frac{1}{\hbar}\right)^3\left(\frac{e}{m_0}\right)^4\frac{-1}{\omega_a\omega_b\omega_c\omega_d}2\pi\delta(\omega_a-\omega_b-\omega_c-\omega_d)$$
$$\sum_1\sum_2\sum_3\sum_4\sum_5\sum_6\sum_7\sum_8 \langle g|\hat{C}_8^\dagger\hat{C}_7\hat{C}_6^\dagger\hat{C}_5\hat{C}_4^\dagger\hat{C}_3\hat{C}_2^\dagger\hat{C}_1|g\rangle$$
$$\hat{\mathcal{P}}_{abcd}\frac{\mathrm{P}^a(8,7)\mathrm{P}^b(6,5)\mathrm{P}^c(4,3)\mathrm{P}^d(2,1)}{(\omega_{kg}-\omega_b-\omega_c-\omega_d-i\delta)(\omega_{lg}-\omega_c-\omega_d-i\delta)(\omega_{mg}-\omega_d-i\delta)}. \quad (2.42)$$

The bracket in Eq. (2.42) is easily evaluated by separating the Bloch states into valence- and conduction-band states. As a result of such evaluation, the sum of single-particle Bloch states in Eq. (2.42)is reduced to the sum of twenty-four transition processes, each of which is made of four successive single-particle transitions. Noticing that $\mathrm{P}^a(8,7)$ is proportional to $\Delta(\mathbf{k}_8-\mathbf{k}_7+\mathbf{q}_a)$, eighteen out of these processes cancel out each other. This is mathematically because they have the apparent system-size dependence described by the divergent delta functions $\Delta(\mathbf{0})$ for wavenumbers and physically because each of them is made of non-correlated single-particle transitions [37]. Examples of these processes are displayed in Fig. 2.5. Then, the bracket is replaced by six terms such as

$$\begin{aligned}\langle g|\hat{C}_8^\dagger\hat{C}_7\hat{C}_6^\dagger\hat{C}_5\hat{C}_4^\dagger\hat{C}_3\hat{C}_2^\dagger\hat{C}_1|g\rangle \quad\rightarrow\quad & -\delta^v_{8,1}\delta^v_{6,3}\delta^c_{4,7}\delta^c_{2,5}-\delta^v_{8,3}\delta^v_{6,1}\delta^c_{4,5}\delta^c_{2,7}\\ & +\delta^v_{8,1}\delta^c_{6,7}\delta^c_{4,5}\delta^c_{2,3}-\delta^v_{8,3}\delta^c_{6,7}\delta^v_{4,1}\delta^c_{2,5}\\ & -\delta^v_{8,5}\delta^v_{6,1}\delta^c_{4,7}\delta^c_{2,3}+\delta^v_{8,5}\delta^v_{6,3}\delta^v_{4,1}\delta^c_{2,7},\end{aligned} \quad (2.43)$$

with the definition of $\delta^v_{8,1}\equiv\delta_{n^v_8,n^v_1}\delta_{\mathbf{k}_8,\mathbf{k}_1}$. Here the superscripts, $v$ and $c$, respectively indicate that the states should be the valence- and conduction-band states. Substituting Eq. (2.43) into (2.42) and rearranging the remaining six terms into five,

$\tilde{\chi}^{(3)}_{abcd}(\mathbf{q}_a,\omega_a;\mathbf{q}_b,\omega_b,\mathbf{q}_c,\omega_c,\mathbf{q}_d,\omega_d) =$

$$\frac{1}{|\hat{\mathcal{P}}^{bcd}|}(\frac{1}{\hbar})^3(\frac{e}{m_0})^4\frac{-1}{\omega_a\omega_b\omega_c\omega_d}2\pi\delta(\omega_a-\omega_b-\omega_c-\omega_d)(2\pi)^3\Delta(\mathbf{q}_a-\mathbf{q}_b-\mathbf{q}_c-\mathbf{q}_d)$$
$$\sum_4\sum_3\sum_2\sum_1\hat{\mathcal{P}}^{abcd}\{$$
$$-\frac{\mathrm{P}^a(1^v,4^c)\mathrm{P}^c(4^c,3^v)\mathrm{P}^b(3^v,2^c)\mathrm{P}^d(2^c,1^v)}{[\omega(4^c,1^v)-\omega_b-\omega_c-\omega_d-i\delta][\omega(4^c,3^v)-\omega_c-i\delta][\omega(2^c,1^v)-\omega_d-i\delta]}$$
$$+\frac{\mathrm{P}^a(1^v,4^c)\mathrm{P}^b(4^c,3^c)\mathrm{P}^c(3^c,2^c)\mathrm{P}^d(2^c,1^v)}{[\omega(4^c,1^v)-\omega_b-\omega_c-\omega_d-i\delta][\omega(3^c,1^v)-\omega_c-\omega_d-i\delta][\omega(2^c,1^v)-\omega_d-i\delta]}$$
$$-\frac{\mathrm{P}^c(1^v,4^v)\mathrm{P}^a(4^v,3^c)\mathrm{P}^b(3^c,2^c)\mathrm{P}^d(2^c,1^v)}{[\omega(3^c,4^v)-\omega_b-\omega_c-\omega_d-i\delta][\omega(2^c,4^v)-\omega_c-\omega_d-i\delta][\omega(2^c,1^v)-\omega_d-i\delta]}$$
$$-\frac{\mathrm{P}^d(1^c,2^v)\mathrm{P}^b(2^v,3^v)\mathrm{P}^a(3^v,4^c)\mathrm{P}^c(4^c,1^c)}{[\omega(1^c,2^v)-\omega_d-i\delta][\omega(4^c,2^v)-\omega_c-\omega_d-i\delta][\omega(4^c,3^v)-\omega_b-\omega_c-\omega_d-i\delta]}$$
$$+\frac{\mathrm{P}^d(1^c,2^v)\mathrm{P}^c(2^v,3^v)\mathrm{P}^b(3^v,4^v)\mathrm{P}^a(4^v,1^c)}{[\omega(1^c,2^v)-\omega_d-i\delta][\omega(1^c,3^v)-\omega_c-\omega_d-i\delta][\omega(1^c,4^v)-\omega_b-\omega_c-\omega_d-i\delta]}\Bigg\}, \tag{2.44}$$

where

$$\omega(4^c,1^v) \equiv \omega(n_4^c\mathbf{k}_4)-\omega(n_1^v\mathbf{k}_1), \tag{2.45}$$

$$\frac{1}{[\omega(4^c,1^v)-\omega_b-\omega_c-\omega_d-i\delta]} \equiv \frac{\Delta(\mathbf{k}_4-\mathbf{k}_1-\mathbf{q}_b-\mathbf{q}_c-\mathbf{q}_d)}{\omega(4^c,1^v)-\omega_b-\omega_c-\omega_d-i\delta}. \tag{2.46}$$

Here $1^c$ and $n_1^c$ ($1^v$ and $n_1^v$) indicate that the summation is carried out over only the conduction (valence) bands. $\Delta$ is the three-dimensional delta function having the periods of reciprocal lattice vectors. Therefore, the **k**-integration over the Brillouin zone in Eq. (2.44) is effectively single because each term in Eq. (2.44) has three delta functions with respect to wavenumbers. This fact reflects that four successive one-particle transitions in respective term in Eq. (2.44) should be fully correlated to each other, as shown in Figs. 2.6(a) to 2.6(e). In other words, such four successive transitions are drawn as a closed loop in figures. The overall sign in respective term in Eq. (2.44) is determined by the number of electron exchange; the odd numbers correspond to the negative sign, while the even numbers to the positive one. In the literature [16], the first, the second, and the other terms are sometimes called "three state", "virtual electron", and "virtual hole" processes, respectively.

The imaginary part of $\tilde{\chi}^{(3)}$, $\mathrm{Im}\tilde{\chi}^{(3)}$, represents the energy resonance between the electromagnetic fields and electronic states, which is clarified by using the identities such as

$$\frac{1}{\omega(4^c,3^v)-\omega_c-i\delta} = \mathcal{P}\frac{1}{\omega(4^c,3^v)-\omega_c}+i\pi\delta(\omega(4^c,3^v)-\omega_c), \tag{2.47}$$

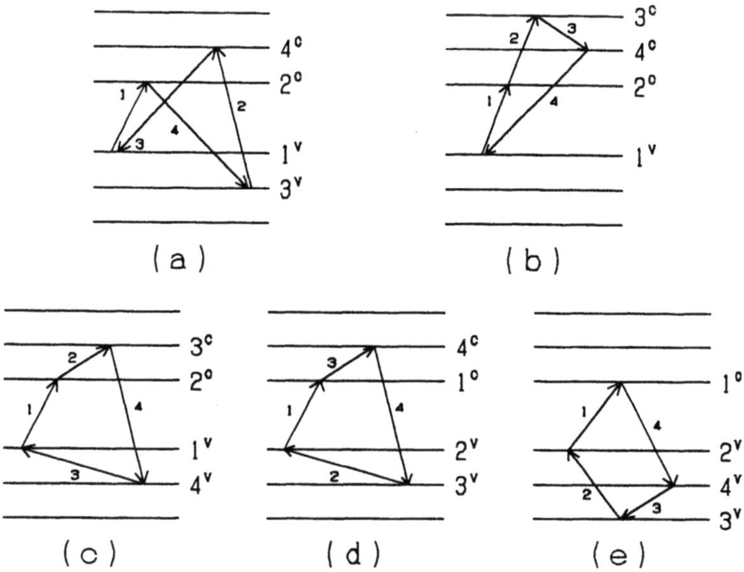


Figure 2.6. Schematic picture of third-order single-particle processes which contribute to the final expression of $\tilde{\chi}^{(3)}$. (a), (b), (c), (d), and (e) respectively correspond to the first, second, third, fourth, and fifth terms in Eq. (2.44).

where $\mathcal{P}$ denotes the principle integration. Using the nature of these delta functions, $\mathrm{Im}\tilde{\chi}^{(3)}$ can be expressed much more compactly by combining five terms in Eq. (2.44). As an example, explicitly displaying the resonant terms with the electromagnetic-field energies, $\omega_b + \omega_c + \omega_d$, $\omega_c + \omega_d$, and $\omega_d$,

$$
\begin{aligned}
&\mathrm{Im}\tilde{\chi}^{(3)}_{abcd}(\mathbf{q}_a, \omega_a; \mathbf{q}_b, \omega_b, \mathbf{q}_c, \omega_c, \mathbf{q}_d, \omega_d) \\
&= \frac{1}{|\hat{\mathcal{P}}^{bcd}|}\left(\frac{1}{\hbar}\right)^3\left(\frac{e}{m_0}\right)^4 \frac{-\pi}{\omega_a\omega_b\omega_c\omega_d}(2\pi)^4\delta[\omega_a - \omega_b - \omega_c - \omega_d]\sum_c\sum_v\sum_l\sum_m \\
&\left\{\mathrm{P}^a(v,c)\left(\hat{\mathcal{P}}^{bcd}\frac{\mathrm{P}^b(c,l)\mathrm{P}^c(l,m)\mathrm{P}^d(m,v)}{[\omega(l,v) - \omega_c - \omega_d][\omega(m,v) - \omega_d]}\right)\delta[\omega(c,v) - \omega_b - \omega_c - \omega_d]\right. \\
&+ \left(\hat{\mathcal{P}}^{ab}\frac{\mathrm{P}^a(v,l)\mathrm{P}^b(l,c)}{[\omega(l,v) - \omega_b - \omega_c - \omega_d]}\right)\left(\hat{\mathcal{P}}^{cd}\frac{\mathrm{P}^c(c,m)\mathrm{P}^d(m,v)}{[\omega(m,v) - \omega_d]}\right)\delta[\omega(c,v) - \omega_c - \omega_d] \\
&+ \left(\hat{\mathcal{P}}^{abc}\frac{\mathrm{P}^a(v,l)\mathrm{P}^b(l,m)\mathrm{P}^c(m,c)}{[\omega(l,v) - \omega_b - \omega_c - \omega_d][\omega(m,v) - \omega_c - \omega_d]}\right)\mathrm{P}^d(c,v)\delta[\omega(c,v) - \omega_d] \\
&\left. + \cdots\right\},
\end{aligned}
\tag{2.48}
$$

where

$$\begin{aligned}&\delta[\omega(c,v)-\omega_b-\omega_c-\omega_d]\\ &\quad\equiv\ \delta(\omega(c,v)-\omega_b-\omega_c-\omega_d)\Delta(\mathbf{k}_c-\mathbf{k}_v-\mathbf{q}_b-\mathbf{q}_c-\mathbf{q}_d).\end{aligned} \tag{2.49}$$

Here $c=(n_c,\mathbf{k}_c)$ and $v=(n_v,\mathbf{k}_v)$, respectively, indicate the conduction- and valence-band states, while $l=(n_l,\mathbf{k}_l)$ and $m=(n_m,\mathbf{k}_m)$ the intermediate states in the valence and/or conduction bands. The equations, (2.44) and (2.48), are the main results in this subsection; the band-state expressions of $\tilde{\chi}^{(3)}$.

### 2.2.3. *Momentum Matrix Element*

#### 2.2.3.1. *Momentum renormalization*

To calculate the nonlinear optical susceptibility microscopically, the momentum matrix element between two band states, *e.g.*, $\mathrm{P}^a(8,7)$ in Eq. (2.40), should be evaluated using the calculated wavefunctions of these states. This momentum matrix element represents the ***optical*** transition-rate amplitude between the states, which should be invariant under the gauge transformation [38]. Therefore, when the equation of motion for the band state includes some kind of nonlocality in the coordinate space, the correction should be added to the local momentum operator, $\hat{\mathrm{p}}^a_{\mathrm{loc}}=\hbar/i\cdot\nabla_a$. This situation is true for all elementary excitations in materials. In this subsection, we consider such correction.

First, let us derive the correction forms by considering the case when both the self-energy and the external potential, $\hat{\Sigma}$ and $\hat{V}_{ext}$, have the nonlocalities. When the electromagnetic field is absent, the self-energy operator works on the state, $|\varphi\rangle$, as

$$\begin{aligned}\langle\mathbf{x}|\hat{\Sigma}|\varphi\rangle &= \int d\mathbf{x}'\langle\mathbf{x}|\hat{\Sigma}|\mathbf{x}'\rangle\langle\mathbf{x}'|\varphi\rangle\\ &= \int d\mathbf{x}'\langle\mathbf{x}|\hat{\Sigma}|\mathbf{x}'\rangle e^{i(\mathbf{x}'-\mathbf{x})\cdot\hat{\mathbf{p}}_{\mathrm{loc}}/\hbar}\langle\mathbf{x}|\varphi\rangle,\end{aligned} \tag{2.50}$$

where we have used the fact that $\exp[i(\mathbf{x}'-\mathbf{x})\cdot\hat{\mathbf{p}}_{\mathrm{loc}}/\hbar]$ acts as a translation operator. Namely, the nonlocal operator can be described in a formal way using the local momentum operator, $\hat{\mathbf{p}}_{\mathrm{loc}}$; symbolically such as $\Sigma(\mathbf{x},\hat{\mathbf{p}}_{\mathrm{loc}})$. When the electromagnetic field is applied, the momentum operator, $\hat{\mathbf{p}}_{\mathrm{loc}}$, should be replaced by $\hat{\mathbf{p}}_{\mathrm{loc}}-e/c\tilde{\mathbf{A}}(\mathbf{x},t)$ through the gauge transformation with the vector potential, $\tilde{\mathbf{A}}$, thus the self-energy operator depending on $\tilde{\mathbf{A}}$. Denoting such dependence as $\hat{\Sigma}_{\tilde{\mathbf{A}}}$, the equation (2.50) is changed to

$$\langle\mathbf{x}|\hat{\Sigma}_{\tilde{\mathbf{A}}}|\phi\rangle = \int d\mathbf{x}'\langle\mathbf{x}|\hat{\Sigma}|\mathbf{x}'\rangle e^{i(\mathbf{x}'-\mathbf{x})\cdot(\hat{\mathbf{p}}_{\mathrm{loc}}-e/c\tilde{\mathbf{A}}(\mathbf{x},t))/\hbar}\langle\mathbf{x}|\phi\rangle. \tag{2.51}$$

Expanding this equation in terms of $\tilde{\mathbf{A}}$ and using the eigenvalue equations, $\langle\mathbf{x}|\hat{\mathbf{x}}=\langle\mathbf{x}|\mathbf{x}$ and $\hat{\mathbf{x}}|\mathbf{x}'\rangle=\mathbf{x}'|\mathbf{x}'\rangle$, we obtain to the first order of $\tilde{\mathbf{A}}(\mathbf{x},t)$ (long-wavelength limit)

$$\langle\mathbf{x}|\hat{\Sigma}_{\tilde{\mathbf{A}}}|\phi\rangle = \langle\mathbf{x}|\hat{\Sigma}|\phi\rangle - \int d\mathbf{x}'\langle\mathbf{x}|[\hat{\Sigma},i\hat{\mathbf{x}}]|\mathbf{x}'\rangle\langle\mathbf{x}'|\phi\rangle\frac{e}{c\hbar}\tilde{\mathbf{A}}(\mathbf{x},t), \tag{2.52}$$

where the first term is the self-energy in the case of no electromagnetic field. This equation indicates that the new interaction between the quasiparticle state and the electromagnetic fields appears due to the presence of nonlocality in the self-energy operator, in addition to the $\hat{\mathbf{p}}_{\rm loc} \cdot \tilde{\mathbf{A}}$ interaction originating from the kinetic-energy term. The similar situation is true for the nonlocal external potential, $\hat{V}_{ext}$.

In these ways, the optical momentum operator coupling to the electromagnetic field is renormalized as follows;

$$\hat{\rm p}^a = \hat{\rm p}^a_{\rm loc} + \frac{m}{\hbar}[\hat{V}_{ext}, i\hat{\rm x}^a] + \frac{m}{\hbar}[\hat{\Sigma}, i\hat{\rm x}^a], \tag{2.53}$$

where $\hat{\rm x}^a$ is the $a$-component of the position operator. This type of momentum renormalization is physically a reflection from the current conservation law, which is often called the Ward-Takahashi-identity in many-body problems. Levine and Allan [18] were the first who pointed out its significance in the realistic *ab initio* calculations of optical properties in materials, proposed the simple recipe to evaluate this correction, and demonstrated its relevancy in the calculations of the lower-energy $\varepsilon_{1,xx}(\omega)$ spectra [19] and the static SHG magnitude of $\chi^{(2)}_{xyz}(0;0,0,0)$ for Si, Ge, AlP, AlAs, GaP, and GaAs [18, 19].

As long as the gauge invariance holds, the calculated results should not depend on the choice of gauge; whether the $\tilde{\mathbf{E}}(\mathbf{x},t)\cdot\hat{\mathbf{x}}$ or $\tilde{\mathbf{A}}(\mathbf{x},t)\cdot\hat{\mathbf{p}}_{loc}$ interaction is adopted [39, 40]. Since the extended Bloch state (*i.e.*, crystal-momentum state) becomes the eigen state in infinite bulks, the $\tilde{\mathbf{A}}\cdot\hat{\mathbf{p}}_{loc}$ interaction is well defined. It should be noticed here that because the approximate model such as the effective mass theory sometimes break the local-gauge-transformation invariance the calculated results strongly depend on the choice of gauge [41]. While the *ab initio* approach starting from the exact Hamiltonian does not suffer from such ambiguity.

### *2.2.3.2. Numerical aspects*

To calculate these nonlocal corrections, it is often useful to evaluate them in the momentum space using the plane-waves as a orthogonal and complete basis [42]. This is because many band-structure calculations are performed for periodic systems and because an error in the evaluation is easily estimated using mathematical properties of the Fourier expansion. The matrix elements between the $|\mathbf{q}\rangle$ and $|\mathbf{q}'\rangle$ plane-waves are given by

$$\langle \mathbf{q}|\frac{m}{\hbar}[\hat{\rm V}_{\rm ext}, i\hat{\rm x}^a]|\mathbf{q}'\rangle = \frac{m}{\hbar}(\frac{\partial}{\partial {\rm q}^a} + \frac{\partial}{\partial {\rm q}'^a})\langle \mathbf{q}|\hat{\rm V}_{\rm ext}|\mathbf{q}'\rangle, \tag{2.54}$$

$$\langle \mathbf{q}|\frac{m}{\hbar}[\hat{\Sigma}, i\hat{\rm x}^a]|\mathbf{q}'\rangle = \frac{m}{\hbar}(\frac{\partial}{\partial {\rm q}^a} + \frac{\partial}{\partial {\rm q}'^a})\langle \mathbf{q}|\hat{\Sigma}|\mathbf{q}'\rangle, \tag{2.55}$$

where the long-wavelength approximation is adopted; the photon phase factor $e^{-i\mathbf{q}_a\mathbf{x}}$ in Eq. (2.40) is omitted in the above corrections.

As an example, let us consider the band-structure calculation in the *ab initio* pseudopotential method in the LDA [22], where the pseudopotentials are nonlocal

[43] and the scissors operator is used as the addition of self-energy to address the problem of the band-gap underestimation associated with the LDA [18]. Such scissors operator is given by

$$\hat{\Sigma}_s = \Delta \sum_{c,\mathbf{k}} |c\mathbf{k}\rangle\langle c\mathbf{k}|, \tag{2.56}$$

where $|c\mathbf{k}\rangle$ is the LDA-calculated conduction-band state having the Bloch wavenumber $\mathbf{k}$ and $\Delta$ is the scissors-shift energy for the conduction-band states. The value of $\Delta$ is determined for respective semiconductor to coincide with the experimental value of direct/indirect fundamental band-gap energy. As explained in Sec. 2.2.1, this treatment is effective and justified for the band-structure study at the present stage.

The pseudopotential correction, $\langle \mathbf{q}|\frac{m}{\hbar}[\hat{\mathrm{V}}_{\mathrm{pseudo}}, i\hat{\mathrm{x}}^a]|\mathbf{q}'\rangle$ [44, 45], can be calculated analytically. While the scissors-operator self-energy correction is applied only to the momentum matrix element between the valence and conduction-band states as

$$\langle v\mathbf{k}|\frac{m}{\hbar}[\hat{\Sigma}, i\hat{\mathrm{x}}^a]|c\mathbf{k}\rangle = \frac{\Delta}{\omega^{\mathrm{LDA}}(c\mathbf{k}) - \omega^{\mathrm{LDA}}(v\mathbf{k})} \langle v\mathbf{k}|\hat{\mathrm{p}}^a_{\mathrm{loc}} + \frac{m}{\hbar}[\hat{\mathrm{V}}_{\mathrm{pseudo}}, i\hat{\mathrm{x}}^a]|c\mathbf{k}\rangle, \tag{2.57}$$

where $\omega^{\mathrm{LDA}}(m\mathbf{k})$ is the LDA energy of the state $|m\mathbf{k}\rangle$. This factorized-replacement form can easily be proved by using the $\mathbf{k} \cdot \mathbf{p}$ perturbation theory [18].

In the rest of this subsection, the numerical relevancy of these two nonlocal corrections in the optical transition-rate calculations is displayed. First, in order to see the overall feature, we show the calculated Im $\chi^{(3)}_{xxxx}(\omega; -\omega, \omega, \omega)$, $\varepsilon_{1,xx}(\omega)$, and $\varepsilon_{2,xx}(\omega)$ for ZnSe in Figs. 2.7(a) to 2.7(c), respectively. Here the values with and without corrections are displayed. It is clearly seen that both the pseudopotential and self-energy corrections change the transition rates in Im $\chi^{(3)}$, $\varepsilon_1$, and $\varepsilon_2$ by definite ratios ($5 \sim 30\%$).

To clarify the material and transition-energy dependence of these corrections, the calculated (partially-corrected-Im$\chi^{(3)}$) /(fully-corrected-Im$\chi^{(3)}$) ratios are shown in Fig. 2.8 for GaAs, ZnSe and Si, as a function of photon energy. We note the following features. (a) The pseudopotential correction strongly depends on the materials; the correction value is negative and large ($20 \sim 30\%$) for Si, while positive and small ($5 \sim 10\%$) for ZnSe and GaAs. This feature is also seen in the calculated momentum-matrix-elements between s- and p-pseudo-orbitals of isolated atoms. Thus, the pseudopotential correction is found to depend on its construction scheme [44]. (b) The effect of the pseudopotential correction is nearly independent of the transition energy, thus of the individual types of transitions ($\mathrm{E}_0$, $\mathrm{E}_1$ etc.) in the Brillouin zone. This may be explained by considering that such transitions have the single character in most semiconductors; the transitions from the bonding-p-like upper-valence bands to the antibonding-s-like lower-conduction ones.

On the other hand, (c) the self-energy correction increases the transition rate in Im $\chi^{(3)}$ for all semiconductors. This is because in the present treatment of self-energy the positive energy-shift of conduction bands always increases the momentum-matrix-elements between valence- and conduction-band states by the factor, $\Delta/(\omega^{\mathrm{LDA}}(c\mathbf{k}) - \omega^{\mathrm{LDA}}(v\mathbf{k}))$. With respect to the absorption-edge transition, such factors are 0.668

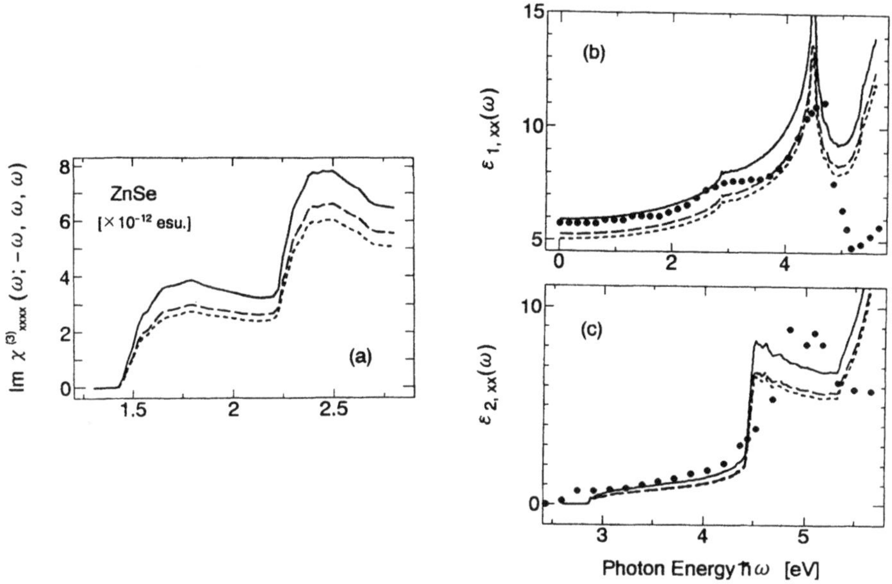


Figure 2.7. Effects of the nonlocal pseudopotential and self-energy corrections on (a) $\mathrm{Im}\chi^{(3)}_{xxxx}(\omega;-\omega,\omega,\omega)$, (b) $\varepsilon_{1,xx}(\omega)$, and (c) $\varepsilon_{2,xx}(\omega)$ for ZnSe. Dotted lines are evaluated by using only matrix elements of local momentum operator, dashed lines with nonlocal pseudopotential correction, and solid lines with full (pseudopotential + self-energy) corrections. Filled circles are experimental data. (After [10]).

eV/2.578 eV $\sim$ 0.26, 0.430 eV/2.390 eV $\sim$ 0.18 and 0.823 eV/0.697 eV $\sim$ 1.18 for Si, ZnSe and GaAs, respectively. Thus the self-energy correction is largest for GaAs. (d) The self-energy correction strongly depends on the transition energy such that it decreases the transition rate with increasing the transition energy. This can easily expected from Eq. (2.57).

The above-mentioned features of two nonlocal corrections are also seen in the linear optical property in Table 2.1, where the calculated and observed static dielectric constants, $\varepsilon_{1,xx}(0)$, are shown. It is seen that the calculated $\varepsilon_{1,xx}(0)$ with full corrections are in good agreement with experiments. Similarly, as seen in the next section, the calculated nonlinear optical spectra with full corrections are also in excellent agreement with experimental observations. All these results clearly indicate that these nonlocal corrections based on the momentum renormalization are essential to the *ab initio* optical transition-rate calculations.

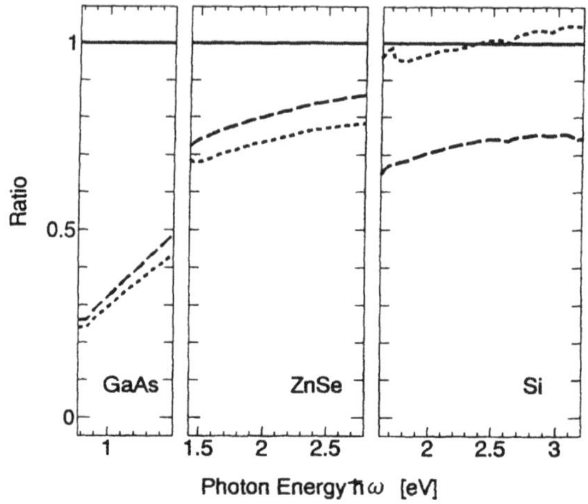


Figure 2.8. Material dependence of the effects of nonlocal pseudopotential and self-energy corrections on $\mathrm{Im}\chi^{(3)}_{xxxx}(\omega; -\omega, \omega, \omega)$ for Si, ZnSe, and GaAs as a function of photon energy. Dotted lines; $\mathrm{Im}\chi^{(3)}$ with only the local momentum operator, dashed lines; $\mathrm{Im}\chi^{(3)}$ with the nonlocal pseudopotential correction, and solid lines as a unit; fully-corrected $\mathrm{Im}\chi^{(3)}$ with the nonlocal pseudopotential and self-energy corrections. (After [10]).

Table 2.1. Static dielectric constants $\varepsilon_{1,xx}(0)$, calculated with and without the nonlocal corrections on the momentum operator, and observed in experiments. The calculated results in left three column are obtained as the static limit of $\varepsilon_{1,xx}(\omega)$ transformed from $\varepsilon_{2,xx}(\omega)$. The results denoted as "direct" are evaluated by directly using the real-part formula of $\varepsilon_{1,xx}(0)$. (After [10]).

| | only $\hat{\mathbf{p}}_{\mathrm{loc}}$ | with $\hat{V}_{\mathrm{pseudo}}$ | with $\hat{V}_{\mathrm{pseudo}} + \hat{\Sigma}$ | direct | experiments |
|---|---|---|---|---|---|
| Si | 10.18 | 8.79 | 11.64 | 11.73 | 12.1, 11.7 |
| ZnSe | 5.02 | 5.24 | 5.90 | 5.90 | 5.6 |
| GaAs | 6.96 | 7.49 | 10.77 | 10.88 | 10.9 |

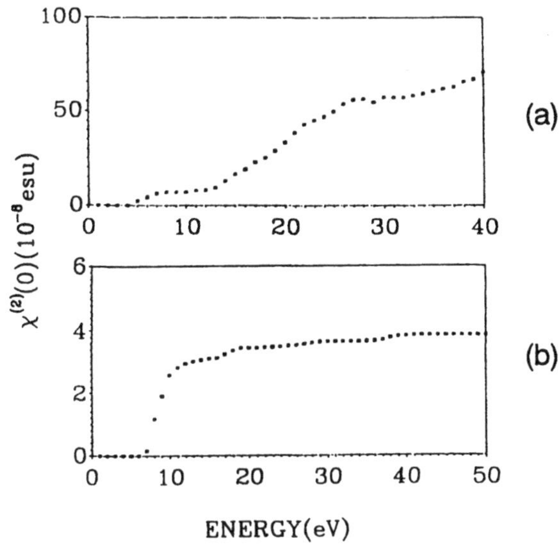


Figure 2.9. Convergence of $\mathrm{Re}\chi^{(2)}_{xyz}(0;0,0)$ as a function of the conduction-band cutoff energy. (a) GaAs and (b) ZnSe. (After [20]).

### 2.2.4. *Numerical Convergence in $\chi^{(n)}$ Calculation*

Finally, the numerical details in the nonlinear-optical-susceptibility calculation are presented. On calculating $\chi^{(n)}$, two kinds of numerical convergence must be checked carefully. The first one is related to the number of bands taken in the summation of Eq. (2.44), while the other is related to the sum over the Brillouin zone. In this subsection, we consider numerical aspects concerning such convergence by viewing the calculated $\mathrm{Re}\chi^{(2)}(0;0,0)$ in the SHG and $\mathrm{Im}\chi^{(3)}(\omega;-\omega,\omega,\omega)$ in the TPA, as examples.

Figures 2.9(a) and 2.9(b) respectively show the calculated static $\mathrm{Re}\chi^{(2)}_{xyz}(0;0,0)$ values for GaAs and ZnSe with zinc-blende structures as a function of a conduction-band cutoff energy, $E_{\mathrm{cf}}$, which is defined as the energy from the valence-band top [20]. In these calculations, the long-wavelength approximation ($\mathbf{q}$=0) is adopted. As seen in these figures, although the way of reaching stable values is different between materials, the static $\chi^{(2)}$ value increases with increasing $E_{\mathrm{cf}}$ and converges for $E_{\mathrm{cf}}$ less than 35 to 40 eV.

This tendency is qualitatively understood from the $\chi^{(2)}$ formula in Eq. (2.13) as follows. Consider the typical transition process, $|\mathrm{VB}\rangle \to |\mathrm{CB}\rangle \to |\mathrm{CB}'\rangle \to |\mathrm{VB}\rangle$, which is made of three successive single-particle transitions and appears in the numerator in Eq. (2.13). Here VB and CB indicate the valence and conduction-band states, respectively. If all these states have eigen energies around $E \sim E_{\mathrm{cf}}$, the momentum matrix elements between these states, such as $\langle \mathrm{CB}|\hat{\mathbf{p}}|\mathrm{VB}\rangle$, are roughly proportional to $E^{1/2}$. Thus the numerator in Eq. (2.13) is estimated being proportional to $E^{3/2}$. On the other hand, in order to estimate the denominator in Eq. (2.13) corresponding to such transition, one must expand it in terms of the electromagnetic-field frequencies, $\omega_a, \omega_b$, and $\omega_c$. This is because the pre-coefficient in $\chi^{(2)}$, $1/\omega_a\omega_b\omega_c$, should be canceled out in the zero-frequency limit with the expanded terms, $\omega_a\omega_b\omega_c/E^3$. (The expanded terms such as $\omega_a\omega_b/E^2$ have zero coefficient, thus being not divergent in the zero-frequency limit [17, 46]). Thus the $\chi^{(2)}$ has the denominator proportional to $E^5$, where the power of two comes from the original denominator while the other power of three from the expansion. In this way, the contribution of the typical transition to the static $\chi^{(2)}$ value is proportional to $E^{-7/2}$, thus it becoming smaller with increasing $E$.

In the TPA case, the imaginary part of $\chi^{(3)}$ should be evaluated. The typical TPA-transition process is written by $|\mathrm{VB}\rangle \rightarrow |\mathrm{IB}\rangle \rightarrow |\mathrm{CB}\rangle \rightarrow |\mathrm{IB}'\rangle \rightarrow |\mathrm{VB}\rangle$, where IB and IB' indicate the intermediate valence- and/or conduction-band states. Since the number of valence bands is finite and the deep core states hardly contribute to the polarization, we have only to check the convergence of Im $\chi^{(3)}$ as a function of the number of intermediate conduction bands included. Table 2.2 shows the convergence of $\mathrm{Im}\chi^{(3)}_{xxxx}(\omega; -\omega, \omega, \omega)$ for zinc-blende ZnSe at two particular photon energies, where the long-wavelength approximation is again adopted. It is seen that more conduction bands should be taken into the calculation when we consider the higher-energy electromagnetic field; the higher-energy transition. In the present case, twenty-six conduction bands are appropriate, which corresponds to the conduction-band cutoff-energy $E_{\mathrm{cf}}$ of about 40 eV. The residual errors due to this truncation is estimated about 1 %. Applying the above-mentioned convergence analysis to the typical TPA process, the numerator in Eq. (2.48)is roughly proportional to $E^2$, while the denominator in the same equation also to $E^2$. On the other hand, the pre-coefficient in $\mathrm{Im}\chi^{(3)}$, $1/\omega_a\omega_b\omega_c\omega_d$, has the $E^{-4}$ dependence. In this way, the contribution of the typical process to $\mathrm{Im}\chi^{(3)}$ is proportional to $E^{-4}$, which indicates that the convergence is faster in $\mathrm{Im}\chi^{(3)}$ than in $\mathrm{Re}\chi^{(2)}$.

It should be noted here that most model-band calculations have adopted one to four conduction bands as the intermediate band states in the TPA spectra calculation. However, as seen in Table 2.2, this treatment may provide a large error in the calculated $\mathrm{Im}\chi^{(3)}$, especially for the TPA for the higher-energy electromagnetic field. This fact clearly demonstrates the necessity of the ***ab initio*** nonlinear-optical-susceptibility calculation coupled with the ***ab initio*** band-structure calculation.

Next, let us examine the numerical convergence with respect to the sum over the Brillouin zone, by considering the TPA described by $\mathrm{Im}\chi^{(3)}$. The standard reliable method to simulate the resonant delta function in the **k**-space integration in $\mathrm{Im}\chi^{(3)}$ is the tetrahedron method [47]. In this method, a whole Brillouin zone is equally divided with n × n × n mesh **k**-points. Then, by evaluating the value of integrand at these **k**-points using the calculated band-state energies and wavefunctions, and by linearly interpolating these values in the respective tetrahedron made of four **k**-points, the integration is performed on the hypersurface defined by the zeros of the function

Table 2.2. Convergence of Im $\chi^{(3)}_{xxxx}(\omega;\omega,-\omega,\omega)$ as a function of the number of conduction bands included. Data are for the case of ZnSe at $\hbar\omega = 1.8$ and 2.6 eV using $36 \times 36 \times 36$ **k**-points in the fcc Brillouin zone. Unit is $10^{-12}$ esu. (After [10]).

| Number of bands | 1.8 (eV) | 2.6 (eV) |
|---|---|---|
| 1 | 3.3324 | 4.8493 |
| 4 | 3.2612 | 6.4866 |
| 8 | 3.7881 | 7.0925 |
| 20 | 3.8363 | 7.2119 |
| 26 | 3.8526 | 7.2367 |
| 38 | 3.8521 | 7.2486 |

Table 2.3. Convergence of Im $\chi^{(3)}_{xxxx}(\omega;\omega,-\omega,\omega)$ as a function of the number of **k**-points in the fcc Brillouin zone. Data are for the case of ZnSe at $\hbar\omega = 1.8$ and 2.6 eV using twenty-six conduction bands. Unit is $10^{-12}$ esu. (After [10]).

| Number of **k** points | 1.8 (eV) | 2.6 (eV) |
|---|---|---|
| $24 \times 24 \times 24$ | 3.7621 | 7.0949 |
| $36 \times 36 \times 36$ | 3.8526 | 7.2367 |
| $48 \times 48 \times 48$ | 3.8703 | 7.2686 |

in the resonant delta function.

Table 2.3 gives the convergence of Im $\chi^{(3)}_{xxxx}(\omega;-\omega,\omega,\omega)$ for zinc-blende ZnSe as a function of the number of **k**-points at particular photon energies. This table shows that $48 \times 48 \times 48$ points are appropriate, which corresponds to 2769 points in the irreducible Brillouin zone and 663 552 tetrahedra in a whole fcc Brillouin zone. This truncation is estimated to have $3 \sim 5$ % errors in the calculated Im $\chi^{(3)}$. It should be noticed here that the error in the **k**-space integration mainly depends on the degree how smoothly the integrand function is described using such **k**-points. The integrand function quickly changes its value at the band-crossing hyperplanes. Because the number of such band-crossing encountered in the $\chi^{(n)}$ calculation increases with increasing the order of nonlinearity, $n$, the appropriate number of **k**-points is generally much larger in the Im$\chi^{(3)}$ calculation than that in the Im$\chi^{(1)}$ calculation.

## 2.3. TWO-PHOTON ABSORPTION IN BULK SEMICONDUCTORS

### 2.3.1. *Two-Photon-Absorption Coefficient*

Two-photon absorption (TPA) is a powerful tool to study the optical and electronic properties of materials. Because the TPA follows the unique selection rules, the TPA provides complementary material information to the one-photon absorption (OPA) [28]. On the other hand, in the application field, the TPA plays a crucial role in limiting the optical transparency of materials and in causing laser-induced damage to compounds [2]. It has also been recognized that nonlinear refraction can be related to TPA through a nonlinear Kramers-Krönig transformation [30, 48].

For the TPA process in cubic-symmetry materials, the third-order absorption coefficient, $\alpha^{(3)}(\omega_1)$, is obtained from Eq. (2.32) as

$$\begin{aligned}\alpha_a^{(3)}(\omega_1) &= \frac{4\pi\omega_1}{cn_a^{(1)}(\omega_1)}\mathrm{Im}\chi_{eff}^{(3)} \\ &= \frac{4\pi\omega_1}{cn_a^{(1)}(\omega_1)}\mathrm{Im}(3\chi_{aaaa}^{(3)}(\omega_1;-\omega_1,\omega_1,\omega_1)|\mathrm{E}_a(\omega_1)|^2 \\ &\quad +6\chi_{abba}^{(3)}(\omega_1;-\omega_2,\omega_2,\omega_1)|\mathrm{E}_b(\omega_2)|^2(1-\delta_{a,b}\delta\omega_1,\omega_2)), \qquad (2.58)\end{aligned}$$

where $n_a^{(1)}(\omega_1)$ is the frequency-dependent linear refractive index and the long- wavelength approximation is assumed. In experiments, however, TPA is often characterized by the TPA coefficient, $\beta$, which is a linearly-dependent part of the third-order absorption coefficient on the light intensities, $\mathrm{I}_a(\omega_1)$ and $\mathrm{I}_b(\omega_2)$, and defined by

$$\alpha_a^{(3)}(\omega_1) = \beta_{aa}(\omega_1;\omega_1)\mathrm{I}_a(\omega_1) + 2\beta_{ab}(\omega_1;\omega_2)\mathrm{I}_b(\omega_2)(1-\delta_{a,b}\delta_{\omega_1,\omega_2}). \qquad (2.59)$$

$\beta_{aa}(\omega_1;\omega_1)$ corresponds to the first self-action term in Eq. (2.58) and is called the degenerate TPA coefficient, while $\beta_{ab}(\omega_1;\omega_2)$ corresponds to the second cross-action term in Eq. (2.58) and is called the non-degenerate one. Together with the degeneracy prefactors in Eq. (2.58), this definition ensures that the degenerate $\beta_{aa}(\omega_1;\omega_1)$ can be obtained continuously from the non-degenerate $\beta_{aa}(\omega_1;\omega_2)$ as $\omega_2$ approaches $\omega_1$.

In the absence of the linear absorption where the probe-light energy is below the band-gap energy ($\hbar\omega_1 \leq \mathrm{E}_g$) of a material, we can exactly express $\beta$ using $\chi^{(3)}$ as

$$\beta_{ab}(\omega_1;\omega_2) = \frac{6(2\pi)^2\omega_1}{n_a^{(1)}(\omega_1)n_b^{(1)}(\omega_2)c^2}\mathrm{Im}\chi_{abba}^{(3)}(\omega_1;-\omega_2,\omega_2,\omega_1), \qquad (2.60)$$

where we have used the familiar relations, $\mathrm{I}_a(\omega_1) = c/2\pi \cdot n_a^{(1)}(\omega_1)|\mathrm{E}_a(\omega_1)|^2$ and $\mathrm{I}_b(\omega_2) = c/2\pi \cdot n_b^{(1)}(\omega_2)|\mathrm{E}_b(\omega_2)|^2$. In this way, the calculation of TPA-coefficient spectra is reduced to the evaluation of $\mathrm{Im}\chi_{abba}^{(3)}(\omega_1;-\omega_2,\omega_2,\omega_1)$, $n_a^{(1)}(\omega_1)$, and $n_b^{(1)}(\omega_2)$, as functions of the photon frequencies and polarizations, $(a,\omega_1)$ and $(b,\omega_2)$. To evaluate the linear refractive index, $n_a^{(1)}(\omega)$, we first calculate the imaginary part of the dielectric function, $\varepsilon_{2,aa}(\omega) = 4\pi\mathrm{Im}\chi_{aa}^{(1)}(\omega;\omega)$, in a similar manner to $\mathrm{Im}\chi^{(3)}$. Then its real part, $\varepsilon_{1,aa}(\omega)$, is obtained from $\varepsilon_{2,aa}(\omega)$ by the Kramers-Krönig transformation [49]. In this way, $n_a^{(1)}(\omega)$ is given by $(\varepsilon_{1,aa}(\omega))^{1/2}$, which is equal to $(1+4\pi\mathrm{Re}\chi_{aa}^{(1)}(\omega;\omega))^{1/2}$.

The general formula of $\mathrm{Im}\chi^{(3)}$ is derived from the microscopic response theory and given in Sec. 2.2.2. Especially in the TPA case, the two-photon energy resonates with the energy difference between the valence- and conduction-band states, and only the term having $\delta(\omega_1+\omega_2-\omega_{cv})$, *i.e.*, the second term in Eq. (2.48), contributes to the spectra. Neglecting the local-field effect and adopting the single-particle approximation,

$$\begin{aligned}
&\mathrm{Im}\chi^{(3)}_{abba}(\omega_1; -\omega_2, \omega_2, -\omega_1) \\
&= \frac{1}{2}\left(\frac{1}{\hbar}\right)^3 \left(\frac{e}{m_0}\right)^4 \frac{\pi}{\omega_1^2\omega_2^2} \int \frac{d\mathbf{k}}{(2\pi)^3} \sum_v \sum_c \\
&\times \Big| \sum_m \Big( \frac{\mathrm{P}^a_{vm}(\mathbf{k})\mathrm{P}^b_{mc}(\mathbf{k})}{\omega_{mc}(\mathbf{k}) + \omega_2} + \frac{\mathrm{P}^b_{vm}(\mathbf{k})\mathrm{P}^a_{mc}(\mathbf{k})}{\omega_{mv}(\mathbf{k}) - \omega_2} \Big|^2 \delta(\omega_1 + \omega_2 - \omega_c(\mathbf{k}) + \omega_v(\mathbf{k})).
\end{aligned} \tag{2.61}$$

Here $v$, $c$ and $m$ are band indices for the valence, conduction and intermediate states, respectively. $\mathrm{P}^a_{vm}(\mathbf{k})$ is the $a$-component momentum matrix element, $\langle v\mathbf{k}|\hat{\mathrm{p}}^a|m\mathbf{k}\rangle$, $\omega_{mv}(\mathbf{k}) = \omega(m\mathbf{k}) - \omega(v\mathbf{k})$ the energy difference between two Bloch states, $|m\mathbf{k}\rangle$ and $|v\mathbf{k}\rangle$. Note that the permutation operator on the sets of incident photon energies and polarizations in Eq. (2.48) is already taken in this formula.

Because the TPA phenomenon is a nonparametric process, the TPA spectra can be evaluated also by means of the transition-rate-calculation scheme based on the Fermi's golden rule in quantum mechanics [50]. The equivalence of the above-mentioned susceptibility-calculation scheme to such transition-rate-calculation one in the TPA-spectra evaluation is easily recognized by noticing that the absolute value in Eq. (2.61) is the second-order transition amplitude between the valence- and conduction-band states, $|v\mathbf{k}\rangle$ and $|c\mathbf{k}\rangle$, having the intermediate states, $|m\mathbf{k}\rangle$. However, it should be noted that the transition-rate-calculation scheme is limited to some phenomena out of various nonparametric phenomena. For example, $\mathrm{Im}\chi^{(3)}_{xxyy}(\omega_1; \omega_1, -\omega_2, \omega_2)$ component detected in the TPA-spectra anisotropy, which will be discussed in the later section, can be calculated not by the transition-rate-calculation scheme but by the susceptibility-calculation scheme. In this sense, the calculation scheme introduced in the present chapter is the most general one for the study of nonlinear optical properties in materials.

### 2.3.2. *Degenerate TPA Spectra*

First, we consider in detail the simplest case, the degenerate TPA spectra in semiconductors, where two photons having the same frequency and polarization are simultaneously absorbed in semiconductors. In experiments, the degenerate TPA can be realized using a single light beam, that works as both the pump and probe-lights.

Let us start with Si, where most characteristics in the TPA spectra can be seen. Figure 2.10(a) shows the calculated degenerate TPA coefficient of Si crystal, $\beta_{xx}(\omega;\omega)$, as a function of the probe-light energy. For comparison, the imaginary part of the dielectric function of the same material, $\varepsilon_{2,xx}(\omega)$, is shown in Fig. 2.10(b). Note that $\varepsilon_{2,xx}(\omega)$ is proportional to OPA coefficient, $\alpha^{(1)}_x(\omega)$, through $\alpha^{(1)}_x(\omega) = \omega\varepsilon_{2,xx}(\omega)/\eta_{xx}(\omega)\mathrm{c}$. The down-arrows indicate the calculated interband energies, $\mathrm{E}_0'(\Gamma_{15c} - \Gamma_{15v} = 3.246$ eV), $\mathrm{E}_1(\mathrm{L}_{1c} - \mathrm{L}_{3v} = 3.431)$, $\mathrm{E}_2(\mathrm{X}_{13c} - \mathrm{X}_{5v} = 4.186)$, and $\mathrm{E}_1'(\mathrm{L}_{3c} - \mathrm{L}_{3v} = 5.247)$. As seen in these figures, $\varepsilon_{2,xx}(\omega)$ spectrum starts to raise the

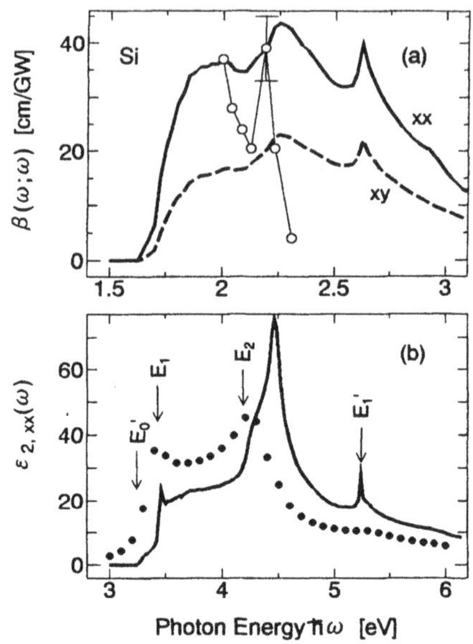


Figure 2.10. (a) Calculated degenerate TPA coefficients, $\beta(\omega;\omega)$, of Si as a function of photon energy. Solid and dashed lines show the different polarizations, $\beta_{xx}(\omega;\omega)$ and $\beta_{xy}(\omega;\omega)$, respectively. Open circles with a guide line are recent experimental data. (b) Photon-energy dependence of the calculated imaginary part of the dielectric function, $\varepsilon_{2,xx}(\omega)$, of Si. Down arrows indicate interband energies. Filled circles are experimental data. (After [10]).

magnitude at the above-mentioned interband energies, while $\beta_{xx}(\omega;\omega)$ does at half of those energies. Their shapes, however, are significantly different between $\beta_{xx}(\omega;\omega)$ and $\varepsilon_{2,xx}(\omega)$ with respect to the structures originating from various critical-point transitions. For example, the $\beta_{xx}(\omega;\omega)$ spectrum has the $E_1$ peak position ($2\hbar\omega \sim 4.0$ eV) higher in energy than that of $\varepsilon_{2,xx}(\omega)$ ($\hbar\omega \sim 3.5$ eV), while the $E_2$ peak position is almost identical between $\beta_{xx}(\omega;\omega)$ and $\varepsilon_{2,xx}(\omega)$. Moreover the $E_1$ and $E_1'$ peaks are sharp in $\varepsilon_{2,xx}(\omega)$, while broad in $\beta_{xx}(\omega,\omega)$.

The energies, $E_1$ and $E_2$, respectively correspond to the *onsets* of the $M_1$ and $M_2$ critical-point transitions associated with van Hove singularities in the joint density of state (JDOS) [53]. Then, in order to qualitatively explain the above features, we derive the scaling law of TPA coefficient as a function of photon energy, for the absorption due to the $M_0$ [8], $M_1$, and $M_2$-transitions [10]. The resulting dependence of $\beta_{xx}(\omega;\omega)$ and $\varepsilon_{2,xx}(\omega)$ on the photon energy $\hbar\omega$ is written for the $M_m$-transition

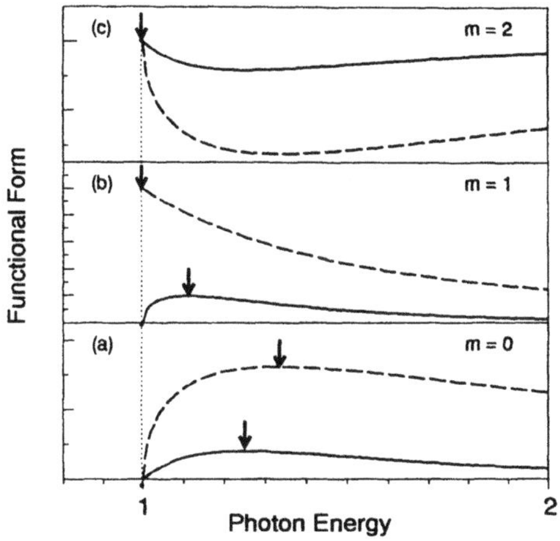


Figure 2.11. Scaling-functional forms of $\beta^m_{xx}(\omega;\omega)$ (solid lines) and $\varepsilon^m_{2,xx}(\omega)$ (dashed lines) for transition-types, $m = 0, 1$ and 2. $\beta^m_{xx}(\omega;\omega)$ is shown as a function of $2\hbar\omega/\mathrm{E}^c_m$, while $\varepsilon^m_{2,xx}(\omega)$ as a function of $\hbar\omega/\mathrm{E}^c_m$. Down arrows indicate the peak positions. The value $x$ in Eqs. (2.62) and (2.63) is taken to be 0.5 here.

above its critical-point energy, $\mathrm{E}^c_m$, as follows,

$$\beta^m_{xx}(\omega) \sim \pm\frac{(2\hbar\omega/\mathrm{E}^c_m - 1)^{|1-m|/2+x}}{(2\hbar\omega/\mathrm{E}^c_m)^5}, \tag{2.62}$$

$$\epsilon^m_{2,xx}(\omega) \sim \pm\frac{(\hbar\omega/\mathrm{E}^c_m - 1)^{|1-m|/2}}{(\hbar\omega/\mathrm{E}^c_m)^2}. \tag{2.63}$$

Here $m = 0$ and 1 correspond to a positive sign, while $m = 2$ to a negative one. The exponential, $|1-m|/2$, comes from JDOS, which is obtained by the usual analysis of the density of states. On the other hand, the other exponential, $x$, comes from the type of the intermediate transition, which is obtained as follows. As seen in Eq. (2.61), the TPA process is caused by two successive transitions, $|v\mathbf{k}\rangle \rightarrow |m\mathbf{k}\rangle \rightarrow |c\mathbf{k}\rangle$. When these two transitions are allowed-allowed (allowed-forbidden) at the critical **k**-point, the numerator in Eq. (2.61) has the energy dependence, nonzero-constant $+(2\hbar\omega - \mathrm{E}^c_m)^{1/2}$ (zero$+(2\hbar\omega-\mathrm{E}^c_m)^{1/2}$). Therefore the value of $x$ is generally between 0 and 1, respectively corresponding to the pure allowed-allowed and allowed-forbidden transitions.

The above consideration with respect to the scaling formula is generalized to the systems of arbitrary dimension, $d$. Because the $\mathrm{M}_m$ critical point in a $d$-dimensional system is equivalent to the $\mathrm{M}_{m-(3-d)}$ critical point in a 3-dimensional one, the corresponding formula are obtained by replacing $m$ in Eqs. (2.62) and (2.62) by $m-(3-\mathrm{d})$.

Figures 2.11(a) to 2.11(c), respectively, illustrate the functional forms of Eqs. (2.62) and (2.63) for $m = 0$, 1, and 2 transitions. From these figure, it is easily recognized that the step-like $\mathrm{E}_1$ peak in $\varepsilon_{2,xx}(\omega)$ at $\hbar\omega = \mathrm{E}_1^c$ shifts to higher energy ($2\hbar\omega = 5/(5-x)\mathrm{E}_1^c$) and broadens in $\beta_{xx}(\omega;\omega)$, while that the $\mathrm{E}_2$ peaks in $\varepsilon_{2,xx}(\omega)$ and $\beta_{xx}(\omega;\omega)$ have the same position at $\mathrm{E}_2^c$. These results clearly explain the calculated spectral features in Si in Figs. 2.10(a) and 2.10(b).

With respect to the spectra magnitude, it is seen in Figs. 2.10(a) and 2.10(b) that the $\mathrm{E}_2$ peak is much larger than the $\mathrm{E}_1$ peak in $\varepsilon_{2,xx}(\omega)$, while that the $\mathrm{E}_1$ and $\mathrm{E}_2$ peaks have comparable magnitude in $\beta_{xx}(\omega;\omega)$. This can be also explained using Eqs. (2.62) and (2.63), where the photon energy is renormalized with the critical-point energy, $\mathrm{E}_m^c$, as a unit. Excluding such renormalization, $\varepsilon_{2,xx}(\omega)$ and $\beta_{xx}(\omega;\omega)$ are roughly proportional to $(\mathrm{E}_m^c)^{-2}$ and $(\mathrm{E}_m^c)^{-5}$, respectively. Thus, the spectra magnitude decreases more sharply in $\beta_{xx}(\omega;\omega)$ than in $\varepsilon_{2,xx}(\omega)$ with increasing the photon energy. In this way, the critical-point transition with lower (higher) energy is generally enhanced (reduced) in $\beta_{xx}(\omega;\omega)$ compared to $\varepsilon_{2,xx}(\omega)$.

Then, the calculated spectra are compared with the experimental observations. Direct TPA spectrum measured by Reitze *et al.* [51] using femtosecond pulses is shown in Fig. 2.10(a) by open circles with a guide line, while the experimentally determined $\varepsilon_{2,xx}(\omega)$ by Aspnes and Studna [52] are plotted by the filled circles in Fig. 2.10(b). Excellent coincidence is found between the theoretical calculation and experimental results regarding the overall magnitude of $\beta_{xx}(\omega;\omega)$ in this energy region and the existence of the $\mathrm{E}_1$ and $\mathrm{E}_2$ absorption peaks below ($2\hbar\omega \sim$) 4.0 eV and around ($2\hbar\omega \sim$) 4.4 eV. Concerning the peak width, it is not understood at present why the calculated $\mathrm{E}_2$ peak is wider than the measured one.

Next, let us consider other semiconductors. Calculated $\beta_{xx}(\omega;\omega)$ and $\varepsilon_{2,xx}(\omega)$ for ZnSe are respectively shown in Figs. 2.12(a) and 2.12(b). The absorption onsets correspond to calculated interband energies; $\mathrm{E}_0$ ($\Gamma_{1c} - \Gamma_{15v} = 2.820$ eV) and $\mathrm{E}_1$ ($\mathrm{L}_{1c} - \mathrm{L}_{3v} = 4.413$). The same analysis of the peak position and its magnitude as that of Si is applicable to the present $\mathrm{E}_0$ and $\mathrm{E}_1$ peaks in $\beta_{xx}(\omega;\omega)$. The result is that, similar to the Si case, the $\mathrm{E}_1$ peak in $\beta_{xx}(\omega;\omega)$ is located at higher energy than that of $\varepsilon_{2,xx}(\omega)$. While the $\mathrm{E}_0$ peak in $\beta_{xx}(\omega;\omega)$ appears at lower energy ($2\hbar\omega = 10/(9-2x)$ $\mathrm{E}_0^c$) than that of $\varepsilon_{2,xx}(\omega)$ ($\hbar\omega = 4/3\ \mathrm{E}_0^c$). Moreover, the spectra-magnitude difference between the $\mathrm{E}_0$ and $\mathrm{E}_1$ peaks is smaller in $\beta_{xx}(\omega;\omega)$ than in $\varepsilon_{2,xx}(\omega)$. Considering that the $\mathrm{E}_0$ peak in $\varepsilon_{2,xx}(\omega)$ is normally hidden in spectra due to other transitions, these features are clearly recognized in Figs. 2.12(a) and 2.12(b).

$\beta_{xx}(\omega;\omega)$ of ZnSe have been measured at particular photon-energies. The calculated $\beta_{xx}(\omega;\omega)$ at $2\hbar\omega = 3.52$ and 4.66 eV are in good agreement with recent accurate experimental data displayed in Fig. 2.12(a). Note that the calculated $\beta_{xx}(\omega,\omega)$ at $2\hbar\omega = 4.66$ eV is mainly made of two kinds of absorption, *i.e.*, the absorption around $\Gamma$ and L points, since the two-photon energy (4.66 eV) is larger than the interband-onset energies ($\mathrm{E}_0 = 2.820$ eV and $\mathrm{E}_1 = 4.413$ eV). The same situation is seen in the calculated and observed $\varepsilon_{2,xx}(\omega)$ at this photon energy.

Figures 2.12(a) and 2.12(b) respectively show the calculated $\beta_{xx}(\omega;\omega)$ and $\varepsilon_{2,xx}(\omega)$

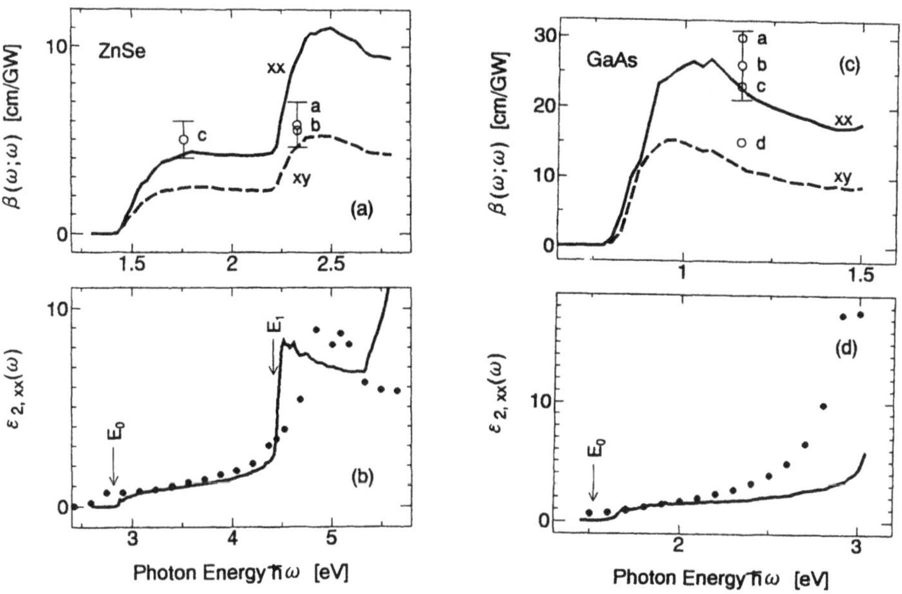


Figure 2.12. (a) Calculated $\beta_{xx}(\omega;\omega)$ (solid line) and $\beta_{xy}(\omega;\omega)$ (dashed line) of ZnSe as a function of photon energy. Open circles (a, b, and c) are experimental data. (b) Photon energy dependence of $\varepsilon_{2,xx}(\omega)$ of ZnSe. Down arrows indicate interband energies. Filled circles are experimental data. (c) Calculated $\beta_{xx}(\omega;\omega)$ (solid line) and $\beta_{xy}(\omega;\omega)$ (dashed line) of GaAs as a function of photon energy. Open circles (a, b, c, and d) are experimental data. (d) Photon energy dependence of $\varepsilon_{2,xx}(\omega)$ for GaAs. Down arrows indicate interband energies. Filled circles are experimental data. (After [10]).

for GaAs, whose interband energy is $E_0(\Gamma_{1c}-\Gamma_{15v}) = 1.520$ eV. GaAs is a typical case having only the transitions around $\Gamma$ point when the photon energy is below the band-gap energy. Similar to ZnSe, we can see the $E_0$ peak not in $\varepsilon_{2,xx}(\omega)$ but in $\beta_{xx}(\omega;\omega)$. The calculated $\beta_{xx}(\omega;\omega)$ is in good agreement with the latest measurements shown in the same figure.

To summarize the conclusions for the degenerate TPA spectra, TPA spectra have abundant structures originating from the critical-point transitions, whose particular shapes can be analyzed using a simple scaling law. Moreover, TPA spectra have the general tendency that the magnitude is larger in the lower-photon-energy region than in the higher-photon-energy region. It should be noted that the degree of coincidence between ***ab initio*** calculations and experiments is closely comparable between the OPA and TPA spectra. Considering that the ***ab-initio***-calculated linear optical susceptibility are often used to analyze and predict the OPA spectra of materials, the

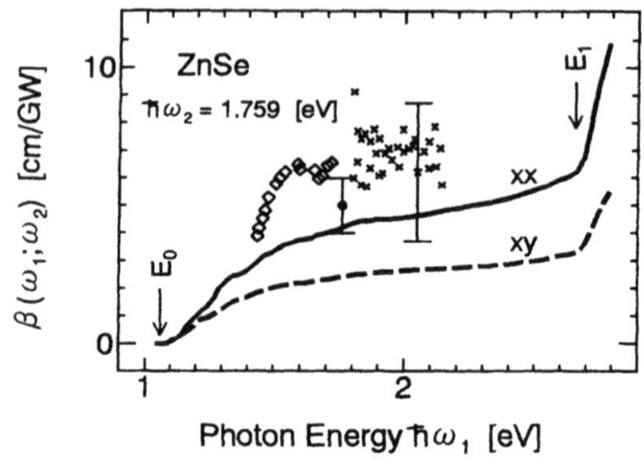


Figure 2.13. Calculated nondegenerate $\beta(\omega_1;\omega_2)$ of ZnSe as a function of probe-light energy $\hbar\omega_1$. Pump-light energy $\hbar\omega_2$ is fixed at 1.759 eV. Solid and dashed lines show the different polarizations, $\beta_{xx}(\omega_1;\omega_2)$ and $\beta_{xy}(\omega_1;\omega_2)$, respectively. Open diamonds and crosses are experimental data with different pump-light intensities, while filled circle is the result of an independent degenerate TPA experiment. (After [10]).

above fact clearly indicates the usefulness of the *ab initio* nonlinear optical susceptibility calculation. As well as the OPA case, the small but definite disagreement of the calculated TPA results with experiments mainly comes from neglecting the local-field effects and excitonic effects, which will be discussed in the last sections.

### 2.3.3. Non-Degenerate TPA Spectra

Off-diagonal tensor elements such as $\beta_{xy}(\omega;\omega)$ are classified as non-degenerate TPA coefficients. In Figs. 2.10(a), 2.12(a), and 2.12(c), we have already shown $\beta_{xy}(\omega;\omega)$ as a function of photon energy for Si, ZnSe, and GaAs, respectively. As seen in these figures, (i) $\beta_{xy}(\omega;\omega)$ is always smaller than $\beta_{xx}(\omega;\omega)$, (ii) but has no remarkable spectral characteristics compared with $\beta_{xx}(\omega;\omega)$. The former feature will be discussed in detail when we consider the anisotropy in the next section.

$\beta_{xx}(\omega_1;\omega_2)$ and $\beta_{xy}(\omega_1;\omega_2)$ belong to another type of nondegenerate TPA coefficient having different probe and pump-light energies. Figure 2.13 shows the calculated nondegenerate $\beta(\omega_1;\omega_2)$ for ZnSe as a function of the probe-light energy, $\hbar\omega_1$, where the pump-light energy is fixed at $\hbar\omega_2 = 1.759$ eV. Down arrows indicate the absorption onsets, $E_0$ and $E_1$, which are 1.061 and 2.654 eV, respectively. As compared with the degenerate TPA results for the same material in Fig. 2.13, (i) the $E_0$ and $E_1$ absorption edges move to lower and higher energies, respectively, and (ii) the spectrum has no apparent peak in the $E_0$-transition region below the $E_1$ edge. The former feature is because the resonance in $\beta(\omega_1;\omega_2)$ occurs at the sum-frequency, $\omega_1 + \omega_2$,

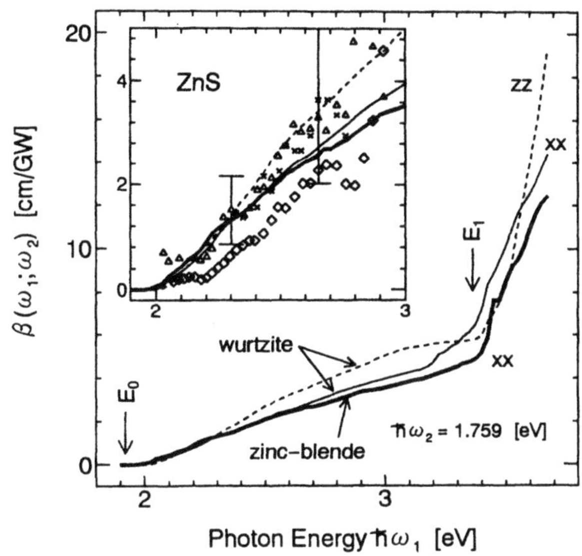


Figure 2.14. Calculated nondegenerate $\beta(\omega_1;\omega_2)$ of zinc-blende and wurtzite ZnS as a function of probe-light energy $\hbar\omega_1$. Pump-light energy $\hbar\omega_2$ is fixed at 1.759 eV. Solid and dashed lines show the different polarizations, $\beta_{xx}(\omega_1;\omega_2)$ and $\beta_{xy}(\omega_1;\omega_2)$, respectively. Open diamonds, open triangles, and crosses are experimental data with different pump-light intensities. (After [10]).

in the nondegenerate TPA case, while the latter one is due to the fact that $\beta(\omega_1;\omega_2)$ has the different $\omega_1$-dependence from that of the degenerate $\beta(\omega_1;\omega_1)$ in Eq. (2.62) [30]. Roughly speaking, $\beta(\omega_1;\omega_2)$ depends on $\omega_1$ such as

$$\beta(\omega_1;\omega_2) \quad \sim \quad \pm\frac{((\hbar\omega_1+\hbar\omega_2)/\mathrm{E}_m^c-1)^{|1-m|/2+x}}{(\hbar\omega_1/\mathrm{E}_m^c)^3}. \tag{2.64}$$

Then the $\mathrm{E}_0$ peak ($\hbar\omega_1 = 6/(5-2x)\cdot(1+\hbar\omega_2/\mathrm{E}_m^c)\mathrm{E}_m^c$) in the nondegenerate $\beta(\omega_1;\omega_2)$ is located at higher energy than that in the degenerate $\beta(\omega_1;\omega_1)$, thus the $\mathrm{E}_0$ peak does not appear in Fig. 2.13. Comparing with the previous measurements [54] shown in the same figure, the calculated results look slightly small but are within the experimental accuracy.

The nondegenerate $\beta_{xx}(\omega_1;\omega_2)$ spectra of ZnS were also measured for a zinc-blende sample [54], which are shown in the inset of Fig. 2.14. However, in this experiment, the possibility of the sample being polycrystals made of zinc-blende and wurtzite structures was suggested. The present *ab initio* approach applies to materials having any crystal structures. Therefore, in order to clarify the TPA-spectra difference between zinc-blende and wurtzite structures, we have evaluated $\beta_{xx}(\omega_1;\omega_2)$ in the zinc-blende structure and $\beta_{xx}(\omega_1;\omega_2)$ and $\beta_{zz}(\omega_1;\omega_2)$ in the wurtzite structure. Note that $\beta_{zz}(\omega_1;\omega_2)$ is equal to $\beta_{xx}(\omega_1;\omega_2)$ in the zinc-blende structure. The results are shown in Fig. 2.14, where down-arrows indicate $E_0$ (1.924 eV) and $E_1$ (3.369 eV) onset energies in the zinc-blende structure. The calculated band-gap energy in the wurtzite structure, $\Gamma_{1c} - \Gamma_{6v}$, is slightly larger than that in the zinc-blende one, $\Gamma_{1c} - \Gamma_{15v}$, by 0.025 eV.

As seen in the inset of Fig. 2.14, the calculated spectra $\beta_{xx}(\omega_1;\omega_2)$ in zinc-blende and wurtzite structures have comparable magnitude around the absorption edges. This is because the band-gap-edge (Γ-point) electronic-structure difference between both crystal structures is small for semiconductors having large ionicity such as ZnS [55]. As a result, the spectra in both structures are within the experimental accuracy, and thus it is difficult to judge the degree of polycrystallization in an experimental sample by comparing the calculated results with experiments. However, the definite difference of magnitude is recognized between the calculated $\beta_{xx}(\omega_1;\omega_2)$ and $\beta_{zz}(\omega_1;\omega_2)$ in a higher-energy region for wurtzite ZnS. Therefore, as one of the way to clarify the polycrystallization, the measurements of both $\beta_{xx}(\omega_1;\omega_2)$ and $\beta_{zz}(\omega_1;\omega_2)$, *i.e.*, some kinds of anisotropy, in a higher-energy region is proposed [10].

#### *2.3.3.1. TPA anisotropy*

For materials having the cubic symmetry such as $T_d$ ($\bar{4}3m$), $\chi^{(3)}$ has four independent nonzero tensor elements, $\chi^{(3)}_{xxxx}$, $\chi^{(3)}_{xyyx}$, $\chi^{(3)}_{xyxy}$, and $\chi^{(3)}_{xxyy}$ [28, 29]. Especially in the *frequency*-degenerate TPA described by $\chi^{(3)}(\omega;-\omega,\omega,\omega)$, $\chi^{(3)}_{xyxy}$ is equal to $\chi^{(3)}_{xyyx}$, thus existing only three independent elements. In this subsection, we consider the anisotropy in such frequency-degenerate TPA.

First we consider the calculated ratios;

$$\mathrm{A} \equiv \frac{\mathrm{Im}\chi^{(3)}_{xyyx}}{\mathrm{Im}\chi^{(3)}_{xxxx}}, \qquad \mathrm{B} \equiv \frac{\mathrm{Im}\chi^{(3)}_{xxyy}}{\mathrm{Im}\chi^{(3)}_{xxxx}}, \tag{2.65}$$

which are shown for GaAs, ZnSe, and Si in Fig. 2.15. The following common features are found among these semiconductors. (i) In the $E_0$-transition regions of GaAs and ZnSe, with increasing the photon energy, the ratio A decreases sharply in the lower-energy side and gradually in the higher-energy side. On the other hand, the ratio B displays the opposite change, thus increasing sharply and gradually in the lower and higher-energy sides, respectively. (ii) In the $E_1$-transition regions of ZnSe and Si, B decreases with increasing the photon energy while the variation of A is small. (iii) In the lower-energy sides of the $E_0'$- and $E_2$-transition regions of Si, both A and B increase with increasing the photon energy.

Some of these anisotropy features are qualitatively explained by considering the symmetry of two-photon transitions. With respect to the $E_0$-transition, we follow the discussion by Dvorak *et al.* [56], which is based on the $\mathbf{k}\cdot\mathbf{p}$ perturbation theory. At the $E_0$ absorption edge, the two-photon transition occurs at Γ point in the fcc Brillouin

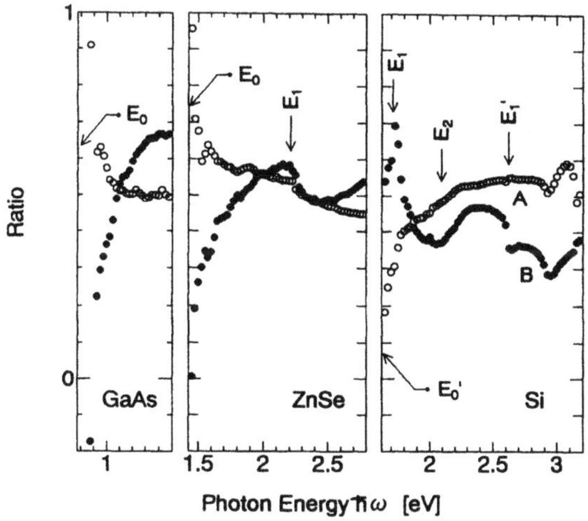


Figure 2.15. Calculated ratios, A = $\mathrm{Im}\chi^{(3)}_{xyyx}/\ \mathrm{Im}\chi^{(3)}_{xxxx}$ (open circles) and B = $\mathrm{Im}\chi^{(3)}_{xxyy}/\ \mathrm{Im}\chi^{(3)}_{xxxx}$ (filled circles), in the frequency-degenerate TPA for GaAs, ZnSe, and Si as a function of photon energy.

zone. Because this point has the high symmetry, only the two-photon transition, $|\Gamma_{15v}\rangle \rightarrow |\Gamma_{15c}\rangle \rightarrow |\Gamma_{1c}\rangle$, with different polarizations, such as $x$ and $y$, is allowed. Thus we expect $\chi^{(3)}_{xyyx} \neq 0$ but $\chi^{(3)}_{xxxx} \sim 0$ at the $\mathrm{E_0}$ absorption edge, which explains why the ratio A has large value at the $\mathrm{E_0}$ absorption edge. Apart from the $\mathrm{E_0}$ absorption edge, the two-photon transition occurs around Γ but at the general **k**-points with no symmetry. Since this transition is allowed for any pairs of polarization directions, $\chi^{(3)}_{xyyx}$ and $\chi^{(3)}_{xxxx}$ have comparable magnitude, which explains why the ratio A sharply decreases apart from the $\mathrm{E_0}$ absorption edge, *i.e.*, in the $\mathrm{E_0}$-transition region. Similar analysis of anisotropy is possible for other critical-point TPA transitions. For example, because the two-photon transition, $|L_{3v}\rangle \rightarrow |L_{3c}\rangle \rightarrow |L_{1c}\rangle$, is allowed for two photons with arbitrary polarizations, $\chi^{(3)}_{xxxx}$ as well as $\chi^{(3)}_{xyyx}$ has non-vanishing values even at the $\mathrm{E_1}$ absorption edge. This explains that the ratio A does not have such large values at the $\mathrm{E_1}$ edge as those at the $\mathrm{E_0}$ edge.

An experimental setup generally possesses the electromagnetic fields with arbitrary propagation and polarization directions. Therefore, from a practical viewpoint, it is useful to define two independent parameters that describe the TPA anisotropy as follows [14],

$$\sigma = \frac{\mathrm{Im}\chi^{(3)}_{xxxx} - 2\mathrm{Im}\chi^{(3)}_{xyyx} - \mathrm{Im}\chi^{(3)}_{xxyy}}{\mathrm{Im}\chi^{(3)}_{xxxx}}, \tag{2.66}$$

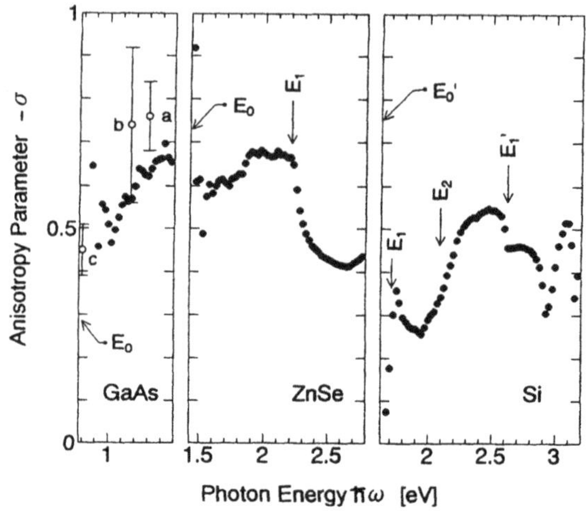


Figure 2.16. Calculated TPA anisotropy parameter, $-\sigma$, for GaAs, ZnSe, and Si as a function of photon energy. Filled circles are the calculated results and open circles (a, b, and c) are experimental data. (After [10]).

$$\delta = \frac{\mathrm{Im}\chi^{(3)}_{xxxx} - 2\mathrm{Im}\chi^{(3)}_{xyyx} + \mathrm{Im}\chi^{(3)}_{xxyy}}{2\mathrm{Im}\chi^{(3)}_{xxxx}}. \tag{2.67}$$

Here $\sigma$ is called the anisotropy parameter while $\delta$ the dichroism parameter. In experiments, $\sigma$ corresponds to the coefficient of the propagation-polarization-angle-dependent part of TPA magnitude, while $\delta$ is relating to the difference of absorption magnitude between the linearly- and circularly-polarized lights. Especially in isotropic materials, $\delta$=$\sigma$=0.

The calculated $-\sigma$ and $\delta$ are shown in Figs. 2.16 and 2.17, respectively, for GaAs, ZnSe, and Si. It is seen that $-\sigma$ gradually increases in the $E_0$-transition region with increasing the photon energy. This is because the compensation occurs between the variations in the ratios, A and B. While $-\sigma$ decreases/increases sharply in the $E_1/E_2$-transition region due to the cooperative variations in A and B. With respect to $\delta$, these features, compensation and cooperation, are found to be opposite to $-\sigma$, thus showing sharp change in the $E_0$-transition region while showing no characteristic change in the $E_1$- and $E_2$-transition regions.

The experimental data are available only for $-\sigma$ in GaAs [57], which are shown in Fig. 2.16. Compared with these experiments, the calculated results show good agreement in the lower-energy region, while they give slightly small values in the higher-energy region. The reason for this discrepancy is not clear at present.

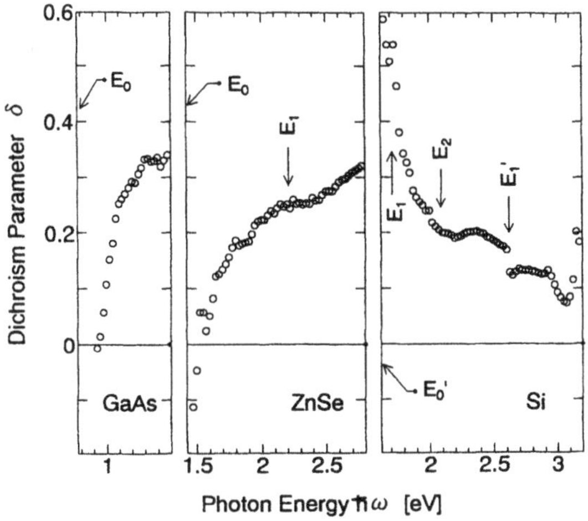


Figure 2.17. Calculated TPA dichroism parameter, $\delta$, for GaAs, ZnSe, and Si as a function of photon energy. (After [10]).

## 2.4. VARIOUS NONLINEAR SPECTRA IN BULKS AND SUPERLATTICES

### 2.4.1. *Second and Third Harmonic Generations in Bulk Semiconductors*

The second and third-harmonic generations (SHG and THG) are the other nonlinear optical properties that have been studied for bulk semiconductors by the recent *ab initio* approaches. In this subsection, the results by Huang and Ching [20, 21] are first reviewed briefly.

#### *2.4.1.1. SHG spectra*

Let us first consider the static limit of the SHG, $\chi^{(2)}(0) \equiv \chi^{(2)}_{xyz}(0;0,0)$. Figure 2.18(a) shows the calculated values vs the measured ones for various semiconductors. Although the overall agreement is satisfactory, there is a tendency that the calculated values are slightly smaller than the measured ones in most semiconductors. One of reasons for this tendency is the neglect of the momentum renormalization effect in this calculation, as shown in Sec. 2.2.3; the momentum renormalization increases the $\chi^{(2)}(0)$ values. It is interesting to see these calculated $\chi^{(2)}(0)$ values as a function of material parameter. Figure 2.18(b) shows the calculated $\chi^{(2)}(0)$ vs the direct-band-gap energy, $E_g$. It is seen that the smaller-band-gap semiconductors have larger $\chi^{(2)}(0)$. This fact had been recognized in the early 1960s and was the basis of many

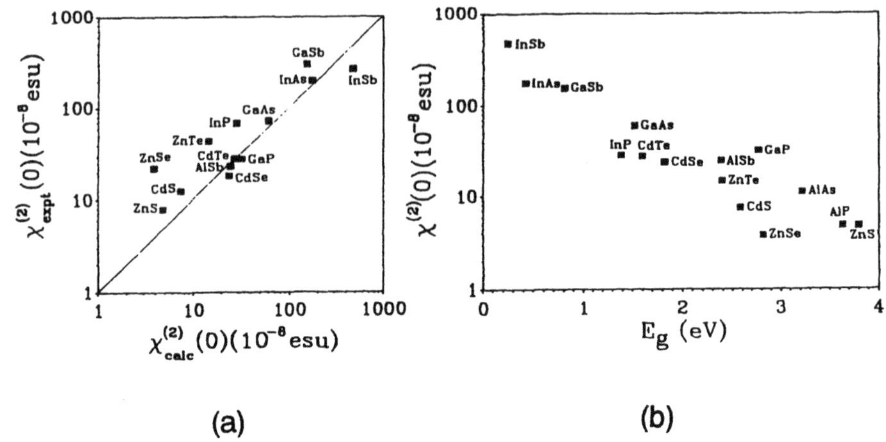


Figure 2.18. (a) Calculated values vs measured ones of the static-limit SHG magnitude, for various semiconductors. (After [20]). (b) Correlation between calculated $\chi^{(2)}(0)$ and direct-band-gap energy, $E_g$. (After [20]).

empirical calculations of $\chi^{(2)}(0)$. However, it should be noted that these values are calculated without any such information. The recent proper success of the scaling laws of various nonlinear optical properties with material parameters and photon energies [8, 30] are strongly supported by only the results of these ***ab initio*** calculations.

The calculated absolute spectra, $|\chi^{(2)}(\omega)| \equiv |\chi^{(2)}_{xyz}(2\omega;\omega,\omega)|$, are plotted in Fig. 2.19(a) for GaAs, together with the observed spectra. The agreement between both results is satisfactory with respect to the spectral magnitude and structures. For example, the peaks at 1.4 and 2.4 eV with a deep valley at 2.0 are well reproduced by the set of experimental data, although it appears that the peak at 2.4 eV is slightly below the measured data. Though the SHG phenomenon manifests itself as the absolute value of $\chi^{(2)}(\omega)$, it is instructive to see the real and imaginary parts of $\chi^{(2)}(\omega)$ separately because such analysis enables to understand the SHG at a more microscopic level. The results are shown for GaAs in Fig. 2.19(b). For the same reason, since the $\mathrm{Im}\chi^{(2)}(\omega)$ is obtained as the sum of the single-photon resonance (the $\omega$ term) and the two-photon resonance(the $2\omega$ term), such two terms are shown in Fig. 2.19(c) for the same material. The following features are clearly seen in the latter figure. (i) The $2\omega$-term usually dominates in the lower-photon-energy region and is positive. It is this feature that determines the sign and magnitude of $\chi^{(2)}(0)$ and the leading peak structures in $|\chi^{(2)}(\omega)|$. (ii) In the higher-photon-energy region, the $\omega$ and $2\omega$ terms have similar magnitude but opposite sign to each other. However,

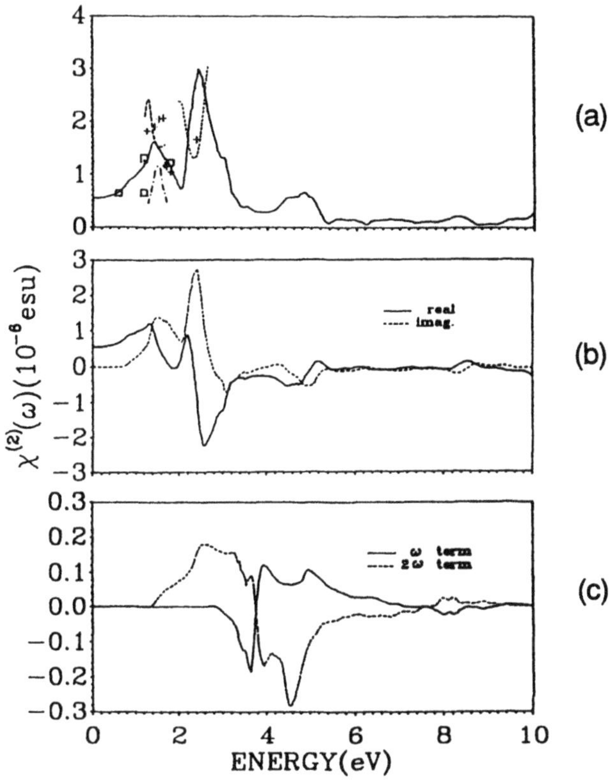


Figure 2.19. Calculated spectra of $\chi^{(2)}_{xyz}(2\omega;\omega,\omega)$ for GaAs as a function of photon energy. (a) Absolute values, $|\chi^{(2)}|$, (b) real (solid line) and imaginary (dashed line) parts of $\chi^{(2)}$, and (c) two contributions to $\mathrm{Im}\chi^{(2)}$, the $\omega$ (solid line) and $2\omega$ (dashed line) resonant terms. All results are from [20].

the detailed analysis for these calculations have not been performed yet.

#### *2.4.1.2. THG spectra*

Next, we consider the THG spectra. Ching and Huang have calculated the THG spectra of 18 cubic semiconductors of group-IV, III-V, and II-VI compounds. Table 2.4 shows some out of their calculated static-limit values, $\mathrm{Re}\chi^{(3)}_{xxxx}(0;0,0,0)$ and $\mathrm{Re}\chi^{(3)}_{xyxy}(0;0,0,0)$, and the corresponding experimental data. Considering that experimental uncertainty renders any results beyond the order of magnitude established

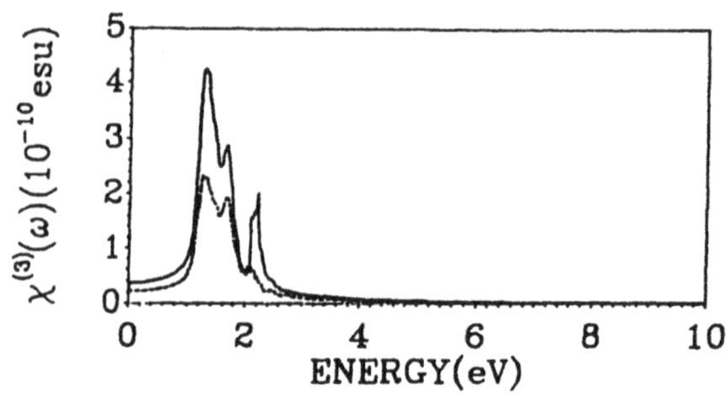


Figure 2.20. Calculated THG spectra for Si. (After [21]). Absolute values of $\chi^{(3)}_{xxxx}$ (solid line) and $\chi^{(3)}_{xyxy}$ (dashed line) are shown here.

Table 2.4. Calculated THG magnitude of semiconductors in static limit. Unit is $10^{-11}$ esu. (After [21]).

| | $\chi^{(3)}_{xxxx}(0;0,0,0)$ | | $\chi^{(3)}_{xyxy}(0;0,0,0)$ | |
|---|---|---|---|---|
| Material | Calc. | Expt. | Calc. | Expt. |
| C | 0.057 | 0.0184 | 0.031 | 0.007 |
| Si | 3.07 | 2.4, 33, 20 | 1.90 | 1.15 |
| Ge | 30.69 | 40, 10, 16, 12, 4.1 | 10.27 | 24, 5.2 |
| GaAs | 6.95 | 3.9, 4.8, 18 | 4.01 | 2.0, 1.2 |
| InSb | 417836 | $10^9$, 8000, $10^{11}$ | 14718 | |

in the data quite meaningless, the calculated results are overall in good agreement with the measured data. Similar to the SHG case, the general feature that $\chi^{(3)}(0)$ increases with decreasing the band-gap energy of the material is found in the table.

The photon-energy-dependent dispersion relations in $\chi^{(3)}_{xxxx}$ $(3\omega;\omega,\omega,\omega)$ and $\chi^{(3)}_{xyxy}$ $(3\omega;\omega,\omega,\omega)$ for various semiconductors have been calculated by the same group. Because there is no available experimental data to compare, the calculated result is displayed only for Si in Fig. 2.20. Although the fine structures and their energy positions in spectra are different between semiconductors, the overall features resemble in most semiconductors. For example, it has been demonstrated that the first major peak in $\chi^{(3)}_{xxxx}(3\omega;\omega,\omega,\omega)$ is closely related to the minimal direct band gap.

### 2.4.2. *TPA and SHG Spectra in Semiconductor Superlattices*

The application of the ***ab initio*** method to study nonlinear optical properties of artificially synthesized complex systems, such as surfaces, interfaces, superlattices, and

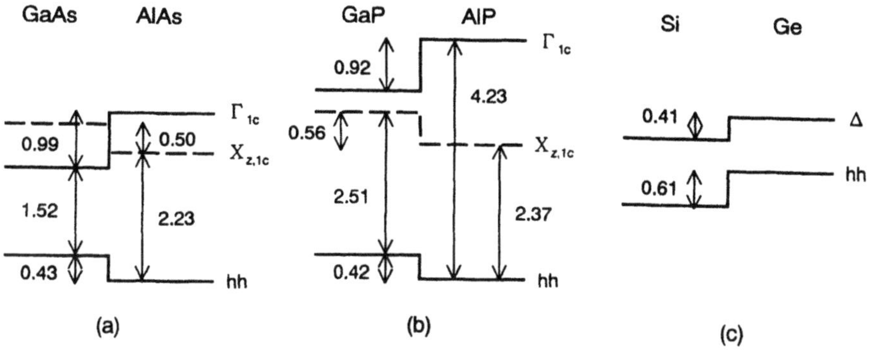


Figure 2.21. Calculated band-offset diagram for GaAs/AlAs (a), GaP/AlP (b), and Si/Ge (on $Si_{0.5}/Ge_{0.5}$ substrate) (c) superlattices. Offsets for Si/Ge are from [58], while others are from [59].

microstructures, is quite fascinating. This is physically because these systems are low-dimensional systems where the electronic and optical properties are often exotic, and computationally because the reliable theoretical calculation of such properties are performed only through the *ab initio* band-structure and susceptibility approaches. However, at the present stage, the calculations of nonlinear optical properties are limited to a few semiconductor superlattice systems, and the comparison with experiments has not been performed yet, *i.e.*, the calculated results still remain the theoretical predictions waiting for the experimental confirmation. In this section, some results out of them are reviewed.

A superlattice, A/B, is an artificially-synthesized layered materials made of different semiconductors, A and B [60]. The electronic structure of a superlattice is well characterized by the band offset, which is normally defined by the energy-level difference between the highest valence-band (lowest conduction-band) states in A and B [61]. Here we concentrate on typical strain-free superlattices, (001)GaP/AlP and (001)GaAs/AlAs, and strained one, (001) Si/Ge, whose calculated band-offset energy diagrams in a real space are schematically shown in Fig. 2.21. (These offsets are calculated in the LDA. Especially for GaAs/AlAs, it is well known that the calculated energy levels in AlAs should be shifted up by about 0.15 eV in order to obtain the correct values observed in experiments. This is due to the insufficient treatment of many-body effects in the LDA calculation [62]. As a result, the calculation for $(GaAs)_2/(AlAs)_2$ shows that the $\Gamma_{1c}$ direct-gap state is located lower than the $X_{z,1c}$ pseudo-direct state contrary to the experiments. However, here, we show the calculated results of GaAs/AlAs as an example of direct-gap superlattice.)

From these figures, one concludes one of characteristics in superlattice electronic structures; the anisotropy. Both the Γ-point electrons and Γ-point holes are confined in the same layers (type-I alignment); in GaP layers for GaP/AlP and in GaAs layers for GaAs/AlAs, respectively. While the X-point electrons are confined in different layers (type-II alignment). As a result, the electron- and hole-states around the fundamental gap have quantized energies along the superlattice direction, thus showing the two-dimensional ***anisotropic*** band structures. Then, how this feature affects on the nonlinear optical spectra ?

### *2.4.2.1. Anisotropy in TPA spectra*

Figures 2.22(a) and 2.23(a) display the calculated degenerate TPA spectra, $\beta_{xx}(\omega;\omega)$ and $\beta_{zz}(\omega;\omega)$, for short-period $(GaAs)_2/(AlAs)_2$ and $(GaP)_2/(AlP)_2$ superlattices, respectively. Here the $z$ axis is along the superlattice direction, while $x$ and $y$ along the interface. It is seen that both spectra have bulk-like $E_0$-type transition (hh (heavy-hole) $\rightarrow \Gamma_{1c}$) around 1.05 ~ 1.55 eV for GaAs/AlAs and around 1.55 ~ 2.0 eV for GaP/AlP, while $E_1$-type transition ($L_{3v} \rightarrow L_{1c}$) above 1.7 eV for GaAs/AlAs and above 2.0 eV for GaP/AlP. In addition, there appears to be the two-photon pseudo-direct transition (hh $\rightarrow X_{z,1c}$) [63], from the hh states to the zone-folded $X_{z,1c}$ state, at 1.17 eV for GaAs/AlAs and at 1.20 eV for GaP/AlP. However, this pseudo-direct transition is extremely small due to the small mixing of the folded $X_{z,1c}$ states with Γ-originated states [64].

In this way, the diagonal $\beta_{xx}$ and $\beta_{zz}$ spectra in these short-period superlattices look having no definite structures intrinsic to superlattices. However, the anisotropy spectra of $A_x = \mathrm{Im}\chi^{(3)}_{xyyx}/\mathrm{Im}\chi^{(3)}_{xxxx}$ and $A_z = \mathrm{Im}\chi^{(3)}_{zxxz}/\mathrm{Im}\chi^{(3)}_{zzzz}$, shown in Figs. 2.22(b) and 2.23(b), have the pronounced feature that $A_z$ has considerably large value while $A_x$ has finite value around the absorption onsets, *i.e.*, both around hh $\rightarrow \Gamma_{1c}$ and hh $\rightarrow X_{z,1c}$. This can be explained by considering the crystal symmetry of superlattices.

Note that in bulks $A_x = A_z$ holds and they have large values around the hh $\rightarrow \Gamma_{1c}$ $E_0$ absorption edge, which has already been shown in Fig. 2.15 for GaAs and ZnSe and explained in Sec. 2.3.3.1 being due to the selective two-photon transitions; $\mathrm{Im}\chi^{(3)}_{xyyx} \neq 0$ and $\mathrm{Im}\chi^{(3)}_{xxxx} = 0$. On the other hand, in superlattices, both the $xyyx$ two-photon transitions from the hh state ($\Gamma_5$ symmetry) to the conduction-band states, $\Gamma_{1c}$ and $X_{z,1c}$ (both $\Gamma_1$ symmetry), are forbidden due to the superlattice symmetry ($D_{2d}$ group), thus $\mathrm{Im}\chi^{(3)}_{xyyx} = \mathrm{Im}\chi^{(3)}_{xxxx} = 0$ and $A_x$ having the finite values. While the $zxxz$ transition is allowed for the above transitions as well as the bulk case, thus $A_z$ having large values. Moreover, because the carrier confinement is much stronger and thus the symmetry breaking from the bulk cubic symmetry is larger near the absorption edge, the above-mentioned anisotropy is remarkable especially near the absorption edge.

### *2.4.2.2. Anisotropy in SHG spectra*

Since IV-family semiconductors such as Si and Ge have the inversion symmetry, $\chi^{(2)}$ tensor always disappears within the dipole approximation [28, 29]. However, when

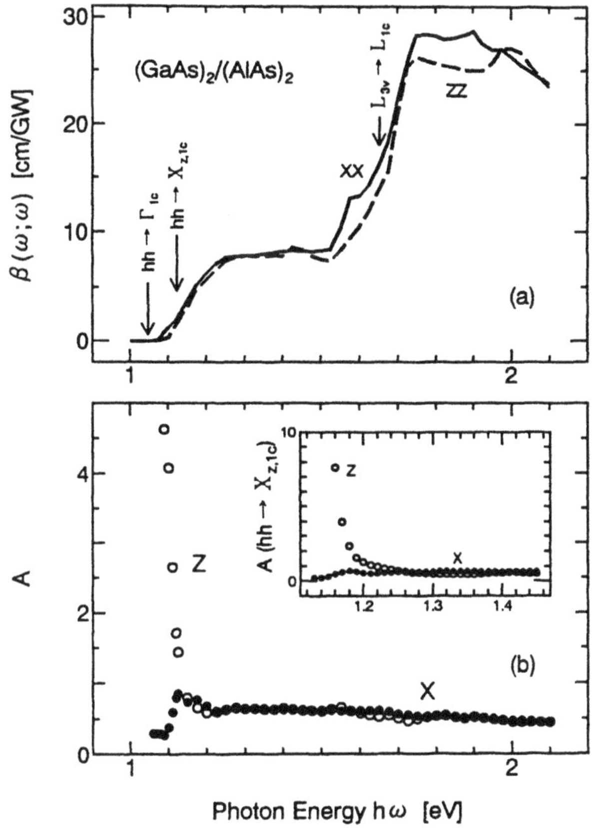


Figure 2.22. (a) Calculated TPA spectra and (b) anisotropy spectra of $(GaAs)_2/(AlAs)_2$. Solid and dashed lines (filled and open circles) respectively correspond to xx and zz polarization, where the superlattice is along the z direction. The down-arrows indicate the onset energies of various transitions. The inset shows the anisotropy spectra originating only from the pseudo-direct transition, hh $\longrightarrow X_{z,1c}$. (After [59]).

the superlattice Si/Ge is produced, such symmetry is broken and the large $\chi^{(2)}$ is expected to appear. Figure 2.24 shows the calculated SHG spectra (the imaginary part of $\chi^{(2)}$, the resonance part) of strained $(Si)_5/(Ge)_5$ (001) superlattice on $Si_{0.4}Ge_{0.6}$ substrate, by Ghahramani and Sipe [65].

The small feature around 0.55 eV in $\mathrm{Im}\chi^{(2)}_{zxy}(2\omega;\omega,\omega)$ is due to $2\omega$ resonance with the transitions between the valence-band-top light-hole states and the zone-folded $X_{z,1c}$ conduction-band states. Even in this Si/Ge system where the mixing between

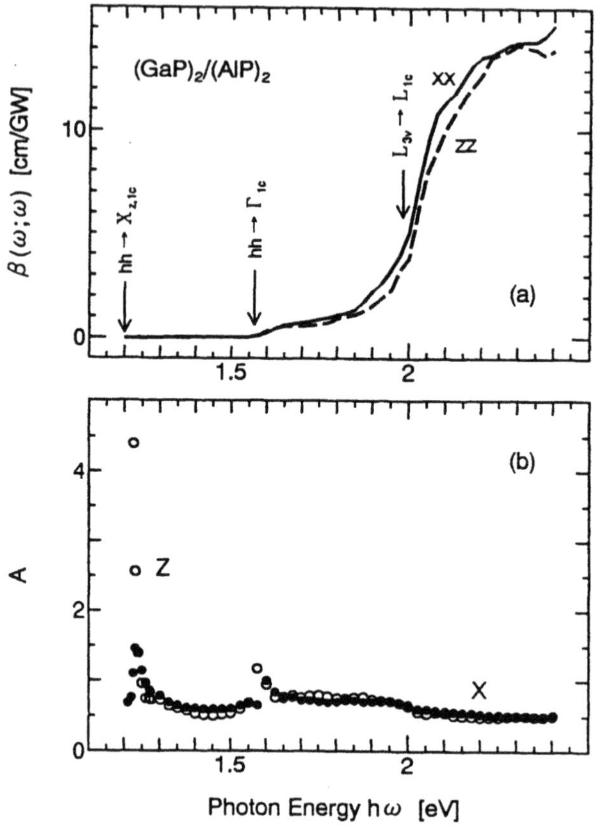


Figure 2.23. (a) Calculated TPA spectra and (b) anisotropy spectra of $(GaP)_2/(AlP)_2$. The abbreviations are same as in Fig. 2.22. (After [59]).

the folded $X_{z,1c}$ and the $\Gamma$ states is relatively large [64], the spectral magnitude of such pseudo-direct transition is small. However, since the photon-energy dependence of $\chi^{(2)}$ ($\sim \omega^{-4}$) is stronger than that of linear $\chi^{(1)}$ ($\sim \omega^{-2}$) and the SHG resonance is located at lower energy than the one-photon resonance, one can expect that the pseudo-direct transition is more prominent in the SHG (by more than an order of magnitude) than in the one-photon absorption [65], which is similar to the energy-dependence feature of TPA explained in Sec. 2.3.2 and thus which is the general feature in nonlinear optical properties.

On the other hand, the major peak around 1.3 eV is due to $2\omega$ resonance with the bulk-like $E_1$ and $E_2$ transitions. Because these transitions does not appear in

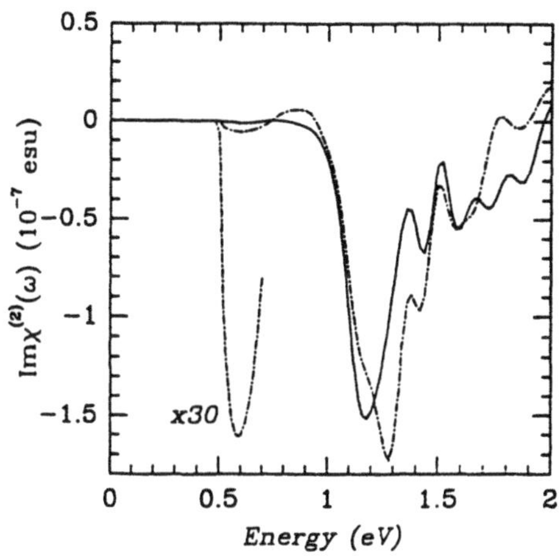


Figure 2.24. Calculated SHG $\chi^{(2)}(2\omega;\omega,\omega)$ for a strained $(Si)_5/(Ge)_5$ superlattice on $Si_{0.4}Ge_{0.6}$ substrate as a function of photon energy, $\hbar\omega$. Solid and dashed lines respectively correspond to $xyz$ and $zyx$ tensor elements. (After [65]).

the bulk SHG, it is concluded that they appear due to the modification of bulk electronic structure associated with the superlattice low symmetry. By changing the layer thickness of superlattices, one can also derive the SHG-spectra component intrinsic to the interface [66]. Compared to the pseudo-direct transition, these $E_1$ and $E_2$ transitions have relatively large magnitude over the wide range of photon energy, thus definitely contributing to the static $\chi^{(2)}$ value. The magnitude is nevertheless so small, *i.e.*, the calculated $\chi^{(2)}_{xyz}(0;0,0)$ is $-2.9 \times 10^{-8}$ esu much smaller than observed anion-cation-type compound-semiconductor values such as $28 \times 10^{-8}$ esu of CdTe.

## 2.5. FUTURE PROSPECTS

### *2.5.1. A Future Prospect*

Now, we have the ***ab initio*** method to quantitatively calculate the optical linear and nonlinear properties of materials without any empirical data, which enables the deep understanding of observed optical responses in materials and the prediction of optical properties of given materials prior to experiments. It is easily expected that many challenges using the ***ab initio*** approach will be performed in the very near future to clarify optical properties of newly-synthesized materials, including 1-, 0-, and 3′-dimensional systems shown in other chapters in this book, as well as 3- and 2-dimensional ones shown in this chapter (see Fig. 2.25).

One example out of such challenges is the calculation of the surface SHG spectra. In the recent surface physics, optical SHG in reflection from surfaces has proven to be a useful tool for studying both semiconductor surfaces and interfaces [67, 68]. This is because the optical response from the surface has the different symmetry (tensor character) from that from the bulk due to the reconstruction of surface atoms and the adsorption of foreign atoms. The surface symmetry is judged qualitatively only by considering the surface atomic configuration [69, 70]. However, in order to identify the spectral structures and to properly account for the observed anisotropy, the microscopic calculation is indispensable because the SHG spectra is natively originated not from the atomic configuration but from the electronic transitions. The *ab initio* approach will work to clarify this surface SHG, which helps identifying the electronic structure, the surface dynamics and the crystal growth mechanism.

From these studies, we obtain the numerous data base for nonlinear optical properties in materials, that enables to induce the microscopic guiding principles of producing state-of-art optically-new-functional materials. On the other hand, from the electromagnetic-field viewpoint, optical nonlinearity in material is a powerful tool to control the photonic state. One example is to produce the exotic photonic states such as the squeezed state [71], which are quite different from the coherent state of classical electromagnetics. Another one is to artificially and effectively modulate the electromagnetic filed through the other electromagnetic fields by using the photonic "superlattice" structures of materials, which corresponds to the many-body effect in photonic band structures [72].

When one proceeds to the optical susceptibility calculations of the above- mentioned lower-dimensional systems, one must include in some cases other many-body effects which is not included in the calculation scheme reviewed in this chapter. One is the optical nonlinearity caused by the non-equilibrium nature of the system, such as nonlinearities due to highly-excited carrier densities, and by the time-developing coherent nature [73], which is beyond the scope of this chapter. In the rest of this section, we consider the relevant effects that should be involved in the future *ab initio* nonlinear-optical-susceptibility calculations; the local field effect and the excitonic effect. These effects are particularly important in low-dimensional systems. This is because the imhomogeneity of electron density is so large in low-dimensional materials that the local field effect is expected to be remarkable. While, as seen in Chapter 1, the excitonic effect is significantly enhanced in low-dimensional systems.

### 2.5.2. *Local Field Effects*

It has long been recognized that the microscopic (short-wavelength, local-field) responses of the electrons caused by the inhomogeneously-varying electronic states on the scale of the interatomic spacings can be different from the macroscopic (long-wavelength, average) responses measured on the scale of the experimental sample size. Thus, in order to compare the measured responses, the microscopic responses normally obtained in the naive theoretical calculation should be corrected, and such correction is called the local filed effect.

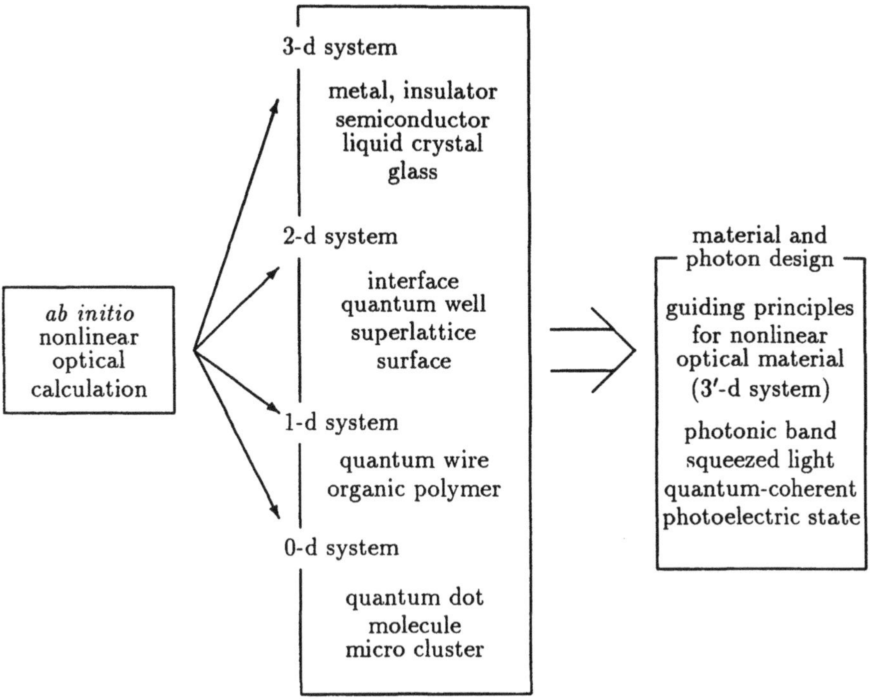


Figure 2.25. Application fields of the *ab initio* nonlinear optical calculation. From these studies, one can obtain the guiding principles for producing nonlinear optical materials and for controlling the quantum photonic states.

Table 2.5. Effect of local field (LF) on the linear and SHG responses in the static limit; $\varepsilon_{1,xx}(\mathbf{q},\omega;\mathbf{q},\omega)$ and $\mathrm{Re}\chi^{(2)}_{xyz}(2\mathbf{q},2\omega;\mathbf{q},\omega,\mathbf{q},\omega)/2$ with $\mathbf{q}=\omega=0$. Unit for $\chi^{(2)}$ is chosen as pm/V. (After [18]).

| | $\varepsilon_{1,xx}$ | | $\mathrm{Re}\chi^{(2)}_{xyz}/2$ | |
|---|---|---|---|---|
| Material | no LF | with LF | no LF | with LF |
| AlP | 7.80 | 7.22 | 16.6 | 14.9 |
| AlAs | 8.65 | 8.08 | 25.5 | 23.6 |
| GaP | 9.30 | 8.78 | 39.5 | 35.1 |
| GaAs | 11.70 | 11.16 | 93.4 | 86.4 |

With respect to the linear response, numerous works have been carried to demonstrate the relevancy of local field effect [45]. For the nonlinear response, however, the study is limited to very few calculations [18, 19] and their formalism is rather complicated. Here, in an analogous manner to the linear response of $\tilde{\chi}^{(1)}_{macro}$, we qualitatively explain how the local field effect is incorporated into the macroscopic $\tilde{\chi}^{(3)}_{macro}$. Namely the following argument is less accurate but physical. The macroscopic $\tilde{\chi}^{(3)}_{macro}$ is the response function of the polarization to the applied electromagnetic field, $\mathbf{E}_{ext}$, while the microscopic $\tilde{\chi}^{(3)}$ calculated using the formula in Sec. 2.2 is that to the microscopic total electromagnetic field, $\tilde{\mathbf{E}}$. Thus we can symbolically write these relations using the tensor product, $\otimes$, as

$$\begin{aligned} \tilde{\mathrm{P}} &\equiv \tilde{\chi}^{(3)}_{macro} : \tilde{\mathbf{E}}_{ext} \otimes \tilde{\mathbf{E}}_{ext} \otimes \tilde{\mathbf{E}}_{ext} \qquad (2.68)\\ &\equiv \chi^{(3)} : \tilde{\mathbf{E}} \otimes \tilde{\mathbf{E}} \otimes \tilde{\mathbf{E}}. \qquad (2.69) \end{aligned}$$

The total electromagnetic field is the sum of the applied field and the field induced by the appearance of the above polarization,

$$\tilde{\mathbf{E}} \otimes \tilde{\mathbf{E}} \otimes \tilde{\mathbf{E}} = \tilde{\mathbf{E}}_{ext} \otimes \tilde{\mathbf{E}}_{ext} \otimes \tilde{\mathbf{E}}_{ext} + (\mathbf{EEE})_{ind}. \qquad (2.70)$$

Since the induced $(\mathbf{EEE})_{ind}$ is a functional of the polarization, it can be formally written using the abstract operator $\hat{\mathcal{F}}$ such as

$$\begin{aligned} (\mathbf{EEE})_{ind} &= \hat{\mathcal{F}} \circ \tilde{\mathbf{P}} \\ &= \hat{\mathcal{F}} \circ \tilde{\chi}^{(3)}_{macro} : \tilde{\mathbf{E}}_{ext} \otimes \tilde{\mathbf{E}}_{ext} \otimes \tilde{\mathbf{E}}_{ext}. \qquad (2.71) \end{aligned}$$

Substituting Eqs. (2.70) and (2.71) into (2.69), we have

$$\tilde{\chi}^{(3)}_{macro,\mathbf{q}_1,\mathbf{q}_2\mathbf{q}_3\mathbf{q}_4} = (1_{\mathbf{q}_a,\mathbf{q}_b} - \tilde{\chi}^{(3)}_{\mathbf{q}_a,\mathbf{q}_c\mathbf{q}_d\mathbf{q}_e} \hat{\mathcal{F}}_{\mathbf{q}_c\mathbf{q}_d\mathbf{q}_e,\mathbf{q}_b} \circ)^{-1}_{\mathbf{q}_1,\mathbf{q}'} \tilde{\chi}^{(3)}_{\mathbf{q}',\mathbf{q}_2\mathbf{q}_3\mathbf{q}_4}, \qquad (2.72)$$

where only the wavenumber dependence are explicitly displayed with the usual summation convention. The experimentally observed macroscopic response is given in the long-wavelength limit such as $\tilde{\chi}^{(3)}_{macro,\mathbf{0},\mathbf{000}}$. Note that $\tilde{\chi}^{(3)}$ in a periodic system has the extended wavenumber conservation as shown in Sec. 2.1.3, thus existing the non-zero components such as $\tilde{\chi}^{(3)}_{\mathbf{G},\mathbf{000}}$ where $\mathbf{G}$ is the reciprocal lattice vector. Therefore it is seen from Eq. (2.72) that not only the wavenumber-diagonal part of $\tilde{\chi}^{(3)}$ such as $\tilde{\chi}^{(3)}_{\mathbf{0},\mathbf{000}}$ but also the off-diagonal part of $\tilde{\chi}^{(3)}$ such as $\tilde{\chi}^{(3)}_{\mathbf{G},\mathbf{000}}$ and $\tilde{\chi}^{(3)}_{\mathbf{q}_a,\mathbf{q}_c\mathbf{q}_d\mathbf{q}_e}$ contribute to $\tilde{\chi}^{(3)}_{macro,\mathbf{0},\mathbf{000}}$. Such contribution from the off-diagonal part represents the local field effect.

As seen in Eq. (2.72), the local effect appears through the existence of the feedback interaction, $\tilde{\mathcal{F}}$, the primary contribution to which comes from the direct Coulomb interaction and which is normally the negative operator. Thus the inverse operator in Eq. (2.72) is less than unity, which indicates that the local field effect decreases the microscopic response; $\tilde{\chi}^{(3)}_{macro} < \tilde{\chi}^{(3)}$. Levine and Allan have calculated the static limit of SHG for various semiconductors in an ***ab initio*** way, the results being shown in Table 2.5 [18]. It is seen that the local field decreases both the linear and nonlinear responses by about 10 %. There has been no study concerning the energy dependence of the local field effect on nonlinear responses. Remembering the linear response result [74], the magnitude variation of the local field effect is expected among various transition regions such as $E_0$, $E_1$ and $E_2$.

On deriving the physically observed quantities, one has often to take into account the local field effect on the linear response as well as that on the nonlinear ones. This is because, for example, the TPA coefficient in Eq. (2.60) is obtained by the ratio, $\mathrm{Im}\chi^{(3)}/(n^{(1)}n^{(1)})$, where the linear response in $n^{(1)}$ also being renormalized by the local field effect.

### 2.5.3. *Excitonic Effects*

The excitonic effects on optical properties have often been recognized being relevant in most semiconducting and insulating materials. However the study of such effects in nonlinear optics so far is limited to the evaluation using model-band structure, such as the two-band model based on the effective approximation. Figure 2.26 shows thus calculated excitonic-enhancement-factor spectra of TPA coefficient above the band-gap energy [7], as a function of the energy of bound-state exciton. It is seen that the enhancement is remarkable around the absorption edge where the excitonic energy, that is normally of the size of bound-state-exciton energy, is comparable to the kinetic energy of carriers (electron and hole). For example, in the case of bulk GaAs having 4 meV binding energy for 1s exciton, such range is about 40meV above the band gap. This tendency is also true in linear optics.

The application of the model-band study is limited to some materials and energy region where the band structures are well known and described by the simple formula. In order to predict the excitonic optical spectra, we need the *ab initio* approach. The standard approach is to use the many-body Green's function method, which was performed by Hanke and Sham for the linear optical properties of semiconductors, the result being shown in Fig. 2.27 [74]. In this work, since the excitonic states are the N-body excited states, they are obtained by solving the Bethe-Salpeter equation [26],

$$\mathrm{S}(1,1';2,2') = \mathrm{S}_0(1,1';2,2') + \mathrm{S}_0(1,1';3,3')\mathrm{I}(3,3';4,4')\mathrm{S}(4,4';2,2'), \quad (2.73)$$

where S(1, 1′; 2, 2′) is the part of the two-particle Green's function, $\mathrm{S}_0$ the free-propagator part of S, I the irreducible electron-hole interaction. Each numeral stands for a set of position, time and spin coordinates, and the repeated numerals, 3, 3′, 4 and 4′, indicate approximate summation. Then the response function, $\chi^{(1)}$, is given by

$$\chi^{(1)}(1,2) = S(1,1;2,2). \quad (2.74)$$

Note that, in this formalism, we calculate both the exciton spectra and the dielectric function at the same time, while the latter of which is the assumed input parameter

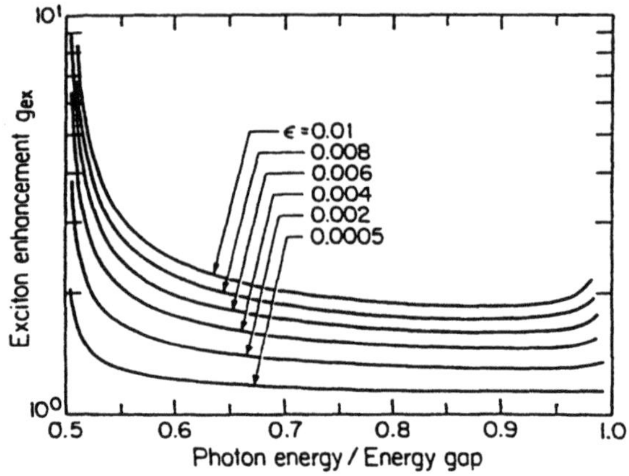


Figure 2.26. Calculated excitonic enhancement factor of TPA coefficient, $\beta_{xx}$, as a function of renormalized photon energy, $\hbar\omega/E_g$. Each curve corresponds to the different value of the ratio $\epsilon$ of the exciton binding energy, $E_b$, to the energy gap, $E_g$, *i.e.*, $\epsilon = E_b/E_g$. (After [7]).

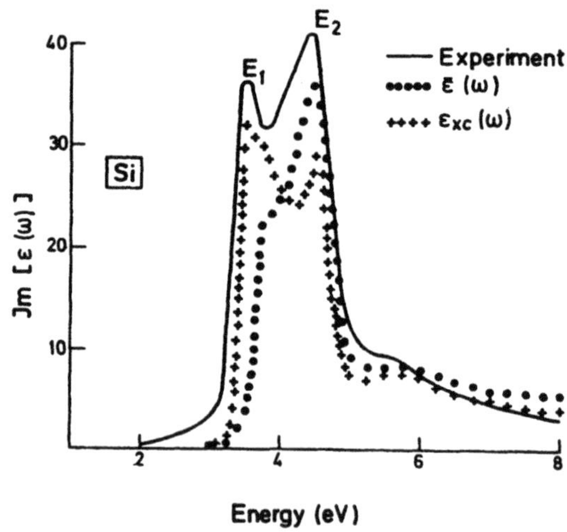


Figure 2.27. Calculated imaginary part of the dielectric function as a function of photon energy; taken from [74]. $\bar{\varepsilon}$ gives the single-particle calculation, and $\varepsilon_{xc}$ the calculation including screened electron-hole attraction and local-field correction.

in the effective-mass exciton theory. As seen in Fig. 2.27, the spectra magnitude is considerably increased in a lower-energy region ($E_1$ region) while decreased in a higher-energy region ($E_2$ region). Moreover, the $E_1$ peak is largely shifted to lower energy due to the electron-hole attraction. All these changes raise the static macroscopic dielectric constant, which is qualitatively opposite to the local-field effect. The development of this kind of approach in nonlinear optics is being desired.

It is a great pleasure to acknowledge the collaboration of Dr. Misao Murayama on the TPA-calculation works reported in this chapter and her assistance preparing the text and figures.

## REFERENCES

[1] E. W. Van Stryland, H. Hanherzeele, M. A. Woodall, M. J. Soileau, A. L. Smirl, S. Guha, and T. F. Boggess, *Opt. Eng.* **24**, 613 (1985).

[2] V. Nathan, A. H. Guenther, and S. S. Mitra, *J. Opt. Soc. Am. B* **2**, 294 (1985).

[3] A. A. Said, M. Sheik-Bahae, D. J. Hagan, T. H. Wei, J. Wang, J. Young, and E. W. Van Stryland, *J. Opt. Soc. Am. B* **9**, 405 (1992).

[4] S. S. Mitra, L. M. Narducci, R. A. Shatas, Y. F. Tsay, and A. Vaidyanathan, *Appl. Opt.* **14**, 3038 (1975).

[5] I. M. Catalano and A. Cingolani, *Solid State Commun.* **43**, 213 (1982).

[6] H. S. Brandi and Cid B. de Araujo, *J. Phys. C* **16**, 5929 (1983).

[7] M. H. Weiler, *Solid State Commun.* **39**, 937 (1981).

[8] B. S. Wherrett, *J. Opt. Soc. Am. B* **1**, 67 (1984).

[9] D. C. Hutchings and E. W. Van Stryland, *J. Opt. Soc. Am. B* **9**, 2065 (1992).

[10] M. Murayama and T. Nakayama, *Phys. Rev. B* **49**, 5737 (1994); **52**, 15 August (1995).

[11] M. Sheik-Bahae, A. A. Said, T. H. Wei, D. J. Hagan, and E. W. Van Stryland, *IEEE J. Quantum Electron.* **QE-26**, 760 (1990).

[12] For example, M. E. Lines, *Phys. Rev. B* **41**, 3383 (1990).

[13] G. D. Mahan, *Phys. Rev. Lett.* **20**, 332 (1968).

[14] D. C. Hutchings and B. S. Wherrett, *Phys. Rev. B* **49**, 2418 (1994).

[15] D. J. Moss, J. E. Sipe, and H. M. van Driel, *Phys. Rev. B* **36**, 9708 (1987).

[16] D. J. Moss, E. Ghahramani, J. E. Sipe, and H. M. van Driel, *Phys. Rev. B* **41**, 1542 (1990).

[17] E. Ghahramani, D. J. Moss, and J. E. Sipe, *Phys. Rev. B* **43**, 9700 (1991); **43**, 8990 (1991); **43**, 9269 (1991).

[18] Z. H. Levine and D. C. Allan, *Phys. Rev. Lett.* **63**, 1719 (1989); *Phys. Rev. B* **43**, 4187 (1991).

[19] Z. H. Levine and D. C. Allan, *Phys. Rev. Lett.* **66**, 41 (1991); *Phys. Rev. B* **44**, 12781 (1991).

[20] M. -Z. Huang and W. Y. Ching, *Phys. Rev. B* **47**, 9449 (1993); **47**, 9464 (1993).

[21] W. Y. Ching and M. -Z. Huang, *Phys. Rev. B* **47**, 9479 (1993).

[22] M. T. Yin and M. L. Cohen, *Phys. Rev. B* **26**, 5668 (1982).

[23] S. Fahy, X. W. Wang, and S. G. Louie, *Phys. Rev. B* **42**, 3503 (1990).

[24] M. S. Hybertsen and S. G. Louie, *Phys. Rev. B* **34**, 5390 (1986).

[25] L. D. Landau and E. Lifshitz, *Quantum Mechanics: Non Relativistic Theory* (Addison-Wesley, Reading, Mass., 1958).

[26] G. D. Mahan, *Many-Particle Physics* (Plenum Press, New York and London, 1981).

[27] B. Honig, J. Jortner and A. Szöke, *J. Chem. Phys.* **46**, 2714 (1967).

[28] R. W. Boyd, *Nonlinear Optics* (Academic, San Diego, 1992).

[29] Y. R. Shen, *The Principles of Nonlinear Optics* (John Wiley & Sons, New York, 1984).

[30] D. C. Hutchings, M. Sheik-Bahae, D. J. Hagan, and E. W. Van Stryland, *Opt. Quantum Electron.* **24**, 1 (1992).

[31] K. -E. Peiponen, *Phys. Rev. B* **37**, 6463 (1988).

[32] See for example the review paper; J. Callaway and N. H. March, *Solid State Physics,* vol. 38, 135 (1984).

[33] See for example the text book; J. C. Inkson, *Many-Body Theory of Solids* (Plenum Press, New York, 1984).

[34] D. Ceperley and B. J. Alder, *Phys. Lett.* **45**, 566 (1980).

[35] J. P. Perdew and A. Zunger, *Phys. Rev. B* **23**, 5048 (1981).

[36] R. W. Godby, M. Schlüter, and L. J. Sham, *Phys. Rev. B* **36**, 6497 (1987); **37**, 10159 (1988).

[37] H. Ishihara and K. Cho, *Phys. Rev. B* **42**, 1724 (1990).

[38] A. F. Starace, *Phys. Rev. A* **3**, 1242 (1971).

[39] F. Bassani, J. J. Forney and A. Quattropani, *Phys. Rev. Lett.* **39**, 1070 (1977).

[40] R. Girlanda, A. Quattropani and P. Schwendimann, *Phys. Rev. B* **24**, 2009 (1981).

[41] A. Pasquarello and A. Quattropani, *Phys. Rev. B* **42**, 9073 (1990).

[42] J. Ihm, A. Zunger, and M. L. Cohen, *J. Phys. C* **12**, 4409 (1979).

[43] G. W. Bachelet, D. R. Hamann, and M. Schlüter, *Phys. Rev. B* **26**, 4199 (1982).

[44] A. J. Read and R. J. Needs, *Phys. Rev. B* **44**, 13071 (1991).

[45] M. S. Hybertsen and S. G. Louie, *Phys. Rev. B* **35**, 5585 (1987).

[46] F. V. Bunkin, *Soviet Phys. JETP* **23**, 1121 (1966).

[47] G. Lehman and M. Taut, *phys. stat. sol. B* **119**, 9 (1983).

[48] M. Sheik-Bahae, D. C. Hutchings, D. J. Hagan, and E. W. Van Stryland, *IEEE J. Quantum Electron.* **QE-27**, 1296 (1991).

[49] C. S. Wang and B. M. Klein, *Phys. Rev. B* **24**, 3417 (1981).

[50] R. Loudon, *The Quantum Theory of Light* (Oxford University Press, 1973).

[51] D. H. Reitze, T. R. Zhang, Wm. M. Wood, and M. C. Downer, *J. Opt. Soc. Am. B* **7**, 84 (1990).

[52] D. E. Aspnes and A. A. Studna, *Phys. Rev. B* **27**, 985 (1983).

[53] M. L. Cohen and J. R. Chelikowsky, *Electronic Structure and Optical Properties of Semiconductors* (Springer-Verlag, Berlin Heidelberg, 1988).

[54] J. A. Bolger, A. K. Kar, B. S. Wherrett, R. DeSalvo, D. C. Hutchings, and D. J. Hagan, *Opt. Commun.* **97**, 203 (1993).

[55] M. Murayama and T. Nakayama, *Phys. Rev. B* **49**, 4710 (1994).

[56] M. D. Dvorak, W. A. Schroeder, D. R. Andersen, A. L. Smirl, and B. S. Wherrett, *IEEE J. Quantum Electron.* **QE-30**, 256 (1994).

[57] R. DeSalvo, M. Sheik-Bahae, A. A. Said, D. J. Hagan, and E. W. Van Stryland, *Opt. Lett.* **18**, 194 (1993).

[58] C. G. Van de Walle and R. M. Martin, *Phys. Rev. B* **34**, 5621 (1986).

[59] M. Murayama and T. Nakayama, unpublished.

[60] For example, E. L. Ivchenko and G. E. Pikus, *Superlattices and Other Heterostructures* (Springer-Verlag, Berlin Heidelberg, 1995).

[61] T. Nakayama, *Physica B* **191**, 16 (1993).

[62] S. B. Zhang, M. S. Hybertsen, M. L. Cohen, S. G. Louie and D. Tománek, *Phys. Rev. Lett.* **63**, 1495 (1989); *Solid State Commun.* **66**, 585 (1988).

[63] For example, G. Shibata, T. Nakayama and H. Kamimura, *Jpn. J. Appl. Phys.* **33**, 6121 (1994).

[64] M. E. Lazzouni and L. J. Sham, *Appl. Phys. Lett.* **63**, 3253 (1993).

[65] E. Ghahramani and J. E. Sipe, *Appl. Phys. Lett.* **62**, 2245 (1993).

[66] E. Ghahramani, D. J. Moss, and J. E. Sipe, *Phys. Rev. B* **43**, 8990 (1991); B**43**, 9269 (1991).

[67] J. F. McGilp, *Condens. Matter* **2**, 7985 (1990).

[68] M. E. Pemble, S. R. Armstrong, S. M. Curry, R. D. Hoare, G. Legothetis, I. M. Povey, A. Stafford, and A. G. Taylor, *Faraday Discuss.* **95**, 199 (1993).

[69] J. E. Sipe, D. J. Moss, and H. M. van Driel, *Phys. Rev. B* **35**, 1129 (1987).

[70] C. Yamada and T. Kimura, *Phys. Rev. B* **49**, 14372 (1994).

[71] O. Hirota, *Squeezed Light* (Elsevier, 1992).

[72] E. Yablonovitch, *J. Phys. Condens. Matter* **5**, 2443 (1993).

[73] H. Haug and S. W. Koch, *Quantum Theory of the Optical and Electronic Properties of Semiconductors* (World Scientific, Hong Kong, 1990).

[74] W. Hanke and L. J. Sham, *Phys. Rev. B* **21**, 4656 (1980); **12**, 4501 (1975).

# *CHAPTER 3*

# WANNIER-STARK LOCALIZATION IN SEMICONDUCTOR SUPERLATTICES

Masaaki Nakayama
*Department of Applied Physics, Osaka City University*

## 3.1. INTRODUCTION

In solid state physics, the effects of a uniform electric field on Bloch electrons in a periodic potential of a crystal have been a topic of interest and controversy over the past 60 years. In the history of the investigation of the electric field effects, "Bloch oscillations" and "Wannier-Stark (WS) localization" have been especially controversial and attractive. The Bloch oscillation conceptually corresponds to the following phenomenon. In the absence of scattering processes, the wave vector of a Bloch electron varies linearly in time at a rate proportional to a uniform electric field, $F$, which is the so-called acceleration theorem. When the wave vector reaches at the Brillouin-zone boundary, $\pi/a$, along the direction of the electric field, the Bloch electron undergoes the Bragg reflection resulting in the transform of its wave vector from $\pi/a$ to $-\pi/a$ as a consequence of the crystal periodicity, where $a$ is the lattice constant of a crystal. Thus, the Bloch electron performs oscillations with a time constant of $\tau_B = h/eFa$ in the momentum and real space. This is called the Bloch oscillation originally discussed by Bloch [1] and Zener [2] almost 60 years ago.

The Bloch oscillation in the time domain translates itself into the frequency (energy) domain as a spectrum of equally spaced energy levels, $h/\tau_B = eFa$, which is called a Stark ladder. The concept of the Stark ladder was extensively discussed by Wannier in 1960 [3] by developing an effective Hamiltonian for one-band Bloch electrons in a uniform electric field. It is noted that James [4], Katsura, Hatta and Morita [5], and Feuer [6] considered the solution of the Schrödinger equation in a periodic potential perturbed by a uniform electric field before the discussion by Wannier [3]. The formation of the Stark-ladder states results from the split of a continuous energy band. From the viewpoint of the spatial properties of the eigenfunctions, the Stark-ladder formation corresponds to the electric-field-induced localization, which has been traditionally called the WS localization. The existence of the WS localization has been continuously discussed in theory [3-18]. The controversial points were whether the Stark ladders exist in an infinite system, and whether they survive with the inclusion of interband interactions. The details in theory have been nicely reviewed by Nenciu [17]. Until now, there has been no clear experimental evidence for the WS localization and Bloch oscillation in bulk crystals. The necessary condition for the observation of the WS localization (Bloch oscillation) is that the coherent length (scattering time) of the state is longer than the localization length $L$ (Bloch-oscillation time $\tau_B$). Since the electronic band width and the lattice constant in bulk semiconductors are of the order of eV and Å, respectively, an electric field required for the WS localization is estimated to exceed a breakdown field of a crystal.

In 1970, Esaki and Tsu [19] proposed an artificial superlattice (SL) consisting of alternating nano-meter-thick layers of semiconductors in order to obtain the Bloch oscillation leading to a THz-oscillator device emitting a coherent electromagnetic radiation. In a SL with its period $D$ longer than the lattice constants of constituents, miniband structures are formed in the mini-Brillouin zone along the growth direction, $-\pi/D \le k_z \le \pi/D$. This is easily recognized in the framework of a well-known Kronig-Penney model for a one-dimensional periodic potential [20]. From the sense of quantum mechanics, it is evident that the miniband width is tailored from zero to a few 100 meV by choosing the individual layer thicknesses and the SL-potential height in the crystal growth of the SL. The proposal of the SL introduced the concept of the design of electronic and optical properties of semiconductors based on quantum mechanics. This opened a novel research field of physics and engineering of semiconductors, and research works in this field have been performed extensively and growing up [21].

For the WS localization and Bloch oscillation in SL's, we reasonably expect that the electric field required for them is much lower than that in a bulk semiconductor because of the much narrower band width and much longer period; however, a long silent period had passed since the proposal of the SL in 1970. In 1988, Mendez *et al.* [22] and Voisin *et al.* [23] reported the first clear experimental evidence for the existence of the WS localization in GaAs/Al$_x$Ga$_{1-x}$As SL's. They observed interband optical transitions with the nature of Stark ladders, $E_m = E_0 + meFD \quad (m = 0, \pm 1, \pm 2, \ldots)$, and the photocurrent spectra in [22] clearly showed the progress of the localization as a function of an applied electric field. Since the first observation in 1988, the WS localization in SL's has been attracting considerable interest in fundamental physics and device applications. It is noted that the controversial points for the existence of the WS localization are settled in the SL because of the finiteness of real samples: The finiteness effect was clearly discussed by Gülseren and Ciraci [18]. The WS localization provides us with a novel concept of the dimensionality crossover of electronic systems controlled by an applied electric field: the transformation from miniband states (3D) to Stark-ladder states (quasi-2D) and vice verse. Furthermore, Schneider *et al.* [24] and Nakayama *et al.* [25,26] demonstrated that Stark-ladder states of a miniband interact with those of a different miniband. If the energy difference between the different Stark-ladder states with the quantum numbers $n=i$ and $j$, $\Delta E_{ij}$, is equal to $meFD$, the $n=i$ Stark-ladder state in a quantum well (QW) resonantly couples with the $n=j$ state in the $m$th nearest neighbor QW. The resonant coupling causes the anticrossing of the related Stark-ladder levels and the wave-function delocalization over the coupling space: This is connected with quantum coherence. For device applications, electro-optic modulators have been considerably studied because of the wide tunability of interband transition energies in the WS localization. Bar-Joseph *et al.* [27] demonstrated for the first time an optical bistability operation in a self-electro-optic effect device at room temperature.

For the observation of Bloch oscillations in SL's, in 1992 Plessen and Thomas [28] theoretically suggested an experimental method by measuring spontaneous photon-echo signals in a time-resolved degenerate-four-wave-mixing experiment. Feldmann *et al.* [29] and Leo *et al.* [30] reported the observation of oscillations of photon-echo signals in a ps-time region. In 1993 Waschke *et al.* [31] reported the direct observation of coherent THz electromagnetic radiation after ultrashort laser-pulse (~100 fs) excitations. In those experiments [29–31], the frequencies of the photon-echo oscillation and electromagnetic radiation vary linearly as a function of an applied electric field. This suggests the occurrence of the Bloch oscillation.

The aim of this chapter is to review the WS localization in SL's from the viewpoint of optical properties. The recent studies of transport properties have been nicely reviewed by Tsu and Esaki [32], Tsu [33], and Beltram and Cappaso [34]. In this chapter, at first, the basic electronic properties of SL's (miniband structures) based on an envelope-function approximation are introduced in Sec. 3.2. The concepts and simple theoretical models of the WS localization are described in Sec. 3.3. In Sec. 3.4 the details of the Stark-ladder states are discussed on the basis of spectroscopic results of interband transitions and the simple theoretical models described in Sec. 3.3. We especially focus on the transformation process from miniband states to Stark-ladder states. In addition, the electro-optical bistability based on the WS localization is described. Section 3.5 covers the resonant coupling between Stark-ladder states. In the above sections, energy states below a barrier potential, so-called below-barrier states, are the subject of the discussion. Section 3.6 deals with electric-field effects on energy states above a barrier potential, above-barrier states. Section 3.7 presents the highlight of the Bloch oscillation corresponding to the dynamical aspect of the WS localization. Finally, the summary of this chapter is described in Sec. 3.8.

## 3.2. MINIBAND STRUCTURES

### 3.2.1. *Band Discontinuity*

In a heterostructure system consisting of two semiconductors $A$ and $B$, one of the most important factors for the electronic and optical properties is the band discontinuity due to a band-gap-energy difference of $\Delta E_g$. The band discontinuity determines QW and SL potentials.

Figure 3.1(a) shows the band structure around $\Gamma$ point in a zincblende semiconductor such as GaAs. The base function of the $\Gamma_6$ conduction band is $s$-like, while those of the $\Gamma_7$ and $\Gamma_8$ valence bands are $p$-like. The $\Gamma_7$–$\Gamma_8$ split results from a spin-orbit interaction. The $\Gamma_8$ valence band is in fourfold degeneracy: The $|J, M_J\rangle$ representation is $|3/2, \pm 3/2\rangle$ and $|3/2, \pm 1/2\rangle$ which are called heavy-hole and light-hole bands, respectively. The split-off $\Gamma_7$ valence band with twofold degeneracy is represented by $|1/2, \pm 1/2\rangle$. Hereafter, we will neglect the split-off $\Gamma_7$ band. We set that the band-offset energy of the $\Gamma_8$ valence ($\Gamma_6$ conduction) band edge in an $A/B$ heterostructure is $\Delta E_v$ ($\Delta E_c$): $\Delta E_g = \Delta E_v + \Delta E_c$. Figures 3.1(b)-(d) show three types of the band discontinuity. Figure 3.1(b) depicts a so-called type-I band discontinuity, which is the most typical: A GaAs/Al$_x$Ga$_{1-x}$As heterostructure [$E_g$(GaAs) $<$ $E_g$(Al$_x$Ga$_{1-x}$As)] is a famous example, and the ratio of the band-offset energies, $\Delta E_v : \Delta E_c$, is now well known to be 0.34 : 0.66 in all the Al concentration range. The band discontinuities in Figs. 3.1.(c) and 3.1.(d) are called type-II. In the history of heterostructures, Fig. 3.1(d) was first called a type-II discontinuity corresponding to the case that the valence band of one semiconductor overlaps with the conduction band of the other: *e.g.*, a GaSb/InAs system [35]. Recently, the type-II system has been also taken to mean a staggered discontinuity shown in Fig. 3.1(c): *e.g.*, a ZnS/ZnSe system. It is noted that the band discontinuity is modified by the lattice-misfit strain of a heterostructure which shifts the energies of the conduction- and valence-band edges and

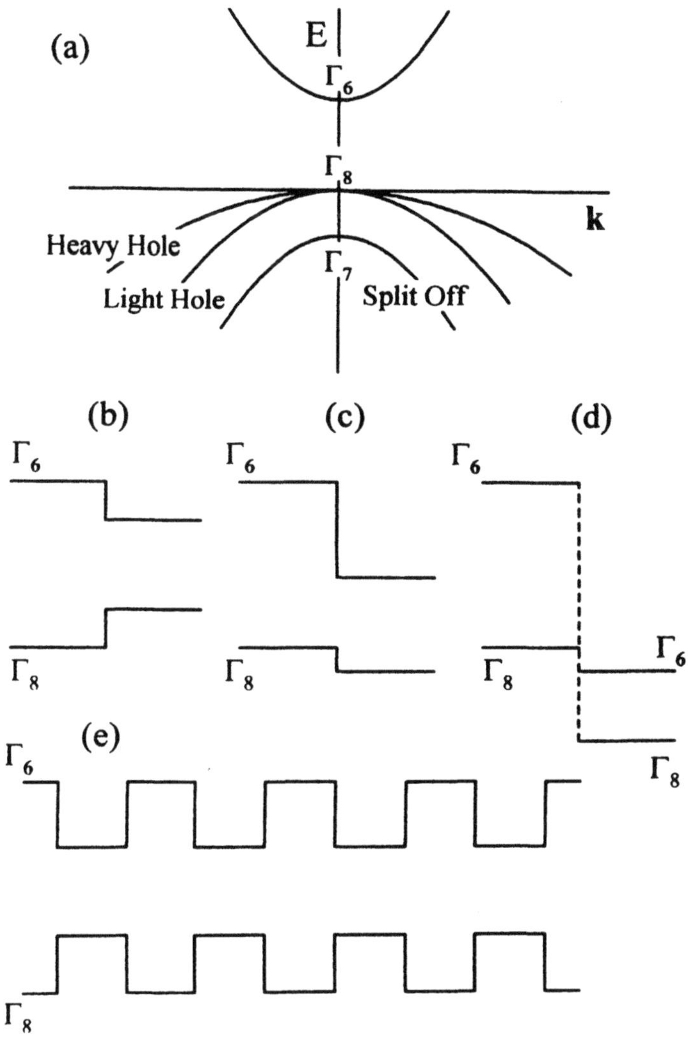


Figure 3.1. (a) Band structure around Γ point in a zincblende semiconductor such as GaAs, (b) type-I band discontinuity, (c) and (d) type-II discontinuity, and (e) a potential structure of the $\Gamma_8$ and $\Gamma_6$ bands in an $A/B$ SL with the type-I discontinuity.

lifts the degeneracy of the heavy-hole and light-hole bands at $\Gamma$ point. Figure 3.1(e) shows schematically the potential structures of the $\Gamma_8$ and $\Gamma_6$ bands in an $A/B$ SL with the type-I discontinuity.

Although the band discontinuity was controversial in the early period of the history of heterostructures, recent theoretical works of first principle calculations have revealed the essence of contributions to the band discontinuity [36–40]. Van de Walle [40] reported the absolute energy levels of the band edges for several important semiconductors and the deformation potentials that describe the strain effects on the electronic bands. Utilizing the reported values (Table II in [40]), we can easily estimate the band discontinuity for several heterostructures including strained-layer systems.

### 3.2.2. *Envelope-Function Approximation*

In a given $A/B$ SL there is a new periodicity along the growth direction $z$. The SL period $D$ (sum of the layer thicknesses of the $A$ and $B$ layers, $d_A + d_B$) is longer than the lattice constants of the constituents, so that a new mini-Brillouin zone is formed in the wave-vector range of $-\pi/D \leq k_z \leq \pi/D$. The mini-Brillouin-zone formation results in generation of miniband structures in the conduction and valence bands. Although it is of course that a first-principles approach is the best way to calculate the band structures, it requires large-scale computation. An envelope-function approximation, which corresponds to an effective-mass approximation, is the most convenient framework, and the effectiveness has been verified by a lot of previously works. The envelope-function approximation was clearly discussed for the first time by Bastard [41].

We consider the miniband structures of the $\Gamma_6$ conduction band and the $\Gamma_8$ valence band (heavy hole, $|J, M_J\rangle = |3/2, \pm 3/2\rangle$ and light hole, $|3/2, \pm 1/2\rangle$) in the framework of the envelope-function approximation [41]. The total wave function in each layer can be written in the form

$$\Psi_{\mathbf{k}}^{(A,B)}(\mathbf{r}) = \sum_{\mu} F_{\mu,\mathbf{k}}^{(A,B)}(\mathbf{r})\, u_{\mu,\mathbf{k}=0}^{(A,B)}(\mathbf{r}), \tag{3.1}$$

where $\mu$ is the band index, $u_{\mu,\mathbf{k}=0}^{(A,B)}(\mathbf{r})$ is the Bloch function at $\Gamma$ point ($\mathbf{k} = 0$) in semiconductors $A$ or $B$, and $F_{\mu,\mathbf{k}}^{(A,B)}(\mathbf{r})$ is the envelope function which slowly varies over the unit cells. The SL potential depends only on the growth direction, $z$, so that the original symmetry survives in the $x$ and $y$ directions: The in-plane wave vectors, $k_x$ and $k_y$, in bulk crystals are good quantum numbers. For that reason, we can use the following form of the envelope function:

$$F_{\mu,\mathbf{k}}^{(A,B)}(\mathbf{r}) = \exp\left(ik_{\mu,x}^{(A,B)}x\right) \exp\left(ik_{\mu,y}^{(A,B)}y\right) f_{\mu,q}^{(A,B)}(z), \tag{3.2}$$

where $f_{\mu,q}^{(A,B)}(z)$ is the envelope function determined by the SL potential and the effective masses, and $q$ is the SL wave vector parallel to $z$. The periodicity of the SL results in the Bloch theorem for the envelope function:

$$f_{\mu,q}^{(A,B)}(z + mD) = \exp(iqmD)\, f_{\mu,q}^{(A,B)}(z), \tag{3.3}$$

where $m$ is integer. According to the Kane model for the **kp** perturbation method [42], the conduction, heavy-hole, and light-hole bands couple each other. In the case of $k_x$, $k_y$=0, the heavy-hole band is completely decoupled from the other two bands. The coupling between the conduction and light-hole bands produces the nonparabolicity of the effective masses. For simplicity, we neglect the band coupling in $k_x$, $k_y$=0, which is a good approximation for GaAs/$Al_xGa_{1-x}As$ SL's [41]. In the case of $k_x, k_y \neq 0$, the dispersion relations of the heavy-hole and light-hole minibands along $k_x$ and $k_y$ are complicated because of the band mixing. The band mixing is described by the Luttinger Hamiltonian [43].

The envelope function of an $A/B$ SL is the solution of the following Schrödinger equation,

$$\left[-\frac{\hbar^2}{2m_\mu(z)}\frac{d^2}{dz^2} + V_\mu(z)\right] f_{\mu,q}(z) = E\, f_{\mu,q}(z), \tag{3.4}$$

where $m_\mu(z)$ is the $\mu$-band effective mass of the SL, and $V_\mu(z)$ is the SL potential determined by the band discontinuity: $m_\mu(z) = m_{\mu,A}(m_{\mu,B})$ and $V_\mu(z) = 0(\Delta E_\mu)$ in the $A(B)$ layer. For convenience we will omit the band index $\mu$ in the following treatment. The form of the envelope function can be written by

$$f_q(z) = \begin{cases} f_q^A(z) = \alpha_q^A \exp(ik_A z) + \beta_q^A \exp(-ik_A z), & \text{in the A layer,} \\ f_q^B(z) = \alpha_q^B \exp(ik_B z) + \beta_q^B \exp(-ik_B z), & \text{in the B layer,} \end{cases} \tag{3.5}$$

$$k_A = \sqrt{2m_A E}/\hbar, \quad k_B = \sqrt{2m_B(E - \Delta E)}/\hbar. \tag{3.6}$$

Considering the continuity of the probability density, $|f_q(z)|^2$, for a stationary state, we obtain the following boundary condition of the envelope function at the $A/B$ interface for $(z_A - z_B) \to 0$:

$$f_q^A(z_A) = f_q^B(z_B) \quad \text{and} \quad \frac{1}{m_A}\frac{df_q^A(z_A)}{dz} = \frac{1}{m_B}\frac{df_q^B(z_B)}{dz}. \tag{3.7}$$

Finally, the miniband dispersion relation, which is similar to the well-known Kronig-Penney relation, is given by

$$\cos(qD) = \cos(k_A d_A)\cos(k_B d_B) - \frac{1}{2}\left(\frac{k_A m_B}{k_B m_A} + \frac{k_B m_A}{k_A m_B}\right)\sin(k_A d_A)\sin(k_B d_B). \tag{3.8}$$

This equation has been widely used to calculate the miniband energies.

Figure 3.2 shows (a) the potential structure and the envelope-function profiles of the electron and heavy hole in a GaAs(3.2 nm)/AlAs(0.9 nm) SL which is a sample used for the discussion in this chapter, and (b) the dispersion relations of

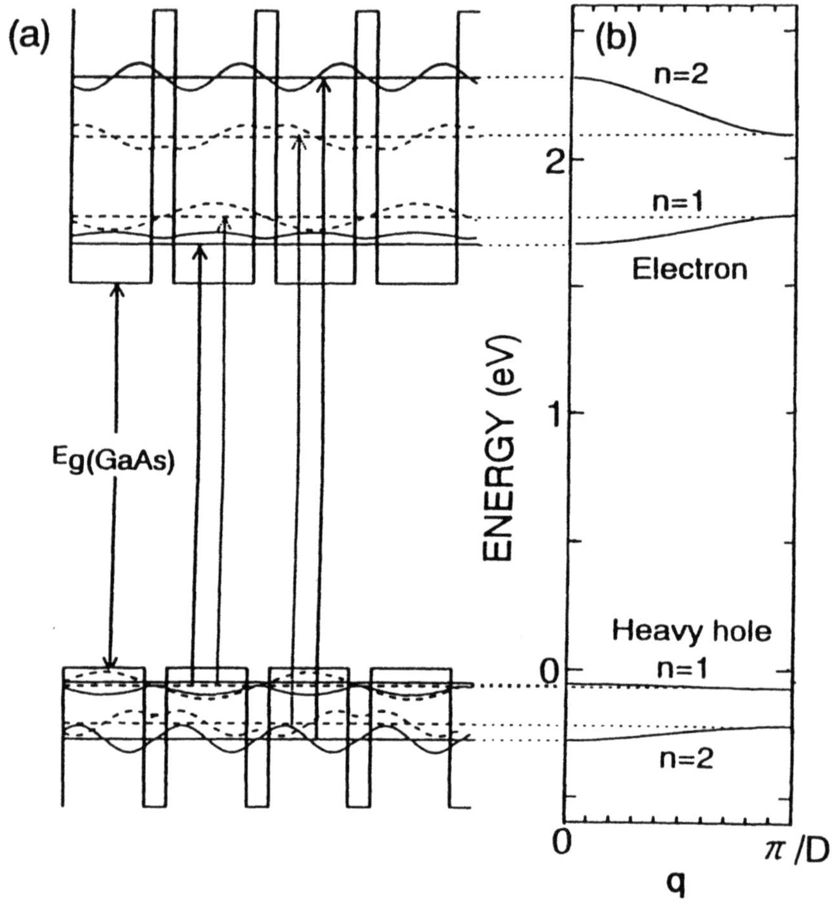


Figure 3.2. (a) Potential structure and the envelope-function profiles of the electron and heavy hole in a GaAs(3.2 nm)/AlAs(0.9 nm) SL which is a sample used for the discussion in this chapter, and (b) the dispersion relations of the electron and heavy-hole minibands calculated using Eq. (3.8). The solid and dashed lines in (a) indicate the envelope functions at the mini-Brillouin-zone center ($q = 0$, $\Gamma$ point) and the mini-Brillouin-zone edge ($q = \pi/D$, $\pi$ point), respectively.

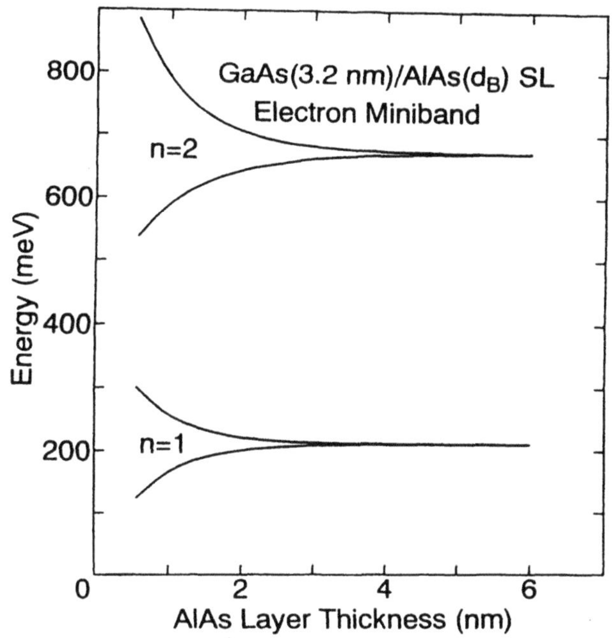


Figure 3.3. Electron-miniband energies at Γ and $\pi$ points in a GaAs(3.2 nm)/AlAs($d_B$ nm) SL as a function of the AlAs-layer thickness.

the electron and heavy-hole minibands calculated using Eq. (3.8). In Fig. 3.2(a) the solid and dashed lines indicate the envelope functions at the mini-Brillouin-zone center ($q = 0$, Γ point) and the mini-Brillouin-zone edge ($q = \pi/D$, $\pi$ point), respectively. In this calculation, the following effective-mass parameters are used [44]: $m_e$=0.0665$m_0$, $m_{HH}$=0.34$m_0$, and $m_{LH}$=0.117$m_0$ for the electron, heavy hole, and light hole of GaAs, and $m_e$=0.15$m_0$, $m_{HH}$=0.4$m_0$, and $m_{LH}$=0.18$m_0$ for AlAs, where $m_0$ is the free-electron mass. The miniband dispersions are tailored by changing the layer thicknesses, potential height, and effective masses which are structural parameters in crystal growth. Figure 3.3 shows the $n$=1 and $n$=2 electron-miniband energies at Γ and $\pi$ points in a GaAs(3.2 nm)/AlAs($d_B$ nm) SL as a function of the AlAs-layer thickness. This figure exhibits a typical example of the miniband tailoring. The miniband width increases with the decrease of the AlAs (barrier) thickness, which results from the increase of the coupling strength of wave functions because of the increase of the tunneling probability. In the case of the disappearance of the miniband width, the electronic state corresponds to a quasi-2D state in an isolated QW.

Next we consider interband optical transitions in the framework of the envelope-function approximation. We define the envelope function of the SL as $f_{\mu,q,n}(z)$, where $n$ is the miniband index corresponding to the quantum number. Assuming that the

Bloch function of semiconductor $A$ is similar to that of semiconductor $B$, $u^A_{\mu,\mathbf{k}=0}(\mathbf{r}) \simeq u^B_{\mu,\mathbf{k}=0}(\mathbf{r}) \rightarrow u_{\mu,\mathbf{k}=0}(\mathbf{r})$, the interband-transition matrix element can be written by

$$M_{c,v} \propto \left\langle u_{c,\mathbf{k}=0} \middle| \mathbf{p} \middle| u_{v,\mathbf{k}=0} \right\rangle \left\langle f_{c,q,n} \middle| f_{v,q,n'} \right\rangle, \tag{3.9}$$

where $\mathbf{p}$ is a momentum operator, and subscripts $c$ and $v$ indicates the conduction and valence bands, respectively. The oscillator strength, $|M_{c,v}|^2$, is determined by the two factors of the Bloch-function term and the overlap between the envelope functions. From the symmetry of the envelope functions as shown in Fig. 3.2(a), we obtain the selection rule of the optical transition for the quantum number: It is allowed in $n - n' = even$, while forbidden in $n - n' = odd$. The value of the overlap integral is almost unity in $n = n'$, so that we mainly observe such a transition. From the Bloch function form [42], the ratio of the Bloch term of the heavy-hole transition to the light-hole one in the light propagation parallel to $z = [001]$ is given by

$$\left|\left\langle u_{c,\mathbf{k}=0} \middle| \mathbf{p} \middle| u_{HH,\mathbf{k}=0} \right\rangle\right|^2 : \left|\left\langle u_{c,\mathbf{k}=0} \middle| \mathbf{p} \middle| u_{LH,\mathbf{k}=0} \right\rangle\right|^2 = 3 : 1. \tag{3.10}$$

The oscillator strength of the heavy-hole transition in $n = n'$ is, therefore, three times larger than that of the light-hole transition: The heavy-hole transitions are dominant in optical measurements of SL samples grown on a (001) substrate which is mainly used in the crystal growth.

## 3.3. THEORETICAL MODELS OF WANNIER-STARK LOCALIZATION

### 3.3.1. *Nearest-Neighbor Tight-Binding Model*

We consider the concepts of the WS localization within a one-band nearest-neighbor tight-binding framework for a one-dimensional crystal proposed by Feuer in 1952 [6]. The same framework was applied to calculate absorption spectra of an SL in the WS localization by Bleuse, Bastard, and Voisin in 1988 [16].

We assume an SL consisting of $(2N + 1)$ QW's with the periodicity of $D$. In this framework the SL wave function for the Schrödinger equation similar to Eq. (3.4) is written by

$$\begin{aligned} \phi_q(z) &= (2N+1)^{-1/2} \sum_{m=-N}^{+N} \exp(iqmD)\, f(z - mD), \\ &\quad \text{with} \quad q = \pi j/ND, \quad j = 0, \pm 1, \pm 2, \ldots, \pm N, \end{aligned} \tag{3.11}$$

where $f(z - mD)$ is the envelope function for an isolated QW state. The eigenenergies at zero electric field are given by

$$\epsilon(q) = \epsilon_0 - \frac{\Delta}{2} \cos(qD), \tag{3.12}$$

where $\epsilon_0$ is the energy of the isolated QW state, and $\Delta$ corresponds to the miniband width which is given by four times the nearest-neighbor interaction energy. It is noted that we neglect an on-site interaction energy which is usually much smaller than $\Delta$. Equation (3.12) gives a quasicontinuous energy spectrum corresponding to a miniband dispersion relation similar to that given by Eq. (3.8).

In the presence of a uniform electric field, $F$, parallel to the SL-growth direction $z$, the potential term of the Schrödinger equation is the combination of the SL potential and the electrostatic potential: $V(z) + eFz$. It is well known that the wave function is generally represented by Airy functions. The details within the framework of the envelope-function approximation will be described in the next subsection. In a high field regime in which the coupling between QW's is broken, the miniband dispersion [Eq. (3.12)] is replaced by the discrete energy levels with an equal energy spacing of $eFD$, Stark ladders:

$$\epsilon_\nu = \epsilon_0 + \nu eFD \quad \text{with} \quad \nu = 0, \pm 1, \pm 2, \ldots, \pm N, \tag{3.13}$$

where $\nu$ is the so-called Stark-ladder index. We note that the SL wave vector $q$ is no longer a good quantum number because $q$ is unstationary due to the acceleration of electrons and holes induced by the electrostatic potential; therefore, we adopt the Stark-ladder index in real space as a quantum number instead of $q$. The present framework gives the following wave-function form for the $\nu$th Stark-ladder state:

$$\phi_\nu(z) = \sum_{m=-N}^{+N} J_{m-\nu}(L/D)\, f(z - mD) \quad \text{with} \quad L = \Delta/2eF, \tag{3.14}$$

where $L$ corresponds to the order of the localization length of the wave function, and $J_{m-\nu}$ is the Bessel function of an integer index $(m - \nu)$. Figure 3.4 depicts schematically the concept of the localization length which is very important to consider the WS localization, where $\Delta$ is the miniband width, and the horizontal lines with the energy spacing $eFD$ indicates the Stark-ladder states. The localization length corresponds to an effective length of the miniband in real space under the electrostatic potential $eFz$, and the definition of $L = \Delta/2eF$ is obtained from a simple geometrical analysis of the potential slope and the miniband width. Figure 3.5 shows the calculated values of $|J_m(L/D)|^2$ as a function of $D/L$ which is proportional to $F$. The value of $|J_m(L/D)|^2$ corresponds to the probability density in the $m$th nearest-neighbor QW. The case of $J_0^2 = 1$ corresponds to the complete localization. It is noted that the localization is not gradually enhanced with the increase of electric field, but exhibits oscillatory profiles.

Figure 3.6 shows the schematic diagram of the Stark-ladder states and the probability density of the localized wave functions, where the interband transitions, which are called Stark-ladder transitions, in a QW are represented by the vertical arrows. It is of course that the same transitions occur in all QW's of the SL. If the localization length of the electron wave function is $pD$, the energies of the observable Stark-ladder transitions are written by

$$E_m = E_0 + meFD \quad \text{with} \quad m = 0, \pm 1, \ldots, \pm p, \tag{3.15}$$

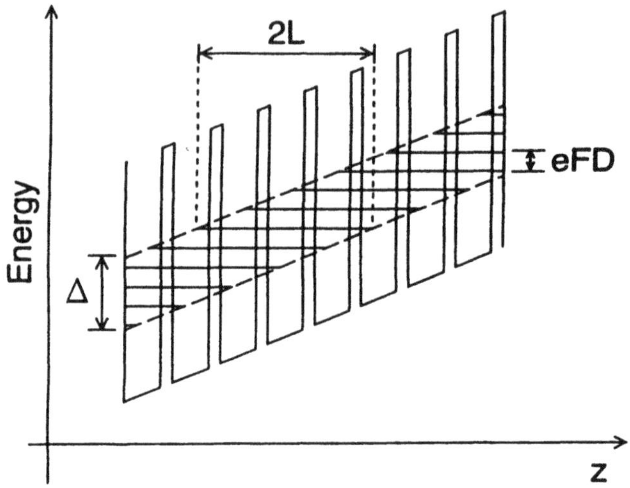


Figure 3.4. Schematic diagram of the concept of the localization length $L$ which is very important to consider the WS localization, where $\Delta$ is the miniband width, the horizontal lines with the energy spacing $eFD$ indicate the Stark-ladder states, and $2L = \Delta/eF$.

where $E_0$ is the interband transition energy of an isolated QW. The energy spectrum of the Stark-ladder transitions as a function of electric field exhibits a fan diagram, which is a direct evidence for the existence of the Stark-ladder states from the viewpoint of spectroscopy. Within the framework of this tight-binding model, the envelope-function term of the interband-transition matrix element [Eq. (3.9)] for the $m$th Stark-ladder transition is represented by

$$\sum_{\nu=-N}^{+N} J_{m-\nu}(\Delta_e/2eFD)J_\nu(\Delta_h/2eFD) = J_m((\Delta_e + \Delta_h)/2eFD), \tag{3.16}$$

where $\Delta_e$ ($\Delta_h$) is the electron (hole) miniband width. Assuming that the Bloch-function term is not changed by the electric field, the oscillator strength of the $m$th Stark-ladder transition is

$$I_m \propto \left|J_m((\Delta_e + \Delta_h)/2eFD)\right|^2. \tag{3.17}$$

This indicates that the oscillator strength of the Stark-ladder transition depends on $eFD$ and the sum of the electron- and hole-miniband widths and that the electric-field dependence is similar to the profiles of $|J_m(L/D)|^2$ shown in Fig. 3.5.

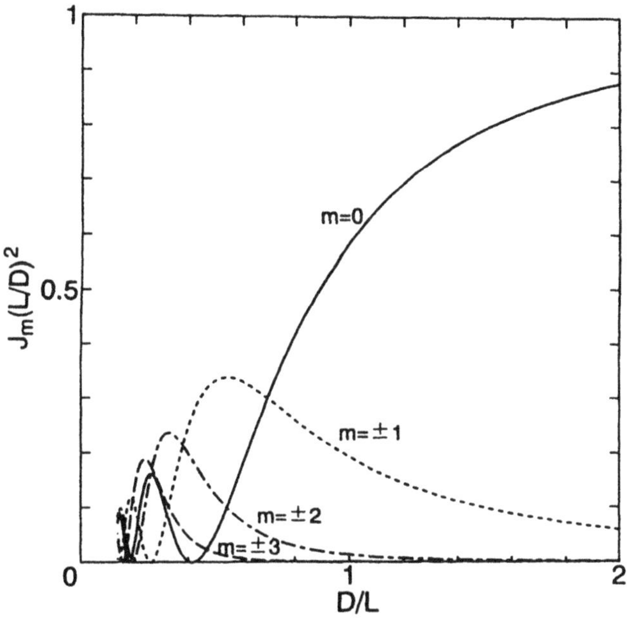


Figure 3.5. Calculated values of $|J_m(L/D)|^2$ as a function of $D/L$ which is proportional to $F$. The value of $|J_m(L/D)|^2$ corresponds to the probability density in the $m$th nearest-neighbor QW.

### 3.3.2. *Transfer-Matrix Method*

The above-described tight-binding model is not useful to calculate precisely the eigenstates of an SL as a function of electric field. Here, we consider a transfer-matrix method within the framework of the envelope-function approximation. As described in the above section, in the presence of an electrostatic potential, the SL wave vector $q$ is no longer a good quantum number; therefore, we adopt a layer-position index $j$, which is analogous to the Stark-ladder index $\nu$, as a quantum number. The Schrödinger equation of the envelope function [Eq. (3.4)] without the band index $\mu$ is written by

$$\left[-\frac{\hbar^2}{2m(z)}\frac{d^2}{dz^2} + V(z) + qFz\right] f_j(z) = E\, f_j(z), \tag{3.18}$$

where $q = -e(+e)$ for electrons (holes). We transform the SL coordinate $z$ to the following dimensionless coordinate $Z_j$:

$$Z_j = -\left[2m_j/(e\hbar F)^2\right]^{1/3} (E - V_j + qFz), \tag{3.19}$$

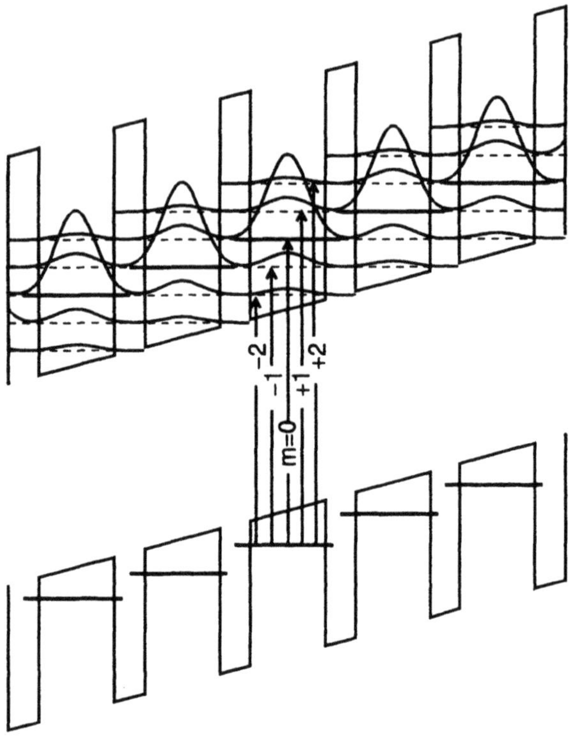


Figure 3.6. Schematic diagram of the Stark-ladder states and the probability density of the localized wave functions, where the interband transitions, which are called Stark-ladder transitions, in a QW are represented by the vertical arrows.

where $m_j$ and $V_j$ are the effective mass and the potential in the $j$th layer. From this transformation, Eq. (3.18) becomes

$$\frac{d^2}{dZ_j^2} f_j(Z_j) - Z_j f_j(Z_j) = 0. \tag{3.20}$$

The general solution of this differential equation is given by [45]

$$f_j(Z_j) = a_j \mathrm{Ai}(Z_j) + b_j \mathrm{Bi}(Z_j), \tag{3.21}$$

where Ai and Bi are the Airy functions.

From the matching condition of the envelope function at every interface [Eq. (3.7)], we obtain the following transfer-matrix relation at the interface between the $j$th and

$(j+1)$th layer:

$$\begin{pmatrix} \mathrm{Ai}(Z_j) & \mathrm{Bi}(Z_j) \\ m_j^{-2/3}\mathrm{Ai}'(Z_j) & m_j^{-2/3}\mathrm{Bi}'(Z_j) \end{pmatrix} \begin{pmatrix} a_j \\ b_j \end{pmatrix} = \begin{pmatrix} \mathrm{Ai}(Z_{j+1}) & \mathrm{Bi}(Z_{j+1}) \\ m_{j+1}^{-2/3}\mathrm{Ai}'(Z_{j+1}) & m_{j+1}^{-2/3}\mathrm{Bi}'(Z_{j+1}) \end{pmatrix} \begin{pmatrix} a_{j+1} \\ b_{j+1} \end{pmatrix}. \tag{3.22}$$

From this transfer-matrix relation, we can obtain the relative wave-function amplitudes in every layer. Here we postulate that a modeled SL with a finite number of periods is sandwiched by barrier layers with a semi-infinite thickness. Assuming that the boundary condition in the final barrier layer is $(a_f, b_f) = (1, 0)$ because of $\mathrm{Ai}(\infty) \to 0$ and $\mathrm{Bi}(\infty) \to \infty$, we can easily calculate the energy spectrum of the transmittance of $T = 1/(a_i^2 + b_i^2)$, where $(a_i, b_i)$ is the wave-function amplitudes in the initial barrier layer. Huchings [46] suggested that the eigenenergies correspond to the energies at the transmittance maxima in a double QW structure. Nakayama *et al.* [25,26] demonstrated that eigenenergies calculated by this transfer-matrix method well explain the observed energies in real SL's in a wide range of electric field.

In a very low electric field and zero field, we cannot apply the above transfer-matrix formulation because of the divergence of the Airy function Bi. This condition is important when we consider the transformation process from miniband states (zero and low field regime) to Stark-ladder states (high field regime). In this case, we use the following approximation: The electrostatic potential $eFz$ is replaced by a step-like potential whose value is the average of the real potential in every layer (see the inset of Fig. 3.9). In the step-like-potential approximation, the eigenfunctions are represented by the linear combination of plane waves in every layer as described by Eq. (3.5). The transfer-matrix relation is given by

$$\begin{pmatrix} \exp(ik_jz) & \exp(-ik_jz) \\ (ik_j/m_j)\exp(ik_jz) & -(ik_j/m_j)\exp(-ik_jz) \end{pmatrix} \begin{pmatrix} a_j \\ b_j \end{pmatrix} = \begin{pmatrix} \exp(ik_{j+1}z) & \exp(-ik_{j+1}z) \\ (ik_{j+1}/m_{j+1})\exp(ik_{j+1}z) & -(ik_{j+1}/m_{j+1})\exp(-ik_{j+1}z) \end{pmatrix} \begin{pmatrix} a_{j+1} \\ b_{j+1} \end{pmatrix} \tag{3.23}$$

This approximation of the step-like potential is valid for SL's with thin layers in which the change of the electric field is small. The two types of the transfer-matrix method will be applied to analyze experimental results of the WS localization in the following sections.

## 3.4. STARK-LADDER STATES

### 3.4.1. *Sample Structures and Spectroscopic Methods*

Before discussing Stark-ladder states in the WS localization, one explains a typical sample structure and spectroscopic methods to detect interband Stark-ladder transitions. The samples, of which optical spectra will be mainly discussed in this review,

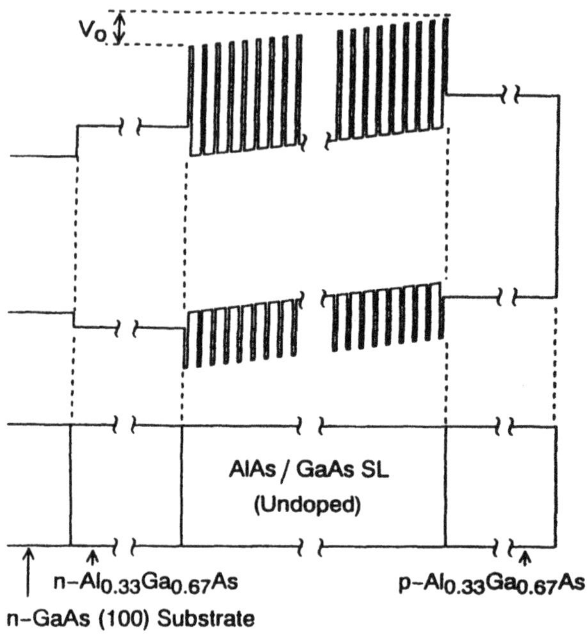


Figure 3.7. Schematic diagram of the sample structure including the potential profile: $p$-$Al_xGa_{1-x}As$/$i$-GaAs/AlAs SL/$n$-$Al_xGa_{1-x}As$. Most samples used in studies of the WS localization have been a similar $p$-$i$-$n$ diode structure.

are an undoped GaAs(3.2 nm)/AlAs(0.9 nm) SL with 80 periods and an undoped GaAs(6.4 nm)/AlAs(0.9 nm) SL with 20 periods grown on an $n$-type (Si-doped) (001)-GaAs substrate by molecular-beam epitaxy (MBE). The SL is sandwiched by $p$- and $n$-$Al_{0.33}Ga_{0.67}As$ layers with the thickness of the order of sub-$\mu$m, so that the total sample structure is a $p$-$i$-$n$ diode. Figure 3.7 shows the schematic diagram of the sample structure including the potential profile: $p$-$Al_xGa_{1-x}As$/$i$-GaAs/AlAs SL/$n$-$Al_xGa_{1-x}As$. Most samples used in studies of the WS localization have been a similar $p$-$i$-$n$ diode structure because the electric field in the SL layer is easily controlled by an applied bias $V_b$. The electric field is given by

$$F = (V_0 - V_b)/L_{SL} \tag{3.24}$$

where $V_0$ is a built-in voltage due to the $p$-$n$ junction, and $L_{SL}$ is the total thickness of the SL layer. In order to observe optical spectra, a semi-transparent contact or a windowed ohmic contact is formed over the sample surface.

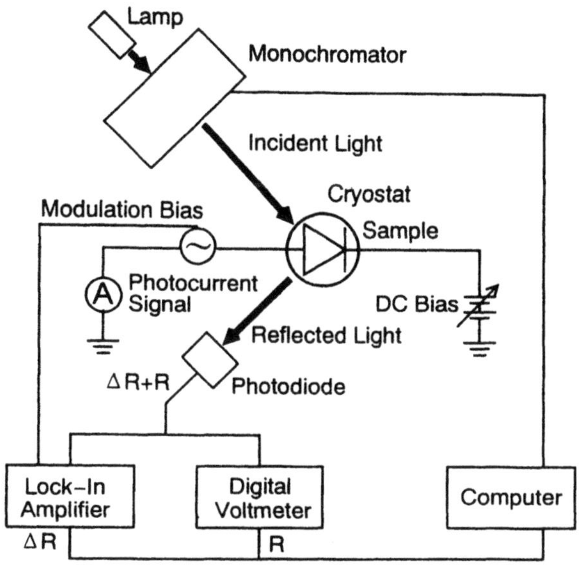


Figure 3.8. Schematic diagram of the measurement system of ER spectroscopy.

Photocurrent (PC) spectroscopy is the most convenient and major method to detect interband Stark-ladder transitions in the *p-i-n* diode structure because the sample itself is a photodiode. Although it is usually accepted that the PC spectrum is similar to an absorption spectrum, the PC properties are affected by dissociation processes from excitons to carriers and by the conductivity of carriers. It is well known that reflectance spectroscopy is a direct method to detect optical transitions. Nakayama *et al.* [25,47] clearly demonstrated that electroreflectance (ER) spectroscopy, which corresponds to an electric-field-modulation spectroscopy of reflectance [48], is much more sensitive to detect Stark-ladder transitions than PC spectroscopy. Figure 3.8 shows the schematic diagram of the measurement system of ER spectroscopy: We can simultaneously observe ER and PC spectra in such a system.

In ER measurements, the applied bias is weakly modulated around a dc bias which determines an applied electric field, and the reflected light synchronizing with the field modulation is detected with a lock-in amplifier. The power of the probe beam which is usually produced by the combination of a halogen lamp and a monochromator is of the order of $\mu W/cm^2$. It is evident from the ER measurement system shown in Fig. 3.8 that ER spectroscopy is also a convenient method.

The ER signal, $\Delta R/R$, is related to the change of the dielectric function $\epsilon = \epsilon_1 + i\epsilon_2$

[49]:

$$\Delta R/R = \alpha(\epsilon_1, \epsilon_2)\Delta\epsilon_1 + \beta(\epsilon_1, \epsilon_2)\Delta\epsilon_2, \tag{3.25}$$

where $\Delta R/R$ is the relative change of reflectance due to an electric-field modulation, and $\alpha$ and $\beta$ are called Seraphin coefficients which remarkably depend on the real and imaginary parts of the dielectric function. For SL's and QW's the electric-field-induced change in the dielectric constant can be given by the first derivative form because of the excitonic transition properties [50–54]:

$$\Delta\epsilon(E_g, \Gamma, I) = \left[\frac{\partial\epsilon}{\partial E_g}\frac{\partial E_g}{\partial F} + \frac{\partial\epsilon}{\partial\Gamma}\frac{\partial\Gamma}{\partial F} + \frac{\partial\epsilon}{\partial I}\frac{\partial I}{\partial F}\right]\Delta F, \tag{3.26}$$

where $E_g$ is a transition energy, $\Gamma$ is a phenomenological broadening factor, and $I$ is an oscillator strength. The first term corresponding to the $E_g$ modulation is usually considered to be a dominant mechanism for ER. The transition-energy modulation in the WS localization is written by

$$\frac{\partial E_g}{\partial F} = \frac{\partial E_0}{\partial F} + meD, \tag{3.27}$$

where $E_g = E_0 + meFD$ for the Stark-ladder transition. The term of $\partial E_0/\partial F$ is due to a quantum-confined Stark effect (QCSE) [55] corresponding to an electric-field-induced energy shift of an interband (exciton) transition associated with electron and hole states confined in a QW. The energy shift due to the QCSE, which is a red shift for the $n$=1 state, results from asymmetry of wave functions confined in a QW under an applied electric field. It was demonstrated that $\partial E_0/\partial F \ll meD$ [47]; therefore, the ER signal of the $m \neq 0$ Stark-ladder transition is remarkably enhanced in comparison with that of the $m$=0 transition. The merit of ER spectroscopy is its high sensitivity, while the demerit is to need numerical calculations of the line-shape analysis in order to determine precisely the transition energy from the derivative line shapes. Gibb *et al.* [56] reported the line-shape analysis of ER signals of Stark-ladder transitions. Since we cannot exactly determine the dielectric function form, which is particular in each sample, the line-shape analysis is phenomenological and arbitrary. For convenience, therefore, we empirically estimate the transition energy from an observed ER spectrum, *e.g.*, peak or dip positions of ER signals, taking account of some uncertainty corresponding to a line width.

If the energy spacing of Stark-ladder levels, $eFD$, agrees with a longitudinal optical (LO) phonon energy of constituents (37 meV for the GaAs-LO phonon) and the multiples, we expect that multiple resonance conditions for Raman scattering processes are achieved. Agulló-Rueda *et al.* [57] and Schneider *et al.* [58] clearly observed doubly and triply resonant Raman scattering in the WS localization. The occurrence of the multiple resonance in a SL induced by an applied electric field demonstrates the Stark-ladder formation.

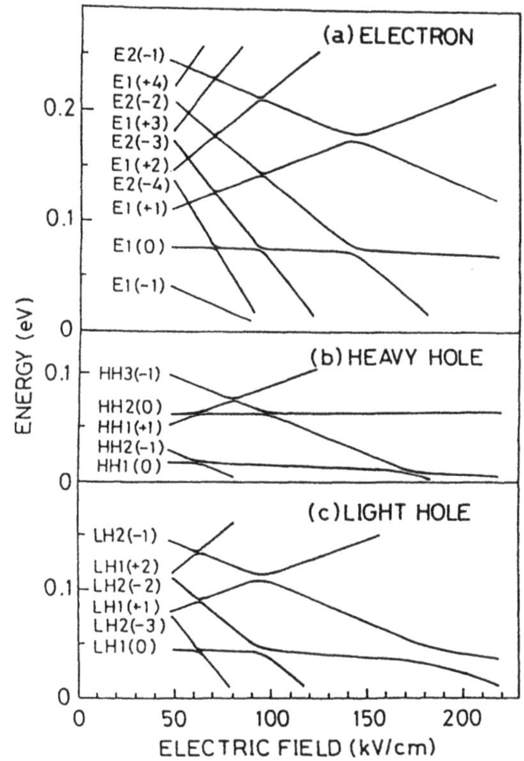


Figure 3.9. Calculated eigenenergies of the $n$=1 electron and heavy-hole states in a GaAs(3.2 nm)/AlAs(0.9 nm) SL as a function of electric field $F$, where the SL is approximated by a system of 19 QW's with a step-like potential as illustrated in the inset, and the energies are measured with respect to the center of the modeled SL. (After [59]).

### 3.4.2. *Transformation Process from Miniband States to Stark-Ladder States*

In this subsection, we consider the transformation process from miniband (quasi-3D) states to Stark-ladder (quasi-2D) states. This corresponds to the electric-field-induced crossover of the dimensionality of electronic systems.

Figure 3.9 shows the calculated eigenenergies of the $n$=1 electron and heavy-hole states in a GaAs(3.2 nm)/AlAs(0.9 nm) SL as a function of electric field $F$ [59], where the SL is approximated by a system of 19 QW's with a step-like potential as illustrated in the inset, and the energies are measured with respect to the center of the modeled SL. The miniband-dispersion relations of the SL at zero field are shown in Fig. 3.2. Since we want to obtain the electric-field dependence of the eigenenergies from zero field, we adapt the step-like potential for the transfer-matrix calculation.

It should be noted that the approximation of the step-like potential is valid in this case because the layer thicknesses of the SL are very thin. Figure 3.10 shows the calculated eigenfunction which belongs to the center QW in the modeled SL with 13 QW's as a function of electric field, where the amplitude is squared. The progress of the wave-function localization is clearly exhibited. In addition, it is evident from Fig. 3.10 that the wave-function amplitude exhibits a peaky profile as a function of electric field, which is similar to the properties of the probability density (Fig. 3.5) represented by the Bessel-function form in the framework of the nearest-neighbor tight-binding model.

Noticing the eigenenergies at $F$=0 in Fig. 3.9, the energy difference between the top and bottom states for the $n$=1 electron (heavy-hole) states is 96 meV (15 meV), which almost corresponds to the miniband width calculated using the envelope-function approximation (see Fig. 3.2): $\Delta_e$=110 meV and $\Delta_{HH}$=15 meV. Focusing on the eigenenergies near the center QW because of neglecting the finite-length effects of the modeled SL, we find that the electric-field dependence of the eigenenergies of the $n$=1 electron (heavy-hole) states changes at $F$=~20 kV/cm (~3 kV/cm). In the lower electric-field regime, the eigenenergies do not depend linearly on the electric field. This behavior is because the electronic states in the individual QW's strongly couple each other. In the higher electric-field regime, the eigenenergies in Fig. 3.9 clearly exhibits the nature of the Stark-ladder states, $\epsilon_0 + \nu eFD$ [Eq. (3.13)], and the calculated eigenfunction of the $n$=1 electron in Fig. 3.10 also indicates the enhancement of the localization with the increase of the electric field above 20 kV/cm. Hence, the electric field of 20 kV/cm (3 kV/cm) is considered to be a critical electric field for the Stark-ladder formation of the $n$=1 electron (heavy-hole) states. For the $n$=1 light-hole state with the miniband width of 88 meV, the critical electric field is estimated to be ~15 kV/cm from the transfer-matrix calculation.

Next we discuss experimental results of the transformation process from miniband states to Stark-ladder states in the GaAs(3.2 nm)/AlAs(0.9 nm) SL whose eigenenergies calculated by the transfer-matrix method are discussed above. Figure 3.11 shows the PC and ER spectra of the GaAs(3.2 nm)/AlAs(0.9 nm) SL at 77 K in the energy range of the H11 and L11 transitions at various electric fields (applied biases) [59], where H$mn$ (L$mn$) denotes the optical transition between the electron state with the quantum number of $m$ and the heavy-hole (light-hole) state with that of $n$. The arrows indicate the H11 and L11 transition energies at zero field calculated using Eq. (3.8), and $\Gamma$ and $\pi$ in the parentheses denote the critical points ($q$=0 and $\pi/D$) of the miniband (see the miniband dispersions shown in Fig. 3.2). The exciton binding energies of 4–5 meV are neglected. The value of the electric field $F$ is estimated from the built-in voltage, 1.77 V, due to the $p$-$n$ junction by using Eq. (3.24). At the lowest electric field, $F$=10 kV/cm, the PC spectrum exhibits two excitonic peaks at 1.71 eV and 1.73 eV originating from the H11- and L11-exciton transitions at $\Gamma$ point, respectively. The exciton transitions in the higher energy range are hidden by the broad background due to continuum interband transitions. Noticing the ER spectrum at $F$=10 kV/cm, we find four signals. The background signal appearing in the PC spectrum is eliminated in the ER spectrum. It is noted that all optical

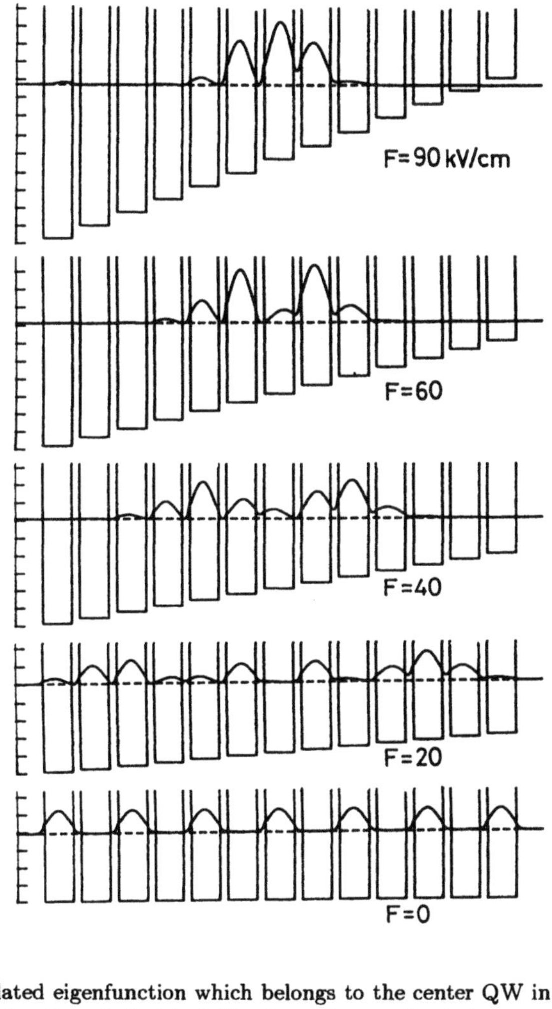


Figure 3.10. Calculated eigenfunction which belongs to the center QW in the modeled SL with 13 QW's as a function of electric field, where the amplitude is squared.

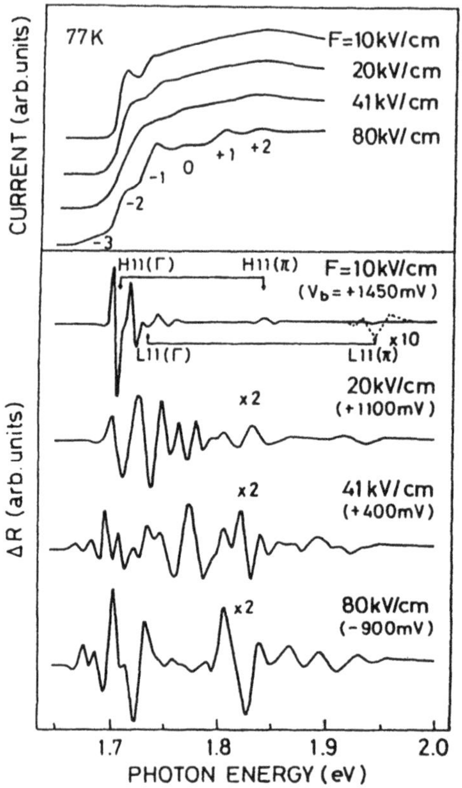


Figure 3.11. PC and ER spectra of the GaAs(3.2 nm)/AlAs(0.9 nm) SL at 77 K in the energy range of the H11 and L11 transitions at various electric fields (applied biases), where $Hmn$ ($Lmn$) denotes the optical transition between the electron state with the quantum number of $m$ and the heavy-hole (light-hole) state with $n$. The arrows indicate the H11 and L11 transition energies at zero field calculated using Eq. (3.8), and $\Gamma$ and $\pi$ in the parentheses denote the critical points ($q$=0 and $\pi/D$) of the miniband. (After [59]).

transitions observed by ER spectroscopy at low temperature correspond to excitonic transitions in the SL. Comparing the energies of the ER signals at 10 kV/cm with those of the calculated miniband transitions indicated by the arrows, we find that the ER spectrum at 10 kV/cm reflects the miniband structures of the SL: the occurrence of the transitions both at $\Gamma$ point and at $\pi$ one.

With the increase of the electric field, the extra structures evolve both to higher and lower energies in the PC and ER spectra. The PC spectrum at 80 kV/cm is a typical example of Stark-ladder transitions in the WS localization. This spectrum consists of six peaks associated with the H11 Stark-ladder transitions with the Stark-ladder index $m$ from $-3$ to $+2$: $E_0 + meFD$. It is noted that the energy of the $m$=0 transition, $E_0$, is almost equal to the center energy between the H11($\Gamma$) and H11($\pi$) transitions observed in the ER spectrum at 10 kV/cm. In much higher electric field, we observe only the $m$=0 transition, resulting from the complete localization. The Stark-ladder behavior can be observed even at room temperature in the same sample [60]. For the ER spectra, they consists of much more signals than in the PC spectra, and exhibit complex profiles owing to the Stark-ladder formation, which will be discussed below with the observed energies as a function of electric field. Since the line-shape analysis of the ER signals of the Stark-ladder transition is complicated and arbitrary as described in Sec. 3.4.1, we simply assume that the transition energies are the peak or dip energies of the ER signals; therefore, the uncertainty of determining the transition energy corresponds to the ER line width of ~10 meV in the present case.

Figure 3.12 shows the transition energies obtained from the ER spectra in the GaAs(3.2 nm)/AlAs(0.9 nm) SL at 77 K as a function of electric field [59], where the closed (open) circles represent the energies of the H11 (L11) transitions, the vertical arrows on the left side indicate the calculated miniband widths, and the number in parentheses means the Stark-ladder index. The solid and dashed lines in Fig. 3.12 indicate the Stark-ladder-transition energies calculated by the transfer-matrix method. From Fig. 3.12, it is obvious that the electric-field dependence of the H11(L11)-transition energies drastically changes at F=~30 kV/cm (F=~45 kV/cm). In the higher electric-field range, we clearly observe the profiles of the Stark-ladder transitions, the so-called fan diagram, with the ladder index from $-6$ to $+5$. On the other hand, in the lower electric-field range, the transition energies are not proportional to the electric field. Such a behavior is considered to be due to the coupling between the electronic states in the individual QW's as discussed with Fig. 3.9. Thus, the critical electric fields for the occurrence of the H11 and L11 Stark-ladder transitions [$F_c$(H11) and $F_c(L11)$] are estimated to be 30 kV/cm and 45 kV/cm, respectively.

The calculated miniband widths of the $n$=1 electron, heavy-hole, and light-hole states are 110, 15, and 88 meV, respectively. The ratio of the total miniband width of the H11 transition to that of the L11 transition, 125 : 198 = 1 : 1.6, agrees with the ratio of $F_c$(H11) to $F_c$(L11), 30 : 45 = 1 : 1.5, obtained from the ER results. Thus, the above indicates that the critical electric field for the Stark-ladder formation depends on the total miniband width. As described in Sec. 3.3.1, the sum of the electron and hole miniband widths is the main parameter to the electric-field dependence of the overlap between the electron and hole envelope functions [Eq. (3.16)] in the framework of the nearest-neighbor tight-binding model. The above experimental results of the critical electric field are consistent with the theoretical prediction. From the transfer-matrix calculation, the critical electric fields for the Stark-ladder formation of the $n$=1 electron, heavy-hole, and light-hole states are 20, 3, and 15 kV/cm, respectively. Assuming that the critical electric field for the occurrence of the interband Stark-ladder transitions corresponds to the sum of those for the Stark-ladder formation of

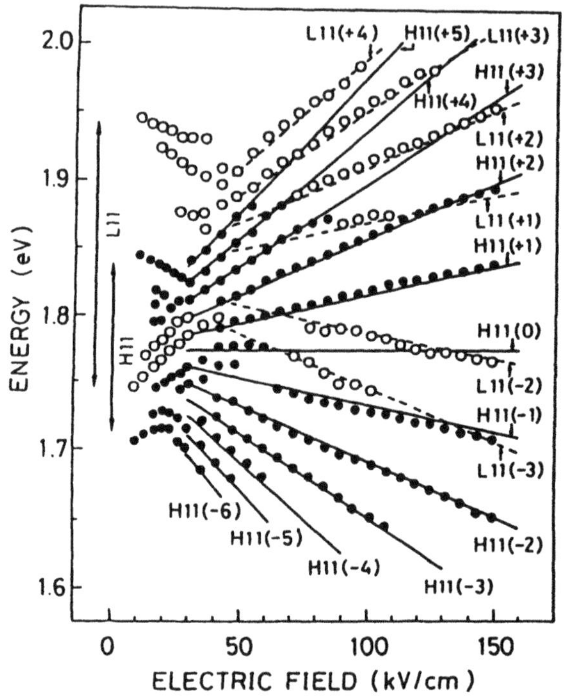


Figure 3.12. Transition energies obtained from the ER spectra in the GaAs(3.2 nm)/AlAs(0.9 nm) SL at 77 K as a function of electric field, where the closed (open) circles represent the energies of the H11 (L11) transitions, the vertical arrows on the left side indicate the calculated miniband widths, and the number in parentheses means the Stark-ladder index. (After [59]).

the electron and hole states, the estimated values of $F_c$(H11) and $F_c$(L11) are 23 kV/cm and 35 kV/cm, respectively. Although the estimated values are smaller than the observed ones, the ratio of the estimated values, $23 : 35 = 1 : 1.5$, is equal to the experimental ratio.

In the above discussion for the Stark-ladder formation, we neglect scattering processes which cause a loss of coherence in the time domain and a homogeneous broadening ($\Gamma_{\mathrm{homo}}$) of optical transitions in the energy domain. For the Stark-ladder transitions to be resolved, the value of the energy spacing $eFD$ must be larger than the value of $\Gamma_{\mathrm{homo}} + \Gamma_{\mathrm{inhomo}}$, where $\Gamma_{\mathrm{inhomo}}$ is an inhomogeneous broadening factor due to imperfections of samples. Plessen *et al.* [61] pointed out that the critical electric field is remarkably affected by the electron-LO-phonon scattering. They summarized the reported results of the Stark-ladder energy spacing at the critical electric field, $eF_cD$, as a function of the miniband width $\Delta$ in GaAs/Al$_x$Ga$_{1-x}$As SL's. It was found that for $\Delta < 36$ meV (GaAs-LO-phonon energy) the value of $eF_cD$ is almost constant, ~5 meV, while for $\Delta > 36$ meV it increases with $\Delta$. From the existence of the threshold energy of the miniband width, Plessen *et al.* concluded that the GaAs-LO-phonon emission is the dominant scattering mechanism in the Stark-ladder formation.

From the viewpoint of the coherent length of wave function, we consider the Stark-ladder transitions. Mendez, Agulló-Rueda, and Hong [22,62,63] pointed out that the number of Stark-ladder transitions observed experimentally is a measure of the coherent length. In the GaAs(3.2 nm)/AlAs(0.9 nm) SL discussed above, we observe the Stark-ladder transitions with the indices from −6 to +5. This indicates that the coherent length of the electron wave function is at least 11 periods (about 45 nm). The coherent length is reduced by scattering due to phonons, impurities, defects, and interface fluctuations. It is of course that an applied electric field reduces the coherent length owing to the localization. They demonstrated that the coherent length decreases with the increase of the period [62] and temperature [63]. Very recently, Bradshaw and Leavitt [64] reported that the electron-wave-function coherent length estimated from observed Stark-ladder indices is 145 nm at 2 K in an In$_{0.53}$Ga$_{0.24}$As(3.0 nm)/In$_{0.52}$Al$_{0.48}$As(2.0 nm) SL and that the value of the coherent length approaches to the theoretical limit.

Under a low electric field condition corresponding to the miniband regime, we have to consider the existence of Franz-Keldysh (FK) oscillations [65–67]. The FK oscillation is a well-known electric-field effect on interband optical transitions in bulk semiconductors [48]. As described in Sec. 3.3.2, the eigenfunctions under an electric field are generally represented by the Airy functions. In the case that the localization length is much larger than the SL period $D$, the oscillatory nature of the Airy function Ai produces a series of oscillations in the energy range higher than the fundamental absorption edge. The oscillation sequence is given by [66]

$$E_j = E_0 + [(eF\hbar)^2/8\mu]^{1/3} X_j \quad \text{with} \quad X_j = [3\pi(j-1/2)/2]^{2/3} \quad (j = 1, 2, \ldots), \quad (3.28)$$

where $E_0$ is the absorption edge of the SL and $\mu$ is the reduced effective mass in the electric-field direction. In the strong localization regime, the FK oscillation should disappear because the quasi-2D state confined in every QW results in infinite $\mu$. The appearance of the FK oscillations, therefore, corresponds to the evidence of the wave-function delocalization. In the GaAs(3.2 nm)/AlAs(0.9 nm) SL discussed in this section, it is difficult to pick up the FK oscillations from the ER spectra (Fig. 3.11) because the narrow energy spacing between the heavy-hole and light-hole transitions makes the complexity of the spectra. Ribeiro *et al.* [66] reported a step-by-step evolution from the FK oscillation to the WS localization in a In$_{0.12}$Ga$_{0.88}$As/GaAs SL with $\Delta$=37 meV for the $n$=1 electron miniband. Schmidt *et al.* [67] reported the coexistence of the FK oscillation and the WS localization in a GaAs (3.1 nm)/AlAs (0.28 nm) SL with the extremely wide electron-miniband width of 380 meV.

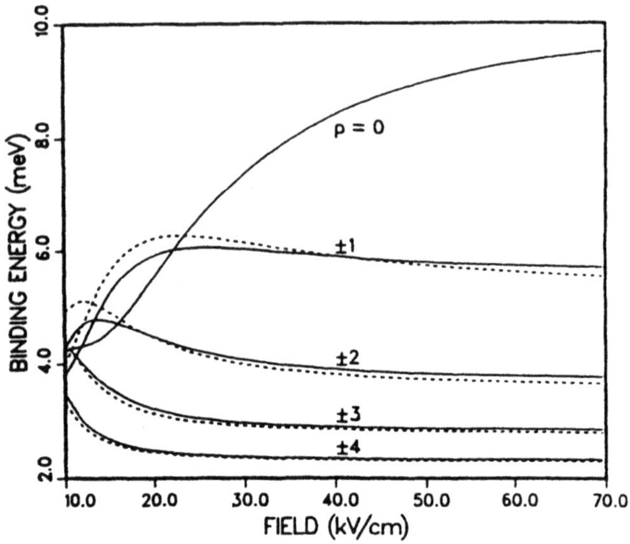


Figure 3.13. Calculated exciton binding energies of the H11-Stark-ladder transitions in a GaAs(3.0 nm)/$Al_{0.35}Ga_{0.65}As$(3.0 nm) SL, where the dashed lines indicate the results for the Stark-ladder transition with a negative index. (After [70]).

### 3.4.3. *Exciton Binding Energy*

In the above discussion, we have neglected excitonic effects on Stark-ladder transitions. Several theoretical studies [68–71] were reported on the excitonic effects. Exciton binding energies and oscillator strengths of Stark-ladder transitions as a function of electric field are precisely described in [70] by Leavitt and Little and [71] by Dignam and Sipe. Figure 3.13 shows the calculated exciton binding energies of the H11-Stark-ladder transitions in a GaAs(3.0 nm)/$Al_{0.35}Ga_{0.65}As$(3.0 nm) SL [70], where the dashed lines indicate the results for the Stark-ladder transition with a negative index. For the $m$=0 transition, the exciton binding energy increases from 4 meV to 9.5 meV with increasing electric field. The increase of the exciton binding energy results from the enhancement of the localization. The binding energy of 4 (9.5) meV almost corresponds to that of a bulk (an isolated 3-nm QW, quasi-2D) exciton. On the other hand, the exciton binding energies of $m \neq 0$ transitions initially increase, go through a maximum, and then decrease to a constant value. The field at which the maximum occurs depends on $m$, and is lower for larger $|m|$. From the viewpoint of spectroscopic experiments, it is difficult to detect clearly such a small change of the binding energy within ~5 meV because spectral line widths are usually comparable to or slightly larger than the amount of the change. It is noted that the excitonic effect remarkably affects the Stark-ladder formation in SL's with very narrow miniband widths comparable to or less than the exciton binding energy. Fox *et al.* [72] reported that the suppression of the observation of Stark-ladder transitions is caused by the excitonic effect in SL's with the miniband widths of 6 and 11 meV.

The exciton binding energy is mainly determined by the degree of the overlap between the corresponding electron and hole envelope functions. Therefore, the electric-field dependence of the exciton binding energies reflects the spatial properties of the wave-function localization which is roughly expected from the Bessel-function properties given by Eq. (3.17). This consideration is closely connected with the oscillator strengths of the Stark-ladder transitions discussed in the next subsection.

### 3.4.4. *Oscillator Strengths of Stark-Ladder Transitions*

As described in Sec. 3.3.1, within the framework of the nearest-neighbor tight-binding model, the oscillator strength of a Stark-ladder transition is proportional to the square of the overlap integral between the hole and electron envelope functions, which is given by the Bessel-function form of Eq. (3.17): The profiles of $|J_m|^2$ are depicted in Fig. 3.5. From this viewpoint, we consider the electric-field dependence of the oscillator strengths of the Stark-ladder transitions.

Figure 3.14 shows (a) ER spectra and (b) PC spectra of the GaAs(3.2 nm)/AlAs (0.9 nm) SL at 77 K in the energy range of the H11-Stark-ladder transitions at various electric fields $F$ [73], where the numbers attached to the ER and PC signals indicate the Stark-ladder index of the H11 transition. The sample is the same used in the discussion in Sec. 3.4.2. Hereafter, we focus only on the H11 transitions because the intensities of the L11 transitions are weak, and the energy range overlaps that of the H11 transitions with the plus Stark-ladder index. The observed intensity in each spectrum is normalized by that at $F$=53 kV/cm in order to compare the relative intensity. Here we assume that the ER and PC intensities are proportional to the oscillator strengths. Figure 3.15 shows the relative peak intensities of (a) ER and (b) PC signals of the H11-Stark-ladder transitions with $m$=0, -1, -2, -3, and -4 as a function of electric field [73], where the lines are guides to the eye. The ER intensities are divided by the Stark-ladder index $m$ following the discussion of the ER mechanism described in Sec. 3.4.1: Eqs. (3.26) and (3.27). The ER intensity of the $m$=0 transition is not plotted because the dominant modulation mechanism is difference from that of the $m \neq 0$ transition. For the PC intensities, the PC spectrum has the intense background of continuum transitions as shown in Fig. 3.14(b); therefore, it is difficult to estimate precisely the PC intensities of the Stark-ladder transitions. Here, we empirically assume the background and subtract it from the spectrum. The electric-field dependence of the PC intensities almost agrees with that of the ER intensities. The intensities of $m \neq 0$ transitions initially increase, go through a maximum, and then decrease to zero. The field at which the maximum occurs depends on $m$, and is lower for larger $|m|$. On the other hand, the intensity of the $m$=0 transition reaches a minimum at ~70 kV/cm, and then increases. Similar results obtained by PC spectroscopy were reported in [63,74]. It is noted that the above electric-field

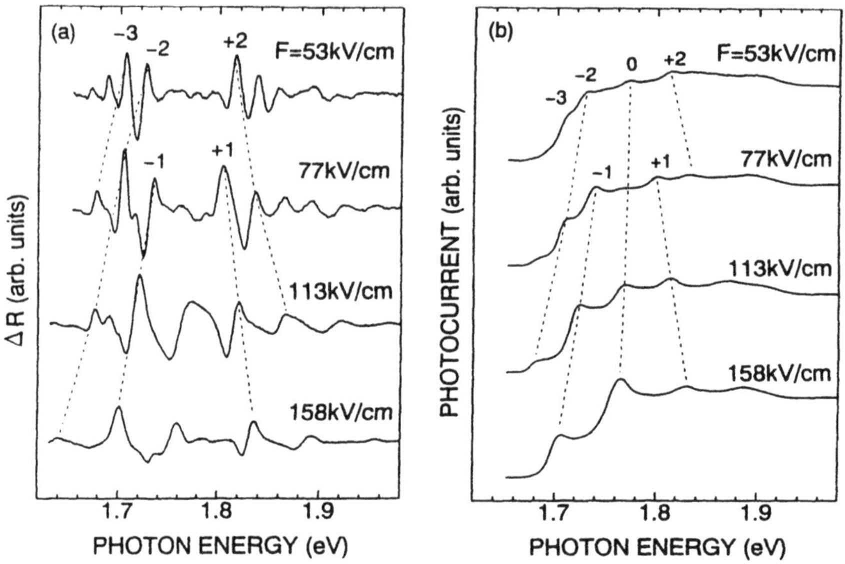


Figure 3.14. (a) ER spectra and (b) PC spectra of the GaAs(3.2 nm)/AlAs(0.9 nm) SL at 77 K in the energy range of the H11-Stark-ladder transitions at various electric fields $F$, where the number attached to the ER and PC signals indicate the Stark-ladder index of the H11 transition. (After [73]).

dependence of the ER and PC intensities of the Stark-ladder transitions is analogous to that of the exciton binding energies shown in Fig. 3.13.

As described in Sec. 3.3.1, the oscillator strength is proportional to the square of the overlap integral of the electron and hole envelope functions. Within the framework of the nearest-neighbor tight-binding model, the oscillator strength of the Stark-ladder transition is proportional to $|J_m[(\Delta_e + \Delta_h)/2eFD]|^2$ [Eq. (3.17)]. The values of $|J_m|^2$ as a function of $D/L \propto F$ shown in Fig. 3.5 is quite similar to the electric-field dependence of the ER and PC intensities (oscillator strengths) shown in Fig. 3.15. From the nature of the Bessel function, the ratio of the electric field at which the value of $|J_m|^2$ reaches a maximum is always $1(m = \pm 1) : 0.60(m = \pm 2) : 0.45(m = \pm 3) : 0.35(m = \pm 4)$. In Fig. 3.15, the intensities of the Stark-ladder transitions occur a maximum at 130, 75, 55, 45 kV/cm for $m = -1, -2, -3$, and $-4$, respectively. The ratio of $130 : 75 : 55 : 45 = 1 : 0.58 : 0.42 : 0.34$ agrees with that expected from the nature of the Bessel function. Thus, the simple tight-binding model well explains the oscillator strengths of the Stark-ladder transitions as a function of electric

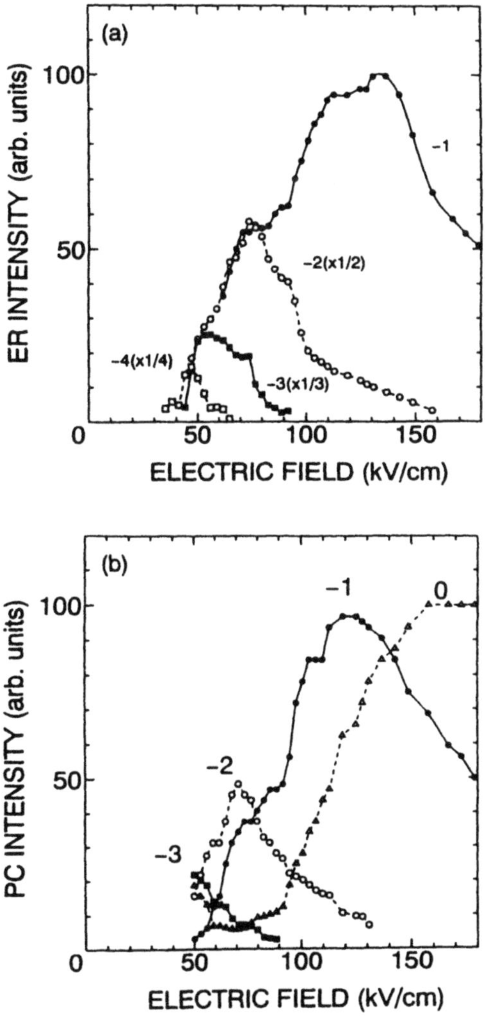


Figure 3.15. Relative peak intensities of (a) ER and (b) PC signals of the H11-Stark-ladder transitions with $m = 0, -1, -2, -3$, and $-4$ as a function of electric field, where the lines are guides to the eye. (After [73]).

field: This indicates that the spatial properties of the wave-function localization is reasonably represented by the Bessel-function form. In more accurate calculations including excitonic effects [70,71], it is suggested that the electric-field dependence of the oscillator strength exhibits a slightly asymmetric behavior with respect to the sign of $m$; however, the profiles are quite similar to those represented by the Bessel-function form. In conclusion, the oscillator strengths of the Stark-ladder transitions as a function of electric field are reasonably explained by the overlap integral of the electron and hole envelope functions given by the Bessel-function form in the framework of the nearest-neighbor tight-binding model.

### 3.4.5. *Stark-Ladder Transitions in Type-II Superlattices*

In a type-II potential structure as shown in Fig. 3.1(c), electron and hole envelope functions are spatially separated. For the Stark-ladder transitions in a type-II SL, the Stark-ladder sequence is given by the fractional form of $\pm eFD/2$, $\pm 3eFD/2$, etc. because the basic distance between the center positions of the electron and hole wave functions is a half of the period, $D/2$.

In $In_xGa_{1-x}As/GaAs$ strained-layer SL's, the lattice-misfit strain lifts the degeneracy of the heavy-hole and light-hole bands at $\Gamma$ point, and produces a type-II potential structure between the electron and light-hole bands: The InGaAs (GaAs) layer is the QW for electrons and heavy holes (light holes) [75]. The Stark-ladder behaviors in $In_xGa_{1-x}As/GaAs$ strained-layer SL's were reported [76-78], and Saker *et al.* [77] and Kajikawa [78] clearly demonstrated the fractional Stark-ladder behavior of the L11 transition. In GaAs/AlAs SL's, the AlAs layer, which is the barrier for the $\Gamma$ potential, is the QW for the electron potential at X point [79]. The fractional Stark-ladder behavior of the transition between the AlAs-X electron state and GaAs-$\Gamma$ heavy-hole one was observed by Shields *et al.* [80]. It should be noted that the observation of the fractional Stark-ladder behavior is a direct evidence for the existence of the type-II potential structure.

### 3.4.6. *Electro-Optical Bistability*

The WS localization has been of interest in the context of electro-optic modulators such as a self-electro-optic effect device (SEED) [27,81–84], a Fabry-Perot reflector [85], and an electroabsorption waveguide modulator [86] because of the large energy shift of Stark-ladder transitions induced by an applied electric field. Before the confirmation of the existence of the WS localization, only the QCSE [55], which is briefly described in Sec. 3.4.1, had been considered to be applied to electro-optic modulators using QW systems. Here, an electro-optical bistability operation in the SEED based on the WS localization is presented.

Miller *et al.* [87] introduced the concept of the SEED based on the QCSE on exciton absorption energies in 1984. Figure 3.16 shows the schematic diagram of the SEED configuration using a *p-i-n* diode, where the $i$ layer is a SL or a multiple QW for an active region. The general principle of the SEED under illumination is that the

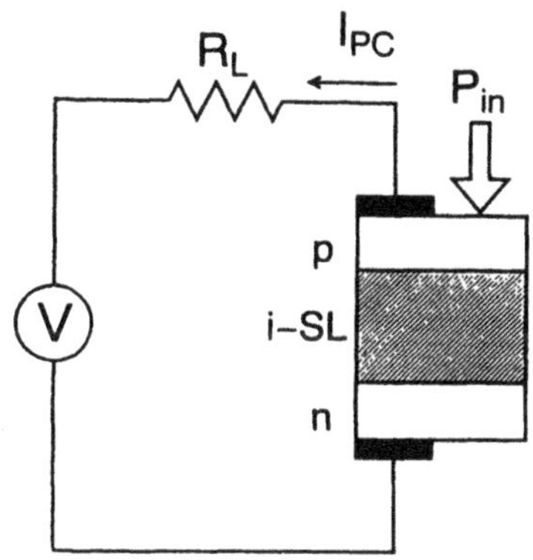


Figure 3.16. Schematic diagram of the SEED configuration using a *p-i-n* diode, where the *i* layer is a SL or a multiple QW for an active region.

PC passing through the electric circuit influences the voltage across the *p-i-n* diode; the voltage (electric field) across the diode influences the absorption of light because of the change of exciton energies, and hence influences the PC. Thus, an electro-optic feedback is established. Under a positive feedback condition we obtain a bistability operation. In the WS localization, the energy shift of the Stark-ladder transition induced by change of an electric field is much larger than that caused by the QCSE in an isolated QW: This is of advantage to the SEED based on the WS localization.

Figure 3.17(a) shows the PC vs. reverse bias at 295 K in a GaAs(3.2 nm)/AlAs(0.9 nm) SL (40 periods) sandwiched by *p*- and *n*-$Al_{0.33}Ga_{0.67}As$ layers [81], where the illumination-light wavelength fits the energy range of the Stark-ladder transitions of H11(−1) and H11(−2). We find the regions of negative differential PC, which are caused by the energy shifts of the H11(−1) and H11(−2) transitions. If we choose an appropriate load resistance and bias voltage in the SEED, we can obtain a bistable operation of PC by using the negative differential PC. Figure 3.17(b) shows the bistability operation of the PC vs. illumination power (100 mW range), where the load resistance is 1 MΩ, and the reverse bias is 6 V. We can also obtain the bistability of the output-light power. It is noted that the wavelength range for the bistability based on the WS localization is much more flexible than that based on the QCSE.

## 3.5. RESONANT COUPLING BETWEEN STARK-LADDER STATES

### *3.5.1. Anticrossing between Stark-Ladder Levels*

In the case of a multiple miniband system, the Stark-ladder states of a miniband should interact with those of a different miniband. Here, we consider the condition that the energy difference between different Stark-ladder states with the quantum

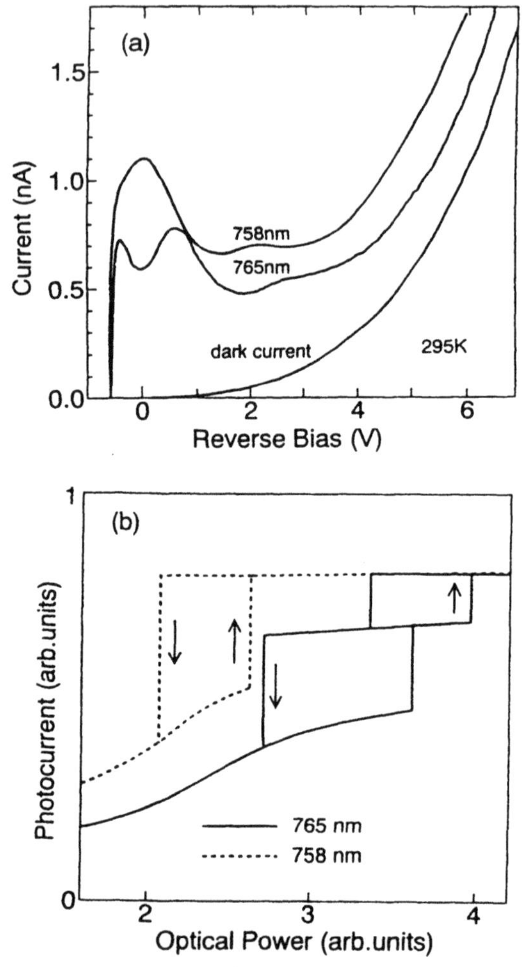


Figure 3.17. (a) Photocurrent vs. reverse bias at 295 K in a GaAs(3.2 nm)/AlAs(0.9 nm) SL (40 periods) sandwiched by *p*- and *n*-$Al_{0.33}Ga_{0.67}As$ layers, where the illumination-light wavelengths fit the energy range of the Stark-ladder transitions of H11(−1) and H11(−2), (b) the bistability operation of the PC vs. illumination power (100 mW range), where the load resistance is 1 MΩ, and the reverse bias is 6 V. (After [81]).

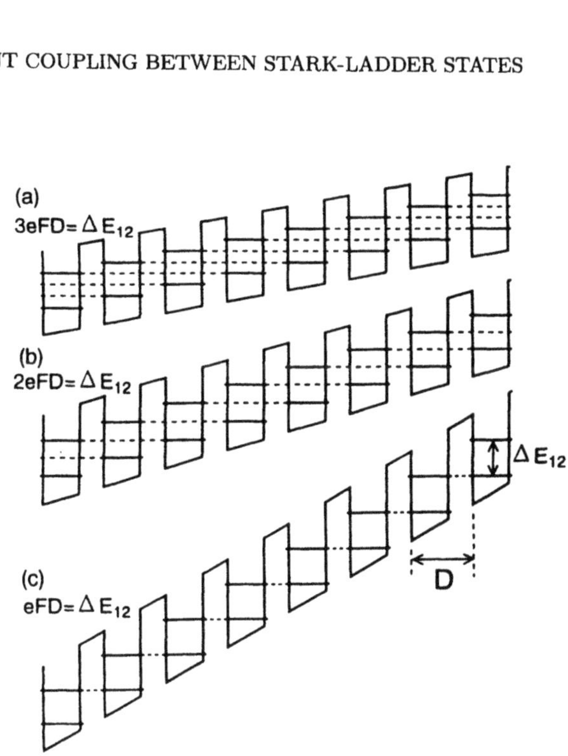


Figure 3.18. Schematic diagrams of the condition of the resonant coupling between Stark-ladder states.

numbers $n$=1 and 2, $\Delta E_{12}$, is equal to $meFD$ as shown in Fig. 3.18: This is called a resonant tunneling condition in the field of transport phenomena. In this case, it is expected that the $n$=1 Stark-ladder state localized in a QW resonantly couples with the spatially-separated $n$=2 state in the $m$th nearest neighbor QW. Schneider *et al.* [24] and Nakayama *et al.* [25,26] revealed the nature of the resonant coupling which corresponds to a phenomenon related to quantum coherence. In this subsection we consider the resonant coupling from the viewpoint of eigenenergies.

Figure 3.19 shows the calculated energies of the Stark-ladder states of (a) electrons, (b) heavy holes and (c) light holes in a GaAs(6.4 nm)/AlAs(0.9 nm) SL as a function of electric field [26], where the energies are measured with respect to the center QW in the modeled SL consisting of eleven QW's. The calculation is based on the transfer-matrix method with the Airy functions (Sec. 3.3.2). The notations of E$n(\pm m)$, HH$n(\pm m)$, and LH$n(\pm m)$ mean the $n$th electron, heavy-hole, and light-hole states in the $m$th nearest-neighbor QW from the center (sixth) QW, and the + (−) sign indicates the QW in the higher (lower) potential side from the center QW. In Fig. 3.19, we find a number of anticrossing behaviors between different Stark-ladder levels. Noticing the electron Stark ladders around 140 kV/cm, the anticrossing between the E1(0) and E2(−2) states and that between the E1(+1) and E2(−1) states occur. This

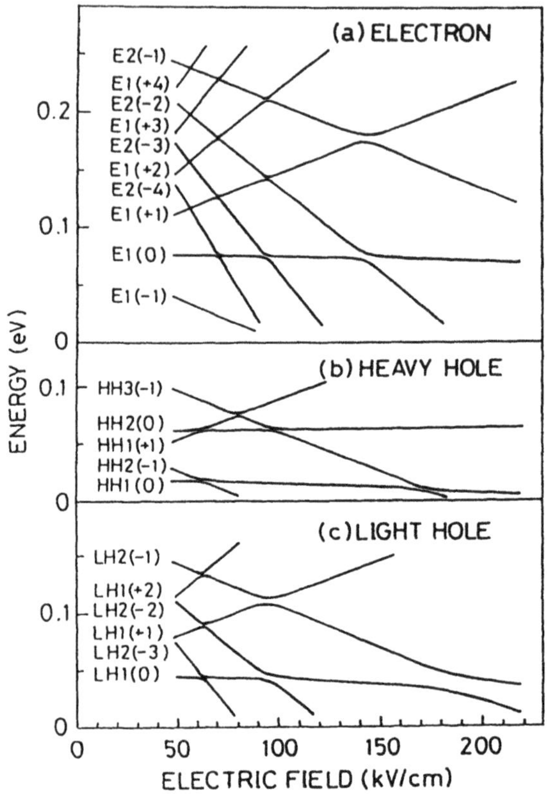


Figure 3.19. Calculated energies of the Stark-ladder states of (a) electrons, (b) heavy holes and (c) light holes in a GaAs(6.4 nm)/AlAs(0.9 nm) SL as a function of electric field, where the energies are measured with respect to the center QW in the modeled SL consisting of eleven QW's. The calculation is based on the transfer-matrix method with the Airy functions. (After [26]).

results from the second-nearest neighbor resonant coupling under the condition of $\Delta E_{12} = 2eFD$ [Fig. 3.18(b)]. Hereafter, the $m$th nearest-neighbor resonant coupling will be denoted by E1(0)–E2($-m$). The occurrence of the anticrossing indicates the mixing between the different Stark-ladder states. Figure 3.20 shows the calculated eigenfunctions under the E1(0)–E2($-2$) resonant coupling. The envelope-function form clearly exhibits the characteristic of the mixing between the E1 and E2 states: the formation of the bonding and antibonding states. The anticrossings around 90 and 70 kV/cm results from the E1(0)–E2($-3$) (third nearest-neighbor) and E1(0)–

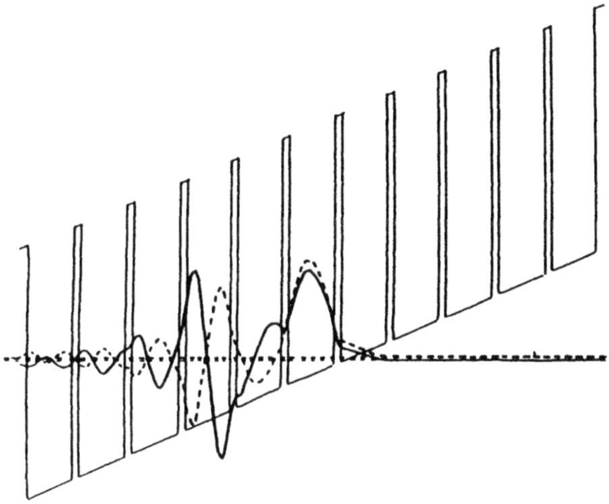

Figure 3.20. Calculated eigenfunctions under the E1(0)–E2(−2) resonant coupling.

E2(−4) (fourth nearest-neighbor) resonant couplings, respectively. We find similar anticrossing behaviors for the Stark ladders of the hole states. Bastard *et al.* [88] proposed an analytical model for such an anticrossing within the framework of the nearest-neighbor tight-binding model.

Under the resonant coupling condition, we expect that the optical transitions associated with the corresponding Stark-ladder states are remarkably modified by the anticrossing behavior. From this viewpoint, we discuss the experimental results of Stark-ladder transitions. Figure 3.21 shows the ER spectra of a GaAs(6.4 nm)/AlAs(0.9 nm) SL at 77 K in the energy range of the H11 and L11 transitions at various electric fields (applied biases) [26], where the ER intensity in each spectrum is normalized by that at 126 kV/cm. The SL has 20 periods, and is sandwiched by $p$- and $n$-$Al_{0.33}Ga_{0.67}As$ layers (see Fig. 3.7 of the sample structure). The miniband width of this sample is about one-third of that of the GaAs(3.2 nm)/AlAs(0.9 nm) SL discussed in Sec. 3.4, $\Delta_e$=30 meV, $\Delta_{HH}$=3 meV, and $\Delta_{LH}$=25 meV for the $n$=1 minibands, so that the degree of the WS localization is much larger. In the electric field range from 53 to 126 kV/cm, the ER intensities of the $m=\pm1$ Stark-ladder transitions, H11($\pm$1) and L11($\pm$1), decrease remarkably with increasing electric field: This clearly results from the enhancement of the WS localization. Above 120 kV/cm, we observe only the H11(0) and L11(0) transitions because the wave functions are completely localized in the individual QW's.

Noticing the ER spectra at 62, 95, 143, and 188 kV/cm in Fig. 3.21, we observe the splitting features of the H11(0) and L11(0) signals. It should be noted that the

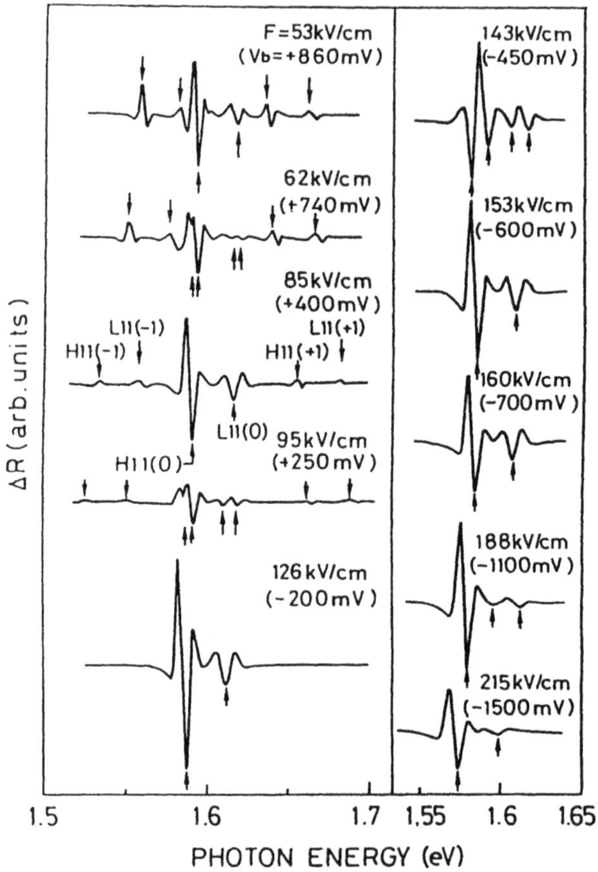


Figure 3.21. ER spectra of a GaAs(6.4 nm)/AlAs(0.9 nm) SL at 77 K in the energy range of the H11 and L11 transitions at various electric fields (applied biases), where the ER intensity in each spectrum is normalized by that at 126 kV/cm. (After [26]).

splitting features cannot be observed by PC spectroscopy. From the above discussion of the anticrossing behaviors between the Stark-ladder states shown in Fig. 3.19, we consider that the appearance of the split of the ER signals indicates the occurrence of the anticrossing due to the resonant coupling. Figure 3.22 shows the calculated and observed energies of the H11(0), H21(−2), L11(0), and L21(−2) transitions as a function of electric field around 140 kV/cm under the E1(0)–E2(−2) coupling [25]. The energies of the splitting ER signals clearly exhibit the anticrossing behaviors

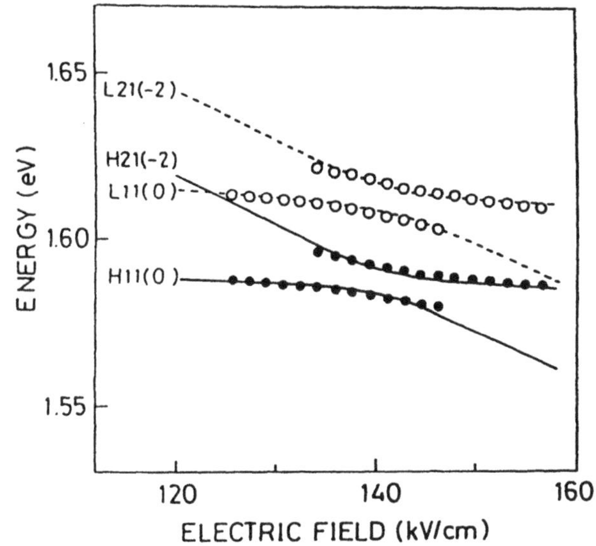


Figure 3.22. Calculated and observed energies of the H11(0), H21(−2), L11(0), and L21(−2) transitions as a function of electric field around 140 kV/cm under the E1(0)–E2(−2) coupling. (After [25]).

predicted by the transfer-matrix calculation. The calculated results in Fig. 3.19 indicate that the split of the ER signals at 95 kV/cm in Fig. 3.21 originates from the E1(0)–E2(−3) coupling. In the ER spectrum at 188 kV/cm in Fig. 3.21, only the L11(0) signal exhibits the splitting feature. This demonstrates the occurrence of the light-hole resonant coupling: The transfer-matrix calculation indicates that the split of the L11 signal results from the first-nearest-neighbor resonant coupling between the light-hole states, LH1(0)–LH2(−1). The split of the H11(0) and L11(0) signals at 62 kV/cm in Fig. 3.21 is due to the simultaneous occurrence of the HH1(0)–HH2(−2) and LH1(0)–LH2(−3) couplings. Thus, the appearance of the line-shape split of the ER signal indicates the occurrence of the resonant coupling which results in the anticrossing between the related Stark-ladder levels. The strength of the resonant coupling corresponds to the anticrossing gap. As shown in Fig. 3.19, the anticrossing gap decreases with the increase of the resonance index. This is demonstrated by the split widths of the ER signals shown in Fig. 3.21 under the various resonant couplings.

### 3.5.2. *Wave-Function Delocalization*

Under the $m$th nearest-neighbor resonant coupling condition, the wave functions of related Stark-ladder states are delocalized over the coupling space of $mD$. It is considered that the delocalization reduces the degree of the overlap between electron and

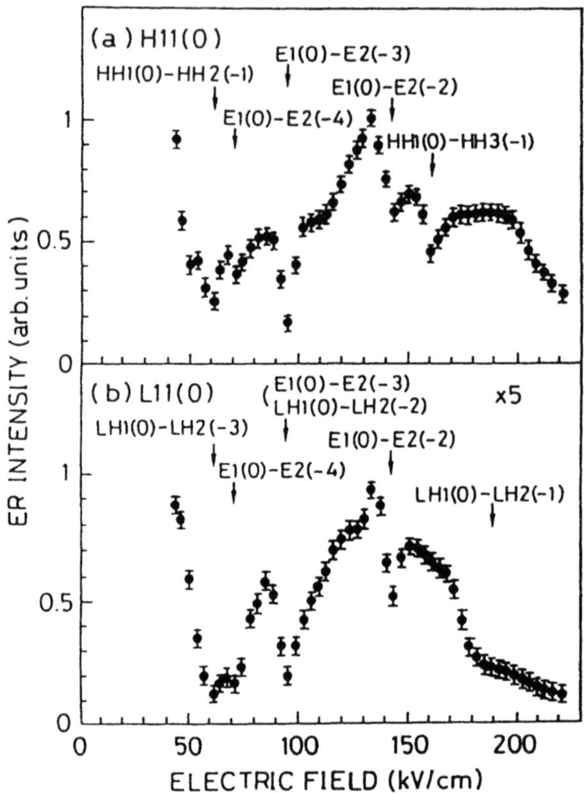


Figure 3.23. ER intensities of the H11(0) and L11(0) transitions as a function of electric field in the GaAs(6.4 nm)/AlAs(0.9 nm) SL at 77 K, where the arrows indicate the calculated electric fields for the various resonant couplings. (After [26]).

hole envelope functions, which results in the reduction of the oscillator strength of interband transitions. From this viewpoint, we discuss the resonant coupling.

Noticing the intensities of the ER signals in Fig. 3.21, we find that they are reduced under the resonant coupling conditions, where the ER signals exhibit the splitting feature. Figure 3.23 shows the ER intensities of the H11(0) and L11(0) transitions as a function of electric field in the GaAs(6.4 nm)/AlAs(0.9 nm) SL at 77 K, where the arrows indicate the calculated electric fields for the various resonant couplings [26]. We notice the dips of the ER intensities under the resonant coupling conditions. It is accepted that the ER intensity is proportional to the oscillator strength. The intensity (oscillator-strength) reduction results from the wave-function delocalization induced by the resonant coupling. In Fig. 3.23 the simultaneous intensity reductions of the H11(0) and L11(0) transitions at 143, 95, and 71 kV/cm result from the resonant coupling between the electron states of E1(0)–E2(−2), E1(0)–E2(−3), and E1(0)–E2(−4), respectively. Since the $m$th-nearest-neighbor resonant coupling between the $n$=1 and $n$=2 states should occur at $meFD = \Delta E_{12}$, the electric field at which the resonant coupling occurs is proportional to the inverse of the resonance index, $1/m$. The ratio of the above electric fields for the electron resonant couplings, 143 : 95 : 71 = 1 : 0.66 : 0.50 is in agreement with the expected ratio of 1/2 : 1/3 : 1/4 = 1 : 0.67 : 0.50. For the resonant coupling between the heavy-hole (light-hole) states, it is expected that only the H11(0) [L11(0)] intensity is reduced: We find this situation at 160 (185) kV/cm for the HH1(0)–HH3(−1) [LH1(0)–LH2(−1)] coupling. The simultaneous intensity reduction of the H11(0) and L11(0) transitions at 60 kV/cm results from the simultaneous occurrence of the HH1(0)–HH2(−1) and LH1(0)–LH2(−3) couplings. The LH1(0)–LH2(−2) coupling overlaps the E1(0)–E2(−3) one. Thus, the electric fields for the light-hole resonant couplings of LH1(0)–LH2(−1), LH1(0)–LH2(−2), and LH1(0)–LH2(−3) are 185, 95, and 60 kV/cm, respectively: They are almost in the ratio of 1 : 1/2 : 1/3 expected from the resonance condition. From the viewpoint of the excitonic effect, it is considered that the wave-function delocalization induced by the resonant coupling causes the reduction of the exciton binding energy. There has been no report of the theoretical analysis for the excitonic effect under the resonant coupling. The occurrence of the intensity reduction and split of the ER signal discussed in this section correspond to the optical detection of resonant tunneling. As described above, we can classify the type of the resonance from the type of detected interband transitions such as H11 and L11.

## 3.6. ABOVE-BARRIER STATES

### 3.6.1. *Existence of Above-Barrier States*

In the above sections, electronic states below a barrier potential (below-barrier states) have been the subject of the discussion. The existence of electronic states above a barrier potential (above-barrier states) is easily recognized in the framework of quantum mechanics; however, only several works of optical transitions associated with above-barrier states (above-barrier transitions) have been reported [89–95]. Very recently, Nakayama *et al.* [95] reported experimental and theoretical studies of the electric-field effects on above-barrier minibands in a GaAs/$Al_{0.1}Ga_{0.9}As$ SL from the viewpoint of the WS localization. In this subsection, we consider the existence of the above-barrier miniband in a SL, and then will discuss the electric-field effects in the next subsection.

In the framework of the envelope-function approximation, we usually calculate the below-barrier miniband dispersion of a SL by using Eq. (3.8). It should be noted that this equation is applicable for the above-barrier miniband. Figure 3.24 depicts the calculated miniband dispersions of electrons, heavy holes, and light holes in the below- and above-barrier energy ranges in a GaAs(7.0 nm)/$Al_{0.1}Ga_{0.9}As$(3.5 nm) SL, where the horizontal dashed lines indicate the band-offset energies of the conduction and valence bands. The calculated results indicate the existence of the above-barrier minibands, and we expect the observation of the optical transitions associated with the above-barrier minibands. Figure 3.25 shows the ER spectra of the GaAs(7.0 nm)/$Al_{0.1}Ga_{0.9}As$(3.5 nm) SL at 77 K at various electric fields [95], where the dashed line indicates the band-gap energy of $Al_{0.1}Ga_{0.9}As$. The SL consists of 70 periods, and is placed in the center of a *p-i-n* diode structure. The *n* layer is a Si-doped ($1\text{x}10^{18}$ $cm^{-3}$) $Al_{0.3}Ga_{0.7}As$ ($\sim$1 $\mu$m) following to a Si-doped (001) GaAs substrate, and the *p* layer is a Be-doped ($5\text{x}10^{18}$ $cm^{-3}$) $Al_{0.3}Ga_{0.7}As$ ($\sim$1 $\mu$m). The arrows located on the ER spectrum at 5 kV/cm indicate the energies of the interband transitions at the $\Gamma$ and $\pi$ (mini-Brillouin-zone edge, $\pi/D$) points at zero field estimated from the mini-

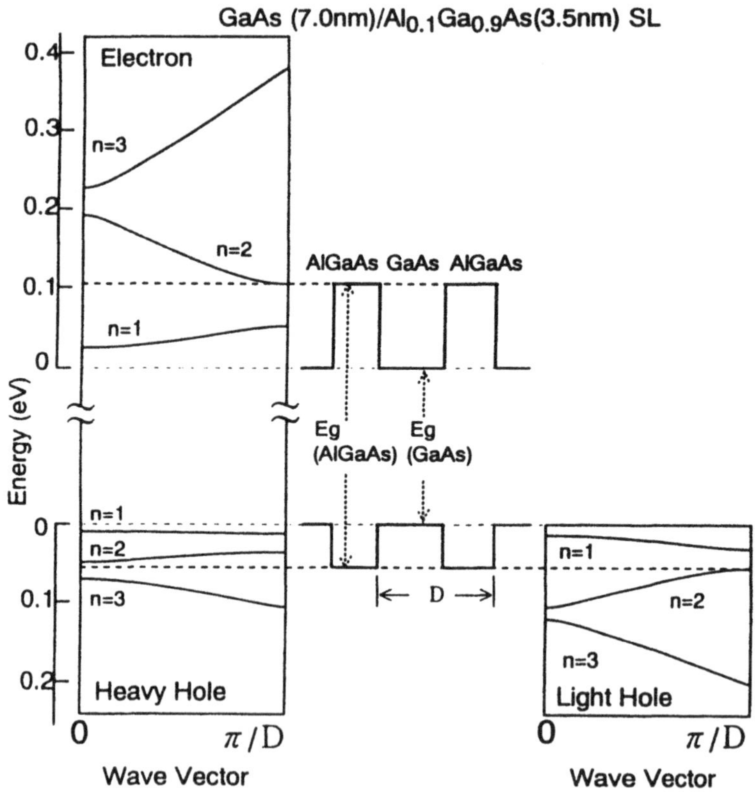


Figure 3.24. Calculated miniband dispersions of electrons, heavy holes, and light-holes in the below- and above-barrier energy ranges in a GaAs(7.0 nm)/$Al_{0.1}Ga_{0.9}As$(3.5 nm) SL.

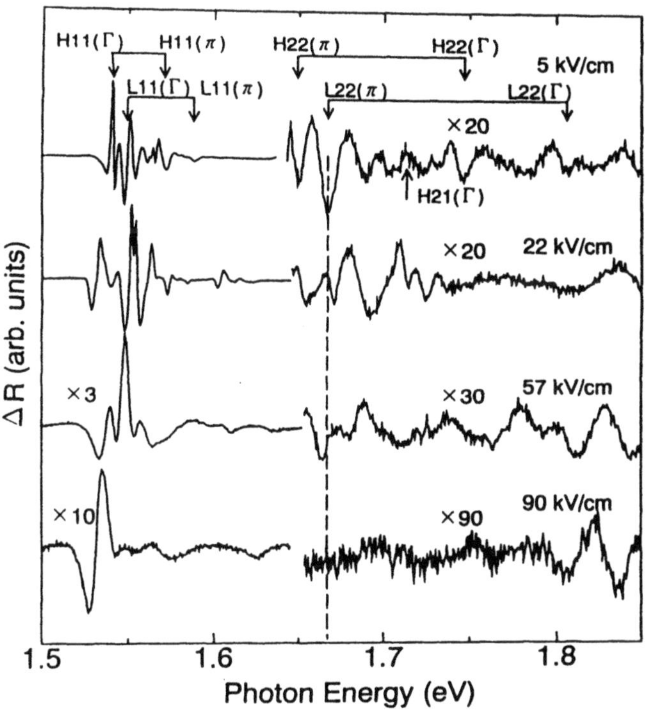


Figure 3.25. ER spectra of the GaAs(7.0 nm)/$Al_{0.1}Ga_{0.9}As$(3.5 nm) SL at 77 K and at various electric fields, where the dashed line indicates the band-gap energy of $Al_{0.1}Ga_{0.9}As$. The arrows located on the ER spectrum at 5 kV/cm indicate the calculated energies of the interband transitions at the $\Gamma$ and $\pi$ (mini-Brillouin-zone edge, $\pi/D$) points at zero field. (After [95]).

band dispersions shown in Fig. 3.24. We neglect the exciton-binding energy of 4–5 meV. The ER spectrum at 5 kV/cm reflects the above-barrier miniband structures in addition to the below-barrier minibands. The oscillatory structures in the energy range below 1.6 eV are mainly due to FK oscillations under the miniband condition discussed in Sec. 3.4.2. Thus, the theoretical and experimental results clearly demonstrate the existence of the above-barrier minibands. Furthermore, it is obvious that the line shapes and energies of the ER signals remarkably change with increasing the electric field, which will be discussed in the next subsection.

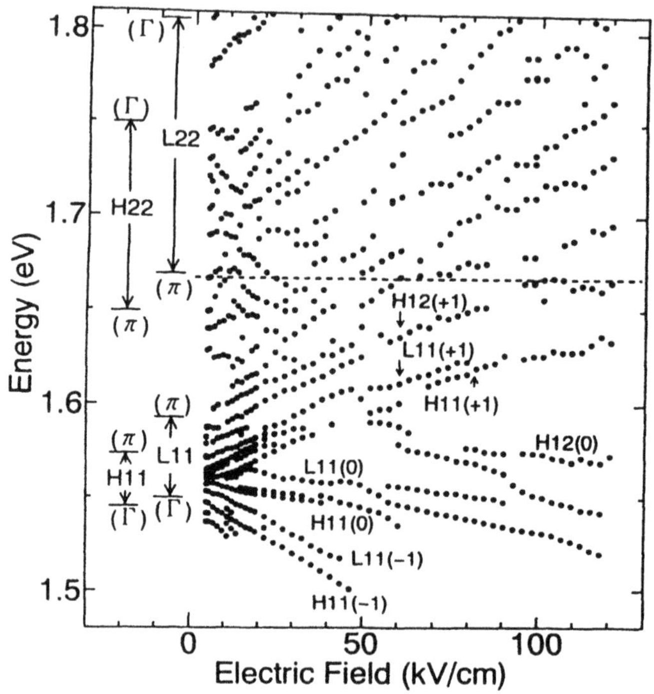


Figure 3.26. Electric-field dependence of the energies of the ER signals in the GaAs(7.0 nm)/$Al_{0.1}Ga_{0.9}As$(3.5 nm) SL, where the dashed line indicates the band-gap energy of $Al_{0.1}Ga_{0.9}As$, and the vertical arrows on the left side indicate the calculated energy ranges of the miniband transitions at $F = 0$. (After [95]).

### 3.6.2. *Stark-Ladder Formation and Resonant Coupling*

The electric-field dependence of the energies of the ER signals in the GaAs(7.0 nm)/$Al_{0.1}Ga_{0.9}As$(3.5 nm) SL is shown in Fig. 3.26 [95], where the dashed line indicates the band-gap energy of $Al_{0.1}Ga_{0.9}As$, and the vertical arrows on the left side indicate the energy ranges of the miniband transitions at $F$=0 estimated from the miniband dispersions shown in Fig. 3.24. From Fig. 3.26, it is obvious that the electric-field dependence of the energies of the below-barrier transitions labeled H11($m$) and L11($m$) clearly shows the properties of the Stark-ladder transitions, the fan diagram as a function of electric field. For the above-barrier transitions labeled H22 and L22, the electric-field dependence of the energies exhibits the similar feature of that of the Stark-ladder transitions below the barrier potential. The Stark-ladder behaviors of the H22 and L22, the downward and upward shifts of their energies vs. the electric field, are remarkable in the electric-field range from 15 to 30 kV/cm. The downward

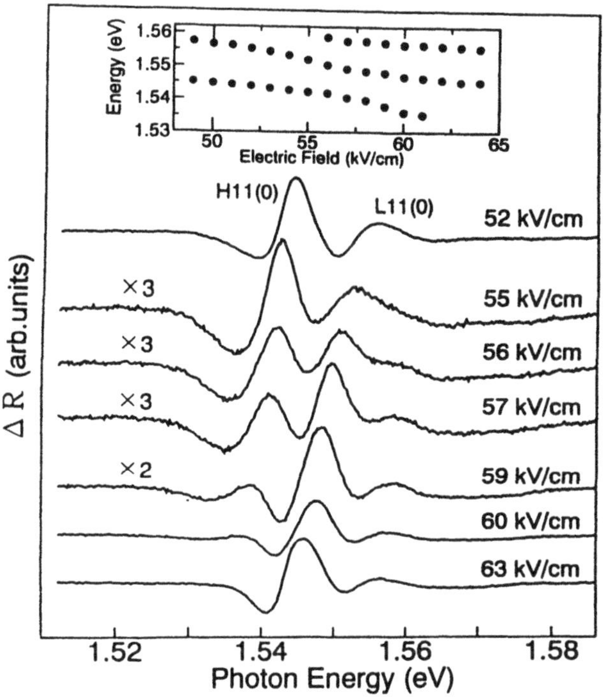


Figure 3.27. ER spectra of the H11(0) and L11(0) transitions around 57 kV/cm, where the inset depicts the electric-field dependence of the ER-signal energies. (After [95]).

shifts above 30 kV/cm are hidden by the upward ones of the below-barrier transitions. The electric-field dependence of the H22 and L22 energies suggests that the Stark-ladder states are formed in the above-barrier region.

Noticing the electric-field dependence of the energies of the H11(0) and L11(0) transitions around 55 and 110 kV/cm in Fig. 3.26, we clearly find the anticrossing behaviors. As discussed in Sec. 3.5, the anticrossing indicates the occurrence of a resonant coupling between different Stark-ladder states. Figure 3.27 shows the ER spectra of the H11(0) and L11(0) transitions around 57 kV/cm, where the inset depicts the electric-field dependence of the ER-signal energies. Both the H11(0) and L11(0) signals exhibit the splitting feature at 56, 57, 59, and 60 kV/cm, and their energies show the anticrossing as depicted in the inset. Since the energy difference between the H11(0) and L11(0) signals is small, ~10 meV, the higher-energy part of the doublet H11(0) signals overlaps with the lower-energy part of the doublet L11(0) ones. The simultaneous change of the H11(0) and L11(0) signals indicates that the Stark-ladder state of the $n$=1 electron miniband resonantly couples with another electron state in a spatially-separated QW. The calculated miniband dispersions (Fig. 3.24) indicate that the energy range of the $n \geq 2$ electron minibands is above the barrier potential; therefore, it is expected that the resonant coupling is related to the above-barrier Stark ladder. Kita *et al.* [96] reported a similar resonant coupling between below-barrier and above-barrier states of the light hole in an InGaAs/GaAs strained multiple QW.

We analyze the above experimental results by using the transfer-matrix method with Airy functions. In the present calculation, the sample structure of the GaAs(7.0 nm)/$Al_{0.1}Ga_{0.9}As$(3.5 nm) SL is approximated by a system of eleven QW's sandwiched by semi-infinite $Al_{0.3}Ga_{0.7}As$ layers. Figure 3.28 shows the electric-field dependence of the eigenenergies of the electron states in the modeled SL calculated by using the transfer-matrix method [95], where the energies are measured with respect to the center of the modeled SL, and the dashed line indicates the barrier potential of electrons. The vertical lines on the left side indicate the calculated energy ranges of the minibands at zero field. Although the electric-field dependence exhibits the complicated behavior because of the various resonant-coupling-induced anticrossings, we find that both the below- and above-barrier minibands transform into the Stark-ladder states with increasing the electric field. This is consistent with the theoretical results reported by Ritze, Horing, and Enderlein from the viewpoint of the density of states of above-barrier states [97]. They suggested that the above-barrier region is dominated by two kinds of structures: One is similar to the Stark-ladder behavior, and the other resembles FK oscillations. Thus, both the experimental and theoretical results demonstrate the formation of the Stark-ladder states of the E2 miniband above the barrier potential.

As described above with Figs. 3.26 and 3.27, we clearly observe the anticrossing due to the resonant coupling between the $n$=1 electron state and the above-barrier state, which results in the split of the H11(0) and L11(0) signals around 57 kV/cm. In the calculated results of the electron eigenenergies (Fig. 3.28), there are two types of the anticrossing between the below-barrier (E1) state and the above-barrier (E2 and E3) states in this electric-field range: For the E1(0) state in the center QW, one around 48 kV/cm is due to the E1(0)–E2(−2) resonant coupling, and the other around 55 kV/cm is the E1(0)–E3(−5) coupling. On the other hand, we observe only one type of the anticrossing as shown in Fig. 3.26. The transfer-matrix calculation (Fig. 3.28) predicts that the E3-related anticrossings of E1(0)–E3(−4) and E1(0)–E3(−3) occur around 68 and 80 kV/cm, respectively; however, we cannot observe them in the energies of the ER signals as a function of electric field (Fig. 3.26). Therefore, the observed anticrossing around 57 kV/cm is attributable to the E1(0)–E2(−2) coupling. The missing of the E3-related coupling in the experiments indicates that the transfer-matrix calculation of the E3 states is quite inaccurate. The following possible reasons for the inaccuracy of the E3 states are considered: The boundary condition is inappropriate for the E3 states whose energies extend over the potential (~280 meV) of the $Al_{0.3}Ga_{0.7}As$ clad layer, and/or the boundary between the clad layer and the modeled SL remarkably modifies the calculated eigenenergies because of the limited number of the QW's. In [95] various types of the resonant coupling related

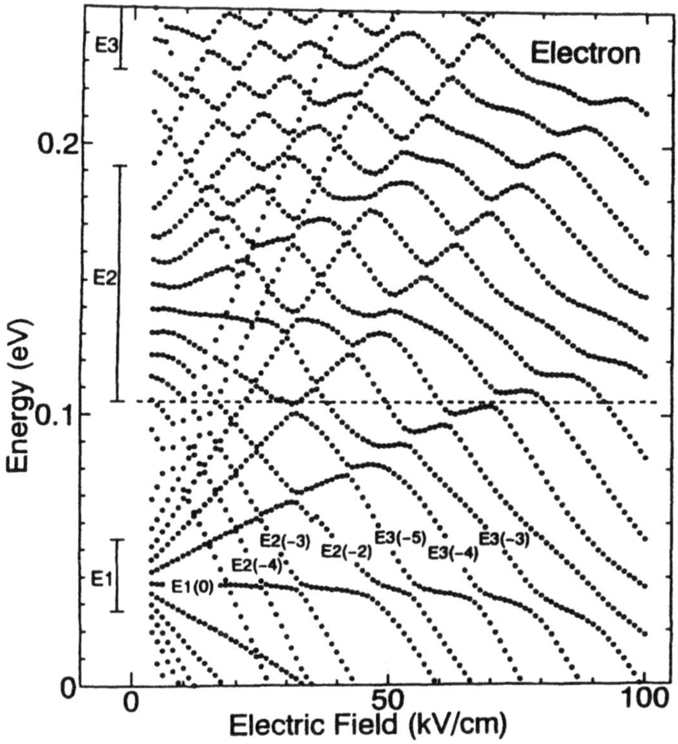


Figure 3.28. Electric-field dependence of the eigenenergies of the electron states in the modeled GaAs(7.0 nm)/$Al_{0.1}Ga_{0.9}As$(3.5 nm) SL calculated by using the transfer-matrix method with the Airy functions [95], where the energies are measured with respect to the center of the modeled SL, and the dashed line indicates the barrier potential of electrons. The vertical lines on the left side indicate the calculated energy ranges of the minibands at zero field.

to the above-barrier states of the electron, heavy hole, and light hole are reported. In conclusion, it is demonstrated that the above-barrier minibands are transformed into the Stark-ladder states and that the above-barrier Stark-ladder state resonantly couples with the below-barrier state.

## 3.7. BLOCH OSCILLATIONS

### 3.7.1. *Concepts of Bloch Oscillations*

The dynamical (time-domain) counterpart of the Stark-ladder formation resulting from the WS localization corresponds to the Bloch oscillation. In a semiclassical treatment with neglecting of scattering processes, an applied electric field changes the wave vector of a Bloch electron along the direction of the field according to the so-called acceleration theorem:

$$k(t) = k_0 + \frac{eF}{\hbar}t, \tag{3.29}$$

where $k_0$ is an initial wave vector. When a Bloch electron in a SL with a period $D$ reaches a boundary of the mini-Brillouin zone ($\pm\pi/D$), it is reflected to the opposite-side boundary: This results in the Bloch oscillation with a time constant $\tau_B$ (frequency $\omega_B$) of

$$\tau_B = h/eFD; \quad \omega_B = eFD/\hbar. \tag{3.30}$$

Although the above semiclassical description is intuitive, it is not accurate because the wave vector, which is unstationary owing to the acceleration, is not a good quantum number in the WS localization. Dignam, Sipe, and Shah [98] reported the theoretical study of the time evolution of electron and exciton wave packets with the nature of the Bloch oscillation in a SL within the framework of the nearest-neighbor tight-binding model described in Sec. 3.3.1. According to [98], the wave packet of electrons moving along the SL direction are represented by a superposition of eigenfunctions with amplitudes $C_\nu$ of the Stark-ladder states ($\epsilon_0 + \nu eFD$):

$$\varphi(z,t) = \sum_\nu C_\nu \exp\left(-i\frac{\nu eFD}{\hbar}t\right)\sum_m J_{m-\nu}(L/D)f(z-mD), \tag{3.31}$$

where $L$ is the so-called localization length ($L = \Delta/2eF$) which has been explained in Sec. 3.3.1. Using Bessel-function recursion relations and sum rules, we can obtain that the expectation value of $z$ in $\varphi(z,t)$ is given by

$$\langle z(t)\rangle = D\sum_\nu C_\nu^2\nu + \cos(\omega_B t)L\sum_\nu C_{\nu-1}C_\nu. \tag{3.32}$$

We find that the motion of the wave packet is periodic with the frequency of $\omega_B$ and the amplitude of $A_z \equiv L|\sum_\nu C_{\nu-1}C_\nu|$ which ranges from zero to $L$. The semiclassical treatment gives the upper limit of $A_z = L$. It should be noted that the realistic amplitude of the Bloch oscillation remarkably depends on the condition of the superposition of eigenfunctions. Dignam, Sipe, and Shah [98] precisely discussed the Bloch oscillation initiated by optical excitations with ultrashort laser pulses. Here, we assume that the excitation optical field is given by [98]

$$E(t) = E_{(-)}\exp(-\Omega^2t^2/2)\exp(-i\omega t) + \text{c.c.}, \tag{3.33}$$

where $\hbar\omega$ and $\hbar\Omega$ are the central energy and the energy width of the excitation laser, respectively. In [98] it is pointed out that the Bloch-oscillation amplitude depends on $L/D$, $\Omega/\omega_B$, and $\omega_c/\omega_B$, where $\omega_c = \omega - E_0/\hbar$ ($E_0$ is the energy of the Stark-ladder transition with $m$=0). For example, if $\Omega$=0, $\Omega/\omega_B \gg 1$, or $\omega_c$=0, the Bloch oscillation is not observable, corresponding to $A_z$=0. In $\Omega$=0, the only one level is excited, so that the wave packet is stationary. In $\Omega/\omega_B \gg 1$ or $\omega_c$=0, the oscillation amplitude is in a symmetrically breathing mode which results in canceling the radiation amplitude. The calculated amplitude exhibits a maximum for $\Omega/\omega_B$ between 1 and 2. The above theory takes account of only a single miniband, though there are several minibands in a real SL. Bloch-electron dynamics in a multiband condition has been controversial because of the existence of interband interactions. Recently, a multiband theory of the Bloch-electron dynamics including the Bloch oscillation and Zener tunneling was extensively discussed by He and Iafrate [99].

### 3.7.2. *Observation of Bloch Oscillations*

The observation of the Bloch oscillation had been expected since the proposal of SL in 1970. For the optical observation, Plessen and Thomas [28] theoretically suggested an experimental method by measuring spontaneous photon-echo signals in a time-resolved degenerate-four-wave-mixing (DFWM) experiment. Feldmann *et al.* [29] and Leo *et al.* [30] reported the observation of oscillations of photon-echo signals in a sub-ps time region. Waschke *et al.* [31] directly observed a coherent THz electromagnetic radiation generated by ultrashort laser pulses. Dekorsy *et al.* [100] demonstrated an electro-optic sampling method for the detection of the temporal change of coherent interband polarization induced by the Bloch oscillation. Very recently, Leisching *et al.* [101] reported a comprehensive study of the Bloch oscillation by using the time-resolved DFWM method. They demonstrated the excitation-condition dependence of the Bloch oscillation as described in the above subsection.

Figure 3.29 shows (a) schematic diagram of the time-resolved DFWM for the observation of photon-echo signals, where the signal in the direction $\mathbf{k}_3$ is detected as a function of a delay time $\tau$, and (b) the semilogarithmic plot of the DFWM signal at 8.5 K as a function of the delay time for various electric fields in a GaAs(10.0 nm)/$Al_{0.3}Ga_{0.7}As$(1.7 nm) SL (40 periods) grown on an $n$-type (001) GaAs substrate [30]. The envelope-function approximation gives the miniband width of the $n$=1 electron (heavy hole) of about 19 (2) meV. In this experiment of the DFWM, a coherent optical polarization is produced in the sample at $t$=0 by the first laser pulse (~100 fs) with the wave vector $\mathbf{k}_1$. This coherent polarization develops in time. After the delay time $\tau$, the second pulse with the wave vector $\mathbf{k}_2$ is applied. The second pulse forms a grating with the polarization left over from the first pulse. A part of the second pulse is diffracted into the direction $\mathbf{k}_3 = 2\mathbf{k}_2 - \mathbf{k}_1$ by this grating. The decay of the DFWM signal is determined by the loss of the coherence of the excited polarization with the so-called dephasing time $T_2$ related to scattering processes resulting in a homogeneous broadening. The photon-echo signal related to the Bloch oscillation (an inhomogeneous broadening) occurs at $t = 2\tau$ in the direction $\mathbf{k}_3$. In Fig. 3.29 the

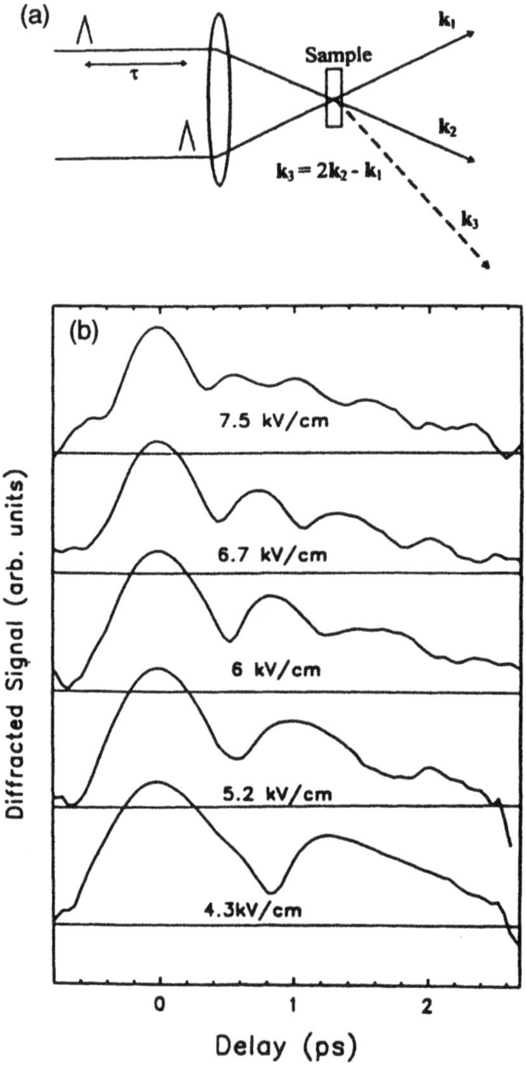


Figure 3.29. (a) Schematic diagram of the time-resolved DFWM for the observation of photon-echo signals, where the signal in the direction $\mathbf{k}_3$ is detected as a function of a delay time $\tau$, and (b) the semilogarithmic plot of the DFWM signal at 8.5 K as a function of the delay time for various electric fields in a GaAs(10.0 nm)/$Al_{0.3}Ga_{0.7}As$(1.7 nm) SL. (After [30]).

DFWM signal exhibits a periodic modulation with a period which remarkably depends on the electric field. The overall decay of the DFWM signal reflects the dephasing of the excited polarization: $T_2$=1.5 ps in this case. The energy estimated from the observed oscillation period agrees with $eFD$: This demonstrates that the modulation of the DFWM signal is caused by the Bloch oscillation. Waschke *et al.* [31] reported the direct observation of the electromagnetic radiation in the THz-frequency region resulting from the Bloch oscillation in a GaAs(9.7 nm)/$Al_{0.3}Ga_{0.7}As$(1.7 nm) SL (40 periods) grown on an $n$-type (001) GaAs substrate: The sample structure is a quite similar to that used in [30]. In the experiment, the sample was excited by 100-fs laser pulses, and the electromagnetic radiation was monitored in the time domain. The observed frequency of the radiation is proportional to the electric field in the WS-localization regime.

The Bloch oscillation is broken by the optically-excited-state dephasing induced by scattering processes: the decrease of the oscillation amplitude in time, only up to four or five oscillations as shown in Fig. 3.29. Plessen *et al.* [61] pointed out that the Bloch oscillation cannot be observed in a GaAs/AlGaAs SL with the miniband width larger than the GaAs-LO phonon energy (~37 meV) because the electron-LO-phonon scattering suppresses the Bloch oscillation. On the other hand, Leisching *et al.* [101] reported that there is no sign of the LO-phonon threshold for the Bloch oscillation. One considers that the observation of the Bloch oscillation in a sample with a large miniband width is remarkably affected by the sample quality because the layer-thickness fluctuation, which considerably modifies the miniband width, reduces the coherency of wave packets. It is noted that the Bloch oscillation has been elusive in transport experiments, save for the reports by Sibille *et al.* [102–104] and Beltram *et al.* [105] of the observation of negative differential conductance ascribed to Bloch-oscillatory phase breaking due to the onset of scattering.

## 3.8. SUMMARY

A review of the WS localization in semiconductor SL's has been presented from the viewpoint of the optical properties. Research works of this subject including device applications have been performed extensively and growing up. The WS localization provides us with a novel concept of the dimensionality crossover of electronic systems induced by an applied electric field: the transformation from extended miniband states (3D) to localized Stark-ladder states (quasi-2D) and vice verse. In addition, the resonant coupling between spatially-separated Stark-ladder states results in the wave-function mixing and delocalization over the coupling space. The wide controllability of the Stark-ladder energy and the localization length enables us to develop new classes of electro-optic devices. The coherent dynamics of the wave packets of the Stark-ladder states corresponds to the Bloch oscillation, which is one of the most attractive phenomena in solid state physics nowadays. The investigation of the Bloch oscillation leads to a coherent THz-radiation oscillator proposed at the start of the history of the SL.

## REFERENCES

[1] F. Bloch, *Z. Phys.* **52**, 555 (1928).

[2] C. Zener, *Proc. R. Soc. London Ser. A* **145**, 523 (1934).

[3] G. H. Wannier, *Phys. Rev.* **117**, 432 (1960); *Rev. Mod. Phys.* **34**, 645 (1962).

[4] H. M. James, *Phys. Rev.* **76**, 1602 (1949).

[5] S. Katsura, T. Hatta, and A. Morita, *Sci. Rep. Tohoku Imp. Univ., Ser.I,* **34**, 19 (1950).

[6] P. Feuer, *Phys. Rev.* **88**, 92 (1952).

[7] J. Zak, *Phys. Rev. Lett.* **20**, 1477 (1968).

[8] G. H. Wannier, *Phys. Rev.* **181**, 1364 (1969); J. Zak, *Phys. Rev.* **181**, 1366 (1969).

[9] A. Rabinovitch and J. Zak, *Phys. Rev. B* **4**, 2358 (1971).

[10] M. Saitoh, *J. Phys. C* **5**, 914 (1972).

[11] H. Fukuyama, R.A. Bari, and H.C. Fogedby, *Phys. Rev. B* **8**, 5579 (1973).

[12] A. Rabinovitch, *Phys. Lett. A* **59**, 475 (1977).

[13] S. Nagai and J. Kondo, *J. Phys. Soc. Jpn.* **49**, 1255 (1980).

[14] F. Bentosela, V. Grecchi, and F. Zironi, *Phys. Rev. B* **31**, 6909 (1985).

[15] D. Emin and C.F. Hart, it Phys. Rev. B **36**, 7353 (1987): Comments for this paper and the reply to comments; D. A. Page and E. Brown, **43**, 2423 (1991); D. Emin and C. F. Hart, **43**, 2426 (1991); J. Zak, **43**, 4519 (1991); C. F. Hart and D. Emin, **43**, 4521 (1991); J. Leo and A. MacKinnon, **43**, 5166 (1991); D. Emin and C. F. Hart, **43**, 5169 (1991).

[16] J. Bleuse, G. Bastard, and P. Voisin, *Phys. Rev. Lett.* **60**, 220 (1988).

[17] G. Nenciu, *Rev. Mod. Phys.* **63**, 91 (1991).

[18] O. Gülseren and S. Ciraci, *Phys. Rev. B* **46**, 7621 (1992).

[19] L. Esaki and R. Tsu, *IBM J. Res. Dev.* **14**, 61 (1970).

[20] See *e.g.*, C. Kittel, in *Introduction to Solid State Physics* (John Wiley Sons, New York, 1986), p. 163–172.

[21] For a review, in *Semiconductors and Semimetals* vol.24, ed. by R. Dingle (Academic Press, San Diego, 1987).

[22] E. E. Mendez, F. Aguulló-Rueda, and J. M. Hong, *Phys. Rev. Lett.* **60**, 2426 (1988).

[23] P. Voisin, J. Bleuse, C. Bouche, S. Gaillard, C. Alibert, and A. Regreny, *Phys. Rev. Lett.* **61**, 1639 (1988).

[24] H. Schneider, H. T. Grahn, K. v. Klitzing, and K. Ploog, *Phys. Rev. Lett.* **65**, 2720 (1990).

[25] M. Nakayama, I. Tanaka, H. Nishimura, K. Kawashima, K. Fujiwara, *Phys. Rev. B* **44**, 5935 (1991).

[26] I. Tanaka, M. Nakayama, H. Nishimura, K. Kawashima, and K. Fujiwara, *Phys. Rev. B* **46**, 7656 (1992).

[27] I. Bar-Joseph, K. W. Goossen, J. M. Kuo, R. F. Kopf, D. A. B. Miller, and D. S. Chemla, *Appl. Phys. Lett.* **55**, 340 (1989).

[28] G. v. Plessen and P. Thomas, *Phys. Rev. B* **45**, 9185 (1992).

[29] J. Feldmann, K. Leo, J. Shah, D. A. B. Miller, J. E. Cunningham, T. Meier, G. v. Plessen, A. Schulze, P. Thomas, and S. Schmitt-Rink, *Phys. Rev. B* **46**, 7252 (1992).

[30] K. Leo, P. Haring Bolivar, F. Bruggemann, and R. Schwedler, *Solid State Commun.* **84**, 943 (1992).

[31] C. Waschke, H. G. Roskos, R. Schwedler, K. Leo, H. Kurz, and K. Köhler, *Phys. Rev. Lett.* **70**, 3319 (1993).

[32] R. Tsu and L. Esaki, *Phys. Rev. B* **46**, 5204 (1991).

[33] R. Tsu, in *Semiconductor Interfaces, Microstructures, and Devices*, ed. by Z.C. Feng (Institute of Physics Publishing, Bristol, 1993), p. 3.

[34] F. Beltram and F. Cappaso, in *Physics of Low-Dimensional Semiconductor Structures*, ed. by P. Butcher, N.H. March, and M.P. Tosi (Plenum Press, New York, 1993), p. 539.

[35] G. A. Sai-Halasz, R. Tsu, and L. Esaki, *Appl. Phys. Lett.* **30**, 651 (1977).

[36] J. Tersoff, *Phys. Rev. B* **30**, 4874(1984).

[37] C. G. Van de Walle and R. M. Martin, *Phys. Rev. B* **35**, 8154(1987).

[38] D. A. Baldereschi, S. Baroni, and R. Resta, *Phys. Rev. Lett.* **61**, 734 (1988).

[39] N. E. Chiristensen, *Phys. Rev. B* **37**,4528(1988).

[40] C. G. Van de Walle, *Phys. Rev. B* **39**, 1987(1989)

[41] G. Bastard, *Phys. Rev. B* **24**, 5693 (1981); *Phys. Rev. B* **25**, 7584 (1982).

[42] E. O. Kane, in *Semiconductors and Semimetals* vol.**1**, ed. by R. K. Willardson and A. C. Beer (Academic Press, New York, 1966), p. 75.

[43] For a review, M. Altarelli, in *Heterojunctions and Semiconductor Superlattices*, ed. by G. Allan, G. Bastard, N. Boccara, M. Lannoo, and M. Voos, (Springer, Berlin, 1986), p. 12.

[44] D. F. Nelson, R. C. Miller, C. W. Tu, and S. K. Sputz: *Phys. Rev. B* **36**, 8063 (1987).

[45] H. A. Antosiewicz, in *Handbook of Mathematical Functions*, ed. by M. Abramowitz and I.A. Stegun (Dover, New York, 1970), p. 435–478.

[46] D. C. Huchings, *Appl. Phys. Lett.* **55**, 1082 (1989).

[47] M. Nakayama, I. Tanaka, T. Doguchi, H. Nishimura, K. Kawashima, and K. Fujiwara, *Solid State Commun.* **77**, 303 (1991).

[48] For a review, D. E. Aspnes, *Surface Sci.* **37**, 418 (1973).

[49] B. O. Seraphin and N. Bottka, *Phys. Rev.* **145**, 628 (1966).

[50] P. C. Klipstein and N. Apsley, *J. Phys. C* **19**, 6461 (1986).

[51] B. V. Shanabrook, O. J. Glembocki, and W. T. Beard, *Phys. Rev. B* **35**, 2540 (1987).

[52] H. Shen, S. H. Pan, F. H. Pollak, M. Dutta, and T. R. AuCoin, *Phys. Rev. B* **36**, 9384 (1987).

[53] W. M. Theis, G. D. Sanders, C. E. Leak, K. K. Bajaj, H. Morkoç, *Phys. Rev. B* **37**, 3042 (1988).

[54] U. Behn, H. T. Grahn, K. Ploog, and H. Schneider, *Phys. Rev. B* **48**, 11827 (1993).

[55] D. A. B. Miller, D. S. Chemla, T. C. Damen, A. C. Gossard, W. Wiegmann, T. H. Wood, and C. A. Burrus, *Phys. Rev. B* **32**, 1043 (1985).

[56] K. Gibb, M. M. Dignam, J. E. Sipe, and A. P. Roth, *Phys. Rev. B* **48**, 8156 (1993).

[57] F. Agulló-Rueda, E. E. Mendez, and J. M. Hong, *Phys. Rev. B* **38**, 12720 (1988).

[58] H. Schneider, J. Wagner, K. Fujiwara, and K. Ploog, *Phys. Rev. B* **42**, 11430 (1990)

[59] I. Tanaka, M. Nakayama, H. Nishimura, K. Kawashima, and K. Fujiwara, *Phys. Rev. B* **48**, 2787 (1993).

[60] K. Fujiwara, *Jpn. J. Appl. Phys.* **28**, L1718 (1989).

[61] G. v. Plessen, T. Meier, J. Feldmann, E. O. Göbel, P. Thomas, K. W. Goosen, J. M. Kuo, and R. F. Kopf, *Phys. Rev. B* **49**, 14058 (1994).

[62] F. Agulló-Rueda, E. E. Mendez, and J. M. Hong, *Phys. Rev. B* **40**, 1357 (1989).

[63] E. E. Mendez, F. Agulló-Rueda, and J. M. Hong, *Appl. Phys. Lett.* **56**, 2545 (1990).

[64] J. L. Bradshaw and R. P. Leavitt, *Phys. Rev. B* **50**, 17666 (1994).

[65] H. Schneider, A. Fischer, and K. Ploog, *Phys. Rev. B* **45**, 6329 (1992).

[66] E. Ribeiro, F. Cerdeira, and A. P. Roth, *Phys. Rev. B* **46**, 12542 (1992).

[67] K. H. Schmidt, N. Linder, G. H. Döhler, H. T. Grahn, K. Ploog, and H. Schneider, *Phys. Rev. Lett.* **72**, 2769 (1994).

[68] R. H. Yan, F. Laruelle, and L. A. Coldren, *Appl. Phys. Lett.* **55**, 2002 (1989).

[69] D. M. Whittaker, *Phys. Rev. B* **41**, 3238 (1990).

[70] R. P. Leavitt and J. W. Little, *Phys. Rev. B* **42**, 11784 (1990).

[71] M. M. Dignam and J. E. Sipe, *Phys. Rev. B* **43**, 4097 (1991).

[72] A. M. Fox, D. A. B. Miller, J. E. Cunningham, W. Y. Jan, C. Y. P. Chao, and S. L. Chuang, *Phys. Rev. B* **46**, 15365 (1992).

[73] I. Tanaka, M. Nakayama, H. Nishimura, K. Kawashima, and K. Fujiwara, *Solid State Commun.* **92**, 385 (1994).

[74] Y. Zhang, D. Jiang, F. Li, J. Zhou, and X. Mei, *J. Appl. Phys.* **72**, 3209 (1992).

[75] For a review, D. Gershoni and H. Temkin, *J. Lumin.* **44**, 381 (1989).

[76] B. Soucail, N. Dupuis, R. Ferreira, P. Voisin, A.P. Roth, D. Morris, K. Gibb, and C. Lacelle, *Phys. Rev. B* **41**, 8568 (1990).

[77] M. K. Saker, D. M. Whittaker, M. S. Skolnick, M. T. Emeny, and C. R. Whitehouse, *Phys. Rev. B* **43**, 4945 (1991).

[78] Y. Kajikawa, *Superlattices and Microstruc.* **9**, 391 (1991).

[79] For a review, L. J. Sham and Y.T. Lu, *J. Lumin.* **44**, 207 (1989).

[80] A. J. Shields, P. C. Klipstein, M. S. Skolnick, G. W. Smith, and C. R. Whitehouse, *Phys. Rev. B* **42**, 5879 (1991).

[81] H. Schneider, K. Fujiwara, H. T. Grahn, K. v. Klitzing, and K. Ploog, *Appl. Phys. Lett.* **56**, 605 (1990).

[82] K. K. Law, R. H. Yan, L. A. Coldren, and J. L. Merz, *Appl. Phys. Lett.* **57**, 1345 (1990).

[83] K. Kawashima, K. Fujiwara, T. Yamamoto, M. Shigeta, and K. Kobayashi, *Jpn. J. Appl. Phys.* **60**, L1542 (1991).

[84] K. Kawashima, M. Hosoda, and K. Fujiwara, *Appl. Phys. Lett.* **62**, 184 (1993).

[85] K. K. Law, R. H. Yan, J. L. Merz, and L. A. Coldren, *Appl. Phys. Lett.* **56**, 1886 (1990).

[86] F. Devaux, E. Bigan, M. Allovon, J. C. Harmand, F. Huet, M. Carre, and J. Landreau, *Appl. Phys. Lett.* **61**, 2773 (1992).

[87] D. A. B. Miller, D. S. Chemla, T. C. Damen, A. C. Gossard, W. Wiegmann, T. H. Wood, and C. A. Burrus, *Appl. Phys. Lett.* **45**, 13 (1984); for a review, D. A. B. Miller, D. S. Chemla, T. C. Damen, T. H. Wood, C. A. Burrus, A. C. Gossard, and W. Wiegmann, *IEEE J. Quantum Electron.* **QE-21**, 1462 (1985).

[88] G. Bastard, R. Ferreira, S. Chelles, and P. Voisin, *Phys. Rev. B* **50**, 4445 (1994).

[89] J. E. Zucker, A. Pinczuk, D. S. Chemla, A. C. Gossard, W. Wiegmann, *Phys. Rev. B* **29**, 7065 (1984).

[90] J. J. Song, Y. S. Yoon, P. S. Fedtowsky, Y. B. Kim, J. N. Schulman, C. W. Tu, J. M. Brown, D. Huang, and H. Morkoç *Phys. Rev. B* **34**, 8958 (1986).

[91] S. H. Pan, H. Shen, Z. Hang, F. H. Pollak, W. Zhuang, Q. Xu, A. P. Roth, R. A. Masut, C. Lacelle, and D. Morris, *Phys. Rev. B* **38**, 3375 (1988).

[92] F. C. Zhang, N. Dai, H. Luo, N. Samarth, M. Dobrowolska, J. K. Furdyna, and L. R. Ram-Mohan, *Phys. Rev. Lett.* **68**, 3220 (1992).

[93] C. Parks, K. Ramdas, M. R. Melloch. and L. R. Ram-Mohan, *Phys. Rev. B* **48**, 5413 (1993).

[94] M. Nakayama, R. Sugie, H. Ohta, and S. Nakashima, *Jpn. J. Appl. Phys.* **34** *Suppl.* 34-1, 80 (1995).

[95] M. Nakayama, M. Ando, I. Tanaka, H. Nishimura, H. Schneider, and K. Fujiwara, *Phys. Rev. B* **51**, 4236 (1995).

[96] T. Kita, T. Inazumi, H. Nakayama, and T. Nishino, *Phys. Rev. B* **50**, 2420 (1994).

[97] M. Ritze, N. J. M. Horing, and R. Enderlein, *Phys. Rev. B* **47**, 10437 (1993)

[98] M. Dignam, J. E. Sipe, J. Shah, *Phys. Rev. B* **49**, 10502 (1994).

[99] J. He and G. J. Iafrate, *Phys. Rev. B* **50**, 7553 (1994).

[100] T. Dekorsy, P. Leisching, K. Köhler, and H. Kurz, *Phys. Rev. B* **50**, 8106 (1994).

[101] P. Leisching, P. Haring Bolivar, W. Beck, Y. Dhaibi, F. Brüggemann, R. Schwedler, H. Kurz, K. Leo, and K. Köhler, *Phys. Rev. B* **50**, 14389 (1994).

[102] A. Sibile, J. F. Palmier, F. Mollot, H. Wang, and J. C. Esnault, *Phys. Rev. B* **39**, 6272 (1989).

[103] A. Sibille, J. F. Palmier, H. Wang, and F. Mollot, *Phys. Rev. Lett.* **64**, 52 (1990).

[104] A. Sibille, J. F. Palmier, and F. Mollot, *Appl. Phys. Lett.* **60**, 457 (1992)

[105] F. Beltram, F. Capasso, D. L. Sivco, A. L. Hutchinson, S. G. Chu, and A. Y. Cho, *Phys. Rev. Lett.* **64**, 3167 (1990).

## *CHAPTER 4*

# ULTRAVIOLET LASER EMISSION FROM ZnS-BASED QUANTUM WELLS

Yoichi Yamada*

*Institute of Physics, University of Tsukuba*



### 4.1. INTRODUCTION

Since the pioneering work of Esaki and Tsu [1] in the early seventies, we have seen an enormous increase in research in the field of semiconductor heterostructures, and a considerable amount of results have been obtained as regards the physical and device properties of semiconductor superlattices and quantum-well structures. The research on alternately layered semiconductor heterostructures has expanded rapidly as a result of the advent of high precision growth techniques, mainly molecular beam

*Present affiliation: Department of Electrical and Electronic Engineering, Yamaguchi University.

epitaxy (MBE) and metalorganic chemical vapor deposition (MOCVD). The emergence of new semiconductor structures has facilitated a number of approaches to producing systems with certain features in their electronic structures which can lead to unique or interesting properties.

While the growth technique has continued to make progress, II-VI compound semiconductors are becoming of increasing interest. Since the early eighties, the growth techniques such as MBE and MOCVD have successfully been extended to the growth of II-VI compound semiconductors. The range of the II-VI materials, which can be combined to produce heterostructures, provides a wealth of opportunity for physical investigations and a new flexibility in the design of novel device structures. From the point of view of optical properties, the II-VI compound semiconductors are unmatched in their range of bandgaps from ultraviolet to far infrared. Most of the interest here is in wide-bandgap II-VI materials such as zinc sulfide (ZnS), zinc selenide (ZnSe), cadmium sulfide (CdS), and their alloys. The current exciting optoelectronic applications for wide-bandgap II-VI compound semiconductors are injection laser diodes and light-emitting diodes operating in short wavelength regions. The first successful report of blue-green injection laser diodes based on $Cd_xZn_{1-x}Se$-ZnSe quantum-well structures embedded in a ZnSe p-n junction [2, 3] has significantly triggered much interest in this research field of both basic physics and device applications. It is through the successful development of doping techniques for p-type conductivity control that the device application has been achieved. Most recently, the one-hour-long operation of blue-green injection laser diodes at room temperature under continuous wave mode has been reported in $Cd_xZn_{1-x}Se$ multiple quantum-well structures combined with $ZnS_ySe_{1-y}$ ternary alloys as optical guiding layers and $Zn_xMg_{1-x}S_ySe_{1-y}$ quarternary alloys as cladding layers [4]. Furthermore, the high-brightness operation of blue and green light-emitting diodes grown homoepitaxially on ZnSe substrates has also been reported [5]. Besides the successful reports mentioned above, many kinds of wide-bandgap II-VI semiconductor heterostructures have been proposed and fabricated for realization of optoelectronic devices operating in the short wavelength regions [6, 7, 8, 9, 10, 11].

Among wide-bandgap II-VI compound semiconductors, ZnS has a large bandgap energy of about 3.73 eV at room temperature, so that ZnS-based quantum-well structures exhibit fundamental absorption edges in the ultraviolet region of the spectrum. Moreover, ZnS has a large exciton binding energy of about 40 meV. This value is approximately one order of magnitude larger than that of GaAs, and is also comparable to the LO phonon energy ($\hbar\omega_{LO}$ = 42 meV). Therefore, it is expected that quasi-two-dimensional excitons in ZnS-based quantum wells have very large binding energies owing to the effect of quantum confinement and that the exciton binding energies exceed the LO phonon energy. Such a situation makes it possible to reduce an exciton-LO phonon coupling [12]. As a result, pronounced excitonic effects can be observed even at room temperature in spite of the much stronger exciton-LO phonon coupling in the wide-bandgap II-VI compound semiconductors compared to that in III-V compound semiconductors.

In this chapter, optical properties of ZnS-based quantum-well structures are reviewed on the basis of our recent experimental results. The subject of this article concentrates on two physical aspects in a new type of wide-bandgap II-VI semiconductor strained-layer quantum-well structures consisting of alternate layers of $Cd_xZn_{1-x}S$ and ZnS. One focus is to understand band lineups at heterointerfaces. Most of the heterostructures consisting of wide-bandgap II-VI semiconductors include large elastic layer strains due to a lattice constant mismatch. These strains directly affect the band lineups and consequently modify the optical properties. The effects of strains on the band lineups and electronic structures are analyzed by semi-quantitative calculations which are based on the model-solid theory [13]. The calculated results agree with the experimental observations of photoluminescence excitation measurements. The agreement shows the validity of the approach. The other focus is to understand the role of quasi-two-dimensional excitons. The excitons, which dominate the optical properties, have the possibility to show the characteristic phenomenon which can never be observed in III-V quantum-well structures because of much larger exciton binding energy and its oscillator strength. The most characteristic phenomenon appears in the process of optical gain formation. It is shown by pump-and-probe experiments that quasi-two-dimensional excitons contribute to the formation of optical gain. This gain mechanism is quite different from that in III-V quantum-well structures where an electron-hole plasma plays a principal role in providing gain. Optical gain spectra are also derived from a phenomenological analysis which is based on a phase-space-filling argument for an inhomogeneously broadened system [14]. This analysis means that excitons are localized at the lower energy tail of the inhomogeneously broadened exciton resonance and that the partial phase-space filling of these localized states takes place. It is shown that experimental results are well explained by the analysis. As a result, it is proposed that the formation of optical gain in the $Cd_xZn_{1-x}S$-ZnS quantum wells originates from the population inversion of localized exciton states. In addition to the above basic investigation, ultraviolet stimulated emission from $Cd_xZn_{1-x}S$-ZnS systems is observed under both optical pumping at room temperature and pulsed current injection at 30 K. The stimulated emission is characterized by the shortest wavelength reported to date among semiconductor laser materials. These observations strongly suggest that this structure is the most fruitful candidate which is capable of providing the semiconductor injection laser diodes operating in the ultraviolet spectral region.

## 4.2. MOCVD GROWTH AND STRUCTURES

We have emphasized so far that the low-pressure metalorganic chemical vapour deposition (MOCVD) method is the most useful and conventional process for fabricating thin layer semiconductor heterostructures. However, in the case of the II-VI compound semiconductors, it should be pointed out that there are several problems concerning the difficulty in precisely controlling the II/VI mole ratio because of the high dissociation pressure of VI elements, and also the premature reaction taking place in the vapour phase prior to epitaxial deposition on a substrate. In order to avoid such problems, we have used new metalorganic sources diluted in a He gas, namely, dimethylzinc (DMZn) with a concentration of 1.06 % in a He gas and dimethylcadmium (DMCd) with a concentration of 0.134 % in a He gas, for the growth of cubic-structured $Cd_xZn_{1-x}S$-ZnS strained-layer multiple quantum wells (MQWs) [15, 16, 17]. A mixture of 10 % $H_2S$ in a $H_2$ gas was used as the sulphur source.

Since DMZn, DMCd, and $H_2S$ gases react in part with W or KA resistance wires in low pressure, it is impossible to use it as a resistance heater in a MOCVD chamber during sufficient periods of the growth. On the other hand, although it is possible to use RF induction coil as an alternative heater, there is a possibility of dissociation of gases which produces radicals due to a plasma discharge in reduced pressures and then it is necessary to employ a large-scaled electric power supply in connection with precise electric circuit. Therefore, in general the MOCVD reaction chamber is basically dependent upon the heating system which fits for long growth run. Our MOCVD apparatus consists of a main reaction chamber and a sample exchange supplemental chamber with a transfer tube. Special feature is that we employ a halogen lamp capable of heating substrates up to 600 °C, by infrared (IR) radiation. Two chambers are constructed by all Al materials and each inner wall is coated by TiN films to avoid accidental impurity contamination.

In order to control the gas flow ratio of II/VI, we used a personal computer connected with a relay box linked to an interface which was capable of switching both the piezo valves of MFC (mass flow controller) and air valves on or off. Then, all of gas sources flowed directly into the MFCs, so that the bubbling system by a carrier gas was unnecessary for the metalorganic (DMZn and DMCd) gas flow as well as the $H_2S$ gas flow [18, 19]. Therefore, our process using the low-pressure MOCVD system with all gaseous sources has proven to be advantageous for precise control of II/VI mole ratio as well as lower-temperature growth (350 °C) of the II-VI compound heterostructures.

The MQW layers were grown on (100)-oriented GaAs substrates, following the deposition of a 1.5-$\mu$m-thick ZnS buffer layer. The MQW structures used in this study consist of 50 periods of 4.1-nm-thick $Cd_xZn_{1-x}S$ well layers separated by 8.1-nm-thick ZnS barrier layers. The Cd composition ratio in the ternary alloy well layer was varied from $x$=0.11 to $x$=0.49. In this case, the ternary alloy well layers are under biaxial compression and the lattice mismatch between the well and barrier layers varies from approximately 0.8 % to 3.8 %. The individual well and barrier layers are believed to be within a critical layer thickness as judged from the argument that the critical layer thickness of a ZnSe-ZnS strained-layer superlattice (SLs) with a large lattice mismatch of about 4.5 % has been determined to be about 10 nm by transmission electron microscopy [20].

## 4.3. BAND LINEUPS IN STRAINED-LAYER SYSTEMS

At a heterointerface consisting of two semiconductors with different lattice constants, all of the lattice constant mismatch can entirely be accommodated by elastic layer strains without the generation of misfit dislocations if the thicknesses of the materials are kept below a certain critical thickness for dislocation generation [21]. Band lineups

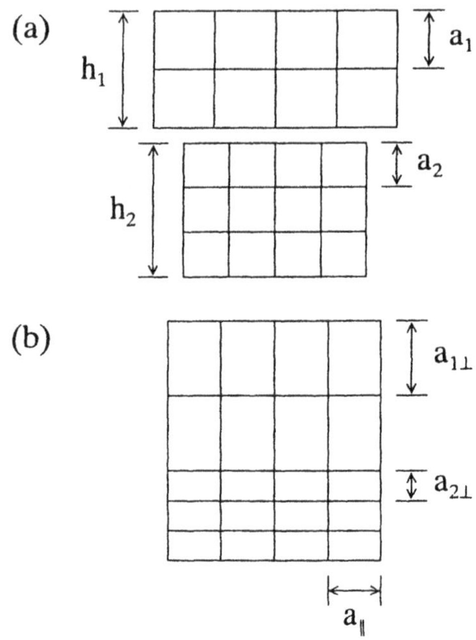


Figure 4.1. Schematic diagram of two-dimensional lattice constants of (a) bulk semiconductors (1 and 2), and (b) a strained-layer interface.

at semiconductor heterointerfaces are therefore affected by the elastic layer strain. In general, the uniaxial component of the strain breaks symmetry of crystals and then removes degeneracy of valence bands. In addition, the hydrostatic component of the strain shifts band edges of both conduction and valence bands. Consequently, electronic structures and optical properties are modified by the changes in the band lineups, especially the band discontinuity at the heterointerface for both conduction and valence bands.

In order to clarify the effects of these strains on the band lineups and electronic structures, semi-quantitative calculations which are based on the "model-solid" theory [13] are presented. In the theory, the center of gravity in the three uppermost valence bands (known as a heavy-hole band, a light-hole band, and a spin-orbital splitting band) is moved with strain and the effects of strain for the conduction- and the valence-band edges can individually be treated. The results can most simply be expressed in terms of an absolute energy level for each semiconductor. The calculations are applied to $Cd_xZn_{1-x}S$-ZnS strained-layer systems. The validity of the calculated results is discussed in comparison with the experimental observations of photoluminescence excitation measurements.

### 4.3.1. *Effects of Strains on Interface Structures*

The derivation of interface properties requires the determination of interface structures. All interfaces are assumed to be ideal, so that the appropriate strains in each of the materials to construct a pseudomorphic interface must be included for strained-layer interfaces. The strains in the pseudomorphic system can be determined by minimizing the macroscopic elastic energy, under the constraint that the lattice constant in the plane, $a_{\parallel}$, is the same throughout the structure as schematically shown in Fig. 4.1. For a system in which $h_1$ and $h_2$ are the respective thicknesses of the unstrained layers of semiconductor 1 and 2, the lattice constants $a_{\parallel}$ and $a_{\perp}$ which are parallel and perpendicular to the plane of the interface can be expressed [13];

$$a_{\parallel} = \frac{a_1 G_1 h_1 + a_2 G_2 h_2}{G_1 h_1 + G_2 h_2}, \tag{4.1}$$

$$a_{i\perp} = a_i \left[1 - D_i \left(\frac{a_{\parallel}}{a_i} - 1\right)\right], \tag{4.2}$$

where $i$ denotes the material (1 or 2), $a_i$ is the lattice constant of the unstrained bulk crystal, and $G_i$ is the shear modulus,

$$G_i = 2\left(c_{11}^i + 2c_{12}^i\right)\left(1 - \frac{D_i}{2}\right). \tag{4.3}$$

The constant $D$ depends on the elastic constants $c_{11}$, $c_{12}$, and $c_{44}$ of the respective materials, and on the interface orientation,

$$D^{001} = 2\frac{c_{12}}{c_{11}}, \tag{4.4}$$

$$D^{110} = \frac{c_{11} + 3c_{12} - 2c_{44}}{c_{11} + c_{12} + 2c_{44}}, \tag{4.5}$$

$$D^{111} = 2\frac{c_{11} + 2c_{12} - 2c_{44}}{c_{11} + 2c_{12} + 4c_{44}}. \tag{4.6}$$

The ratio of $a_{\parallel}$ and $a_{i\perp}$ to the unstrained lattice constants determines the strain components parallel and perpendicular to the interface;

$$\epsilon_{i\parallel} = \frac{a_{\parallel}}{a_i} - 1, \tag{4.7}$$

$$\epsilon_{i\perp} = \frac{a_{i\perp}}{a_i} - 1. \tag{4.8}$$

These formulas determine the strains in the layers and the interface structure. The lattice constant $a_{\parallel}$ parallel to the in-plane of the interface can flexibly be varied by modifying the thickness of each layer in Eq. (4.1), and Eqs. (4.2), (4.7) and (4.8) can be expressed as a function of $a_{\parallel}$.

### 4.3.2. *Derivation of Band Lineups*

First of all, the valence-band offset without strains ($\Delta E_V^0$), which is the energy difference between the top of degenerate valence bands in each of the materials, is determined. The determination of this value is thought to be of great significance in this band lineups calculation because both conduction- and valence-band offsets are under the control of it. Here, Harrison's values in the linear combination of atomic orbitals (LCAO) theory [22] are employed empirically.

From the bandgap energy ($E_g$) and the spin-orbital splitting energy ($\Delta_{so}$) in addition to the above value, the energy values $E_{V,av}^0$ and $E_C^0$ which are the average over the three uppermost valence bands (known as the heavy- and light-hole bands, and the spin-orbital splitting band) and the minimum of the conduction band at Γ-point in the tetragonal Brillouin zone without strains, respectively, can be obtained for each of the materials by the following relations:

$$E_V^0 = E_{V,av}^0 + \frac{\Delta_{so}}{3}, \tag{4.9}$$

$$E_C^0 = E_V^0 + E_g, \tag{4.10}$$

where $E_V^0$ is the maximum of the degenerate valence band at Γ-point without strains. These values indicate the relative energy positions, so that these individual "absolute values" have no physical meaning.

The hydrostatic component of the strain shifts the band edges for both conduction- and valence-bands. One usually considers only the relative shift of the conduction band with respect to the valence band in bulk semiconductors. This shift is expressed by a bandgap deformation potential constant ($a$). For the heterojunction structures, however, values for individual band edges are essential, since they influence the discontinuity at the heterointerface. These effects are expressed as follows,

$$\delta E_C = a_c \frac{\Delta\Omega}{\Omega}, \tag{4.11}$$

$$\delta E_{V,av} = a_v \frac{\Delta\Omega}{\Omega}, \tag{4.12}$$

where $a_c$ and $a_v$ are hydrostatic deformation potential constants for the conduction and valence bands, respectively. The hydrostatic bandgap deformation potential constant ($a = a_c - a_v$) can theoretically be divided into the conduction- and valence-band deformation potential constants ($a_c$ and $a_v$), using the calculated values of the ratio $a_c/a_v$ given by Camphausen *et al.* [23]. $\Delta\Omega/\Omega = (\epsilon_{xx} + \epsilon_{yy} + \epsilon_{zz})$ is the fractional volume change. Then, the energy positions, $E_C$ and $E_{V,av}$, are derived as follows,

$$E_C = E_C^0 + \delta E_C = E_C^0 + a_c \frac{\Delta\Omega}{\Omega}, \tag{4.13}$$

$$E_{V,av} = E_{V,av}^0 + \delta E_{V,av} = E_{V,av}^0 + a_v \frac{\Delta\Omega}{\Omega}. \tag{4.14}$$

The uniaxial component of the strain also splits the uppermost degenerate valence bands (the heavy- and light-hole bands). These bands are labeled by $E_{V,1}$(the light-hole band), $E_{V,2}$(the heavy-hole band), and $E_{V,3}$(the spin-orbital splitting band). The strain splittings themselves are proportional to the magnitude of the strain, and are conventionally expressed in terms of the deformation potentials. For the strain along [001], the following shifts of the individual valence bands with respect to the average $E_{V,av}$ are given [24];

$$\Delta E_{V,2} = \frac{1}{3}\Delta_{so} - \frac{1}{2}\delta E_{001}, \tag{4.15}$$

$$\Delta E_{V,1} = -\frac{1}{6}\Delta_{so} + \frac{1}{4}\delta E_{001} + \frac{1}{2}[\Delta_{so}^2 + \Delta_{so}\delta E_{001} + \frac{9}{4}(\delta E_{001})^2]^{\frac{1}{2}}, \tag{4.16}$$

$$\Delta E_{V,3} = -\frac{1}{6}\Delta_{so} + \frac{1}{4}\delta E_{001} - \frac{1}{2}[\Delta_{so}^2 + \Delta_{so}\delta E_{001} + \frac{9}{4}(\delta E_{001})^2]^{\frac{1}{2}}. \tag{4.17}$$

In these equations, $\delta E_{001}$ can be expressed in terms of the magnitude of the strain and the shear deformation potential constant ($b$) as follows,

$$\delta E_{001} = 2b(\epsilon_{zz} - \epsilon_{xx}). \tag{4.18}$$

From these formulas [Eqs. (4.11)–(4.18)] shown above, the energy positions of the conduction band ($E_C$) and the valence bands ($E_{V,1}$, $E_{V,2}$ and $E_{V,3}$) at Γ-point can be obtained, and the conduction- and valence-band offsets ($\Delta E_C$ and $\Delta E_V$) and the bandgap energies ($E_{g,1}$ and $E_{g,2}$) of each material (1 and 2) can be derived from these energy positions.

### 4.3.3. *Band Offsets in* $Cd_xZn_{1-x}S$-ZnS *Systems*

The calculations mentioned above are performed as regards $Cd_xZn_{1-x}S$-ZnS strained-layer systems. In this case, $Cd_xZn_{1-x}S$ well layers and ZnS barrier layers are subject to biaxial compressive and tensile strains, respectively, along the <100> and <010> directions in the (100) plane. The conduction- and valence-band offsets and bandgap energies modified by the strains are evaluated.

From Eqs. (4.13)–(4.18), the conduction-band offset($\Delta E_C$) and valence-band offsets ($\Delta E_{V,hh}$: heavy-hole-band offset and $\Delta E_{V,lh}$: light-hole-band offset) are derived as follows,

$$\Delta E_C = E_C^{ZnS} - E_C^{CdZnS}, \tag{4.19}$$

$$\Delta E_{V,hh} = E_{V,av}^{CdZnS} + \Delta E_{V,2}^{CdZnS} - (E_{V,av}^{ZnS} + \Delta E_{V,1}^{ZnS}), \tag{4.20}$$

$$\Delta E_{V,lh} = E_{V,av}^{CdZnS} + \Delta E_{V,1}^{CdZnS} - (E_{V,av}^{ZnS} + \Delta E_{V,1}^{ZnS}). \tag{4.21}$$

Here, the valence-band offsets are determined by the energy difference between each of the bands (heavy-hole band and light-hole band) in the $Cd_xZn_{1-x}S$ well layer and the top of the valence band in the ZnS barrier layer (light-hole band). Moreover, the

Table 4.1. Physical parameters of CdS and ZnS used for calculations of band lineups.

| | CdS | ZnS[a)] |
|---|---|---|
| Lattice constant | | |
| $a$ (Å) | 5.832[b)] | 5.4093 |
| Elastic constant | | |
| $C_{11}$ (Mbar) | 0.78[c)] | 1.067 |
| $C_{12}$ (Mbar) | 0.53[c)] | 0.666 |
| Deformation potential constant | | |
| $a_c - a_v$ (eV) | −3.1[d)] | −4.0 |
| $a_c$ (eV) | −2.08[e)] | −2.78 |
| $a_v$ (eV) | 1.02[e)] | 1.22 |
| $b$ (eV) | −4.7[c)] | −0.7 |
| Bandgap energy at 4.2 K | | |
| $E_g$ (eV) | 2.56 | 3.84 |
| Spin-orbital splitting energy | | |
| $\Delta_{so}$ (eV) | 0.07 | 0.072 |
| Exciton binding energy | | |
| $E_B^{1s}$ (meV) | 27 | 40 |
| Offset of the valence band at no strain | | |
| $\Delta E_V^0$ (eV) | 0.28 | |

a)Reference [25].
b)K. Ichino *et al.*, *J. Cryst. Growth* **117**, 527 (1992), and references therein.
c)Reference [27].
d)Reference [26].
e)From our calculations based on [23].

bandgap energies of both $Cd_xZn_{1-x}S$ and ZnS strained layers ($E_g^{CdZnS}$ and $E_g^{ZnS}$) are derived as follows,

$$E_g^{CdZnS} = E_C^{CdZnS} - (E_{V,av}^{CdZnS} + \Delta E_{V,2}^{CdZnS}), \tag{4.22}$$

$$E_g^{ZnS} = E_C^{ZnS} - (E_{V,av}^{ZnS} + \Delta E_{V,1}^{ZnS}). \tag{4.23}$$

Here, Eqs. (4.19)–(4.23) can be expressed as a function of the lattice constant $a_\parallel$ parallel to the in-plane of the interface, which is determined by modifying the respective thicknesses of $Cd_xZn_{1-x}S$ and ZnS layers.

Physical and material parameters of bulk CdS and ZnS [25] are listed in Table 4.1. The hydrostatic deformation potential constants ($a_c$ and $a_v$) of CdS with zincblende structures were calculated to be −2.08 eV and 1.02 eV using $a_c - a_v$=−3.1 eV [26] and $a_c/a_v$=−2.05 [23], respectively. The offset of the degenerate valence band in the

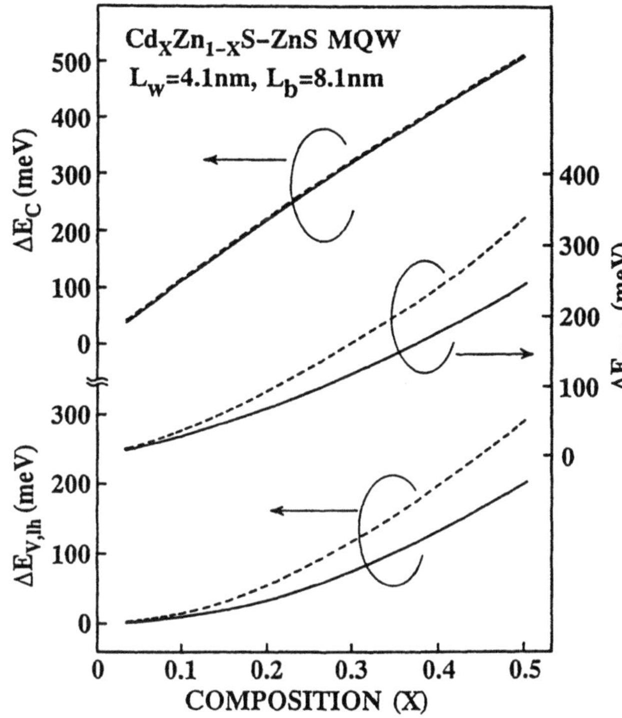


Figure 4.2. Calculated results of a conduction-band offset ($\Delta E_C$) and valence-band (heavy-hole and light-hole bands) offsets ($\Delta E_{V,hh}$ and $\Delta E_{V,lh}$) in a (100) $Cd_xZn_{1-x}S$-ZnS strained-layer quantum-well structure ($L_w$=4.1 nm and $L_b$=8.1 nm) as a function of Cd composition ratio in ternary alloy well layers. The solid curves indicate the results calculated in the case that the quantum-well structures are free-standing, while the dashed curves in the case that the well layers only are affected by strains.

absence of strain ($\Delta E_V^0$=0.28 eV) is estimated from the energy difference between $E_V$(ZnS)=−11.40 eV [22] and $E_V$(CdS)=−11.12 eV [22]. A shear deformation potential constant ($b$) has recently been estimated to be −4.7 eV in a cubic-structured CdS film using high-resolution angle-resolved synchrotron-radiation photoemission spectroscopy [27].

The bandgap energy of the unstrained $Cd_xZn_{1-x}S$ ternary alloy layer varies with composition $x$ and can be described by [28]

$$E_g(x) = E_g^{CdS} + (E_g^{ZnS} - E_g^{CdS} - b)(1-x) + b(1-x)^2, \tag{4.24}$$

where $b$ is a bowing parameter ($b = 0.3eV$). The lattice constant of the midcomposition alloy is calculated from Vegard's law, which states that the lattice constant of the

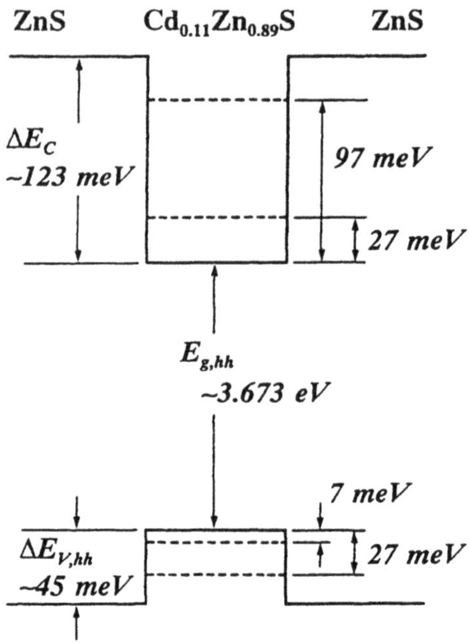


Figure 4.3. Calculated results of band lineups and energy levels for both electron and heavy-hole quantum states on the basis of Kronig-Penney's analysis in a $Cd_{0.11}Zn_{0.89}S$-ZnS strained-layer quantum well.

solid solution varies linearly with composition. Other parameters, such as the elastic constant and the deformation potential constant, are calculated by a linear interpolation from the binary materials data. Figure 4.2 shows the calculated results of the conduction-band offset ($\Delta E_C$), the heavy-hole-band offset ($\Delta E_{V,hh}$), and the light-hole-band offset ($\Delta E_{V,lh}$) as a function of the Cd composition ratio in the $Cd_xZn_{1-x}S$ ternary alloy well layer. In this calculation, the thicknesses of the $Cd_xZn_{1-x}S$ well and ZnS barrier layers are assumed to be 4.1 nm and 8.1 nm, respectively. The solid curves indicate the results calculated in the case that the quantum well structures are free-standing, while the dashed curves indicate the results calculated in the case that the well layers only are affected by the strains. It can be seen from this figure that this type of strained-layer quantum-well structure exhibits a type-I band lineup and that the conduction-band offset is relatively larger than the valence-band offset. These band offsets for both the conduction and valence bands increase with the Cd composition ratio in the ternary alloy well layer.

Figure 4.3 shows a band structure in a $Cd_{0.11}Zn_{0.89}S$-ZnS strained-layer quantum well, which is derived from the calculated results shown as the dashed curves in Fig. 4.2. As for the valence band, a heavy-hole band only is indicated. The

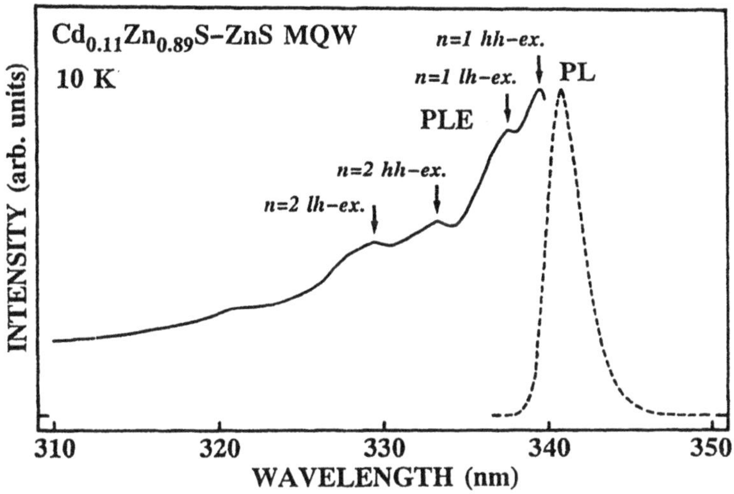


Figure 4.4. Photoluminescence (PL; dashed line) and photoluminescence excitation (PLE; solid line) spectra at 10 K taken from a $Cd_{0.11}Zn_{0.89}S$-ZnS strained-layer multiple quantum-well structure consisting of 50 periods of 4.1-nm-thick $Cd_{0.11}Zn_{0.89}S$ well layers separated by 8.1-nm-thick ZnS barrier layers.

conduction-band offset and the heavy-hole-band offset are estimated to be 123 meV and 45 meV, respectively. (The light-hole-band offset is estimated to be 16 meV.) The effective bandgap energy which is determined by the energy difference between the conduction band and the heavy-hole band at Γ-point in the $Cd_{0.11}Zn_{0.89}S$ well layer is estimated to be 3.673 eV. On the basis of Kronig-Penney's analysis, energy levels for both electron and heavy-hole quantum states are calculated and also shown in this figure. In this calculation, the effective masses of electrons and heavy holes are assumed to be $m_e^*/m_0$=0.39 [29] and $m_{hh}^*/m_0$=1.76 [30], respectively. It is understood from this figure that the quantum states up to n=2 are formed both in the conduction-band offset and in the heavy-hole-band offset. Optical transition energies concerning n=1 and n=2 heavy-hole excitons are, therefore, estimated to be 3.641 eV and 3.731 eV, respectively, on the assumption that the binding energy of the heavy-hole exciton is 66 meV which is derived from the correlation between the well layer thickness and the Bohr radius of bulk excitons [31]. Figure 4.4 shows photoluminescence (PL; dashed line) and photoluminescence excitation (PLE; solid line) spectra at 10 K taken from a $Cd_{0.11}Zn_{0.89}S$-ZnS strained-layer multiple quantum-well structure consisting of 50 periods of 4.1-nm-thick $Cd_{0.11}Zn_{0.89}S$ well layers separated by 8.1-nm-thick ZnS barrier layers. The PL spectrum is dominated by the radiative recombination of n=1 heavy-hole excitons, which shows the peak at 340.8 nm (3.637 eV). In the PLE spectrum, on the other hand, four peaks denoted by arrows are observed and these peaks are attributed to n=1 heavy-hole excitons (3.650 eV), n=1 light-hole excitons (3.672 eV), n=2 heavy-hole excitons (3.720 eV), and n=2 light-hole excitons (3.762 eV). The difference in peak energy between the PL and PLE spectra of the n=1 heavy-hole excitons is about 13 meV. This energy difference, defined as a Stokes shift, results from the localization of excitons. It is found from this figure that the exciton transition energies observed in the PLE spectrum approximately agree with the calculated results shown in Fig. 4.3. The validity of the calculations shown above has also been confirmed in the case of a ZnSe-ZnS strained-layer superlattice with a large lattice constant mismatch of about 4.5 % [32].

## 4.4. FORMATION OF OPTICAL GAIN IN II-VI QUANTUM WELLS

The quantum-well structures consisting of alternate layers of $Cd_xZn_{1-x}S$ and ZnS have the following physical properties: (a) The effective bandgap energy can be tuned within green-blue-ultraviolet regions by varying a Cd composition ratio $x$ or by changing a well layer thickness; (b) since there is a relatively large conduction-band offset as already shown in the preceding section, quantum confinement of electrons in the $Cd_xZn_{1-x}S$ well layers is strong; (c) owing to the large 1s exciton binding energies of bulk materials and to the smaller dielectric constant of the ZnS barrier layer ($\varepsilon_0$=8.32) compared to that of the $Cd_xZn_{1-x}S$ well layer ($\varepsilon_0$=8.58 for CdS), the quantum confinement of excitons may be very pronounced; (d) the quasi-two-dimensional free excitons undergo disorder scattering in the ternary alloy well layers which are under biaxial compression.

In this section, the formation of optical gain in $Cd_xZn_{1-x}S$-ZnS strained-layer multiple quantum-well (MQW) structures is studied both experimentally and theoretically. First of all, experimental evidence for ultraviolet stimulated emission from this type of MQW is demonstrated under both optical pumping at room temperature and pulsed current injection at 30 K. The optical gain spectra are also presented for this type of MQW measured by means of a pump-and-probe technique employing nanosecond excitation pulses. These experimental results suggest that the origin of optical gain is different from that in III-V quantum-well systems in which an electron-hole plasma plays a principal role in providing gain, and that excitons contribute to the formation of optical gain. Moreover, optical gain spectra are derived from the phenomenological analysis which is based on a phase-space-filling effect of excitons in an inhomogeneously broadened system due to interface and alloy fluctuations. Finally, the exciton-related mechanisms of optical gain formation in wide-bandgap II-VI quantum-well structures are discussed.

### 4.4.1. *Experimental Procedures*

For optical pumping experiments, the MQW samples were cleaved to approximately 1-mm-long resonators with uncoated facets and were excited perpendicular to the MQW layer plane. The excitation source was a frequency-tripled Q-switched $Nd^{3+}$:YAG laser (355 nm). The pulse width was approximately 5 ns and the repetition rate was

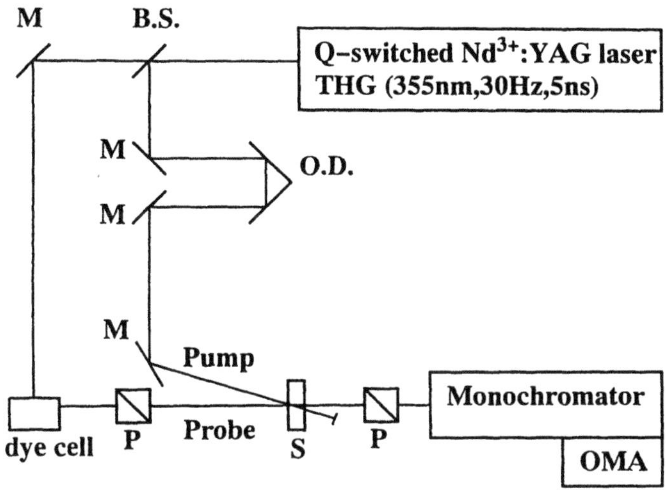


Figure 4.5. Experimental setup for nanosecond pump-and-probe measurement.

30 Hz. The second harmonic light of a dye laser pumped by a frequency-doubled Q-switched $Nd^{3+}$:YAG laser was also employed in order to perform resonant excitation of excitonic absorption peaks. Edge emission from the cleaved facets of the samples was detected using an optical multichannel analyzer (OMA) in conjunction with 75-cm or 25-cm single grating monochromators.

In pump-and-probe experiments, the frequency-tripled Q-switched $Nd^{3+}$:YAG laser was used as a pump source. As a probe source, amplified spontaneous emission from *Exalite 376* dye solution, excited by a portion of the pump source, was used. The probe pulse was possible to cover the broad spectral range from approximately 360 to 390 nm, and was weak enough to induce no absorption change. The spot of the probe pulse on the sample was 200 $\mu$m in diameter and spatially centered in the excitation spot of the pump pulse which was 2 mm in diameter. The time delay between pump and probe pulses was adjusted to be zero. The transmittance of the probe pulse through the sample was detected using the OMA in conjunction with a 25-cm monochrometer. In order to reduce the influence of the luminescence, which was induced by the pump pulse, a pair of polarizers were placed before and after the sample along the probe beam path. The two polarizers had the same polarization which was perpendicular to the polarization of the pump beam. The experimental setup for the pump-and-probe measurement is shown in Fig. 4.5.

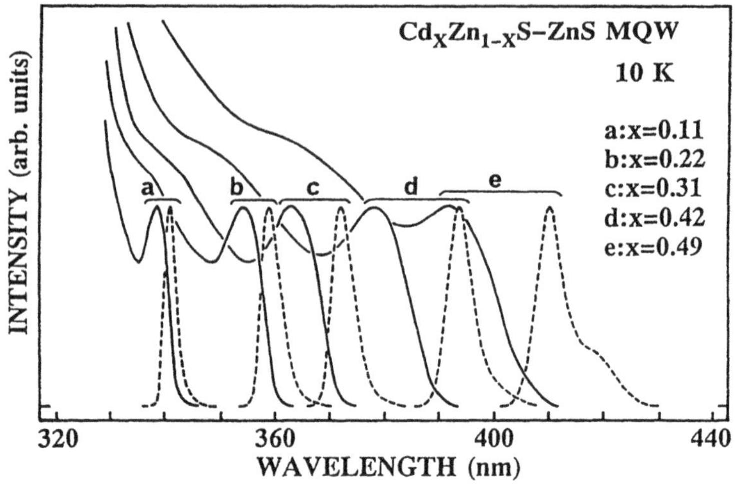


Figure 4.6. Absorption (solid lines) and photoluminescence (dashed lines) spectra taken from $Cd_xZn_{1-x}S$-ZnS MQWs at 10 K with various Cd composition ratios in ternary alloy well layers (a: $x$=0.11, b: $x$=0.22, c: $x$=0.31, d: $x$=0.42, and e: $x$=0.49). (After [33]).

### 4.4.2. *Ultraviolet Stimulated Emission under Optical Pumping*

Figure 4.6 shows absorption (solid lines) and photoluminescence (dashed lines) spectra at 10 K taken from $Cd_xZn_{1-x}S$-ZnS strained-layer MQWs with various Cd composition ratios in the ternary alloy well layer (a: $x$=0.11, b: $x$=0.22, c: $x$=0.31, d: $x$=0.42, and e: $x$=0.49). The absorption spectra were measured after removal of the GaAs substrate by chemical etching. The photoluminescence (PL) spectra were obtained under weak excitation using a continuous-wave He-Cd laser (325 nm) from the front surface of the sample. It can be seen from this figure that the effective bandgap energy of this type of MQW varies widely within the ultraviolet spectral region according to the composition, $x$, of the $Cd_xZn_{1-x}S$ well layer. Moreover, it is found from our separate experiments that this type of MQW exhibits a strong excitonic emission band even at room temperature (RT). It is noted here that the band lineup of this type of MQW is found to be always type-I on the basis of the semi-quantitative calculations as shown in Sec. 4.3. The dominant excitonic emission band in each MQW shown in Fig. 4.6 is attributable to the radiative recombination of n=1 heavy-hole excitons which are localized owing to interface roughness and statistic compositional fluctuations in the ternary alloy well layer. With increasing Cd mole fraction, the Stokes shift becomes larger as well as the linewidth of both the absorption and emission spectra. This may be due to the effects of the interface fluctuations, compositional fluctuations, and lattice mismatch strain. The dominant cause is not clearly understood at the present time, however, as will be shown later,

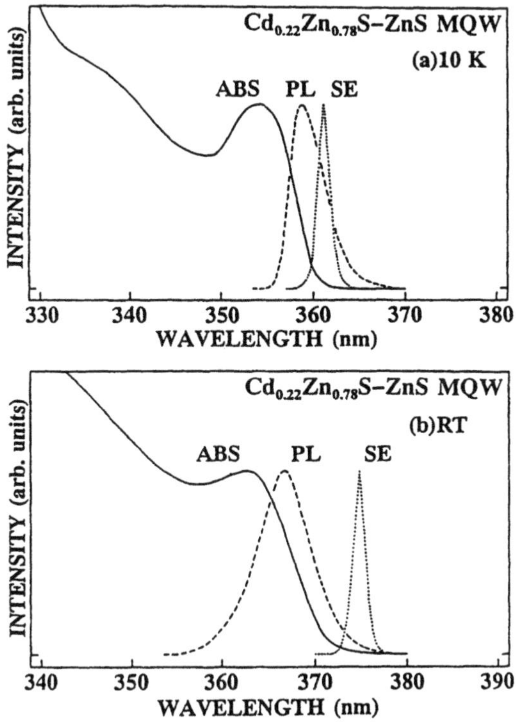


Figure 4.7. Stimulated emission spectra (SE; dotted lines) from the edge facet of a $Cd_{0.22}Zn_{0.78}S$-ZnS MQW at 10 K in (a) and at room temperature in (b). Absorption spectra (ABS; solid lines) and photoluminescence spectra (PL; dashed lines) from the front face of the sample at each temperature are also shown. (After [33]).

the discussion on the formation of optical gain will be done by assuming that the compositional and interface fluctuations are in the quantum well.

For the sample with a Cd composition of $x$=0.22 in the ternary alloy well layer, stimulated emission spectra (SE; dotted lines) observed from the cleaved facet at 10 K and RT are shown in Figs. 4.7(a) and (b), respectively [33]. In this figure, absorption spectra (ABS; solid lines) and photoluminescence spectra (PL; dashed lines) observed from the front face of the sample are also shown. The peak position of the stimulated emission is located at 361.2 nm at 10 K and is shifted to 374.9 nm at RT. Figure 4.8 shows the edge emission intensity at 10 K, 80 K, 150 K, 225 K, and RT as a function of excitation power density. The threshold excitation power density ($I_{th}$) for the stimulated emission is estimated to be approximately 7 kW/cm$^2$ at 10 K. This value increases gradually with temperature, and the room temperature threshold is

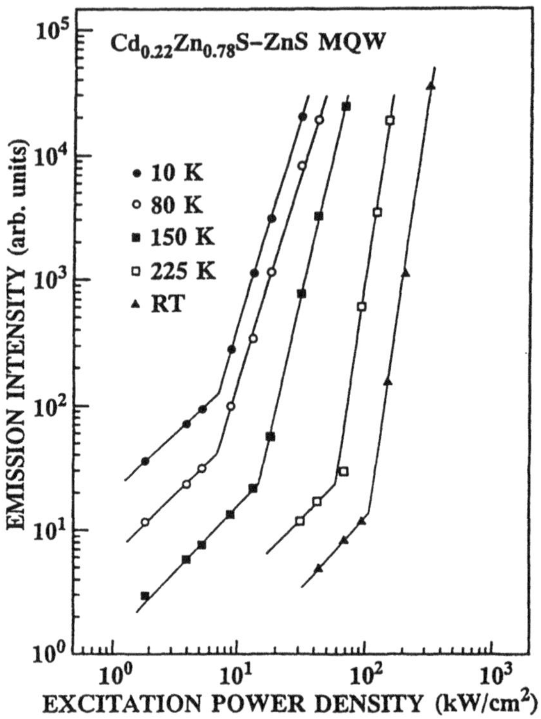


Figure 4.8. Edge emission intensity of a $Cd_{0.22}Zn_{0.78}S$-ZnS MQW as a function of excitation power density at different temperatures. (After [33]).

approximately 110 $kW/cm^2$. The stimulated emission spectra at 10 K and at RT shown in Fig. 4.7 were obtained at the excitation power densities of approximately 9 $kW/cm^2$ ($1.3I_{th}$) and 210 $kW/cm^2$ ($1.9I_{th}$), respectively. These linewidths (full widths at half maximum) were approximately 1.4 nm at 10 K and 1.7 nm at RT. As shown in Fig. 4.2, the conduction- and valence-band offsets in the sample were estimated to be $\Delta E_C$=240 meV and $\Delta E_V$=107 meV (heavy-hole band), respectively, including the effects of the strain due to the lattice mismatch (1.7 %). These values are thought to be sufficient to confine both electrons and heavy holes in the quasi-two-dimensional potential wells up to room temperature. It is noted here that a Stokes shift can be observable even at RT as shown in Fig. 4.7(b). This fact clearly indicates that the excitons in the quantum wells are still localized at RT.

Figure 4.9 shows the polarization dependence of edge emission intensity of (a) the polarization of light parallel to the heterointerface where a TE mode is dominant, and (b) the polarization of light perpendicular to the heterointerface where a TM mode

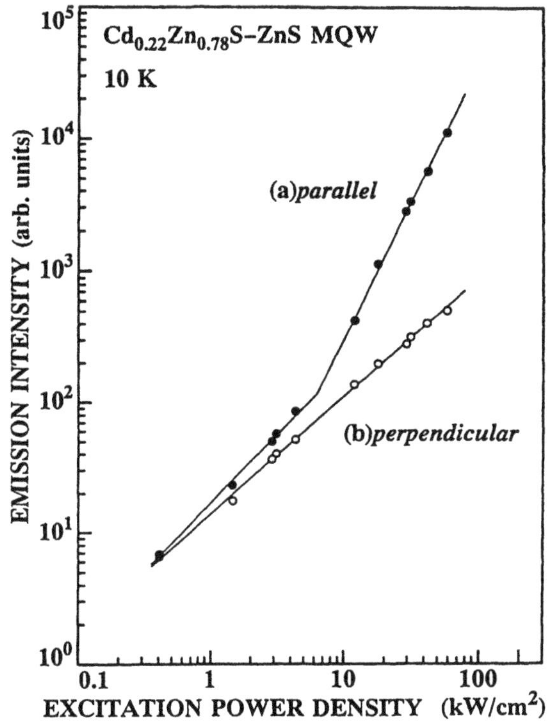


Figure 4.9. Polarization dependence of edge emission intensity of (a) the polarization of light parallel to the heterointerface where a TE mode is dominant and (b) the polarization of light perpendicular to the heterointerface where a TM mode is dominant as a function of excitation power density in a $Cd_{0.22}Zn_{0.78}S$-ZnS MQW at 10 K.

is dominant as a function of excitation power density. The result shows that the stimulated emission occurs at the TE mode. In general, it is reported that the gain of TE mode is mainly supported by an electron transition from conduction to heavy-hole bands, while the gain of TM mode is mainly supported by an electron transition from conduction to light-hole bands [34]. In fact, it is expected that the electronic transition between n=1 conduction electrons and n=1 heavy holes is dominant in this MQW structure because the well layers are under compressive strain. The gain of TM mode has been reported on ZnSe-$Zn_xMn_{1-x}Se$ systems [35], where the electronic transition between n=1 conduction electrons and n=1 light holes was dominant because the well layers were under tensile strain.

Figure 4.10 demonstrates the room-temperature ultraviolet stimulated emission spectra (SE; dotted lines) taken from three $Cd_xZn_{1-x}S$-ZnS strained-layer MQWs (a:

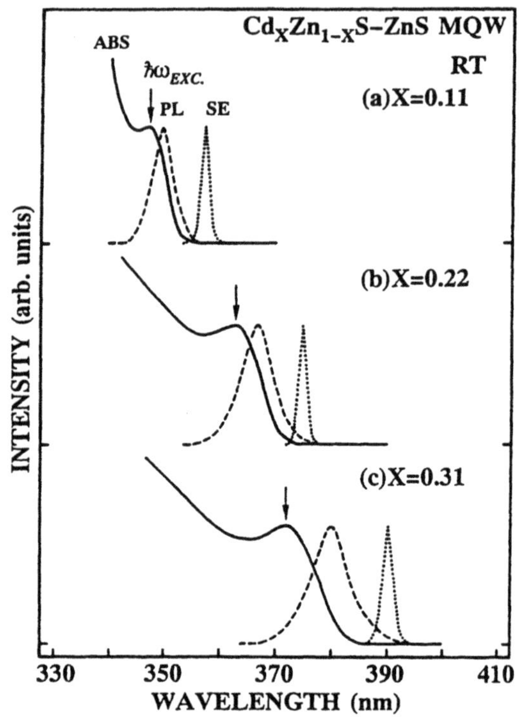


Figure 4.10. Room-temperature ultraviolet stimulated emission spectra (SE; dotted lines) observed under resonant excitation of each excitonic absorption peak shown as the vertical arrow in $Cd_xZn_{1-x}S$-ZnS MQWs (a: $x$=0.11, b: $x$=0.22, and c: $x$=0.31). Absorption spectra (ABS; solid lines) and photoluminescence spectra (PL; dashed lines) from the front face of the sample are also shown. (After [46]).

$x$=0.11, b: $x$=0.22, and c: $x$=0.31). In this case, each stimulated emission spectrum was observed under resonant excitation of the excitonic absorption peak shown as the vertical arrow. Absorption spectra (ABS; solid lines) and photoluminescence spectra (PL; dashed lines) observed from the front face of the sample are also shown in this figure. The peak position of each stimulated emission is located at (a) 357.2 nm, (b) 374.9 nm, and (c) 390.2 nm and the threshold excitation power density for each stimulated emission is estimated to be approximately (a) 46 kW/cm$^2$, (b) 64 kW/cm$^2$, and (c) 79 kW/cm$^2$. It can be seen from this figure that the stimulated emission taken from the $Cd_xZn_{1-x}S$-ZnS strained-layer MQW is characterized by the shortest wavelength reported to date among semiconductor laser materials. These experimental results strongly suggest that this type of MQW is the most fruitful

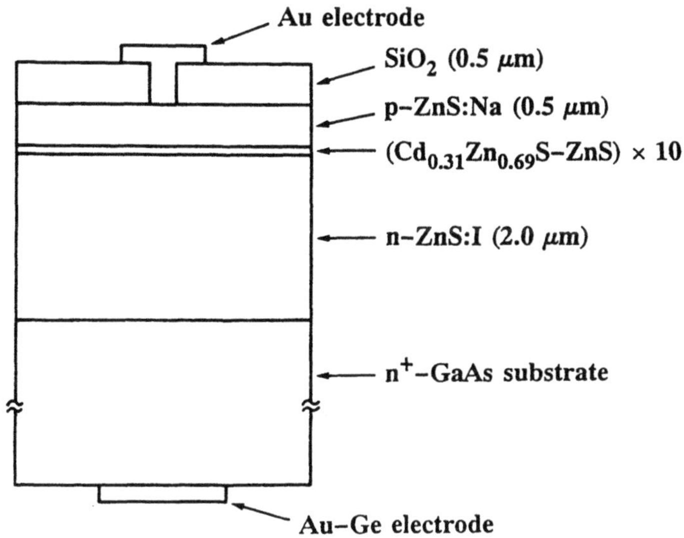


Figure 4.11. The schematic structure of an ultraviolet injection laser diode based on a $Cd_{0.31}Zn_{0.69}S$-ZnS MQW.

candidate capable of providing the semiconductor injection laser diodes operating in the ultraviolet spectral region.

### *4.4.3. Ultraviolet Injection Laser Diode*

The laser diode structure was fabricated on (100)-oriented $n^+$-GaAs substrates, following the deposition of a 2-μm-thick n-type ZnS layer doped with iodine. The active layer was a MQW structure consisting of 10 periods of 4.1-nm-thick $Cd_{0.31}Zn_{0.69}S$ well layers separated by 8.1-nm-thick ZnS barrier layers. This MQW structure was embedded in a p-n junction made from ZnS layers doped with sodium acceptors and iodine donors. The carrier concentration of sodium acceptors and iodine donors was estimated to be approximately $1\times10^{16}$ and $1\times10^{19}$ $cm^{-3}$ at room temperature, respectively, both of which were confirmed by means of Hall measurement. The schematic structure is shown in Fig. 4.11. Gain-guided lasers were fabricated using evaporated Au electrodes as the ohmic contact on the p-type ZnS:Na layer. Current confinement was achieved with 15-μm-wide stripes in a $SiO_2$ insulator layer and 1-mm-long cleaved resonator. A Au-Ge electrode was used as the ohmic contact on the $n^+$-GaAs substrate.

Figure 4.12 shows the voltage-versus-current characteristics at 30 K for the laser diode structure mentioned above. The pulse width and the repetition rate were 300 ns and 4 Hz, respectively. The inset shows the spontaneous emission [(a); dashed line]

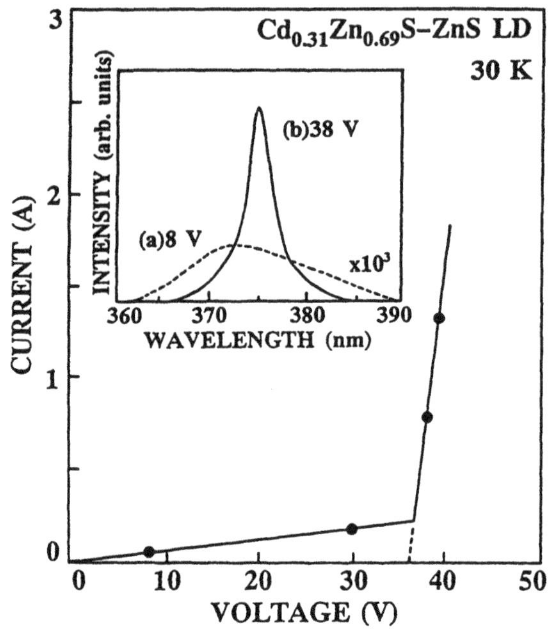


Figure 4.12. Voltage-versus-current characteristics at 30 K for an injection laser diode based on a $Cd_{0.31}Zn_{0.69}S$-ZnS MQW. The inset shows the spontaneous emission [(a); dashed line] and stimulated emission [(b); solid line] spectra at 8 and 38 V, respectively.

and stimulated emission [(b); solid line] spectra. The spontaneous electroluminescence was obtained at about 372 nm at 8 V and its linewidth was estimated to be about 15 nm. From the voltage-versus-current characteristics, the built-in voltage for causing stimulated emission was found to be about 36 V and the threshold current density was estimated to be about 1.5 kA/cm$^2$. Above 36 V, the emission intensity became strong and the peak position moved towards the longer-wavelength side. The stimulated electroluminescence of linewidth less than 3 nm was clearly observed at the wavelength of 375 nm at 38 V. The present energy position is close to that of the stimulated emission observed at 10 K under optical pumping. This result strongly demonstrates the possibility of lasing at the ultraviolet wavelength [36, 37].

### 4.4.4. *Optical Gain Spectra*

It has been reported recently that excitons play a central role in the formation of optical gain in wide-bandgap II-VI MQW structures [14]. In order to study the formation of optical gain in this $Cd_xZn_{1-x}S$-ZnS MQW, pump-and-probe experiments with nanosecond excitation pulses were performed. Figure 4.13(a) shows absorption

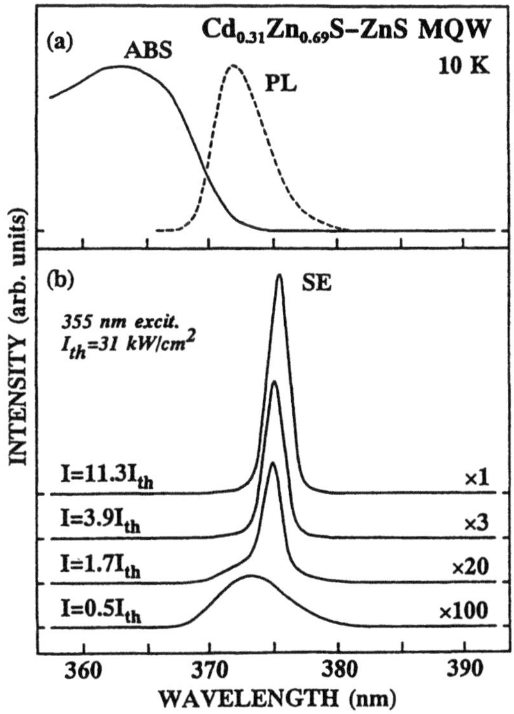


Figure 4.13. Absorption (ABS) and photoluminescence (PL) spectra in (a) and edge emission spectra both below and above the threshold for stimulated emission (SE) in (b) taken from a $Cd_{0.31}Zn_{0.69}S$-ZnS MQW at 10 K.

(ABS; solid line) and photoluminescence (PL; dashed line) spectra at 10 K taken from a $Cd_{0.31}Zn_{0.69}S$-ZnS MQW used in this pump-and-probe measurement. Edge emission spectra both below and above the threshold for stimulated emission are also shown in Fig. 4.13(b). The peak position of the stimulated emission just above the threshold is located at 374.9 nm and the threshold excitation power density for stimulated emission is estimated to be approximately 31 kW/cm$^2$.

Figure 4.14 shows pump-and-probe absorption spectra at 10 K taken from the $Cd_{0.31}Zn_{0.69}S$-ZnS MQW [33]. The dashed line shows the absorption spectrum without pump pulses, while the solid line shows the spectrum with an excitation power density of approximately 253 kW/cm$^2$. The inset shows the enlarged spectra at the tail region of the excitonic absorption peak at excitation power densities of (a) 18, (b) 65, (c) 143, (d) 193, and (e) 253 kW/cm$^2$. With increasing pump intensity, absorption around 375 nm gradually decreases. Above 143 kW/cm$^2$, negative absorption

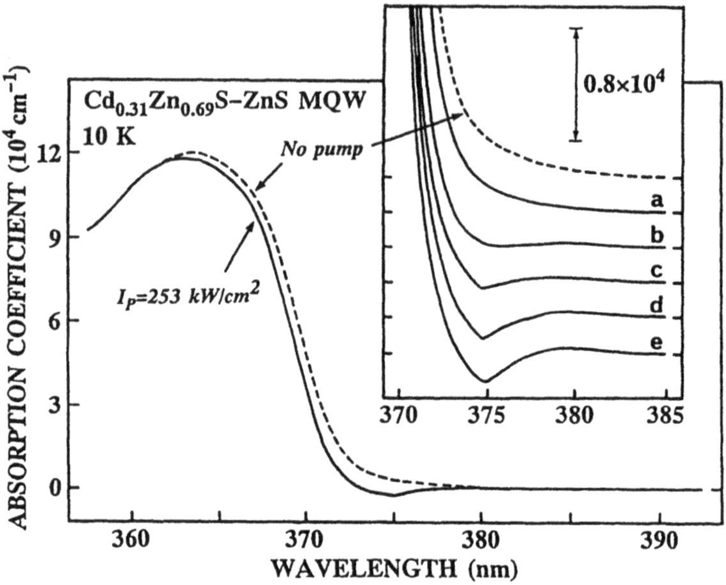


Figure 4.14. Pump-and-probe absorption spectra taken from a $Cd_{0.31}Zn_{0.69}S$-ZnS MQW at 10 K. The dashed line shows the absorption spectrum without pump pulses. The solid line shows the spectrum with an excitation power density of approximately 253 kW/cm$^2$. The inset shows the enlarged spectra at the tail part of the excitonic absorption peak at excitation power densities of (a) 18, (b) 65, (c) 143, (d) 193, and (e) 253 kW/cm$^2$.

is observable and then increases with pump intensity. This observation unambiguously indicates the presence of gain. The stimulated emission observed in this sample was naturally located at 374.9 nm at 10 K as shown in Fig. 4.13(b), and the threshold excitation power density was estimated to be approximately 31 kW/cm$^2$. The pump intensity at which negative absorption appears is several times larger than the threshold for stimulated emission. This is due to the inhomogeneous excitation over 50 quantum wells (QWs) along the growth direction. In fact, the transmittance of the excitation pulse through the 50 QWs is estimated to be approximately 20 %. Several QWs near the surface of the sample dominantly contribute to the stimulated emission. On the other hand, all of the 50 QWs, on the average, contribute to the optical gain which is observed in the pump-and-probe configuration. Figure 4.15 shows differential transmission spectra obtained at each excitation condition shown in the inset of Fig. 4.14. As the excitation power density increases, the peak observed around 370 nm increases gradually up to $\Delta T/T \approx 0.2$. This indicates an increase in the transmission and is attributed to the bleaching of exciton absorption. Moreover, the other peak is observable around 375 nm at the excitation power densities above

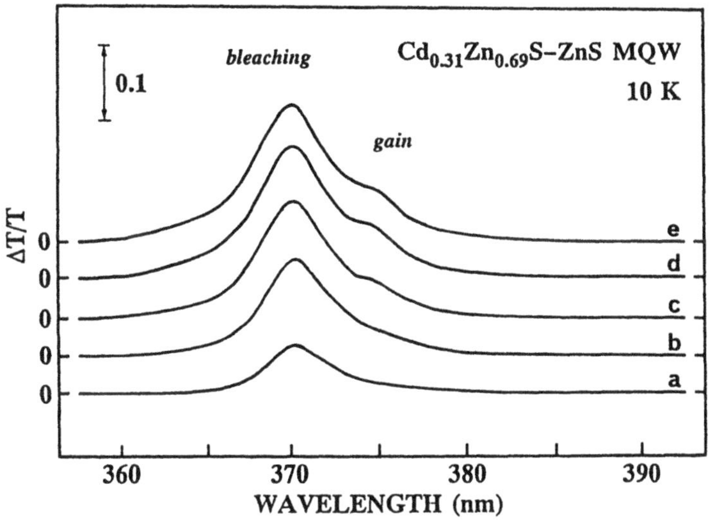


Figure 4.15. Differential transmission spectra taken from a $Cd_{0.31}Zn_{0.69}S$-ZnS MQW at 10 K. Each spectrum was obtained at the excitation power densities of (a) 18, (b) 65, (c) 143, (d) 193, and (e) 253 kW/cm$^2$.

143 kW/cm$^2$. This is attributed to the gain.

The main feature of the experimental findings is the observation of optical gain even in the presence of the clear excitonic absorption peak which is slightly bleached at the lower energy side. This observation suggests that excitons contribute to the formation of optical gain and, as a result, there is no evidence for the presence of an electron-hole plasma.

### *4.4.5. Discussion*

For III-V compound semiconductor lasers with quantum-well structures or double heterostructures, it has well been understood that the origin of optical gain is due to the population inversion of a degenerate electron-hole plasma. In comparison with II-VI compound semiconductors, the plasma regime is much more rapidly reached under increasing excitation densities in III-V compound semiconductors. This is because the effective masses in both the conduction and the valence bands of III-V compounds are much smaller than those of II-VI compounds. In other words, excitons in II-VI compounds have larger binding energies as well as smaller Bohr radii than those in III-V compounds. Then, the excitons in II-VI compounds exist stably at higher densities than those in III-V compounds.

Since the seventies, the possibility of excitonic gain and related stimulated emission has been suggested for wide-bandgap II-VI compound semiconductors by several

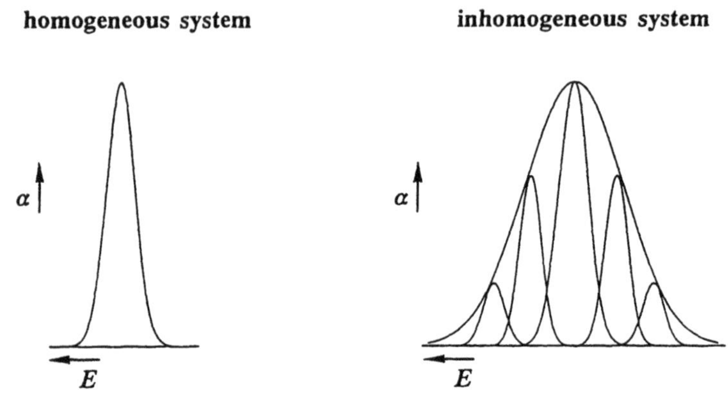


Figure 4.16. Schematic illustration of exciton density of states (exciton absorption) for both homogeneous and inhomogeneous systems. The inhomogeneous broadening of exciton resonance results from alloy composition fluctuations in ternary alloy well layers and interface fluctuations between well and barrier layers.

workers [38, 39, 40, 41, 42, 43, 44]. Their experiments for bulk materials argued that optical gain was formed by inelastic exciton-exciton or exciton-LO phonon scattering processes at low temperatures, and by an inelastic exciton-free carrier (electron or hole) scattering process at higher temperatures where a considerable number of excitons were thermally ionized.

As for quantum-well structures based on the wide-bandgap II-VI compound semiconductors, on the other hand, the excitonic gain process was first proposed by Ding *et al.* in 1992 [14]. They explained the excitonic gain in $Cd_xZn_{1-x}Se$-ZnSe quantum wells by considering the partial phase-space-filling effect of excitons at the lower energy states of inhomogeneously broadened exciton resonance. This circumstance derives from the large enhancement in the exciton binding and its oscillator strength owing to the effect of quantum confinement, so that quasi-two-dimensional excitons are stable against ionization by optical phonons up to room temperature. It was considered that these excitons were subjected to localization owing to alloy composition fluctuations in ternary alloy well layers and interface fluctuations between well and barrier layers. Therefore, the excitons are localized at the lower energy states of the inhomogeneously broadened exciton resonance as schematically shown in Fig. 4.16, and the partial phase-space-filling of these localized exciton states occurs. The localized exciton states are considered to be easily saturated because of the low density of states. As a result, it was explained that the population of the localized exciton states can easily be inverted at the electron-hole pair (or exciton) densities below the onset to an electron-hole plasma. The gain/absorption coefficient was then given as

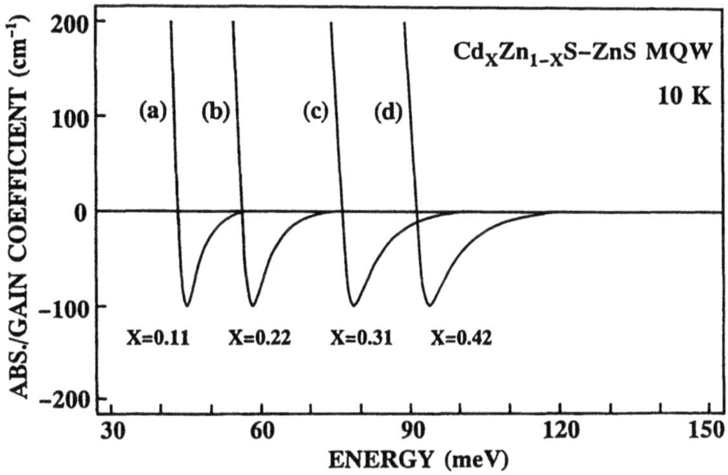


Figure 4.17. Calculated results of optical gain spectra of $Cd_xZn_{1-x}S$-ZnS MQWs (a: $x$=0.11, b: $x$=0.22, c: $x$=0.31, and d: $x$=0.42) at 10 K. In this figure, the horizontal axis indicates the photon energy measured from the center of the n=1 heavy-hole inhomogeneously broadened exciton resonance. (After [46]).

the following [14, 45]:

$$g(E) = \int_{-\infty}^{\infty} D_i(E')D_h(E-E')\left(\frac{2}{\exp[(E'-\mu)/kT]+1}-1\right)dE', \tag{4.25}$$

where $D_i(E)$ is the inhomogeneous and $D_h(E)$ the homogeneous line shape function, and $\mu$ is the chemical potential. The energies in the above equation are defined with respect to the center of the exciton resonance broadened inhomogeneously. The quantity $g(E)$ is normalized to the peak value of the absorption coefficient in unexcited materials. The density of excitons was also given as follows,

$$n = \int_{-\infty}^{\infty} D_i(E)\frac{1}{\exp[(E-\mu)/kT]+1}dE. \tag{4.26}$$

The energies in the above equation are also defined with respect to the center of the exciton resonance. The density of excitons, $n$, is normalized to the maximum density of excitons $n_{\max}$ for the total phase-space-filling of the inhomogeneously broadened exciton line ($n_{\max} \approx (\pi a_B^2)^{-1}$).

Using the same procedure, we analyzed the optical gain spectra of $Cd_xZn_{1-x}S$-ZnS MQWs [46]. Figure 4.17 shows the calculated results of optical gain spectra as a function of Cd composition ratio in the ternary alloy well layers (a: $x$=0.11, b: $x$=0.22, c: $x$=0.31, and d: $x$=0.42) at 10 K. In this figure, the horizontal axis indicates

Table 4.2. Calculated energy shifts of the gain maximum ($\sim$100 cm$^{-1}$) from the exciton resonance and densities of excitons per quantum well as a function of Cd composition ratio in $Cd_xZn_{1-x}S$-ZnS MQWs at 10 K. Experimental energy shifts which are determined by the energy difference between the n=1 heavy-hole exciton peaks and the stimulated emission peaks, and experimental densities of excitons which are derived from the threshold excitation power densities for stimulated emission are also indicated.

| | Energy shift (meV) | | Density of excitons (cm$^{-2}$) | |
|---|---|---|---|---|
| | Calc. | Exp. | Calc. | Exp. |
| (a) $x$=0.11 | 45 | 44 | $1.0\times10^9$ | $2.4\times10^{10}$ |
| (b) $x$=0.22 | 58 | 55 | $1.1\times10^9$ | $4.8\times10^{10}$ |
| (c) $x$=0.31 | 78 | 73 | $1.2\times10^9$ | $1.9\times10^{11}$ |
| (d) $x$=0.42 | 93 | 88 | $1.3\times10^9$ | $3.7\times10^{11}$ |

the photon energy measured from the center of the n=1 heavy-hole exciton resonance broadened inhomogeneously in each sample. At low temperature, the homogeneous broadening is much smaller than the inhomogeneous one, so that the results shown in Fig. 4.17 were approximately derived from the following equation [14, 45]:

$$g(E) = D_i(E)\left[\frac{2}{\exp[(E-\mu)/kT]+1} - 1\right]. \tag{4.27}$$

With increasing Cd composition ratio in the ternary alloy, the redshift of the gain spectra is observed, that is, the energy difference between the center of exciton resonance and the gain maximum increases. This is due to the increase in the inhomogeneous linewidth with Cd composition. In fact, the absorption linewidths (the full widths at half maximum) are approximately estimated to be 28 meV for (a), 37 meV for (b), 51 meV for (c), and 61 meV for (d). These energy shifts of the gain maximum from each of the exciton resonance are summarized in Table 4.2, and are compared with the experimental results which are determined by the energy difference between the center of n=1 heavy-hole exciton resonance and the peak position of stimulated emission just above the threshold. The electron-hole pair (or exciton) densities which are necessary for the threshold gain for each of the stimulated emission ($\sim$100 cm$^{-1}$) per quantum well are also summarized in Table 4.2. These values are also compared with the experimental results which are estimated from the threshold excitation power densities for stimulated emission in each of the samples. It is noted here that the densities obtained both theoretically and experimentally are below the estimated values of electron-hole-plasma phase transition [$n_{\max} \approx (\pi a_B^2)^{-1}$=$6.3\times10^{12}$ cm$^{-2}$ for (a), $6.1\times10^{12}$ cm$^{-2}$ for (b), $5.9\times10^{12}$ cm$^{-2}$ for (c), and $5.5\times10^{12}$ cm$^{-2}$ for (d)].

Figure 4.18 shows the calculated results of optical gain spectra of a $Cd_{0.22}Zn_{0.78}S$-ZnS MQW as a function of temperature. With increasing temperature up to 80 K,

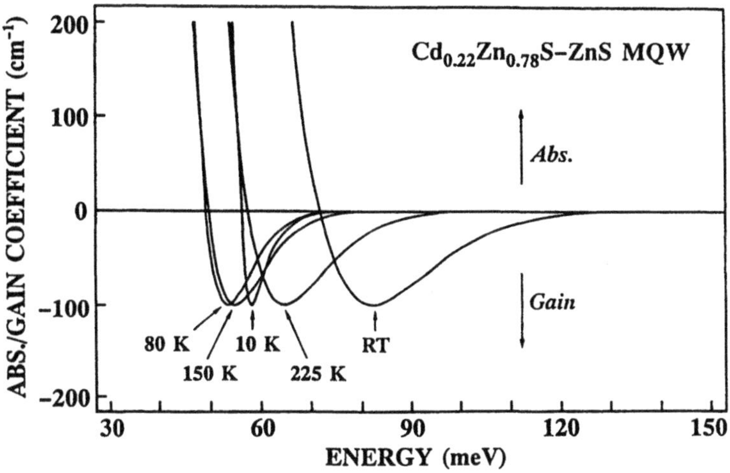


Figure 4.18. Calculated results of optical gain spectra of a $Cd_{0.22}Zn_{0.78}S$-ZnS MQW as a function of temperature. In this figure, the horizontal axis indicates the photon energy measured from the center of the n=1 heavy-hole inhomogeneously broadened exciton resonance. (After [46]).

Table 4.3. Temperature dependence of the n=1 heavy-hole exciton absorption linewidth ($\Delta_{total}$) in a $Cd_{0.22}Zn_{0.78}S$-ZnS MQW and its inhomogeneous ($\Delta_i$) and homogeneous ($\Delta_h$) components which are derived from the following relation: $\Delta_{total}=(\Delta_i^2+\Delta_h^2)^{\frac{1}{2}}$ on the assumption that $\Delta_i$=37 meV.

| | $\Delta_{total}$ (meV) | $\Delta_i$ (meV) | $\Delta_h$ (meV) |
|---|---|---|---|
| 10 K | 37.0 | 37.0 | – |
| 80 K | 37.0 | 37.0 | – |
| 150 K | 40.0 | 37.0 | 15.2 |
| 225 K | 51.0 | 37.0 | 35.1 |
| RT | 69.0 | 37.0 | 58.2 |

the energy shift of the gain maximum from the center of exciton resonance decreases. With further increasing temperature, this energy shift conversely increases. This aspect can be explained by considering the temperature dependence of the distribution function of excitons and the homogeneous broadening due to the exciton-phonon scattering at higher temperatures. Namely, the decrease in the energy shift up to 80 K can be explained only by the temperature-dependent slope around the chemical potential in the distribution function. Above 80 K, the increase in the homogeneous broadening due to the exciton-LO phonon scattering must be considered in addition

Table 4.4. Calculated energy shifts of the gain maximum (~100 $cm^{-1}$) from the exciton resonance and densities of excitons per quantum well as a function of temperature in a $Cd_{0.22}Zn_{0.78}S$-ZnS MQW. Experimental energy shifts which are determined by the energy difference between the n=1 heavy-hole exciton peaks and the stimulated emission peaks, and experimental densities of excitons which are derived from the threshold excitation power densities for stimulated emission are also indicated.

| | Energy shift (meV) | | Density of excitons ($cm^{-2}$) | |
|---|---|---|---|---|
| | Calc. | Exp. | Calc. | Exp. |
| 10 K | 58 | 55 | $1.1\times10^{9}$ | $4.8\times10^{10}$ |
| 80 K | 53 | 50 | $4.6\times10^{10}$ | $4.8\times10^{10}$ |
| 150 K | 55 | 55 | $3.8\times10^{11}$ | $1.0\times10^{11}$ |
| 225 K | 65 | 70 | $1.2\times10^{12}$ | $3.7\times10^{11}$ |
| RT | 82 | 95 | $2.0\times10^{12}$ | $7.2\times10^{11}$ |

to the above circumstance. In fact, the absorption linewidth remains almost constant below 80 K and gradually increases above 150 K. The inhomogeneous linewidth ($\Delta_i$) measured in this sample is about 37 meV which is determined by the linewidth at low temperature (10 K). The homogeneous linewidth ($\Delta_h$) can be obtained by dividing the total linewidth ($\Delta_{\text{total}}$) into the inhomogeneous and the homogeneous components using the following relation: $\Delta_{\text{total}} = (\Delta_i^2 + \Delta_h^2)^{\frac{1}{2}}$ if both the inhomogeneous and homogeneous line shapes are assumed to be a Gaussian distribution. These values are listed in Table 4.3. Then, the energy shift of the gain maximum from the exciton resonance increases with temperature above 80 K. These calculated energy shifts are summarized in Table 4.4 with the experimental results which are determined similarly as mentioned above. The electron-hole pair (or exciton) densities are also summarized in Table 4.4; note also that the densities are below the estimated electron-hole-plasma phase transition ($6.1\times10^{12}$ $cm^{-2}$).

It can be seen from Tables 4.2 and 4.4 that the calculated energy shifts of the gain maximum from the center of exciton resonance approximately agree with the experimental results for both the composition and the temperature dependence. Especially, the temperature dependence of the experimental energy shifts, that is, the decrease in energy shift up to 80 K and the increase in energy shift above 80 K is well supported by the calculated results. On the other hand, the calculated results of the electron-hole pair (or exciton) densities indicate that the densities, which give threshold gain (~100 $cm^{-1}$), increase with Cd composition and also with temperature. This tendency is consistent with the experimental results which are derived from the threshold excitation power densities for stimulated emission. There are some cases that imply a difference of about one order of magnitude between the calculated and the experimental results. It seems that the discrepancy becomes larger at lower temperature as well as at higher Cd composition ratio. However, no further consideration can be done at the present stage as judged from the fact that the estimation of the carrier densities includes somewhat experimental errors.

Next, the possibility of other excitonic gain processes, for example, the inelastic exciton scattering pictures which have been considered in bulk II-VI compounds is examined. It was argued both experimentally and theoretically that at low temperatures inelastic exciton-exciton scattering was the most dominant process to provide gain and stimulated emission in pure binary II-VI materials such as CdS, ZnO, ZnTe, and ZnSe [38, 39, 40, 42, 44]. In our MQW systems, however, inhomogeneous broadening of an exciton density of states is present owing to compositional fluctuations as well as interface fluctuations. Therefore, excitons are easy to be subject to localization. Such localization of excitons does not exist in the above mentioned pure binary bulk materials. In fact, the inelastic exciton-exciton scattering picture does not apply to our MQW systems because the energetic distance between the stimulated emission and the exciton resonance for all of our MQW samples can not consistently be explained by the binding energy of excitons (see Tables 4.2 and 4.4). Therefore, the possibility of inelastic exciton-exciton scattering process to give gain and stimulated emission can completely be ruled out in our MQW systems. Similarly, the possibility of inelastic exciton-LO phonon scattering process can also be eliminated in our MQW systems. While, at higher temperatures (near room temperatures) it was argued that inelastic exciton-free carrier (electron or hole) scattering became more probable. It was explained by considering that a considerable number of excitons were thermally ionized [39, 40]. However, quasi-two-dimensional excitons in the wide-bandgap II-VI MQWs are more stable against ionization because of the large enhancement in the binding energy and oscillator strength owing to the effect of quantum confinement. For example, as already mentioned above, the binding energy of excitons in the $Cd_{0.11}Zn_{0.89}S$-ZnS MQW system is estimated to be as large as 66 meV on the basis of the correlation between the well layer thickness and the Bohr radius of bulk excitons [31]. Therefore, the quasi-two-dimensional excitons are thought to be much more stable even at room temperature. Furthermore, as shown in Table 4.4, the large energy shift of the stimulated emission peak from the center of exciton resonance can not be explained by the inelastic exciton-free carrier scattering picture. As a result, it is hard to apply the exciton-free carrier scattering picture to our MQW systems.

In addition, there is another excitonic gain process, which is due to the decay of a biexciton (excitonic molecule) into an exciton and a photon. The optical gain and stimulated emission due to the transition from the biexciton states to the longitudinal exciton states have been observed in CuCl bulk crystals [47] and microcrystals [48]. On the other hand, in the case of our MQW systems, it was found that the optical gain was not due to the biexciton decay process [49]. Figure 4.19 shows the photoluminescence spectra at 10 K taken from the front surface of the $Cd_{0.22}Zn_{0.78}S$-ZnS MQW under various excitation power densities. The lowest spectrum was obtained under He-Cd laser (325 nm) excitation, and the upper five spectra were obtained under frequency-tripled Q-switched $Nd^{3+}$:YAG laser (355 nm) excitation. The luminescence at the low-excitation density is dominated by the radiative recombination of n=1 heavy-hole excitons. With increasing excitation power density, a shoulder on the low-energy side of the exciton line grows more rapidly than the exciton line

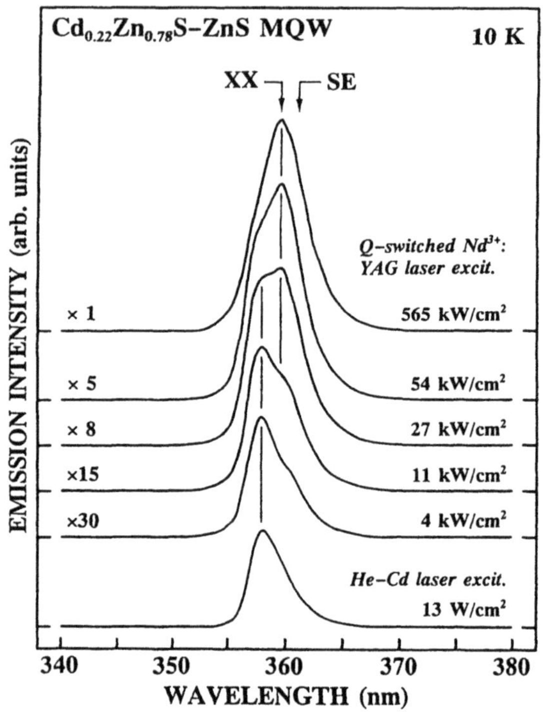


Figure 4.19. Photoluminescence spectra at 10 K taken from a $Cd_{0.22}Zn_{0.78}S$-ZnS MQW under various excitation power densities. The lowest spectrum was obtained under He-Cd laser (325 nm) excitation, and the upper five spectra were obtained under frequency-tripled Q-switched $Nd^{3+}$:YAG laser (355 nm) excitation.

and becomes the dominant radiative-recombination process (denoted by XX), which shows the peak at 359.7 nm (3.446 eV). On the basis of the spectral position, we attribute the line XX to the radiative recombination of biexcitons. This biexciton line follows the expected superlinear excitation-density dependence, and the dependence of the biexciton intensity on the exciton intensity gives approximately $I_{biex} \propto I_{ex}^{1.6}$. This result indicates that the biexciton density does not vary as the square of the exciton density. Such a reduction in the exponent from 2 has also been observed in the case of GaAs-based quantum-well structures [50, 51, 52, 53, 54]. One of the reasons for the reduction was explained by considering the short recombination time of excitons and biexcitons in such direct-bandgap materials [54]. Specifically, owing to the shorter radiative lifetime of biexcitons than that of excitons, it was considered that the biexciton density varied less strongly than quadratically as the exciton density was increased. Then, the peak position of the biexciton luminescence is located at about 14 meV higher energy side of that of the stimulated emission (3.432 eV) as shown in Fig. 4.7(a). Therefore, the possibility of biexciton decay process to provide gain and stimulated emission can also be eliminated in our MQW systems. We have also studied the temporal behaviour of biexcitons under high-density excitation in a $Cd_{0.17}Zn_{0.83}Se$-$ZnS_{0.08}Se_{0.92}$ MQW by means of time-resolved luminescence spectroscopy [55]. It was found from the experimental results that the biexcitons did not contribute to the formation of optical gain in the $Cd_{0.17}Zn_{0.83}Se$ MQW.

In this way, it is reasonable to understand that the formation of optical gain in the $Cd_xZn_{1-x}S$-ZnS MQWs originates from the localization of excitons at the lower-energy states of inhomogeneously broadened exciton resonance and the partial phase-space-filling effect of excitons at the localized exciton states. We only point out that the exciton localization model gives the lowest threshold exciton density for gain and stimulated emission in our MQW systems with inhomogeneous broadening. Therefore, another excitonic gain process such as the inelastic exciton scattering model may be dominant in the "perfectly-controlled" II-VI quantum-well structure where the localization of excitons does not take place.

Very recently, in contrast to the above-mentioned systems with ternary alloy quantum-well layers, we have studied nonlinear optical properties of dense excitonic systems in a ZnSe single quantum-well (SQW) structure [56]. The structure consisted of a 10.5-ML (3-nm-thick) ZnSe well layer sandwiched between 150- and 50-nm-thick $Zn_{0.90}Mg_{0.10}S_{0.18}Se_{0.82}$ barrier layers. The luminescence linewidth of the n=1 heavy-hole excitons was as small as 5 meV, which showed the high quality of the SQW structure. The main result obtained by means of time-resolved nonlinear luminescence spectroscopy, based on an excitation correlation technique, was that the enhancement in the luminescence efficiency of biexcitons was clearly observed in the time-correlation trace of nonlinear biexciton luminescence. Then, the origin of the enhancement was attributed to the higher-order nonlinearity with respect to the density of biexcitons, and was explained by considering the stimulation of biexciton luminescence. This experimental result strongly suggests the possibility of biexciton lasing, that is, the formation of optical gain due to the biexciton decay process in high-quality II-VI semiconductor quantum-well structures with binary quantum-well layers.

In wide-bandgap II-VI semiconductor quantum-well structures, however, general understanding of the formation of optical gain can not be obtained at the present stage. In fact, besides the exciton localization model [14], there have been several different reports to date about the formation of optical gain in the similar $Cd_xZn_{1-x}Se$-ZnSe quantum-well structures: inelastic exciton-exciton scattering process [57], inelastic exciton-LO phonon scattering process [58], localized biexciton decay process [59], and free-carrier recombination process [60]. This fact suggests that the origin of optical gain formation strongly depends on the structural quality of the sample as well as the structural design such as a composition ratio in ternary alloy well layers and a well layer thickness. The former dominates the degree of exciton localization, and the latter dominates the enhancement of exciton binding and its oscillator strength.

We consider that the most important point is to clarify the optimum structure in which some process gives the lowest threshold exciton density for gain and stimulated emission rather than to clarify the process which gives the lowest threshold exciton density for gain and stimulated emission in a given structure. In that case, we must discuss whether such an extrinsic factor as exciton localization is effective or not. Recently, excitonic gain has also been observed in GaAs-based quantum wires [61]. Even in the GaAs-based structures, the contribution of excitons to the formation of optical gain becomes important when the dimensionality of the structure is lowered. This is just due to the large enhancement in the exciton binding and the resultant stability of excitons in the one-dimensional quantum wires.

First experimental evidence for exciton localization by alloy fluctuations has been reported by Lai and Klein [62] for indirect-gap $Ga_{1-x}Al_xAs$. After that, such localization has been studied for direct-gap II-VI alloy $CdS_{1-x}Se_x$ systems [63, 64]. The general picture of exciton motion in semiconductor alloys was well explained as follows [65]: at intraband excitation excitons are created in the region of high density of states and have a kinetic energy far exceeding the amplitude of crystal potential fluctuations. Such excitons do not "see" the potential fluctuations and can be regarded as free particles. Therefore, the excitons have a high probability of spatial migrations. The energy relaxation through the emission of phonons brings the excitons into the energy region of the random potential relief where further energy relaxation and spatial migration are possible only through the tunneling into the localized states with lower energy. Namely, at kinetic energies comparable with the potential fluctuations, the spatial localization of excitons can take place in the wells of crystal potential. The states of localized and delocalized (free or extended) excitons are separated by some characteristic energy, the so-called mobility edge which corresponds to the general Mott-Anderson model in disordered systems. When the thermal energy ($k_BT$) becomes comparable to the localization energy of excitons, the population of excitons changes from the localized states to the delocalized (extended) states, and then the excitons behave similarly to free excitons in binary materials. Such a study on exciton localization has recently been revisited in a cubic $Cd_{0.58}Zn_{0.42}S$ film lattice matched to GaAs [66]. It was shown by means of time-resolved luminescence spectroscopy that the luminescence at 2 K was well explained by the model of exciton localization and also that the luminescence at RT mainly originated from nearly delocalized free excitons. In our MQW systems, however, the localization energies of excitons are considerably large. As already shown in Figs. 4.7(b) and 4.10, the large Stokes shifts were observed even at RT. This observation means that most of the excitons in the quantum wells are still localized at RT. Then, it is considered that the change in the population of excitons from the localized states to the delocalized states hardly occurs in our MQW systems even at RT.

Majumder *et al.* [67] measured optical gain spectra in disordered $CdS_{1-x}Se_x$ systems and concluded that the population of localized excitons in the tail states resulting from alloy fluctuations was inverted at low temperatures resulting in gain and stimulated emission. They also concluded that the many-particle effects of these localized excitons were very weak compared to that of free and extended excitons.

Then, the gain observed for the alloys was thought to be simply due to gradual filling of the localized states and population inversion of these states, without too much influence on the absorption spectra above the mobility edge. This identification clearly supports the contribution of localized excitons to the formation of optical gain in our MQW systems in which the crystal statistic potential fluctuations are induced by the ternary alloy fluctuations and by the interface fluctuations.

## 4.5. CONCLUSIONS

The main objective of this chapter has been to examine band lineups and optical characteristics in one of the novel wide-bandgap II-VI semiconductor strained-layer multiple quantum wells, $Cd_xZn_{1-x}S$-ZnS systems.

The effects of elastic layer strains due to a lattice constant mismatch on band lineups and electronic structures were analyzed on the basis of the modified model-solid theory. The effects of strains on conduction- and valence-band edges can individually be treated and the center of gravity in the three uppermost valence bands is moved with strains in this theory. The calculated results indicated that the $Cd_xZn_{1-x}S$-ZnS strained-layer system exhibited a type-I band structure and that the conduction-band offset was two or three times as large as the valence-band offset. These calculated results were examined in comparison with the experimental results of photoluminescence excitation spectroscopy. The agreement with the experiments was good, showing the validity of the calculations.

The exciton-related mechanism of optical gain formation, which is one of the most characteristic phenomenon in wide-bandgap II-VI compound semiconductors, was studied. First, room-temperature ultraviolet stimulated emission was demonstrated in optically-pumped $Cd_xZn_{1-x}S$-ZnS strained-layer multiple quantum-well structures. It was shown that the stimulated emission observed in this study was tuned from approximately 350 nm to 390 nm by varying a Cd composition ratio in ternary alloy well layers and was characterized by the shortest wavelength reported to date among semiconductor laser materials. Furthermore, a laser diode structure based on the $Cd_xZn_{1-x}S$-ZnS systems has been fabricated using a conventional low-pressure MOCVD method. Spectral narrowing in the ultraviolet electroluminescence above the threshold current was clearly observed at 30 K under pulsed operation. These experimental results strongly suggest that this type of quantum-well structure is the most fruitful candidate which is capable of providing the semiconductor injection laser diodes operating in the ultraviolet spectral region. This possibility was also supported by the calculated results of its band lineups. Next, optical gain spectra were measured by means of a pump-and-probe technique with nanosecond excitation pulses. It was found that gain was observable even in the presence of the clear excitonic absorption peak which was slightly bleached at the lower energy side. These results indicated that the formation of optical gain was excitonic in origin and that there was no evidence for the presence of an electron-hole plasma. A phenomenological analysis for excitonic gain was introduced on the basis of the inhomogeneous broadening of exciton resonance due to alloy fluctuations and interface fluctuations.

It was considered that excitons were localized from initially excited states to lower energy states of the inhomogeneously broadened exciton resonance, and that the partial phase-space filling of these localized exciton states took place resulting in gain in terms of induced emission to ground states. The calculated results based on the above analysis were found to be good agreement with the experimental results. The possibility of other excitonic gain processes, for example, the inelastic exciton scattering pictures which have been considered in bulk II-VI materials was also examined and found not to apply to our quantum-well structures. As a result, it was proposed that the formation of optical gain in the $Cd_xZn_{1-x}S$-ZnS quantum well structures originated from the population inversion of localized exciton states.

This article is based on the collaboration with Professor Tsunemasa Taguchi and Professor Yasuaki Masumoto. The author would like to thank them for valuable discussions. The author would also like to thank Dr. Yoichi Kawakami, Dr. Yasuyuki Endoh, and Dr. John T. Mullins for stimulating discussions.

## REFERENCES

[1] L. Esaki and R. Tsu, *IBM J. Res. Develop.* **14**, 61 (1970).

[2] M. A. Haase, J. Qiu, J. M. DePuydt, and H. Cheng, *Appl. Phys. Lett.* **59**, 1272 (1991).

[3] H. Jeon, J. Ding, W. Patterson, A. V. Nurmikko, W. Xie, D. C. Grillo, M. Kobayashi, and R. L. Gunshor, *Appl. Phys. Lett.* **59**, 3619 (1991).

[4] A. Ishibashi and S. Itoh, in *Proceedings of Lasers and Electro Optics Society* (Boston, 1994).

[5] D. B. Eason, Z. Yu, W. C. Hughes, W. H. Roland, C. Boney, J. W. Cook, Jr., J. F. Schetzina, G. Cantwell, and W. C. Harsch, *Appl. Phys. Lett.* **66**, 115 (1995).

[6] R. N. Bylsma, W. M. Becker, T. C. Bonsett, L. A. Kolodziejski, R. L. Gunshor, M. Yamanishi, and S. Datta, *Appl. Phys. Lett.* **47**, 1039 (1985).

[7] Y. Kawakami, S. Yamaguchi, Y.-h. Wu, K. Ichino, Sz. Fujita, and Sg. Fujita, *Jpn. J. Appl. Phys.* **30**, L605 (1991).

[8] K. Ichino, Y.-h. Wu, Y. Kawakami, Sz. Fujita, and Sg. Fujita, *J. Cryst. Growth* **117**, 527 (1992).

[9] K. Nakanishi, I. Suemune, Y. Fujii, Y. Kuroda, and M. Yamanishi, *Appl. Phys. Lett.* **59**, 1401 (1991).

[10] G. Sun, K. Shahzad, J. M. Gaines, and J. B. Khurgin, *Appl. Phys. Lett.* **59**, 310 (1991).

[11] D. Lee, A. M. Johnson, J. E. Zucker, C. A. Burrus, R. D. Feldman, and R. F. Austin, *Appl. Phys. Lett.* **60**, 739 (1992).

[12] N. T. Pelekanos, J. Ding, M. Hagerott, A. V. Nurmikko, H. Luo, N. Samarth, and J. K. Furdyna, *Phys. Rev. B* **45**, 6037 (1992).

[13] C. G. Van de Walle, *Phys. Rev. B* **39**, 1871 (1989).

[14] J. Ding, H. Jeon, T. Ishihara, M. Hagerott, A. V. Nurmikko, H. Luo, N. Samarth, and J. Furdyna, *Phys. Rev. Lett.* **69**, 1707 (1992).

[15] Y. Endoh and T. Taguchi, *Mater. Res. Soc. Symp. Proc.* **161**, 211 (1990).

[16] T. Taguchi and Y. Endoh, *Jpn. J. Appl. Phys.* **30**, L952 (1991).

[17] T. Taguchi, Y. Endoh, and Y. Nozue, *Appl. Phys. Lett.* **59**, 3434 (1991).

[18] Y. Yamada, T. Taguchi, and A. Hiraki, *Tech. Rep. Osaka University* **39**, 211 (1989).

[19] T. Taguchi, Y. Kawakami, and Y. Yamada, *Physica B* **191**, 23 (1993).

[20] Y. Kawakami, T. Taguchi, and A. Hiraki, *J. Cryst. Growth* **89**, 331 (1988).

[21] G. C. Osbourn, *IEEE J. Quantum Electron.* **QE-22**, 1677 (1986).

[22] W. A. Harrison, *"Electronic Structure and Properties of Solids" — The Physics of Chemical Bond —* (W. H. Freeman and Company, San Francisco, 1980).

[23] D. L. Camphausen, G. A. N. Connell, and W. Paul, *Phys. Rev. Lett.* **26**, 184 (1971).

[24] F. H. Pollak and M. Cardona, *Phys. Rev.* **172**, 816 (1968).

[25] T. Taguchi and Y. Yamada, *Mater. Res. Soc. Symp. Proc.* **161**, 199 (1990).

[26] A. Blacha, H. Presting, and M. Cardona, *phys. stat. sol. (b)* **126**, 11 (1984).

[27] D. W. Niles and H. Höchst, *Phys. Rev. B* **44**, 10965 (1991).

[28] L. G. Suslina, A. G. Plyukhin, D. L. Fedorov, and A. G. Areshkin, *Sov. Phys. Semicond.* **12**, 1331 (1978).

[29] M. Cardona, *J. Phys. Chem. Solids* **24**, 1543 (1963).

[30] P. Lawaets, *Phys. Rev. B* **4**, 3460 (1971).

[31] Y. Shinozuka and M. Matsuura, *Phys. Rev. B* **28**, 4878 (1983); **29**, 3717(E) (1984).

[32] Y. Yamada, Y. Masumoto, T. Taguchi, and K. Takemura, *Phys. Rev. B* **44**, 1801 (1991).

[33] Y. Yamada, Y. Masumoto, J. T. Mullins, and T. Taguchi, *Appl. Phys. Lett.* **61**, 2190 (1992).

[34] M. Yamada, S. Ogita, M. Yamanishi, K. Tabata, N. Nakaya, M. Asada, and Y. Suematsu, *Appl. Phys. Lett.* **45**, 324 (1984).

[35] R. L. Gunshor and L. A. Kolodziejski, *IEEE J. Quantum Electron.* **QE-24**, 1744 (1988).

[36] T. Taguchi, C. Onodera, Y. Yamada, and Y. Masumoto, *Jpn. J. Appl. Phys.* **32**, L1308 (1993).

[37] Y. Yamada, Y. Masumoto, and T. Taguchi, *J. Cryst. Growth* **138**, 570 (1994).

[38] J. M. Hvam, *Solid State Commun.* **12**, 95 (1973).

[39] S. W. Koch, H. Haug, G. Schmieder, W. Bohnert, and C. Klingshirn, *phys. stat. sol. (b)* **89**, 431 (1978).

[40] C. F. Klingshirn, W. Maier, B. Hönerlage, H. Haug, and S. W. Koch, *Solid-State Electron.* **21**, 1357 (1978).

[41] C. Klingshirn and H. Haug, *Phys. Rep.* **70**, 315 (1981).

[42] I. M. Catalano, A. Cingolani, M. Ferrara, and M. Lugara, *Solid State Commun.* **43**, 371 (1982).

[43] B. Hönerlage, R. Lévy, J. B. Grun, C. Klingshirn, and K. Bohnert, *Phys. Rep.* **124**, 161 (1985).

[44] P. R. Newbury, K. Shahzad, and D. A. Cammack, *Appl. Phys. Lett.* **58**, 1065 (1991).

[45] J. Ding, M. Hagerott, T. Ishihara, H. Jeon, and A. V. Nurmikko, *Phys. Rev. B* **47**, 10528 (1993).

[46] Y. Yamada, Y. Masumoto, and T. Taguchi, *Physica B* **191**, 83 (1993).

[47] K. L. Shaklee, R. F. Leheny, and R. E. Nahory, *Phys. Rev. Lett.* **26**, 888 (1971).

[48] Y. Masumoto, T. Kawamura, and K. Era, *Appl. Phys. Lett.* **62**, 225 (1993).

[49] Y. Yamada, Y. Masumoto, and T. Taguchi, (unpublised).

[50] R. C. Mıller, D. A. Kleinman, A. C. Gossard, and O. Munteanu, *Phys. Rev. B* **25**, 6545 (1982).

[51] S. Charbonneau, T. Steiner, M. L. W. Thewalt, E. S. Koteles, J. Y. Chi, and B. Elman, *Phys. Rev. B* **38**, 3583 (1988).

[52] D. C. Reynolds, K. K. Bajaj, C. E. Stutz, R. L. Jones, W. M. Theis, P. W. Yu, and K. R. Evans, *Phys. Rev. B* **40**, 3340 (1989).

[53] D. J. Lovering, R. T. Phillips, G. J. Denton, and G. W. Smith, *Phys. Rev. Lett.* **68**, 1880 (1992).

[54] R. T. Phillips, D. J. Lovering, G. J. Denton, and G. W. Smith, *Phys. Rev. B* **45**, 4308 (1992).

[55] Y. Yamada, T. Mishina, Y. Masumoto, Y. Kawakami, S. Yamaguchi, K. Ichino, Sz. Fujita, Sg. Fujita, and T. Taguchi, *Phys. Rev. B* **51**, 2596 (1995).

[56] Y. Yamada, T. Mishina, Y. Masumoto, Y. Kawakami, J. Suda, Sz. Fujita, and Sg. Fujita, *Phys. Rev. B* **52**, (1995) in press.

[57] J. Y. Jen, T. Tsutsumi, I. Souma, Y. Oka, and H. Fujiyasu, *Jpn. J. Appl. Phys.* **32**, L1542 (1993).

[58] Y. Kawakami, I. Hauksson, H. Stewart, J. Simpson, I. Galbraith, K. A. Prior, and B. C. Cavenett, *Phys. Rev. B* **48**, 11994 (1993).

[59] F. Kreller, M. Lowisch, J. Puls, V. Rossin, A. Schülzgen, J. Griesche, N. Hoffmann, M. Rabe, and F. Henneberger, in *Proceedings of 22nd International Conference on the Physics of Semiconductors* (Vancouver, 1994).

[60] R. Cingolani, R. Rinaldi, L. Calcagnile, P. Prete, P. Sciacovelli, L. Tapfer, L. Vanzetti, G. Mula, F. Bassani, L. Sorba, and A. Franciosi, *Phys. Rev. B* **49**, 16769 (1994).

[61] W. Wegscheider, L. N. Pfeiffer, M. M. Dignam, A. Pinczuk, K. W. West, S. L. McCall, and R. Hull, *Phys. Rev. Lett.* **71**, 4071 (1993).

[62] S. Lai and M. V. Klein, *Phys. Rev. Lett.* **44**, 1087 (1980).

[63] S. Permogorov, A. Reznitskii, S. Verbin, G. O. Müller, P. Flögel, and M. Nikiforova, *phys. stat. sol. (b)* **113**, 589 (1982).

[64] E. Cohen and M. D. Sturge, *Phys. Rev. B* **25**, 3828 (1982).

[65] S. Permogorov and A. Reznitsky, *J. Lumin.* **52**, 201 (1992).

[66] Y. Kawakami, M. Funato, Sz. Fujita, Sg. Fujita, Y. Yamada, and Y. Masumoto, *Phys. Rev. B* **50**, 14655 (1994).

[67] F. A. Majumder, S. Shevel, V. G. Lyssenko, H. E. Swoboda, and C. Klingshirn, *Z. Phys. B* **66**, 409 (1987).

*CHAPTER 5*

# LUMINESCENCE FROM SILICON NANOSTRUCTURES

Yoshihiko Kanemitsu
*Institute of Physics, University of Tsukuba*



## 5.1. INTRODUCTION

Recently, in silicon science and technology, the desire for the integration of opto-electronics devices with silicon microelectronics has led to the search for Si-based materials and structures that emit light with a high quantum efficiency. Crystalline silicon (c-Si) is the dominant material in microelectronics and is one of the best-studied materials. Crystalline silicon does not show efficient light emission at room temperature, because of its band structure with an indirect band gap of $\sim$ 1.1 eV and a small exciton binding energy ($\sim$ 15 meV). One promising approach to overcoming the indirect nature of optical transition in silicon is the relaxation of the

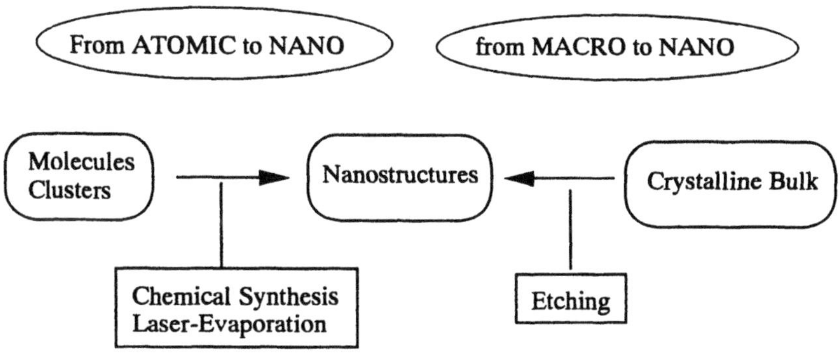


Figure 5.1. Our experimental approaches for the understanding of electronic and optical properties of Si nanostructures. Si nanostructures have hybrid electronic properties between the molecular and crystalline bulk limits.

k-selection rule due to the spatial confinement in low-dimensional Si nanostructures. In low-dimensional semiconductor systems, three categories are usually considered: two-dimensional (2D) quantum wells, one-dimensional (1D) quantum wires, and zero-dimensional (0D) quantum dots. Efficient visible light emission has been observed from Si polymers as 1D systems and Si nanocrystallites as 0D systems, at room temperature [1]. The discovery of visible luminescence from these Si nanostructures opens a new possibility for use of group IV indirect-gap semiconductors as materials for optoelectronic applications. In this chapter, we discuss optical properties of light-emitting Si nanostructures.

The first synthesis of Si polymers by Kipping [2] can be back to the early 1920s. In 1949, Burkhard [3] synthesized dimethylpolysilanes. Yajima *et al.* [4] produced silicon carbide fibers from polysilanes in 1975. However, it is fair to say that the success in soluble polysilanes in 1979 [5] has created the silicon polymers science. After the synthesis of soluble Si polymers, chemical, physical and electronic properties of Si polymers have been actively studied [6, 7]. Chemically synthesized Si polymers are often regarded as natural quantum wires whose unique properties are primarily attributed to the quantum confinement effect on the conjugated electrons delocalized in polymer backbones. A sharp luminescence in 1D Si chainlike polymers is observed near the absorption edge with high quantum efficiencies, at room temperature [1].

In 1984, Pickering *et al.* [8] reported visible photoluminescence (PL) from electrochemically etched porous silicon at low temperature. They considered the origin of the visible luminescence as amorphous silicon or impurities. In 1988, Furukawa and Miyasato [9] prepared nanometer-sized Si crystallites terminated by hydrogen atoms by means of a reactive sputtering technique and first suggested that the blueshift of the optical band gap is explained by the quantum confinement effect in 0D Si nanocrystallites. In 1990, Takagi *et al.* [10] reported the room-temperature visible

PL from silicon nanocrystallites prepared by a microwave plasma decomposition of $SiH_4$ gas. From the size dependence of the PL peak energy, they pointed out that the quantum confinement effects play an important role in the visible luminescence process in Si nanocrystallites. Similar visible luminescence from porous silicon was observed by Canham [11]. The blueshift of the band gap energy was also observed in porous silicon by Lehman and Gosele [12]. After this, there have been numerous reports describing efficient visible luminescence from porous silicon. Now, we can obtain visible luminescence ranging from red to violet in color from porous silicon fabricated under carefully controlled conditions. Furthermore, we can observe visible photo- and electro-luminescence from nanocrystallites made from indirect-gap semiconductors such as Ge [13, 14] and SiC [15, 16]. However, the mechanism of visible luminescence from Si, Ge, and SiC nanocrystallites is not yet clear. There are several possibilities for the origin of luminescence in porous silicon, Si nanocrystallites, and related materials.

In order to understand electronic and optical properties of Si nanostructures, we have two different experimental approaches, as illustrated in Fig. 5.1.
(a) *From molecules to nanostructures*: The nanometer-sized materials are chemically synthesized by polymerization techniques of molecules. The size dependence of the optical properties is explained by the bathochromic effect due to conjugated electrons.
(b) *From bulk to nanostructures*: The nanometer-sized structures are fabricated by epitaxial growth and etching techniques. The optical properties are discussed in term of the quantum confinement effects.

In this chapter, we review current experimental studies on optical properties of Si nanostructures on the basis of the above two approaches. We describe luminescence properties of 1D Si polymers and 0D Si clusters and nanocrystallites and their applications to optical logic gates and light-emitting devices.

## 5.2. CLUSTERS AND POLYMERS: FROM MOLECULES TO NANOSTRUCTURES

In exploring new optoelectronic materials and devices, a great deal of research effort is focused on reducing the dimensionality of the electronic structures in semiconductors. Natural analogues of low-dimensional silicon structures such as wires and sheets are chemically synthesized clusters and polymers. Modern organic synthesis techniques allow us to produce Si materials with controlled structures. In this sense, we have an approach for the understanding of microscopic nature of silicon nanostructures: *From molecules to nanostructures* [17, 18]. Studies of optical properties of chemically synthesized Si materials (small Si clusters, Si based polymers, siloxenes, etc.) are needed to understand the electronic structures and the microscopic PL mechanisms in porous silicon, Si nanocrystallites and other light-emitting silicon materials, because the local Si structures can be controlled in chemically synthesized Si materials.

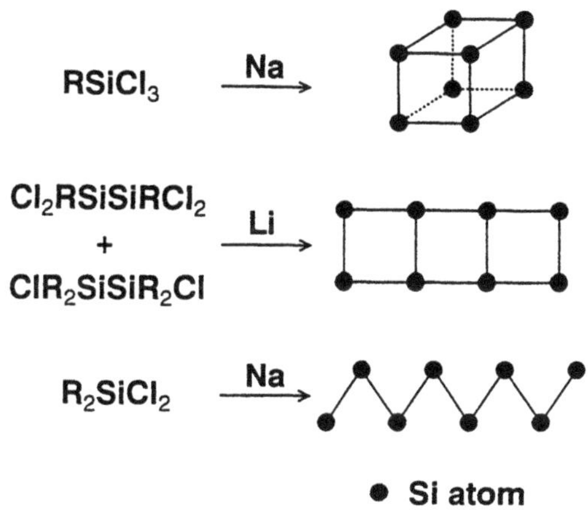


Figure 5.2. Chemically synthesis procedures of $Si_8$ clusters with cubic, ladder, and chain structures. Modern organic synthesis and purification techniques produce silicon materials with controlled structures. The dangling bonds are terminated by the organic substituents, $R$.

### 5.2.1. *Small Clusters*

The study of the structures and properties of silicon clusters has been an extremely active area of current research. The small Si clusters have been usually produced by pulsed laser evaporation methods [19, 20, 21]. Recent laser-evaporation methods combined with mass spectrometer produce size-selected Si clusters. However, since small clusters are short-lived intermediate species and there are few experimental studies of optical properties in small clusters, we have little information on the optical properties of vapor-phase small clusters [19]. We need to contrive new experimental procedures for the study of the detailed structure and properties of Si clusters.

Organic synthesis and purification techniques can produce small silicon materials with controlled structures. We discuss the shape dependence of the optical properties in chemically synthesized silicon clusters. Chemically synthesis techniques have some advantages over the widely-used laser-evaporation methods: For example, the size and shape of Si clusters can be exactly controlled [22, 23, 24, 25, 26]. The study of the chemically-synthesized Si clusters helps us to understand the electronic properties of vapor-phase Si clusters and to provide new information for the ongoing discussion on the origin of the strong visible luminescence of nanometer-sized Si materials. Here, we discuss the optical properties of $Si_8$ clusters and the size dependence of luminescence properties of small Si clusters.

The synthesis procedures of $Si_8$ clusters are illustrated in Fig. 5.2. Synthetic,

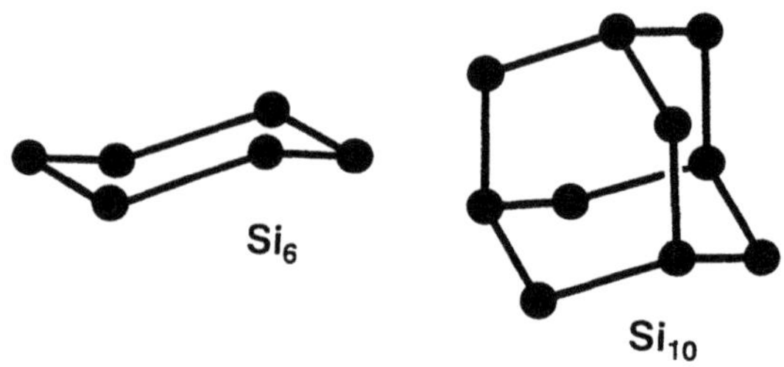


Figure 5.3. $sp^3$-hybridized "crystalline" clusters: chair-like cyclohexane $Si_6$ and adamantane $Si_{10}$ structures.

purification, and characterization methods were described in [22, 27, 28]. The dangling bonds are terminated by organic substituents $R$. In each structures, the optical properties are not sensitive to the kind of the organic terminations. The optical absorption and luminescence properties of these clusters reflect the electronic structures of the central silicon skeleton structures. Saturatedly bonded $Si_8$ clusters terminated by an organic substituent have no dangling bonds and become a new model material for small silicon clusters.

For example, in laser-evaporation experiments, magic number clusters are observed, where magic number is the number of atoms in the relatively stable clusters. Magic-number $Si_6$ and $Si_{10}$ clusters are prototypes of silicon clusters. In Fig. 5.3, we illustrate the $sp^3$-hybridized structures, chair-like cyclohexane $Si_6$ and adamantane $Si_{10}$ structures. These structures are obtained from the diamond structure of c-Si. These crystalline-series magic numbers were found to show good correspondence with the observed magic numbers [29]. However, the stable structures of small $Si_6$ and $Si_{10}$ clusters are different from those of $sp^3$-hybridized crystalline structures. As an example, a structural study [21] of size-selected $Si_4$, $Si_6$, and $Si_7$ clusters in $N_2$ matrix shows that $Si_4$ is a planar rhombus, $Si_6$ a distorted octahedron, and $Si_7$ a pentagonal bipyramid, as illustrated in Fig. 5.4. Since the structures of small Si clusters are very complicated and depend strongly on the number of Si atoms in the clusters, the synthesis of Si clusters with different geometry is very important for the understanding of the presence of the magic numbers and of unique electronic properties of small Si clusters.

The optical absorption and PL spectra of $Si_8$ clusters with different structures at room temperature are summarized in Fig. 5.5. The Si-backbone structures are illustrated in this figure. In the chain and ladder $Si_8$ structures, the absorption spectra are observed in the ultraviolet spectral region. On the other hand, in the cubic cluster, a small absorption band around 2.4 eV and a large one around 3.5 eV are observed.

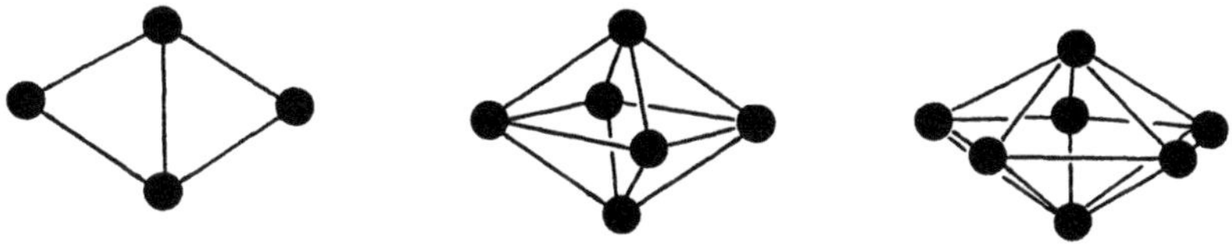

Figure 5.4. Stable structures of laser-evaporated $Si_4$, $Si_6$ and $Si_7$ clusters.

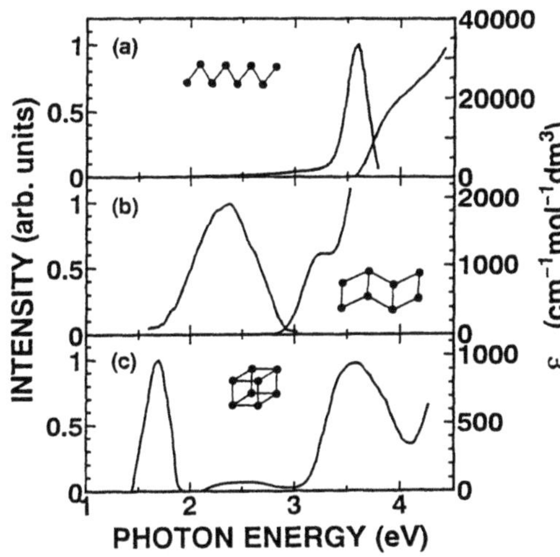


Figure 5.5. Optical absorption and photoluminescence of chemically synthesized $Si_8$ clusters at room temperature: (a) chain, (b) ladder, and (c) cubic structures. The lowest absorption peak energy in the cubic structure is low compared with those in the chain and ladder structures. (After [24]).

The absorption edge energy of the cubic structure is very small compared with those of the chain and ladder structures.

In the chain structure, a sharp and strong PL band is observed near the absorption edge. In the ladder structure, a broad PL band is observed with a large Stokes shift in the visible spectral region. However, in the cubic structure, PL is not clearly observed

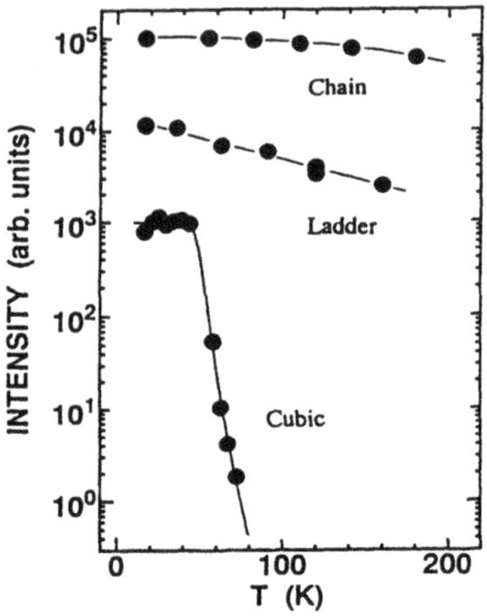


Figure 5.6. Temperature dependence of the luminescence intensity in the chain, ladder and cubic $Si_8$ clusters. The luminescence intensity in the cubic structure abruptly decreases with increasing temperature above 40 K, while the luminescence intensity in the chain and ladder structures gradually decreases with increasing temperature. (After [24]).

at room temperature, but observed only at low temperatures. Figure 5.6 shows the temperature dependence of the PL intensity in $Si_8$ clusters. The PL intensity in the cubic cluster abruptly decreases with increasing temperature above 40 K, while the PL intensities in the chain and ladder clusters gradually decrease with increasing temperature. The PL decay properties of the cubic $Si_8$ cluster are entirely different from those of the $Si_8$ chain and $Si_8$ sheet: The lifetimes of the chain and ladder $Si_8$ structures are 110 ps and 700 ps, respectively, at room temperature. The lifetime of the cubic structure is 3.1 ms at low temperatures. As seen in Figs. 5.5 and 5.6, the optical properties of small Si clusters are very sensitive to the Si backbone structures. In particular, the optical properties of the cubic structure are quite different from those of the chain and ladder structures.

The luminescence characteristics (the slow-decay and temperature-sensitive luminescence) in the cubic structure are different from those in the chain and ladder structures. The optical absorption edge energy (the lowest energy gap or the optical band-gap) of the cubic cluster is very small compared with those of the chain and ladder clusters. The Si-Si bond angle in the cubic structure is about 90°, which is different from the bond angle of tetrahedral Si bonds ($sp^3$-hybridized orbital). On the other hand, the Si-Si bond angle in the chain and ladder structures is nearly equal to that of tetrahedral Si bonds. The molecular orbital characteristics in the cubic clusters are "$p$-like" and are different from those in the $sp^3$-hybridized chain and ladder structures. Therefore, we speculate that the difference of the optical properties between the cubic and the other clusters is caused by that of the orbital hybridization and bonding in the clusters. The study of the electronic properties of Si compounds with different geometry and different orbital hybridization is very important for the understanding of the presence of the magic numbers of small Si clusters and remarkably size-independent optical properties of small Si clusters [20].

Optical studies of clusters and wires containing a few to several thousand atoms help to understand how molecules evolve into solids. We studied the size dependent optical properties of confined excitons in chains of Si atoms [25]. The oscillator strength per Si atom, $f_1$ is linearly proportional to the number of Si atoms and then saturates to a constant value: In Si chains of $N < 20$, $f_1$ is approximately given by $f_1 \propto N$ [25]. If chains of Si atoms are solids in nature, the oscillator strength per atom is almost independent of the number of atoms. Therefore, we believe that the size dependent region is the continuous transition of Si chains from molecular to solid forms and long chains become solid-like in optical properties. To our knowledge, there is no theoretical work of the size dependence of the oscillator strength of excitons in small "molecular-like" clusters and chains. The observation of the size dependence of radiative decay time confirms the size dependence of the oscillator strength of the lowest exciton, $f_1$. Moreover, with increasing chain length, the PL spectrum becomes narrower, the PL intensity increases, and the PL lifetime decreases. These size dependence show that in chains having 20 Si atoms or less, the exciton is coherently excited over the chain of Si atoms.

In 1D chainlike Si polymers, a sharp and strong PL band was observed with essentially no Stokes shift: Excitons are delocalized on Si-backbone chains and luminescence is due to free-exciton recombination. In Si sheets, broad luminescence spectra with large Stokes shifts were observed at the visible region, and the exciton localization determines the luminescence properties. The size dependence of the luminescence properties is not clearly observed in Si materials other than 1D Si chains [26].

We can chemically synthesize Si clusters with unique structures. Figure 5.7 shows some examples of small Si clusters. The study of these clusters helps us to understand the electronic properties of laser-evaporated Si clusters, because laser-evaporated Si clusters have complicated structures but similar structures will be produced by chemically synthesis and purification methods. Our group are trying to clarify the optical properties of small Si clusters with various structures as illustrated in Figs. 5.3 and 5.7.

### 5.2.2. *Polymers*

In order to understand the nature of Si nanostructures, we need to fabricate Si materials having large number of Si atoms and to study their optical properties. In the approach from molecules to nanostructures, a class of large Si materials is Si-based polymers. Here, we summarize the optical properties of Si polymers with different backbone structures and discuss the origin of visible luminescence in silicon materials. We studied the optical properties of various Si-backbone polymers: (a) polymethylphenylsilane, (b) poly(dimethyl-phenylsilyl)-phenylsilane, (c) dodecaiso-

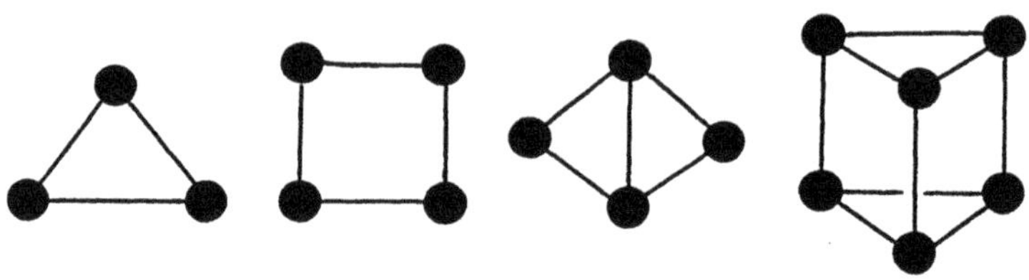

Figure 5.7. Chemically synthesized Si clusters. The dangling bonds are terminated by the organic substituents.

propyltetracyclodesilane, (d) network polysilynes, and (e) planar siloxene. The synthesis procedures were described in the literature [7, 27, 28, 30, 31]. The Si-backbone structures of these polymers are illustrated in Fig. 5.8 and hereafter, we call these structures as the chain [Fig. 5.8(a)], branch [Fig. 5.8(b)], ladder [Fig. 5.8(c)], network [Fig. 5.8(d)], and siloxene [Fig. 5.8(e)] structures. In the chain structure, the organosilicon units on the polymer backbone have two Si-Si bonds. In the other structures, the polymer backbones are constructed by the organosilicon units having three Si-Si bonds. Using these samples, we discuss the effects of the "lateral" dimension and backbone geometry on the optical properties of Si polymers.

Figure 5.9 shows luminescence and absorption spectra of the chain, branch, ladder, network, and siloxene structures at room temperature. In the chain structure [Fig. 5.9(a)], sharp luminescence and sharp absorption bands are observed. A Stokes shift between luminescence and lowest absorption peaks is relatively small. It is well accepted that the sharp absorption and PL bands are attributed to 1D excitons delocalized on the backbone chain. The structures of chainlike Si polymers take a variety of conformations, depending on organic substituents attached to the polymer backbones [32]. However, the luminescence band is very sharp in all conformations including disordered forms, as shown in Fig. 5.10. Therefore, the sharp luminescence band is the most important feature of the optical properties of the chain structure [33].

In the branch and ladder structures, broad luminescence spectra are only observed in the visible region. The intensities of the lowest absorption peak per organosilicon unit constructing the polymer backbones in the branch and ladder structures are very small compared with that in the chain structure. Time-resolved PL measurements showed that excitons are strongly localized on 2-3 Si atoms in the branch and ladder structures, but excitons are delocalized over about 20 Si atoms in the chain structure [25, 26]. Si backbone polymers consisting of Si atoms with three Si-Si bonds have

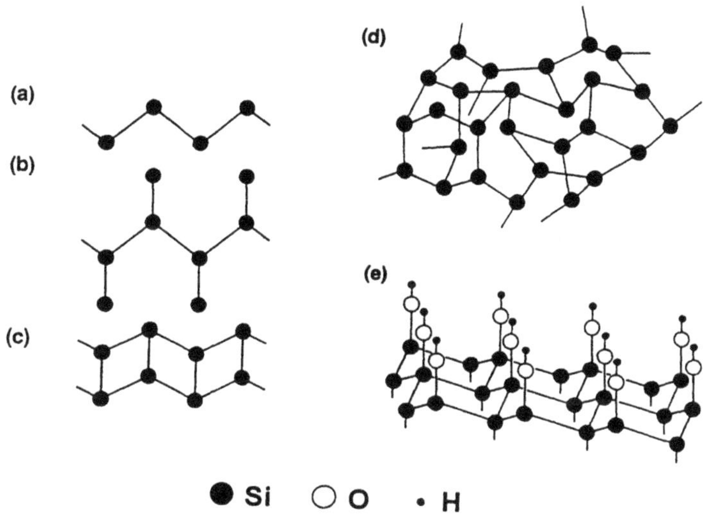


Figure 5.8. Schematics of Si polymers with different backbones: (a) chain, (b) branch, (c) ladder, (d) network, and (e) planar siloxene structures. In the chain structure, the organosilicon units in the polymer backbone have two Si-Si bonds. In other structures, the polymer backbone are constructed by the organosilicon units having three Si-Si bonds.

spectroscopic characteristics that excitons are strongly localized, the PL spectrum is broad, and the quantum efficiency is low. The broad visible PL in Si polymers is caused by the introduction of Si units with three Si-Si bonds into the polymer backbone.

In the network structure, the broad PL is observed in the visible region. The luminescence properties are determined by the disordered network Si-backbone structures. In order to discuss the 2D characteristics of the network polymers, we compare optical absorption and PL spectra in the network structure and planar siloxene. The planar siloxene $[Si_6H_6O_3)_n]$ is a direct-gap semiconductor because of the orbital mixing of delocalized electrons in 2D-Si layer and non-bonding $n$ electrons in O atoms [34]. In planar siloxene, the sharp PL and absorption bands are observed. A Stokes shift between the PL and absorption peaks is small. The broad PL spectrum in the network structure is different from the sharp PL spectrum due to 2D excitons in planar siloxene.

We discuss the origin of the broad luminescence spectra in the branch and ladder structures from the electronic band structures of Si polymers. The band calculations were performed by Takeda and Shiraishi with the first-principle local density functional method [35, 36]. In theoretical calculations, the organic substitutions without Si atoms in polymers are replaced by H atoms and the trans-planar form is considered as the chain structure and the branch and ladder structures are constructed

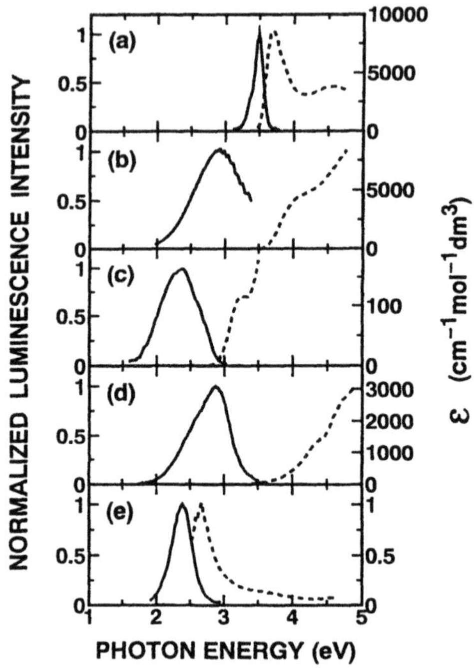


Figure 5.9. Optical absorption and normalized photoluminescence spectra in Si polymers: (a) chain, (b) branch, (c) ladder, (d) network, and (e) planar siloxene structures. In the chain structure, a sharp luminescence band is observed in the ultraviolet region. In the other structures, broad luminescence spectra are observed in visible region.

using the chain structure of the trans-planar form. The chain-like polysilane has a directly allowed-type band structure with a band gap of 3.89 eV. Both the highest occupied valence band (HOVB) and the lowest unoccupied conduction band (LUCB) states are formed by the skeletal Si orbital and are delocalized toward the skeleton direction. Therefore, these chain-like polysilanes can be described as "$\sigma$–conjugated one-dimensional structures."

2D Si-backbone polymers have intermediate electronic properties between 1D chainlike Si polymers of the direct-gap band structure and 3D bulk Si of the indirect-gap band structure. Although 2D crystalline planar polymers still a hypothetical polymer, disordered planar and network polymers are of scientific importance because they serve as a conceptual bridge between the properties of the linear chains and fully coordinated amorphous or crystalline silicon. The 2D structure has two characteristic band gaps: an indirect one of 2.48 eV and a direct one of 2.63 eV [35]. The 2D structure has an energy structure similar to the indirect gap one of 3D bulk

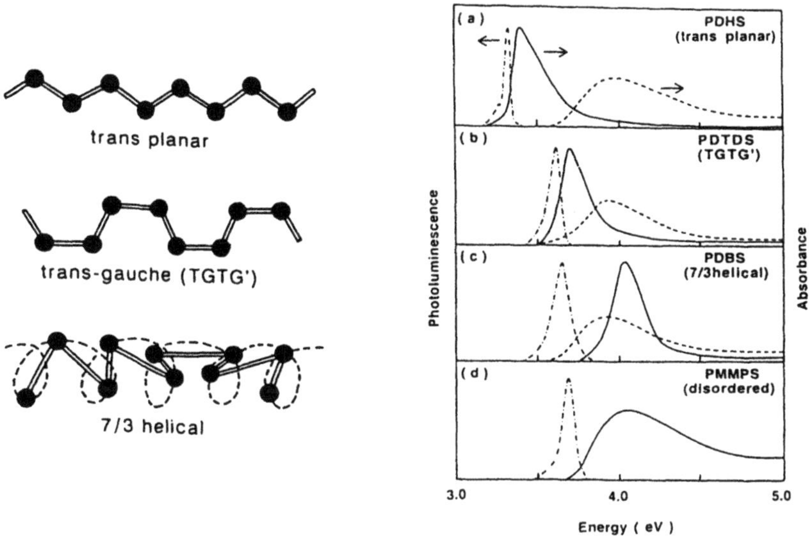


Figure 5.10. Various backbone conformation and optical properties of one-dimensional Si chains. (After [32]).

c-Si. The addition of oxygen atoms to the 2D structure causes the strong modification of the band structure [34]. The oxidized 2D sheet (planar siloxene structure) has a direct gap structure of the band gap of 1.7 eV at the point Γ. Oxygen atoms are in out-of-plane sites and oxygen atoms do not cut the electron and hole delocalization in 2D sheet. The lifting of the HOVB due to the $\sigma$-n mixing and the stabilization of the LUCB due to the oxygen hybridization effectively reduce the band gap at the point Γ. The planar siloxene is the optically allowed direct gap structure.

The calculated band structures are consistent with experimental facts that strong PL is observed in the chain and siloxene structures and weak broad PL spectra at the visible region are observed in the branch, ladder, and network structures. However, the polymers have a noncrystalline characteristics and the exciton localization in the structural disorders determines the luminescence properties and carrier dynamics in of branch, ladder, and network structures, as discussed below.

In ladder and network silicon polymers, the luminescence temperature dependence is described by the inverse Arrhenius equation [33]. This relation is usually observed in the noncrystalline semiconductors with low coordination numbers; *e.g.*, chalcogenide glasses and highly hydrogenated silicon alloy. The inverse Arrhenius temperature dependence is a feature of noncrystalline materials constructed by the low-coordination groups. The polymers with three Si-Si bonds exhibit the noncrystalline nature and there is a broad distribution of activation energies for nonradiative process. In the network polymers, the branching points and the random network structures give rise to localized electronic states known as the band tail of the both conduction and valence bands [37]. We consider that broad PL bands observed in many Si-based materials such as the branch, ladder, and network Si polymers and a-Si:H are closely related to the existence of Si atoms having three or four Si-Si bonds. The disorder and strain of Si backbone structures constructed by Si units with three or four S-Si bonds bring about the localized electronic states like amorphous semiconductors and semiconducting glasses.

The branch, ladder, and network Si polymers constructed by the organosilicon units with three Si-Si bonds exhibit broad luminescence spectra with long decay times. The broad luminescence properties of these structures are entirely different from the sharp luminescence properties of the chain structure constructed by two Si-Si bonds. In the chain structure, the sharp absorption and luminescence bands are mainly observed in the ultraviolet spectral region. These sharp bands due to the quasi-1D excitons are observed even in disordered form. In the disordered network structures, the excitons are strongly localized and the noncrystalline nature dominates the optical properties. The luminescence properties of branch, ladder, and network Si polymers resemble those in amorphous semiconductors and semiconducting glasses. The disorder and spatially inhomogeneous strain of Si-backbone structure with Si units of three or four Si-Si bonds bring about the localized electronic states which control the luminescence properties of Si materials. Since exciton localization determines the luminescence properties in ladder and network polymers, we cannot clarify the 2D characteristics of excitons using these polymers.

Figure 5.11 summarizes the size dependence of the peak energy of the lowest absorption band. The absorption spectrum depends strongly on the size of Si-skeleton chains or sheets. This suggests the quantum size effect on the lowest exciton energy. In small Si clusters, the gap energy is sensitive to the shape of clusters, rather than the size of clusters. In order to clarify the electronic structures of Si nanocrystallites, we need to fabricate very large Si materials having $N > 500$. The structures and properties of Si chains are actively studied and the microscopic structure of Si chains is controlled [25]. However, the structures of large ladder and network polymers are not well characterized and the exciton localization determines the luminescence properties in Si-based polymers. Moreover, by conventional organic synthesis methods, it is difficult to produce the crystalline Si materials with controlled size and microstructure at the present stage. The optical and electronic functionality of chemically synthesized Si polymers are still unclear (*e.g.*, electroluminescence is not observed in Si polymers). Since Si clusters and nanocrystallites have hybrid electronic properties between the molecular and crystalline limits, consecutive research from molecules to the bulk crystalline is required to understand electronic properties of Si nanostructures. However, there is a large gap in experimental study between Si clusters fabricated from molecules and Si nanocrystallites fabricated from bulk c-Si.

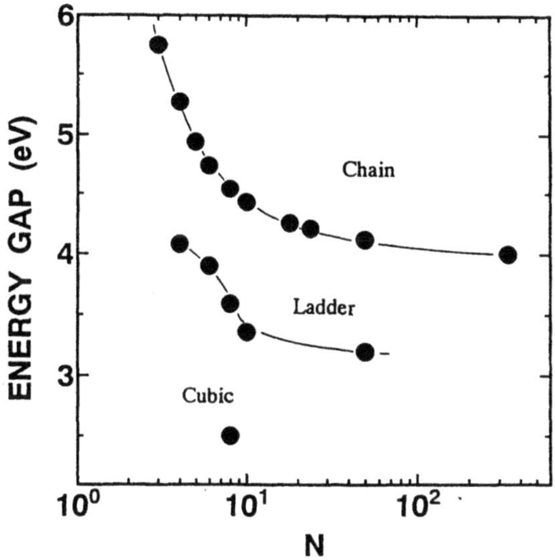


Figure 5.11. The absorption peak energy of the lowest excitation state as a function of the number of Si atoms in chain, ladder and cubic clusters. The solid lines are guides to the eye. The absorption spectrum depends on the size of Si materials.

## 5.3. NANOCRYSTALLITES: FROM BULK TO NANOSTRUCTURES

In the previous section, we described one approach for the understanding of the microscopic nature of silicon nanostructures: ***From molecules to nanostructures.*** However, by chemical synthesis techniques, it is difficult to produce crystalline silicon materials with large number of Si atoms. On the other hand, Si nanocrystallites are fabricated by electrochemical etching of bulk c-Si wafer and their electronic properties are considered on the basis of the band structures: ***From bulk to nanostructures.*** In this section, we discuss optical properties of Si nanocrystallites.

Bulk c-Si has an indirect gap which is unfavorable to efficient light emission. To obtain the efficient light emission, low-dimensional quantum structure are introduced into silicon materials: short-period Si/Ge superlattices, Si nanocrystallites and highly porous silicon. The efficient visible luminescence from porous silicon initiates remarkable theoretical studies in low-dimensional silicon materials. Low-dimensionality plays an essential role in optical transition of Si materials. The theoretical studies show the widening of the band gap energy and the enhancement of the oscillator strength with decreasing size of wires or dots: The quantum confinement effects plays an important role in the efficient visible luminescence in nanometer-sized Si crystallites [38, 39, 40, 41, 42, 43]. However, electronic structures of surface states cannot be calculated theoretically.

On the other hand, a drastic size reduction to a few nanometers is required for the observation of strong visible light emission. A large fraction of the silicon atoms appears on the surface: About 30% Si atoms are at the surface in 3-nm Si nanocrystallites. It is considered that the surface states in the Si nanocrystallites play a significant role. Thus, in spite of the intensive experimental and theoretical studies, no conclusive argument has been given yet on the mechanism of the efficient light-emission from porous silicon and related materials. There have been several possibilities for the origin of luminescence from porous silicon: (a) the "pure" quantum confinement effect in Si wires and dots, (b) siloxene-based molecular compounds, (c) silicon hydride complexes, (d) polysilanes, and (e) surface localization states on Si nanocrystallites.

However, recent experimental studies seem to rule out molecular-based materials such as siloxene, polysilanes, and silicon hydride complex. We reported that the luminescence properties of porous silicon are very similar to those of isolated c-Si spheres [44, 45]. Recent studies also support that porous silicon consisted of nanometer sized crystalline dots rather than wires [46, 47, 48, 49]. The isolated Si nanocrystallites are interpreted as the source of efficient luminescence from porous silicon. Therefore, we discuss the optical properties of porous silicon as nanometer-sized c-Si quantum dots.

### 5.3.1. *Fabrication of Nanocrystallites*

Porous silicon was discovered by Uhlir in 1956 [50]. The porous layer is created by electrochemical etching of Si wafers in HF-based electrolytes. Despite many years of basic research on the chemical properties of porous silicon, the formation mechanism is still not well understood [51]. However, it is generally agreed that the etching reaction requires holes injected from c-Si substrate. Since the band gap in porous silicon (Si nanocrystallites in the porous layer) increases compared with that of bulk c-Si due to the quantum confinement effect, holes need the additional energy $\Delta E_v$ to penetrate into the porous layer. $\Delta E_v$ is the valence-band offset between the porous layer and Si substrate. If $\Delta E_v$ is larger than the bias, the porous layer becomes depleted of holes and further dissolution is stopped. Since $\Delta E_v$ is a function of the size of nanocrystallites, one can conclude that an increase of the formation of bias results in an increase of the band gap energy and a decrease of the crystallite size in the porous layer. This process is self-adjusting. The quantum confinement effect limits the size of the nanocrystallites [51].

A high-resolution transmission electron microscopy (TEM) image in porous silicon shows that Si nanocrystallites with different sizes and shape are dispersed in the amorphous Si-based material matrix. The size and shape of Si nanocrystallites depend on experimental conditions. The pore structure also depends on the kinds of Si wafer (n- or p-type resistivity) and anodization parameters (HF concentration, current density, light illumination, etc.). The spatial structure on macroscopic scale reflects the available paths for the etching current. Si nanocrystallites in porous silicon retain the substrate crystallinity, but with a small lattice expansion compared to bulk c-Si [52].

Light illumination generates holes in the porous layer. Light illumination under the applied field promotes further dissolution of the porous structure. Then, we can control the size of nanocrystallites under light illumination [53]. Light illumination techniques produce the porous silicon layer even on n-type silicon wafer. By changing the illumination wavelength during electrochemical etching, we can also control the photoluminescence wavelength from n-type porous silicon layer. These light illumination effects support that the quantum size effect and self-adjusting effect are important in the formation process of the porous silicon layer.

Light-emitting Si nanocrystallites were prepared by using microwave plasma decomposition of $SiH_4$ silane gas. Si nanocrystallites were produced by means of microwave plasma decomposition of $SiH_4$ with $H_2$ and/or Ar gas mixture in a resonant cavity. Dissociated species of $SiH_4$ coalesced with each other and formed Si clusters or nanocrystallites. Altering the mixture ratio of gases and the power of microwave, the size of Si nanocrystallites was varied from 2.5 to 20 nm in diameter. The surface amorphous layer was easily oxidized in a humid atmosphere. After oxidation, the Si nanocrystallites were covered by $SiO_2$ matrices. The visible luminescence was not observed in hydrogen-terminated Si nanocrystallites. The strong visible photoluminescence in the red spectral region was observed after surface oxidation [10].

Si nanocrystallites were also produced by laser breakdown of silane ($SiH_4$) gas [54]. Pure $SiH_4$ gas was introduced into the vacuum chamber (a background pressure of$\sim 10^{-6}$ Torr). Average diameter of the Si nanocrystallites was controlled by this pressure. Laser pulses ($\sim$ 200 mJ per pulse at 1.06 $\mu$m, 10 ns pulse duration) from a $Nd^{3+}$:YAG laser system were focused into the chamber and a spark was observed at the focal point. The Si nanocrystallites were deposited on quartz or Ge wafer substrates. The samples oxidized at room temperature in a clear-air box exhibited strong visible PL. Fourier-transform infrared (FTIR) spectroscopy, x-ray photoemission spectroscopy (XPS), and TEM examinations indicate that the oxidized Si nanocrystallites consist of a c-Si core and an 1.6-nm-thick a-$SiO_2$ surface layer. The size distributions of oxidized Si nanocrystallite samples are described by a log-normal function [12,40]. In these sample, after surface oxidization, the efficient visible PL was observed in the red spectral region.

Si or Ge nanocrystallites in $SiO_2$ matrices were fabricated by using rf magnetron cosputtering techniques. This method is simple, but it is hard to control the surface structure of nanocrystallites. The oxides (*e.g.*, $SiO_2$, $SiO_x$, $GeO_2$, or $GeO_x$) are formed at the surface of nanocrystallites. The nanocrystallite samples were prepared by a method of rf-magnetron cosputtering of semiconductor wafers and $SiO_2$. Thin films of the mixture of semiconductors and $SiO_2$ were deposited onto substrates, and then annealed in an Ar gas atmosphere at high temperature (300–800 °C) in order to grow nanocrystallites in the $SiO_2$ glassy matrix and to control the size of nanocrystallites. The TEM images show that as-deposited samples are glassy structure, while in the annealed samples, nanocrystallites exist in the glassy $SiO_2$ matrix. Luminescence spectra of Si nanocrystallites in $SiO_2$ in the red spectral region are also similar to those of porous silicon. Moreover, luminescence properties of Ge nanocrystallites in $SiO_2$ [13, 14] are similar to those of oxidized porous germanium [55].

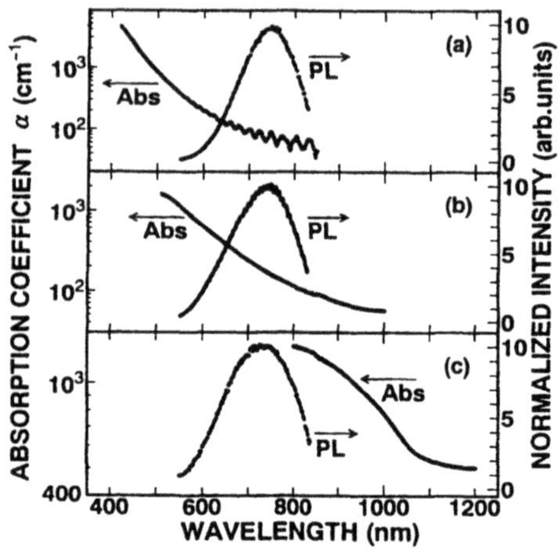


Figure 5.12. Typical optical absorption and photoluminescence spectra of porous silicon films at room temperature: (a) $L \sim 2$ nm, (b) $L \sim 3.5$ nm, and (c) $L \sim 9$ nm. The exponential absorption tail is observed. No significant size dependence of the luminescence spectrum is observed. (After [47]).

As mentioned above, most experiments show that the strong visible luminescence ("RED" luminescence) is observed when the surface of the Si nanocrystallite is covered by silicon oxides. A drastic size reduction to a few nanometers is needed for the observation of strong visible light emission. Since an appreciable fraction of the silicon atoms appears on the surface, the surface states on the Si nanocrystallites play a significant role. Here, we will discuss the mechanism of visible luminescence from Si nanocrystallites and porous silicon.

### 5.3.2. *Red Photoluminescence*

The widely studied luminescence in Si nanocrystallites is in the "red" spectral region. First of all, we discuss the mechanism of the red luminescence.

To understand the quantum confinement effect on electronic processes in nanocrystallites, we need to know the size dependence of both optical absorption and luminescence properties of nanocrystallites. The blueshift of absorption spectra due to the quantum confinement effects is observed in a variety of direct-gap semiconductor nanocrystallites. It is very important to identify the optical absorption properties and compare absorption spectra with luminescence spectra in Si nanocrystallites.

Figure 5.12 shows typical optical absorption spectra of free-standing porous silicon films with different crystallite sizes [(a) $L \sim 2$ nm, (b) $L \sim 3.5$ nm, and (c) $L \sim 9$ nm] [47], where the average diameter $L$ of the crystallites was determined from Raman measurements and TEM observations. A monotonical increase in the optical absorption coefficient is observed with increasing photon energy. This also indicates that the absorption edge has an exponential tail (the Urbach tail) in the visible spectral region. The exponential tail states are caused by the disordered surface states in the crystallites and the size distribution of the crystallites. It is found that the blueshift of optical absorption spectrum occurs with a decrease in the average diameter of Si crystallites. The size dependence of optical band edge $E_\alpha$ is shown in Fig. 5.13. The blueshift of the absorption spectrum indicates that the quantum confinement effect plays a key role in the absorption process. These optical absorption measurements provide strong evidence for the quantum confinement effects in porous silicon.

Figure 5.12 also shows the PL spectra of porous silicon films with different crystallite sizes. The PL spectrum was measured under cw 325 nm laser excitation. No significant size dependence of the PL peak energy is observed. The PL peak energy in samples with small average diameters [see Figs. 5.12(a) and 5.12(b)] is much smaller than the band gaps determined from the experiment $E_\alpha$ and the theoretical calculations (see Fig. 5.13). The above PL characteristics and the blueshift of the absorption spectrum imply that the site for the photogeneration of carriers is different from that for radiative recombination of carriers. Moreover, in porous silicon films having large Si crystallites, there appears an unusual relationship between the absorption and PL spectra: The PL peak energy is higher than the absorption edge energy as shown in Fig. 5.12(c). In porous silicon samples with large crystallite diameters, the PL excitation spectrum entirely differs from the optical absorption spectrum. TEM studies also showed that there are Si crystallites of various sizes in the sample. Thus, the unusual spectroscopic relation in Fig. 5.12(c) strongly suggests that only small crystallites contribute to the efficient visible PL.

The size-dependence of the PL peak energy has been also studied in oxidized Si nanocrystallites. The size dependence of the band gap energy (theoretical calculations) and the PL peak energy (experimental results) are summarized in Fig. 5.13. The size dependence of the PL peak energy is observed in oxidized Si nanocrystallites. However, the size dependence of the PL peak energy is very small compared with that of the absorption spectrum. Theoretical calculations can not explain the size dependence of the PL peak energy: This size dependence is not explained by the quantum confinement effect only. The above experimental observations show that some energy-loss occurs in the electronic processes from the absorption to light emission. The carriers or excitons relax from the high-energy absorption state to the lower-energy emission state [47, 56]. It is considered that this interface region between the Si core and $SiO_2$ layer play a role in the size-insensitive luminescence in oxidized Si nanocrystallites [54].

Considering that the surface of light-emitting Si nanocrystallites and porous silicon is terminated by H atoms, OH groups or $SiO_2$, we believe that the electronic properties of the surface state in the crystallite are different from those of the c-Si core in the crystallite. With large surface-to-volume ratios in nanometer-sized crystallites, the surface effects become more enhanced on decreasing the size of nanometer-sized crystallites. The present of the crystallite surface as a boundary and source of active

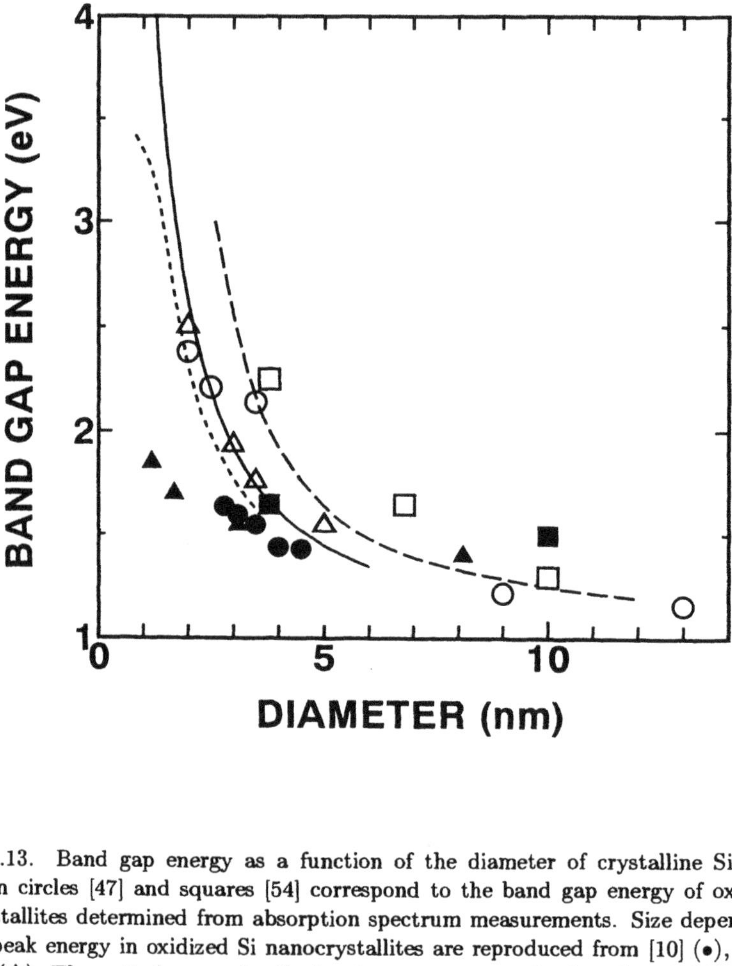


Figure 5.13. Band gap energy as a function of the diameter of crystalline Si spheres. The open circles [47] and squares [54] correspond to the band gap energy of oxidized Si nanocrystallites determined from absorption spectrum measurements. Size dependence of the PL peak energy in oxidized Si nanocrystallites are reproduced from [10] (•), [54] (■), and [48] (▲). Theoretical curves for the band-gap energy are also from by solid [39], broken [40], and dot [43] lines. The size dependence of the luminescence peak energy is very small compared with that of the theoretically calculated band-gap energy.

surface states makes crystallites different from epitaxial low-dimensional structures. In particular, oxygen atoms critically modify the electronic structure of Si crystallites [44, 45, 54], because the stable red luminescence is clearly observed after surface oxidation. The surface plays a role in the luminescence other than passivation (the reduction of the nonradiative recombination centers on the crystallite surface).

Moreover, the temperature dependence of the PL intensity shows a complicated behavior. This behavior implies that the carriers generated in the core state are rapidly localized at the lower-energy luminescent state with increasing temperature. At higher temperatures, the nonradiative recombination process increases. For the non-radiative recombination process, we consider the tunneling and thermally activated escape processes of carriers from the nanocrystallites to the outer materials. At higher temperatures, the thermally activated nonradiative recombination causes a decrease of the PL intensity. At lower temperatures below 50 K, the nonradiative recombination is tunneling process and the long radiative lifetime enhances this process. Many physical processes including the tunneling and thermally-activated nonradiative recombination, the Auger recombination process, and the surface localization of excitons determine the curious temperature dependence of the PL intensity in porous silicon and oxidized Si nanocrystallites.

The above experimental facts (the size-sensitive optical absorption spectrum, the size-insensitive luminescence spectrum, and the temperature dependence of the luminescence intensity) can be explained by a simple picture that the photogeneration of carriers occurs in the c-Si core with the size-dependent band gap and the radiative recombination in the localized state on the nanocrystallite surface causes the strong visible PL at room temperature. In our model, the PL efficiency in the surface state increases with a decrease in the crystallite size, because the surface-to-volume ratio increases with a decrease in the size of crystallites. Nonradiative processes are both of the escape processes of carriers or excitons from the nanocrystallites into the outer surrounding materials and nonradiative recombination on the surface defects. The strong red luminescence comes from the surface states of crystallites with sizes less than a certain diameter. The sample having a sharp size distribution will show more efficient luminescence. Next, we will discuss the electronic structures of the luminescent surface states.

In order to clarify the importance of the surface localized states, we studied the size dependence of the luminescence spectrum of oxidized Si nanocrystallites. Hereafter, $\langle D_{\text{core}} \rangle$ is the average diameter of the crystalline core in the oxidized nanocrystallites and was determined from TEM observations and a log normal function. The PL peak energy of Si nanocrystallites is $1.6 \sim 1.7$ eV, and is almost independent of the average diameter of the c-Si core. The PL spectra of oxidized Si nanocrystallites are very similar to those of porous silicon. The PL intensities of the oxidized Si nanocrystallites depend strongly on the crystallite size. The size dependence of PL intensity is very clear compared with that of the PL peak energy. Since the samples have a broad distribution of the crystallite size, we need to consider the effect of the size distribution on the PL intensity. Then, we first find that the PL intensity increases with the number of the oxidized Si nanocrystallites that have a small $\langle D_{\text{core}} \rangle$:

The PL intensity is approximately proportional to the volume fraction of the oxidized Si nanocrystallites of $\langle D_{core}\rangle \leq 7$ nm, indicating that only smaller nanocrystallites contribute efficiently to the strong PL. This finding is consistent with the blueshift of the absorption spectrum in nanocrystallites. The band gap widening due to the quantum confinement effect plays an essential role in the absorption and luminescence processes in Si nanocrystallites.

To evaluate the possible origins of the strong PL at $1.6 \sim 1.7$ eV for Si nanocrystallites, we consider the electronic and optical properties of the crystalline core. For core sizes of nanometer order, the band-gap energy depends sensitively on the core size. In the case of the c-Si core, according to theoretical calculations, the lowest (*1s*) exciton energy of the c-Si core changes from 1.3 eV for 10 nm to 2.0 eV for 4 nm diameter [39, 40]. This is inconsistent with the size-insensitive PL peak energy. Moreover, the electronic properties of a c-Si core more than several nanometers in diameter are thought to be similar to those of bulk Si. Therefore, the indirect-to-direct transition in the band gap may not occur in 3 nm-Si crystallites. The optical transition strength would then be small because of the indirect character, and this would result in no strong PL at room temperature.

The outer surface $SiO_2$ layer does not itself contribute to the visible PL because the band-gap energy is out of the visible range ($> 8$ eV). Because the size-insensitive PL peak energy implies that visible PL processes have an origin other than the quantum confinements in the crystalline core, we stress the importance of the interfacial region between the crystalline core and the surface oxide layer. We then propose a "three-region model" for oxidized nanocrystallites composed of (a) a crystalline core with diameter $\langle D_{core}\rangle$, (b) an outer oxide layer, and (c) an interfacial layer between (a) and (b). The a-$SiO_2$ surface region form a high-potential barrier to confine the exciton inside. In this model, the interfacial region (surface state on nanocrystallites) plays the most important role in the PL process as illustrated in Fig. 5.14. In Si nanocrystallites, the strong red PL observed after the surface is covered by silicon oxides [10, 54]. It is considered that the oxidized surface state becomes the localized states exhibiting strong and stable luminescence at the red spectral region.

The composition ratio of silicon-oxygen compound in this interfacial region should be intermediate between that of the oxide layer [a-$SiO_2$ (Si:O = 1:2)] and that of the crystalline core [c-Si (Si:O = 1:0)]. That is, this region contains nonstoichiometric amounts of oxygen atoms. In incompletely oxidized Si layers, oxygen atoms play important roles in electronic structures. To understand the most essential feature of the electronic properties of this thin region, we use the result of the *ab initio* calculations for the one-sided oxidized Si planar sheets shown in Fig. 5.14, because this compound has an intermediate composition ratio (Si:O) and a thin layer structure. The most characteristic feature is that this Si layer has the direct-allowed transition of 1.7 eV at the point. These calculated transition energies are nearly equal to the observed PL peak energies.

In the above calculation, we consider the large crystalline Si sheet as the surface of the Si nanocrystallite. The surface roughness, the disorder of the nanocrystallite surface, the size of the Si surface region, and the composition fluctuation (*e.g.*, Si:O ratios) cause the broadening of the surface states near 1.7 eV. The transition energy in the interfacial region is not strongly sensitive to the core size $\langle D_{\rm core}\rangle$, whereas that in the crystalline core region depends on $\langle D_{\rm core}\rangle$. Even if the surface states depend on the crystalline size, the size-dependence is very weak, as shown in Fig. 5.13. The disordered potential of the surface states causes the broad band tail states. Moreover, in order to clarify that the oxygen atoms critically affects the optical properties of Si nanostructures, Matsumoto *et al.* prepared the Si sheet as illustrated in Fig. 5.15. In the out-of-Si-skeleton case, the PL intensity in the oxidized compound is two orders in magnitude higher than that in non-oxidized Si sheet.

When the transition energy of the interfacial layer (the near-surface region of the nanocrystallites) is lower than that of the c-Si core region, excitons or carriers are subject to a confinement potential and is localized on the interfacial spherical-shell region. The condition for exciton confinement causing efficient visible luminescence is roughly determined by the relative magnitude of the transition energy of each region. Theoretical calculations indicate that the energy gaps of the c-Si core of $\langle D_{\rm core}\rangle \leq$ 5–7 nm are higher than the transition energies of the interfacial layers. As a result, when $\langle D_{\rm core}\rangle \leq$ 5–7 nm for Si nanocrystallites, the exciton is confined in the interfacial shell region. A schematic diagram of the energy gap is drawn in Fig. 5.16. Therefore, the exciton confinement (or exciton localization) in the interfacial layer is a possible origin of the strong red PL from oxidized Si nanocrystallites. The experimental results show that only the oxidized Si nanocrystallites of $\langle D_{\rm core}\rangle \leq 7$ nm contribute to the strong PL at 1.65 eV, and the theoretical evaluations of $\langle D_{\rm core}\rangle \leq$ 5–7 nm for Si nanocrystallites agree with this finding.

According to our three-region model, the process of radiative recombination proceeds as follows. The photogeneration of excitons mainly occurs within the c-Si core. Some of the excitons in the core are localized into the interfacial layer, where the localized states on the nanocrystallite surface causes the band tail state. The strong PL is then caused by the radiative recombination of excitons confined in the interfacial layer. In this scenario, the localization of excitons (the exciton transfer) from the core to the interfacial layer increases with increasing temperature. This surface exciton localization picture also explains why the PL intensity depends strongly on the size of the c-Si core whereas the PL peak energy is not sensitive to the core size. The interfacial region between the crystalline core and the outer oxides plays a role in strong red luminescence at room temperature.

The surface effects as well as the quantum confinement effects control the optical and electronic properties of nanocrystallites. These two effects also complicate the mechanism of the broad visible PL in Si nanocrystallites. In fact, time-resolved PL studies in Si nanocrystallites [44, 45, 57, 58, 59] indicate that the recombination processes are complex and the PL decay exhibits nonexponential behaviors. A wide range of PL decay times has been reported in picosecond to microsecond time regions [44, 45, 57, 58, 59, 60]. Since the slow decay component in time regions of microseconds or more determines the time-integrated PL intensity, the experimental and theoretical analyses of the slow PL decay are expected to clarify the origin of the strong visible PL and carrier recombination processes in Si nanocrystallites. Here, we discuss the

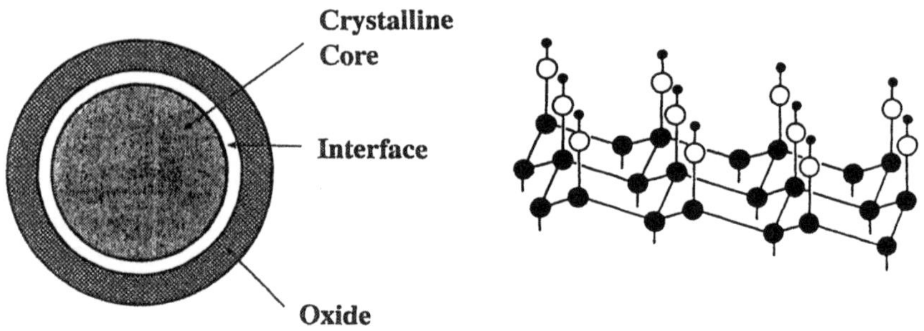


Figure 5.14. Schematic view of oxidized silicon nanocrystallites (left): The oxidized Si nanocrystallite consists of three regions: the core, interfacial, and outer oxide regions. Model structure of luminescent interfacial region (right): one-sided oxidized Si sheet for calculations of the electronic structures of the interface state. The composition ratio of silicon-oxygen in the interfacial region is intermediate between that of the oxide layer and that of the crystalline core. This region contains nonstoichiometric amounts of oxygen atoms.

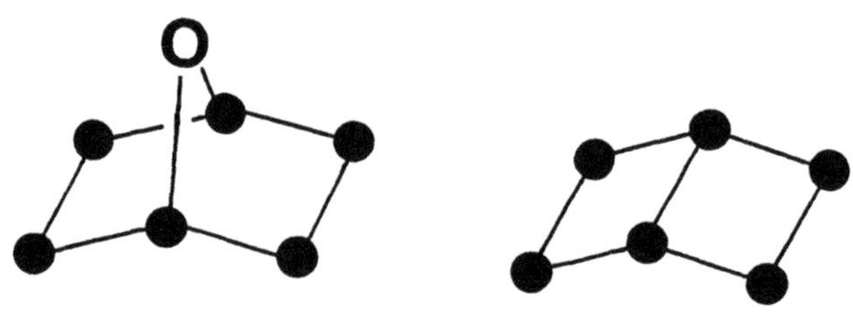


Figure 5.15. Oxidized (left) and non-oxidized (right) small Si sheets. The luminescence intensity in the oxidized sheet is two orders of magnitude higher than that in non-oxidized sheet.

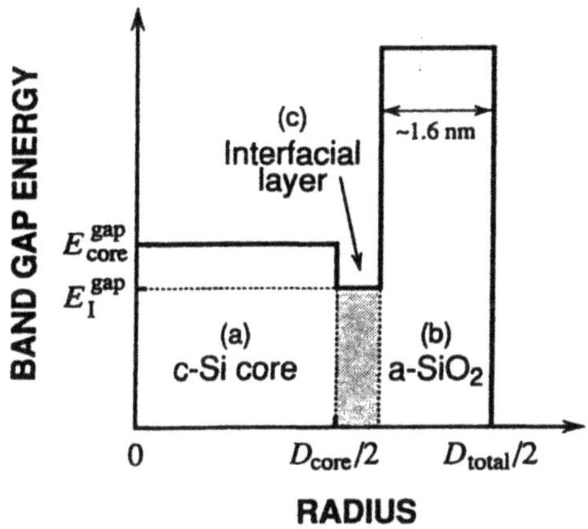


Figure 5.16. Energy-gap diagram of the three-region model. The direct-allowed transition energy of the interfacial Si layer is 1.6 eV, respectively, including the exciton binding energy. The energy gap of the crystalline core increases with decreasing core diameter. In Si nanocrystallites of $\langle D_{core} \rangle \leq$ 5–7 nm, the exciton is confined in the interfacial region between the crystalline core and the surface oxide layer.

PL decay dynamics in the $10^{-7}$ to $10^{-2}$ s time regions in Si nanocrystallites with different surface structures, electrochemically-etched porous silicon and oxidized Si crystallites.

In order to compare the carrier dynamics of porous silicon with isolated Si nanocrystallites, we studied the PL decay dynamics in two different samples: (a) porous silicon of the average crystallite diameter $\langle D_{core} \rangle \sim 3.5$ nm and (b) oxidized Si nanocrystallites of $\langle D_{core} \rangle \sim 3.7$ nm. In both samples, the PL peaks are around 750 nm under steady state laser excitation: The PL spectrum of porous silicon is very similar to that in the oxidized Si nanocrystallites. This results show that the outer surface materials surrounding the Si crystallites, a-$SiO_2$ or Si:O:H compounds, do not contribute themselves to the strong visible luminescence. The carriers in nanocrystallites contribute to the strong PL.

Figure 5.17 shows the PL decay profiles at different wavelength in the broad PL band at room temperature. The PL decay profiles are nonexponential, and they are well described by a stretched exponential function [44]:

$$I(t) = I_0 \left(\frac{t}{\tau}\right)^{\beta-1} \exp\left[-\left(\frac{t}{\tau}\right)^{\beta}\right], \tag{5.1}$$

where $\tau$ is an effective decay time, $\beta$ is a constant between 0 and 1, and $I_0$ is a constant. The solid curves in this figure are given by the above function using the least-squares-fitting of the data for determining values of $\tau$ and $\beta$. This stretched exponential decay is usually observed in the PL decay and transport properties of disordered systems.

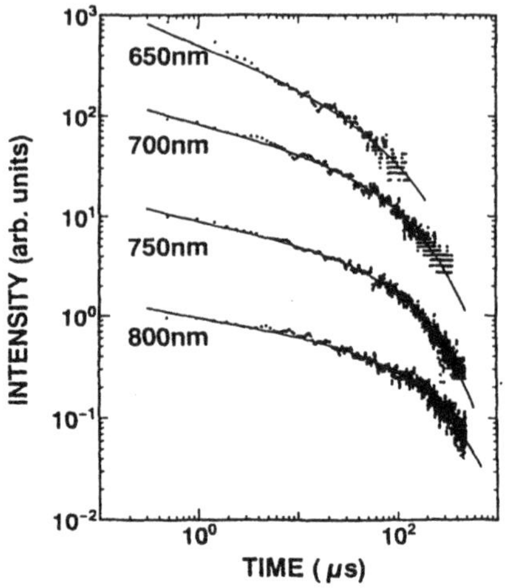


Figure 5.17. Double logarithmic plot of the decay curves at different wavelengths in the broad red-luminescence spectrum at room temperature. The solid lines are theoretical curves given by stretched exponential functions. (After [45]).

This experimental result suggests that the localization and hopping motion of excitons or carriers play an important role in the PL process in Si nanocrystallites.

The effective decay time $\tau$ depends on the temperature. Figure 5.18 shows the PL decay profiles at 750 nm in a porous silicon film at 20, 78, and 290 K. Even at low temperature, the PL decay profiles are nonexponential, and they are well described by the above stretched exponential function. The temperature dependence of in oxidized Si crystallites was very similar to that in porous silicon. This indicates that the slow decay dynamics of red PL is not sensitive to the materials surrounding the c-Si spheres (either a-$SiO_2$ or Si:O:H compounds). Moreover, the PL decay dynamics depended scarcely on the average size of c-Si spheres in porous Si films. The PL decay dynamics of porous silicon was similar to that of isolated Si nanocrystallites. The carrier diffusion between Si nanocrystallites does no play a role in the PL decay processes. These PL properties strongly suggest that the red PL comes from the surface localized states within Si nanocrystallites.

Moreover, it was pointed out that both the competing effect between the radiative and nonradiative recombination processes and the geometrical effects on carrier transport in complex topology cause the nonexponential decay dynamics of the luminescence [59, 60]. As discussed above, the PL decay dynamics of porous silicon is similar to that of isolated oxidized Si nanocrystallites. A large degree of linear polarization of luminescence was observed in porous silicon. These experimental facts indicate that no significant exciton migration and energy transfer occur between nanocrystallites. At low temperatures, the tunneling nonradiative recombination becomes important,

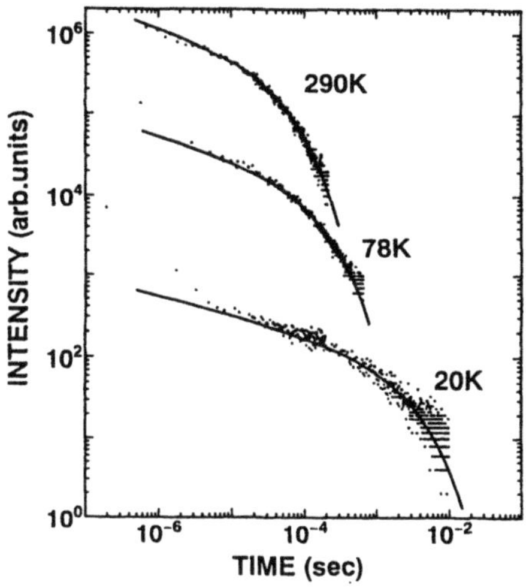


Figure 5.18. Double logarithmic plot of photoluminescence decay curves in a porous silicon film of $\langle D_{\rm core} \rangle \sim 3.5$ nm at 20, 78, and 290 K. The solid lines are theoretical curves given by stretched exponential functions. (After [44]).

because the radiative recombination lifetime becomes very long.

The stretched exponential PL decay and a wide range of PL decay times reflect the inhomogeneity of porous silicon. Porous silicon is inhomogeneous in the sense that it has a distribution of crystallite size and shape and variation in surface structure and surface stoichiometry. The observed PL decay profiles reflect the energy distribution of the recombination sites. We can point out that the slow dynamics reflects the disorder nature and the exciton localization nature. The nanoscopic inhomogeneity will produce the unique optical properties of porous silicon. The luminescence properties of inhomogeneous systems can be controlled by changing the experimental conditions such as laser wavelength, laser pulse width, laser fluence, measurement temperature, and ambient conditions. The selective excitation and fluorescence-line-narrowing spectroscopies are successful to study the optical properties of inhomogeneous nanocrystallite systems. In inhomogeneous systems, the linewidth of single molecule or single dot in the sample (homogeneous width) is several orders of magnitude narrower than the inhomogeneous band. Several techniques such as photon echoes, spectral hole burning, and single-molecular (single-dot) spectroscopies have been developed to study narrow homogeneous structures. These techniques will be used in nanocrystalline systems and will provide important information on electronic properties of nanocrystallites. In particular, the preparation of the size-controlled Si nanocrystallites is needed for detailed discussion on the luminescence mechanisms and for higher emission efficiencies.

In nanometer-sized Si crystallites, the quantum confinement and surface localization effects determine the optical properties. The widening of the band-gap energy occurs with a decrease of the crystallite size. Many experimental results suggest that the surface states influenced by the surface roughness and compositional disorders play an important role in the efficient and stable "red"-luminescence in oxidized Si nanocrystallites and porous silicon at room temperature. Hydrogen and silicon oxides passivation reduce the formation of the surface dangling bonds which act as nonradiative recombination centers. In particular, the oxygen-modified and disordered surface states play an essential role in red luminescence other than passivation. However, porous silicon shows complicated behaviors. For example, in as-prepared porous silicon without air exposure, we observe weak luminescence whose wavelength depends on the crystallite size. After air exposure (surface oxidation), the luminescence shifts to red and becomes stable and efficient. By controlling the sample preparation and measurement conditions, it is possible to obtain visible luminescence originating from the core states in nanocrystallites.

### 5.3.3. *Blue and Green Luminescence*

In the previous section, we discussed red luminescence properties and pointed out that porous silicon is a nanoscopic inhomogeneous system. This section describes that the luminescence wavelengths in porous silicon can be controlled by changing the excitation conditions or the surface structures. First, we show the excitation power dependence of both PL spectra and picosecond PL decay in porous silicon [61, 62]. The porous silicon layers were formed by electrochemical anodization of p-type silicon. After anodization, the sample was oxidized by exposing it to air at room temperature, in order to obtain stable PL with higher quantum efficiency. The average diameter of nanometer-sized Si crystallites was about 3 nm.

Figure 5.19 shows luminescence spectra of porous silicon. The luminescence was measured using the fourth harmonics of a cw mode-locked Ti:$Al_2O_3$ laser. The excitation wavelength was 240 nm, the pulse duration was about 500 fs, and the repetition rate was 81 MHz. The sample was kept in a vacuum to avoid surface oxidation. A computer controlled stepper motor moved the porous silicon sample during the laser illumination. These are necessary to prevent laser-induced fatigue or photo-oxidization caused by the same spot excitation in air. In the case of 325-nm steady-state laser excitation with 640 mW/cm$^2$ excitation intensity, the porous silicon shows an usual red luminescence with 670 nm peak wavelength [Fig. 5.19(d)]. The peak energy of the red PL shifted to higher energy with increasing excitation laser energy. However, using the 500 fs laser pulses with the same excitation intensity, the luminescence spectrum varies from red to blue-green. We observed the blue-green PL from porous silicon using third harmonics (355 nm) of a Q-switched pulse yttrium aluminum garnet (YAG) laser with 10 ns pulse duration. These results suggest that there are two luminescent states in porous silicon, in which the PL intensity depends on excitation conditions such as pulse intensity and pulse duration. With a decrease of the excitation intensity, the PL spectrum became red. When the laser intensity increases again, the PL returned to the blue-green. Cycle shifts in the PL spectra between the

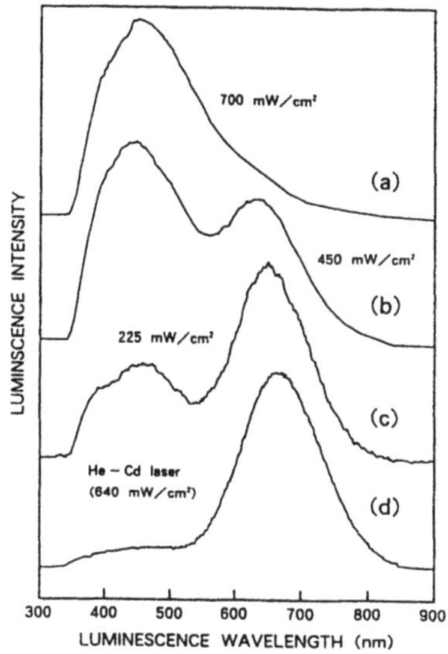


Figure 5.19. Photoluminescence spectra of porous silicon as a function of excitation intensity. Spectra (a), (b) and (c) were obtained by 240-nm and 500-fs laser pulse excitation and spectrum (d) was obtained by high-power cw 325 nm laser excitation. With increasing femtosecond laser intensity, a blue-green luminescence band appears. With decreasing laser intensity, the spectrum returns to red. (After [61]).

blue-green and red were obtained by changing the intensity of short laser pulses.

Figure 5.20 shows the red and blue-green luminescence intensities as a function of the 240-nm and 500-fs laser-excitation intensity. The red PL is observed at less than 100 mW/cm$^2$ excitation intensity and increases sublinearly with increasing laser intensity. Then the red luminescence shows saturation around 500 mW/cm$^2$. On the other hand, the intensity of the blue-green luminescence increases linearly with increasing excitation intensity. These nonlinear luminescence behaviors can be explained from a change of the recombination route from the lower energy states to the higher ones due to the population saturation. Picosecond PL decay measurements clarify the carrier dynamics as a function of the excitation intensity between the blue-green and red luminescence states.

These results indicate that there are two inhomogeneously broadened luminescent states overlapping in the visible spectral region, as illustrated in Fig. 5.21. The fast component is mainly observed in the blue-green region and the slow one is mainly observed in the red region. With increasing the excitation intensity, the luminescence with faster decay components becomes large compared to that with slower ones. Therefore, porous silicon shows a wide range of decay components on the order from

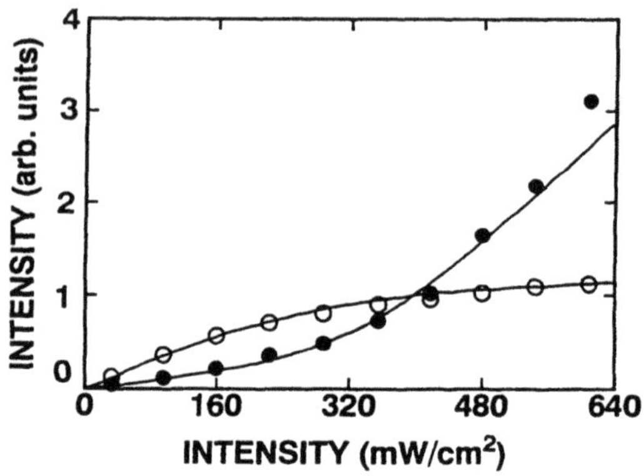


Figure 5.20. Blue-green luminescence (solid circles) and red luminescence (open circles) intensities as a function of excitation intensity. The solid lines represent the theoretically fitted curves. Nonlinear luminescence spectra are caused by a change of the recombination route from a lower-energy state to a higher-energy state due to the population saturation. (After [62]).

picoseconds to milliseconds due to the change in the ratio of the two luminescence intensities, which depends on the excitation conditions such as pulse energy and pulse duration. The calculated curves in Fig. 5.20 (solid lines) describe the nonlinear luminescence behaviors well. The experimental and theoretical results clearly show that there are two different luminescence states in porous silicon in the visible spectral region.

In as-prepared porous silicon without air exposure, a very weak green PL was observed under cw laser excitation. After surface oxidation in air, a stable and strong PL band was observed in the red spectral region [53, 63, 64]. This oxidization effect can be explained by the above discussion in Sec. 5.3.2. The surface oxidation causes a reduction of the band gap energy of the near-surface region: In out-of-Si skeleton systems, $\sigma$-n mixing effectively reduces the band gap [36]. Moreover, the oxidization enhances the oscillator strength. The above calculation explains the experimental observations such as the broad PL spectrum, weak green PL in as-prepared porous silicon, and the strong red PL after surface oxidation. These disordered surface states cause the exponential tail states.

Moreover, cycle shifts in the PL spectra between green and red were obtained by changing the solutions [64]. The as-anodized samples in immersed in HF electrolyte showed the green PL under UV excitation light. When the sample was removed from the electrolyte, rinsed in water and dried brown, the PL spectrum changed from

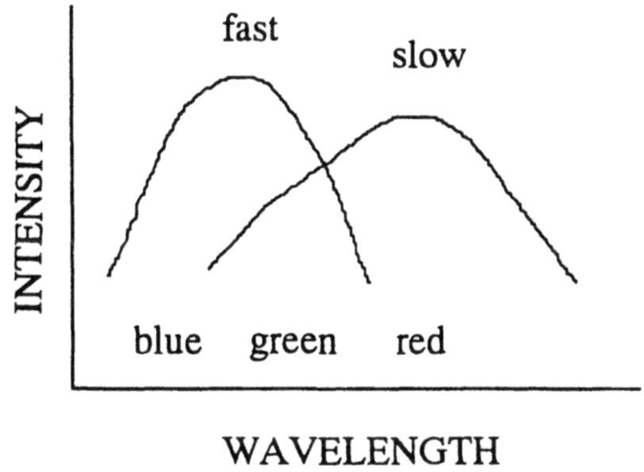


Figure 5.21. Illustration of the fast and slow decay components as a function of luminescence wavelength. There are two luminescence components in the visible spectral region in porous silicon. We can control the luminescence wavelength and the effective luminescence decay time by changing the excitation conditions.

green to red. However, returning the samples to the electrolyte caused the green PL to appear again. When the sample was again dried, the PL returned to red. Modification of the surface state in this process causes the cycling tuning between green and red luminescence. It is well known that the PL wavelength and intensity are very sensitive to the surface chemistry of as-prepared porous silicon. Surface chemistry of porous silicon shows that the surface plays a role other than passivation and the radiative recombination proceeds via low-energy states at the disordered surface. This cycle shift due to the surface modification is similar to our observation of the cycle shift in the spectrum by the femtosecond laser excitation. During the repeated cycling, the luminescence intensity decreased.

Efficient blue PL is also observed in rapid-thermal-oxidized porous silicon. The luminescence properties of usual porous silicon samples are sensitive to the ambient conditions and the luminescence intensity is unstable. In particular, the degradation of luminescence is observed at high temperature. The luminescence quenching is discussed with the loss of the $SiH_2$ surface species [65]. Hydrogen passivation reduces the formation of the dangling bonds which act as nonradiative recombination centers and hydrogen desorption leads to mid-gap states related to dangling bonds which decrease the luminescence intensity. Hydrogen passivation on the surface of porous silicon is not stable in air. For optoelectronic device applications and the ongoing discussion on the mechanism of visible PL, we need to control the surface structure and to preserve the crystalline cores with diameters of several nanometers [66]. Thermal oxidization processes produce a higher quality oxide at the Si surface for microelectronics technology. High quality $SiO_2$ formed on the surface of the Si nanocrystallite play a role [67, 68]:

(i) the mechanical and chemical stability,
(ii) the reduction of the nonradiative recombination centers, and
(iii) optically transparent windows for the study of the c-Si core state.

To clarify the effects of the surface structure on the optical properties in nanostructures and to obtain the stable luminescence, we tried to control the surface structure of Si crystallites by rapid-thermal-oxidization (RTO) processes. The as-prepared porous silicon samples were processed in a RTO apparatus. The heating rate was 200 °C/s and the samples were kept for 35 s at the oxidation temperature $T_{ox}$ ranging from 480 to 1200 °C under 1 atm $O_2$ pressure. The cooling rate was 100 °C/s. By controlling $T_{ox}$, we fabricated the strongly red- and blue-light-emitting porous silicon. By changing the surface structures of the crystallites from silicon oxyhydrides to silicon dioxides, the PL spectrum was changed from the red to the blue region.

Figure 5.22 shows PL spectra from oxidized porous Si under 325 nm excitation at room temperature as a function of $T_{ox}$. In porous silicon oxidized at low $T_{ox}$ ( $<$ 800 °C), the PL peak is around 750 nm. The spectral width of the broad PL is about 0.3 eV (FWHM). Both the PL peak and the PL width in oxidized porous silicon are nearly equal to those in the usual porous silicon. At higher $T_{ox}$ above 800 °C, the strong blue PL near 400 nm appears, while the red PL near 750 nm disappears. The TEM studies indicated that the number of nanocrystallites decreases rapidly with increasing $T_{ox}$ ($T_{ox} > 800$ °C). With increasing $T_{ox}$ above 900 °C, the blue PL intensity decreases. When the number of Si crystallites decreases, the blue PL intensity decreases. These imply that Si nanocrystallites participate in the blue PL process in oxidized porous silicon.

FTIR spectroscopy provides information about the internal surface of porous silicon. In the as-etched porous silicon samples (before RTO processes), the absorption peaks due to Si-H and Si-$H_2$ were observed near 900 and 2100 $cm^{-1}$. After oxidization, these two peaks disappeared. At low $T_{ox}$ below 800 °C, an absorption band due to silicon oxyhydride (Si:O:H) is observed near 3400 $cm^{-1}$. At $T_{ox}$ higher than 800 °C, this broad absorption band disappears. With increasing $T_{ox}$, the amount of silicon oxide monotonically increases. These FTIR spectra strongly suggest that at low $T_{ox}$, the surface of Si crystallites is mainly terminated by Si-O-H, while at high $T_{ox}$ above about 800 °C, the surface of Si crystallites is covered by amorphous $SiO_2$ (a-$SiO_2$). It is found that there is a good relationship between the PL spectrum and the surface-termination condition. During RTO processes, the composition of the internal surface changes from silicon hydrides to silicon oxyhydrides (Si:O:H) for $T_{ox} < 800$ °C or to a-$SiO_2$ for $T_{ox} > 800$ °C. The PL peak energy changes from about 1.65 eV ($T_{ox} < 800$ °C) to 3 eV ($T_{ox} > 800$ °C).

Figure 5.23 shows the temperature dependence of the red PL intensity in the sample of $T_{ox} = 680$ °C and of the blue PL intensity in the sample of $T_{ox} = 880$ °C under 325 nm excitation. The red PL increases with rising the temperature up to about 100 K and then gradually decreases. Similar temperature dependence is also observed in usual porous silicon. This behavior is explained by a picture that carriers

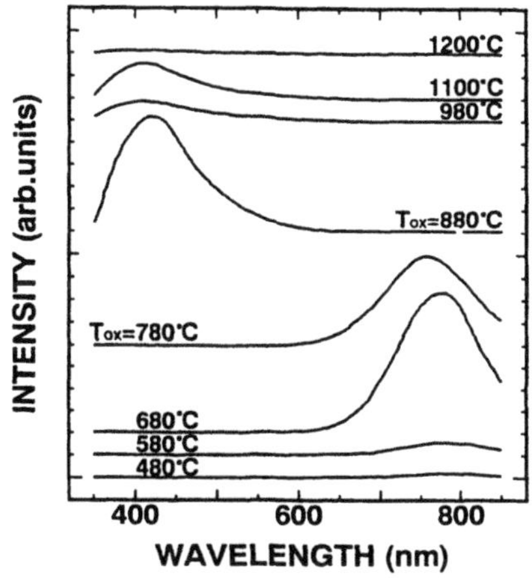


Figure 5.22. Luminescence spectra from oxidized porous Si under 325 nm excitation as a function of oxidation temperature $T_{ox}$. The luminescence spectrum drastically changes at higher than $T_{ox} = 800$ °C. There is a correlation between the luminescence peak wavelength and the surface structure of nanocrystallites. (After [68]).

are confined or localized into the lower-energy state and the PL originates from the lower-energy state (the surface localized state) rather than the band-edge state (the delocalized state). On the other hand, the blue PL intensity does not decrease at low temperatures.

In porous silicon oxidized at low $T_{ox}$, the long and nonexponential components of the red PL are clearly observed in the microsecond time region. These rapid redshift in the picosecond time region and the long nonexponential PL decay suggest that carriers are localized into the long-lived lower-energy state. On the other hand, the blue PL decay is approximately exponential with a time constant of about 650 ps, and the redshift of the PL peak is not observed, where we measured the PL decay up to 5 ns and the PL decay is single-exponential over 5 ns. There is no long component of the blue PL. Moreover, no significant wavelength dependence of the PL decay was observed [67, 68, 69].

In the experiments, both the temperature dependence of the blue PL intensity and the decay dynamics of the blue PL are entirely different from those of the red PL. The blue PL intensity decreases monotonically with increasing temperature, while the unusual temperature dependence of the red PL intensity is observed. The initial decay of the blue PL is approximately given by a single exponential function. The long nonexponential decay of the blue PL is not observed, but that of the red PL is clearly observed. These results suggest that the mechanism of the blue PL is different from that of the red PL originating from the long-lived near-surface state.

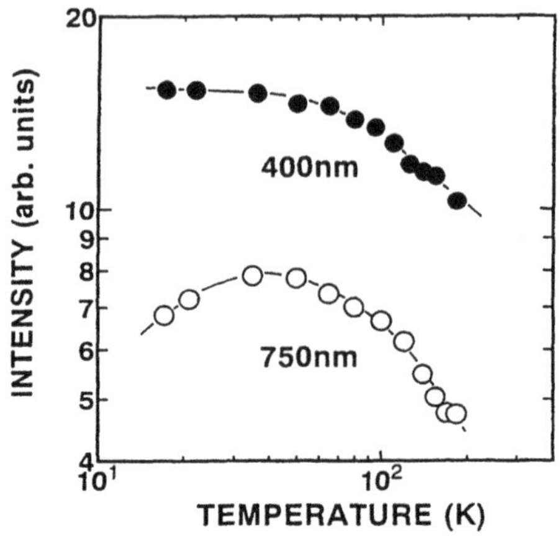


Figure 5.23. Temperature dependence of the blue and red luminescence intensities in rapid-thermal-oxidized porous silicon.

Both the PL dynamics and the temperature dependence of the PL intensity indicate that the mechanism of blue PL is different from that of red PL. The spectroscopic characteristics of the red PL (the picosecond PL dynamics, the long nonexponential PL decay, and the temperature dependence of the PL intensity) suggest that the photogenerated carriers in the c-Si core are rapidly localized into the lower-energy state with small overlap between the electron and hole wavefunctions. The above theoretical considerations and the spectroscopic data lead us to conclude that the photogeneration of carriers occurs in the c-Si core whose band gap is modified by the quantum confinement effect and the radiative recombination occurs at the surface localized state. The red PL is caused by the radiative recombination of excitons localized in the luminescent surface state.

At higher $T_{ox}$, the surface of the c-Si core is covered by an a-$SiO_2$ layer. The a-$SiO_2$ layer does not itself contribute to the visible PL, because the band-gap energy is out of the visible range (> 8 eV). However, this layer creates a novel electronic state in an interfacial region between the c-Si core and the a-$SiO_2$ layer, as discussed in Sec 5.3.2. Since the diameter of the c-Si core decreases at high $T_{ox}$, oxygen atoms migrate into the Si-skeleton layer and oxygen atoms form the bridges or interconnect Si atoms. These structures have wider gaps than 3-4 eV. On the other hand, at high $T_{ox}$ above 800 °C, the decrease of the c-Si core size occurs.

The band-gap energy of the c-Si core with 1-2 nm is estimated to be about 3 eV, which is nearly equal to or less than that of the interfacial region. However, this very small c-Si core has hybrid electronic properties between the molecular and crystalline states. We can not distinguish the surface state from the core state in a very small crystallite like a molecular-like cluster. It is considered that these small Si clusters act similar to Si-rich oxides (nonstoichiometric oxides, $SiO_x$) [69].

Several models are proposed for the origin of the blue luminescence in RTO porous silicon. The nanostructure of the luminescence state is not understood, but experimental results suggest that the Si-rich oxides including very small Si clusters play a role in the strong blue luminescent process. In as-prepared porous silicon without air exposure, a blue-green PL is observed. Under femtosecond laser excitation, the blue-green PL appears near the absorption edge. The spectrum and intensity of the blue-green luminescence are sensitive to the sample preparation conditions and laser excitation conditions. In RTO porous silicon, the surface localized state plays an essential role in strong blue and red PL process: The surface structure and excitation condition changes the origin of visible PL.

Next, we discuss the origin of the higher-energy state exhibiting the blue-green luminescence from usual porous silicon under high-power ultrashort laser excitation. The comparison between the blue PL observed in RTO porous silicon and the blue-green PL in porous silicon gives an important hint for understanding the origin of the blue-green PL in usual porous silicon. A correlation between the blue PL spectrum and FTIR spectrum was reported in RTO porous silicon. An oxygen-related origin for blue PL band, small Si clusters similar to Si-rich oxides, is proposed in RTO porous silicon. However, porous silicon under these high-power femtosecond laser excitation showed the high value ($\sim 0.4$) of the linear polarization of blue-green PL, and the degree of luminescence polarization of porous silicon is higher than that of RTO porous silicon. The lifetime of the blue-green PL in porous silicon depended strongly on the monitored wavelength: The lifetime decreased with decreasing monitored PL wavelength. This wavelength dependence of the PL lifetime may imply the size dependence of the radiative lifetime from the crystalline core state. On the other hand, the lifetime of the blue PL in the RTO porous silicon does not depend on the monitored wavelength. The wavelength dependence of the PL lifetime in porous silicon is different from that in RTO porous silicon. Theoretical calculations indicate that spherical Si nanocrystallites with 2-3 nm-diameter have 2.4–3.0 eV-bandgap energy due to the quantum confinement effect. This is supported by the absorption spectrum measurements and the blue-green PL appears near the absorption edge [45]. One possibility is that the blue-green PL is evidence of the quantum confinement effect in the nanometer-sized Si crystallites.

At the present stage, the origin of blue-green luminescence from porous silicon remains a mystery. The power-dependence of the luminescence spectrum and intensity shows that the blue-green luminescence is caused by the change of carrier recombination routes due to the population saturation of lower-lying surface localized states. The blue-green PL properties in porous silicon are similar to those in Si nanocrystallites, but different from the blue PL properties in RTO porous silicon. The reversible cycle tuning in the PL spectrum between blue-green and red and the power dependence of the luminescence can not be explained by the irreversible photochemical reaction such as photo-oxidization at the surface. It is usually observed in other semiconductor nanocrystallites (CdSe, CdS, CuCl, etc.) that the fast decay PL at higher energy regions comes from the crystalline core state and the slow decay at the lower energy regions comes from the surface localized states. Our experimental results clearly show that there are two different luminescent state in porous silicon in the visible spectral region, like CdSe and CdS nanocrystallites. Further experimental and theoretical studies are needed for the understanding of the microscopic origin of the blue-green PL in porous silicon.

There are two different luminescence state in the visible region in porous silicon. Carrier recombination occurs from the lower-energy states in the case of weak excitation while in the case of strong excitation, it occurs from higher-energy states. The power-dependent luminescence spectrum can be explained from the change of carrier recombination routes due to the population saturation of lower-lying surface localized states. The luminescence spectrum from two different states is overlapped, because porous silicon has a distribution of the crystalline size and shape and a fluctuation of the surface structure and stoichiometry, as illustrated in Fig. 5.22. By changing the excitation conditions, we can control the luminescence components (fast or slow decay components and red or blue-green components). Furthermore, by changing the surface structures, we can obtain the strong blue PL in RTO porous silicon. Porous silicon shows the various optical properties depending on the excitation condition and surface structures.

## 5.4. APPLICATIONS

### *5.4.1. Light-Emitting Diodes*

Silicon light-emitting devices have not been achieved yet, because of the poor radiative efficiency of the indirect band gap structure. The external quantum efficiency of 1.1 $\mu$m electroluminescence (EL) from c-Si pn junctions is less than $10^{-6}$. In contrast, the efficiencies of red GaAs and blue GaN light-emitting diodes exceed several percents. Light-emitting porous silicon has a potential to open the way to Si optoelectronic devices such as EL devices, because it shows strong visible PL at room temperature. Since EL devices can electrically control a light emission, they seem advantageous to display devices and to new optoelectronic integrated circuits (OEICs). In addition, EL studies provide important information on the carrier injection, transport and recombination mechanisms in the porous layer.

In 1992, porous silicon EL devices with pn heterostructures were reported by Namavar *et al.* [70] and Futagi *et al.* [71]. They observed a rectifying junction behavior and red EL at a forward bias voltage less than 10 V. The EL spectra of these devices were similar to the PL spectra of used porous silicon. The recent EL studies are roughly classified into two categories:

(i) Improvement of emission efficiencies of red EL.

(ii) Fabrication of multi-color diodes.

First, we summarize typical fabrication methods of EL devices and their EL properties. Red PL is usually observed in porous silicon and the most widely studied EL is also in the red spectral region. Various device structures for red EL are proposed and demonstrated to improve EL properties such as the threshold voltage, emission efficiency (brightness) and stability. For this purpose, improved injection of electrons and holes into the porous silicon layer (the active recombination layer) is essential. In particular, it is important to develop how to inject minority carriers into a porous silicon layer efficiently. We describe typical attempts to achieve higher emission efficiency.

To improve the injection efficiency of minority carriers into a p-type PS layer, Namavar *et al.* fabricated a quasi-pn junction between p-type PS and n-like material, indium tin oxide (ITO) [72]. ITO films were deposited onto the porous layer by an rf sputtering technique. Solid-state ITO/p-type PS devices showed strong rectifying behaviors. The transparent ITO film allows light emission through the top surface of device under forward electrical bias conditions. At currents below 100 mA, the EL intensity increased as the square of the current, finally becoming linear in current at higher levels. Such behavior is explained as follows: The low-current square-law regime is dominated by recombination at defects in the space-charge region, while injected minority carrier leads to the emission of a photon at higher current levels. The experimental results suggested the presence of a thin interfacial dielectric region between the surface of the PS layer and the ITO, and they estimated that interface state density available to store charge is of the orders of $10^{13}/\text{cm}^2$ eV.

To improve the brightness and stability of porous silicon EL devices, Mimura, Matsumoto and their coworkers, have fabricated a light emitting diode (LED) using a pn heterojunction between n-type microcrystalline silicon carbide ($\mu$c-SiC) and p-type PS [71, 73]. This is because the injection efficiency in a pn heterojunction is expected to be superior to that in a Schottky junction. In a pn heterojunction using porous silicon, the junction material has to satisfy the following requirements. The preparation process must be below 400 °C, because of the disappearance of PL due to the release of surface hydrogen at higher temperatures [65]. In addition, since the band gap energy of porous silicon is considered to be larger than about 2 eV, a highly conducting material with a band gap wider than 2 eV is required for the effective injection of minority carriers. $\mu$c-SiC prepared by electron cyclotron resonance plasma chemical vapor deposition (ECR plasma CVD) [71, 73] meets the above requirements.

After anodization, the samples were transferred into an ECR CVD apparatus, and then n-type a-SiC (2-nm thickness), and n-type $\mu$c-SiC (15-50-nm thickness) were deposited onto the porous layer in sequence. The a-SiC layer served as a buffer layer against hydrogen plasma damage. Finally, ITO was evaporated as a transparent electrode. In this device structure, the $\mu$c-SiC film showed no PL at room temperature, because the n-type $\mu$c-SiC film contained a lot of dangling bonds which originate from the hydrogen content less than several percent. Rectification behavior was clearly observed and the rectification ratio was about 10 at $\pm 5$ V. Some samples showed larger rectification ratios $10^2$–$10^3$ at $\pm 5$ V. Red EL was clearly observed with the naked eye at a forward voltage larger than about 15 V. The light emission was quite uniform over an electrode area of 1 cm$^2$. The EL spectrum from the n-type $\mu$c-SiC/p-type PS pn junction is (700 nm peak) roughly similar to the PL spectrum from p-type PS. The EL intensity increased by a square of the forward current, suggesting that the EL is not dominated by a monomolecular recombination but dominated by a bimolecular recombination. That is, it was found that electrons from n-type $\mu$c-SiC and holes from p-type c-Si are injected into the porous silicon layer, acting as the recombination region. The above results indicate that n-type $\mu$ c-SiC serves as an electron injector to p-type PS, and that fabricating a pn heterojunction is very useful to improve the EL efficiency and stability.

In contrast to pn heterojunctions, it was reported that some light-emitting devices have porous silicon pn homojunctions. Steiner *et al.* [74] fabricated a porous silicon pn homojunction. They employed the selective formation of different kinds of porous silicon substructures caused by the doping level and the illumination during anodization. The device has a nanoporous light-emitting n-type layer between a mesoporous $p^+$-doped capping layer and a macroporous n-type substrate. Ion implantation and thermal annealing were used to achieved the desired doping profile in the p-n junction. The light illumination during the anodization has two important effects. It allows a photocurrent to flow through the reverse biased pn junction and the buried n-type c-Si becomes nanoporous. This procedure makes three kinds of porous substructures formed in the wafer: The top mesoporous $p^+$ layer, the macroporous n-type layer near the n-type c-Si, and the intermediate light-emitting nanoporous n-type layer. The external quantum efficiency of red EL was estimated to be about $10^{-4}$. This value is the highest efficiency reported for porous silicon light-emitting diodes in the solid-state structure and nearly reaches the EL efficiency of porous silicon in electrolyte with liquid contacts. However, the quantum efficiency from porous silicon EL devices in the solid state structures is low ($< 10^{-4}$) compared with other light-emitting diodes. One of the reasons for the poor efficiency is that porous silicon has a large internal surface area and the good contact is not obtained. The liquid/porous silicon junctions have achieved the low threshold voltage for light emission and the high quantum efficiency. The solid-state porous silicon EL devices have greater potential for device applications. However, in solid-state structures, it is difficult to make a sufficient contact area because of a large internal surface structure of porous silicon.

To fabricate porous silicon light-emitting devices in the solid-state structure having uniform carrier injections, polymer/porous silicon junctions are demonstrated: polypyrrole/p-type PS diodes [75] and polyaniline/n-type PS diodes [76]. The EL efficiency and stability in the conducting-polymer/porous silicon devices have significantly improved, compared with those in the Au/porous silicon devices. In the porous silicon diodes, the polymers have two advantages [76]: (i) Polymers are relatively transparent at the red emission wavelengths, and (ii) Polymers permeate the porous silicon layer better than Au or ITO.

The EL and PL exhibit similar spectral characteristics. The above experiments indicate that the electrodes thoroughly permeating the porous silicon layer are very useful for efficient and stable EL operations. These polymer techniques are attractive for the development of efficient porous silicon devices in the solid-state structure.

As mentioned above, we are making much effort to obtain red EL devices with higher efficiencies. Furthermore, we have tried the approaches to fabricate multi-color light emitting devices. "Display" is one of the important applications of a light-emitting device. In the display applications, three primary colors are required. However, almost of the porous silicon EL devices show the red in color. For multi-color EL devices, the porous silicon layers are fabricated with specially treatments during or after anodization or other porous materials are used.

For n-type Si substrates, light illumination is required to produce porous layer, because the etching reaction required hole from c-Si substrate. When the band gap becomes larger than the photon energy of illumination light, the etching process automatically stops because the light cannot continue to generate holes [51]. The short wavelength light illumination produces the porous silicon layer with smaller sized Si crystallites. For blue and green light-emitting diodes, we studied the luminescence properties of UV-illuminated n-type porous silicon (UV-PS) and fabricated green light-emitting devices [53,77]. During anodization, the n-type c-Si wafers were illuminated with only UV light (wavelength, from 240 to 420 nm) produced by filtering light with a 150 W Hg-Xe lamp in the dark room (UV-PS). After anodization, samples for PL measurements were rinsed with ethanol (not water) and then dipped into ethanol in the dark without being exposed to air and light in order to suppress oxidization as much as possible. For EL device samples, ITO was deposited onto UV-PS. Green EL was clearly observed with the naked eye in the dark beyond a current density of about 10 mA for both polarities. Before or during PL measurements, the samples were kept in an vacuum and in the dark room. The sample was not exposed to air and to light. The peak wavelength of the EL spectrum (520 nm) is similar to that of the PL spectrum. However, the EL spectrum was much broader than the PL spectrum. After the sample was exposed to air, the PL peak immediately shifted to a longer wavelength (red) region and the spectrum became broader. At a current below about 8 mA, the EL intensity increased by a square of the current and then showed a linear dependence on the current at high levels. This behavior indicates that at low currents the EL is dominated by a bimolecular recombination, but at high currents it is dominated by a monomolecular recombination. The green EL in these experiments seems to reflect the size of the Si crystallites, because EL is still green after the PL moves to red. UV-PS realizes the green EL devices. The difference in the spectrum between EL and PL imply that blue-green luminescence originates from the "delocalized" states.

We reported that RTO porous silicon shows blue PL. However, the blue EL efficiency in porous silicon is extremely low. This is because the $SiO_2$ surface layer in oxidized porous silicon prevents the efficient injection efficiency. We noticed silicon carbide which is one of the wide gap semiconductors (the band gap of 6H-SiC is 2.86 eV). Due to the indirect band gap character, the efficiency of a light-emitting diode is low ($< 10^{-4}$). For the enhancement of the EL efficiency, we formed porous silicon carbide (porous SiC) by anodizing a single crystalline 6H-SiC substrate in HF solution, and fabricated a junction between ITO and porous SiC [16]. The porous SiC showed strong PL with a peak of about 470 nm at room temperature and that its intensity was about one hundred times stronger than that of the 6H-SiC substrate [15]. Porous silicon carbide is an attracting material to a new blue light emitting device.

In this section, we briefly summarized current studies of porous silicon EL devices in the solid state structure. The light-emitting devices using porous silicon have much progressed in recent three years. Quantum efficiencies up to $\sim 10^{-4}$ are obtained and the way to multi-color devices is also opened [77]. However, device performances are insufficient for practical use, and the details of the EL mechanism in porous silicon is still unclear. For the further improvements, it is necessary to understand the carrier injection and transport mechanisms and the band structure in porous silicon, and to develop good contacts between the injection materials and porous silicon with large internal surfaces. The liquid/porous silicon junctions have achieved the low threshold voltage for light emission and the high quantum efficiency. Moreover, the liquid/porous silicon junctions produced wavelength-tunable EL devices [78]. These efficient and unique EL properties are not obtained in the solid state structure. In the near future, we will obtain efficient EL light emission and unique EL properties even in the solid state structure.

### 5.4.2. *Optical Logic Gates*

The materials studied for future optical processing and optical communication systems are inorganic compound semiconductors, dielectrics, or organic materials. However, LSI fabrication processes for these materials devices are not highly developed compared with those for Si devices. On the other hand, electrochemical anodization can be easily included in a LSI circuit fabrication process, therefore porous silicon would find many applications in both LSI and optoelectronics technologies. We first observed large optical nonlinear absorption and optical bistability in porous silicon at room temperature [79, 80]. This finding will be promising for many applications such as phase conjugations, beam amplifications, and neural networks. We summarize here all-optical logic and switching experiments using optical nonlinear properties of porous silicon. Free-standing porous silicon films were used in nonlinear optical experiments. Transmission-electron-microscopy and Raman-spectroscopy examinations showed that the average diameter of Si crystallites was about 3 nm. All optical logic and switching experiments were performed with 457.9 $\sim$ 514.5 nm light from a cw Ar ion laser at room temperature. The incident laser beam was modulated using an acousto-optic modulator.

In the high power region, the optical absorption coefficient increased with increasing incident laser power. In this power range, the optical absorption $\alpha$ is given by

$$\alpha = \alpha_0 + \alpha_{\mathrm{NL}} I, \tag{5.2}$$

where $\alpha_{\mathrm{NL}}$ is the nonlinear absorption coefficient and $I$ is the incident laser power. The nonlinear absorption $\alpha_{\mathrm{NL}}$ in the porous silicon film is clearly observed at laser wavelengths shorter than about 500 nm. This photon energy was roughly equal to the band-gap energy of the 3-nm crystallite. This implies that the carriers are generated in the silicon nanocrystallites and the nonlinear absorption is caused by these photogenerated carriers.

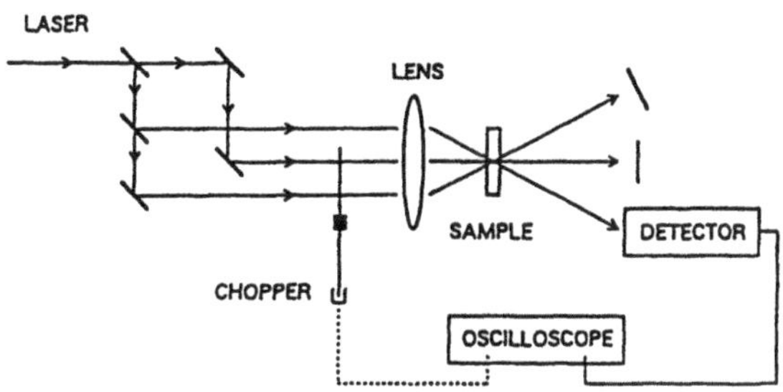


Figure 5.24. Experimental arrangement for optical logic gates. The two-beam configuration is for the optical inverter and the three-beam configuration is for the optical NOR gate. (After [79]).

Using this nonlinear absorption phenomenon, we have demonstrated some all-optical logic and switching experiments needed for the promotion of silicon-based all-optical integrated circuits. The experimental setup for logic gates is shown in Fig. 5.24. The laser beam was split into a strong and a weak beam for inverter-function experiments or into two strong beams and a weak beam for NOR-function experiments. Both the strong writing beams and the weak reading beam were focused onto the same spot of porous silicon films. The spot diameter was about 50 $\mu$m.

Figures 5.25 and 5.26 show the optically inverted transmission (one writing beam) and the optically switched NOR function (two writing beams) signals, respectively. The laser beam was split into a 3-mW writing beam (excitation beam) and 0.2-mW reading beam (probe beam). The relative intensity change of the reading beam ($\Delta T/T$) was more than 0.4. In porous silicon, the photogenerated carriers in the nanocrystallites will separate by diffusion process into the amorphous materials (Si:O:H compounds) surrounding nanocrystallites. A nonuniform space charge distribution in the sample is produced by the carrier diffusion, and this causes nonlinear absorption and spatial modification of the index of refraction. The relaxation time of the optically induced absorption change is about $0.5 \sim 1$ ms at room temperature. This relaxation time is longer than the luminescence lifetime in silicon nanocrystallites. The relaxation time of the nonlinear absorption change is in good agreement with the carrier diffusion lifetime determined from our time-of-flight photocurrent measurements [81]. These slow and wavelength-dependent nonlinear phenomena can be explained by nonuniform space charge causing the photorefractive effect or the

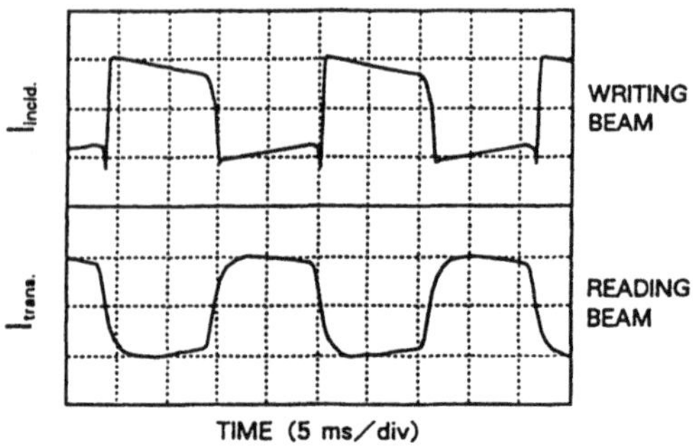


Figure 5.25. Optically inverted transmission signals. The upper and lower figure mean the incident laser intensity $I_{incid}$ and the transmitted laser intensity $I_{trans}$, respectively. The relative intensity change of the reading beam ($\Delta T/T$) is more than 0.4. The relaxation time of the optically induced absorption change is about 1 ms. (After [79]).

thermal effects. In porous silicon, the photogenerated carriers in the nanocrystallites will separate by diffusion process into the amorphous materials (Si:O:H compounds) surrounding nanocrystallites. A nonuniform space charge distribution in the sample is produced by the carrier diffusion, and this causes nonlinear absorption and spatial modification of the index of refraction. Moreover, since porous silicon has a extremely low thermal conductivity, the thermal effect due to laser heating may also play a role in slow nonlinear phenomena.

The nonlinear absorption in free-standing porous silicon is very large. However, it is also very slow (time scales of the order of ms). Therefore, it seems advantageous to consider the possibility of parallel operations. Figure 5.27 shows the experimental arrangement for an image processing application [82]. The three beams transmitted by the mask are focused onto the porous silicon layer, each beam having an intensity of 300 W/cm$^2$. The weak reading beam probes the spatially modulated absorption change of the pattern, and this is recorded by a CCD camera. In Fig. 5.27, we show the CCD image of the spatially modulated reading beam at 457.9-nm. Three circular shadows characteristic of the inverted writing beam pattern are evident. This result could be applied to the multiple processing of images. For example, parallel logic operation such as a NOR image could be obtained by using two writing beams (two image patterns).

The optical nonlinear phenomena were observed in porous silicon fabricated from silicon wafer. The optical bistability behavior was also observed using low-power laser: "moonlight bistability" [83]. This will be promising for many applications based on silicon technology. In particular, the combination of the optical nonlinearity and

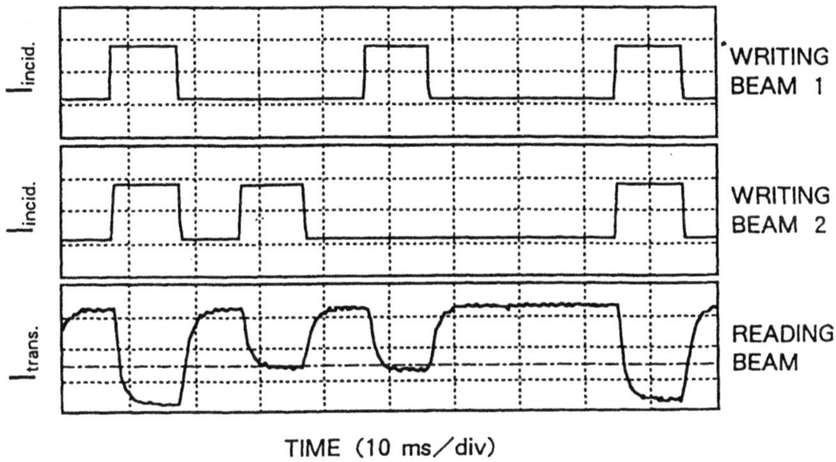


Figure 5.26. Optically switched NOR function signals. $I_{\text{incid}}$ and $I_{\text{trans}}$ mean the incident laser intensity and the transmitted laser intensity, respectively. The relative intensity change of the reading beam ($\Delta T/T$) is more than 0.4. (After [79]).

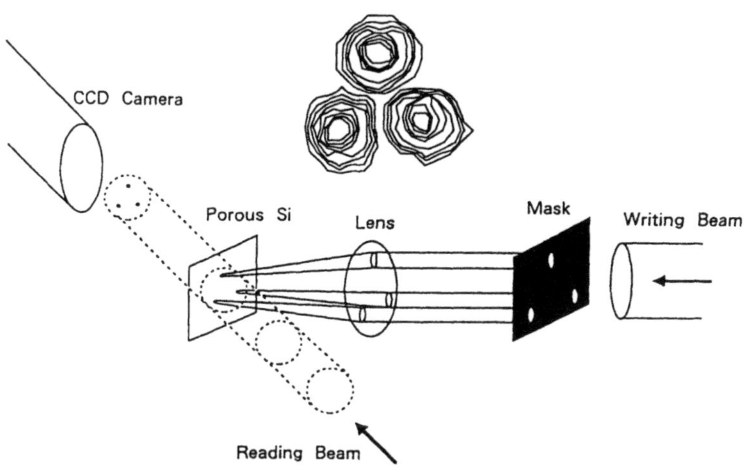


Figure 5.27. Experimental arrangement for an image processing application. The contour CCD image of the spatially modulated pattern of the reading beam. The reading image is made up of three circular shadows, inverted with respect to the writing beam pattern. (After [82]).

the blue-green luminescent devices open a possible way to fabricate silicon-based all-optical integrated circuits. Moreover, we can select the optimized wavelength of the optical bistability devices by changing the size of crystallites in porous silicon films. These characteristics are not observed in bulk semiconductors and semiconductor superlattices. Therefore, optical nonlinear devices fabricated by porous silicon having controllable optical properties seems to be the key elements of all-optical logic and computing systems.

## 5.5. SUMMARY

In this chapter, we present an overview of experimental studies on optical properties in Si nanostructures. The quantum confinement effect causes the widening of the band-gap energy and the enhancement of the oscillator strength in the low-dimensional Si nanocrystallites such as 1D polymers and 0D nanocrystallites.

Natural analogues of low-dimensional silicon structures such as wires and sheets are chemically synthesized clusters and polymers. Modern organic synthesis techniques allow us to produce silicon materials with controlled structures. We studied the optical properties of $Si_8$ clusters with chain, ladder and cubic structures. In small Si clusters, the luminescence properties are very sensitive to the shape of clusters. Chemical synthesis methods make for laser-evaporation experiments. Furthermore, the size dependence of the optical properties of clusters and wires containing a few to several thousand atoms helps us to understand how molecules evolve into solids. In chain structure, the oscillator strength of excitons depends on the size of Si chains. On the other hand, the luminescence properties of branch, ladder, and network polymers are determined by the exciton localization and resemble those of amorphous Si materials. The synthesis of crystalline network polymers is required to study the two-dimensional nature of excitons in Si-based materials. The crystalline network polymers will realize the high efficient and wavelength-tunable luminescence in the visible region.

Luminescence properties of Si nanocrystallites are sensitive to the laser excitation conditions. The power-dependence of the luminescence spectra shows that there are two different luminescence states in the visible region in porous silicon and Si nanocrystallites. Carrier recombination occurs from the lower-energy states in the case of weak excitation. In the case of strong excitation, it occurs from the higher-energy states. By changing the excitation conditions, we can control the luminescence properties or can select either the fast and blue PL component or the slow and red PL component. Luminescence from porous silicon and Si nanocrystallites is also very sensitive to the surface structure of the crystallites. The surface effect plays an important role other than passivation. The fluctuation of surface composition and surface roughness cause the shallow tail states. The size dependence of the optical absorption and luminescence spectra shows that some energy loss occurs during the absorption-emission process and the visible luminescence state is below the delocalized band edge. Many experimental results are more easily understood by a picture that the cooperation of the quantum confinement and surface effects in nanocrystallites causes the efficient and stable red-luminescence at room temperature.

Since Si clusters and nanocrystallites have hybrid electronic properties between the molecular and bulk crystalline limits, consecutive research from molecules to the bulk crystalline is required to understand electronic properties of Si nanostructures. However, at present, there is a large gap in experimental study between molecular-like Si clusters and solid-like Si nanocrystallites.

Finally, we would like to stress that Si nanostructures are complex materials that have unique optical properties. For example, Si nanocrystallites show large optical nonlinearity and efficient luminescence under weak laser excitation. We consider that the combination of electroluminescence and large nonlinear phenomena can open the way to fabricating integrated all-optical logic devices with silicon technology. Si nanostructures open a new frontier field among Physics, Chemistry and Electronics communities.

The author would like to thank H. Mimura (ATR Optical and Radio Communications Research Laboratories), T. Matsumoto (Electronics Research Laboratories, Nippon Steel Corporation), T. Ogawa (Osaka City University), K. Takeda (Waseda University), K. Shiraishi (NTT Basic Research Laboratories), M. Kondo (Electrotechnical Laboratory), T. Ohno (National Research Institute for Metals), H. Matsumoto and S. Kyushin (Gunma University), and H. Uto and K. Suzuki (University of Tsukuba) for discussions and collaborations.

## REFERENCES

[1] *Light Emission from Novel Silicon Materials*, ed. Y. Kanemitsu, M. Kondo, and K. Takeda (Phys. Soc. Jpn., Tokyo, 1994).

[2] F. S. Kipping, *J. Chem. Soc.* **119**, 830 (1921).

[3] C. Burkhard, *J. Am. Chem. Soc.* **71**, 963 (1949).

[4] S. Yajima J. Hayashi and M. Omori, *Chem. Lett.* 931 and 1209 (1975).

[5] P. J. Wesson and T. C. Willams, *J. Polm. Sci., Polym. Chem. Ed.* **17**, 2883 (1979).

[6] R. West, *J. Organometal Chem.* **300**, 327 (1996)

[7] R. D. Miller and J. Michl, *Chem. Rev.* **89**, 1359 (1989).

[8] C. Pickering, M. I. J. Beale, D. J. Robbins, P. J. Pearson, and R. Greef, *J. Phys. C* **17**, 6535 (1984).

[9] S. Furukawa and T. Miyasato, *Phys. Rev. B* **38**, 5726 (1988).

[10] H. Takagi. H. Ogawa, Y. Yamazaki, A. Ishizaki, and T. Nakagiri, *Appl. Phys. Lett.* **56**, 2349 (1990).

[11] L. T. Canham, *Appl. Phys. Lett.* **57**, 1046 (1990).

[12] V. Lehmann and U. Gosele, *Appl. Phys. Lett.* **58**, 856 (1991).

[13] Y. Maeda, N. Tsukamoto, Y. Yazawa, Y. Kanemitsu, and Y. Masumoto, *Appl. Phys. Lett.* **59**, 3168 (1991).

[14] Y. Kanemitsu, H. Uto, Y. Masumoto, and Y. Maeda, *Appl. Phys. Lett.* **61**, 2187 (1992).

[15] T. Matsumoto, J. Takahashi, T. Tamaki, T. Futagi, H. Mimura, and Y. Kanemitsu, *Appl. Phys. Lett.* **64**, 226 (1994).

[16] H. Mimura, T. Matsumoto, and Y. Kanemitsu, *Appl. Phys. Lett.* **65**, 3350 (1994).

[17] Y. Kanemitsu, *Mater. Res. Soc. Symp. Proc.* **358**, 81 (1995).

[18] Y. Kanemitsu, K. Suzuki, H. Uto, Y. Masumoto, T. Matsumoto, S. Kyushin, K. Higuchi, and H. Matsumoto, *Appl. Phys. Lett.* **61**, 2446 (1992).

[19] M. F. Jarrold, *Science* **252**, 1085 (1991).

[20] K. D. Rinnen and M. L. Mandich, *Phys. Rev. Lett.* **69**, 1823 (1992).

[21] E. C. Honea, A. Ogura, C. A. Murray, K. Raghavachai, W. O. Sprenger, M. F. Jarrold, and W. L. Brown, *Nature* **366**, 42 (1993).

[22] H. Matsumoto, K. Higuchi, S. Kyushin, and M. Goto, *Angrew. Chem. Int. Ed. Engl.* **31**, 1354 (1992).

[23] Y. Kanemitsu, K. Suzuki, K. Kondo, and H. Matsumoto, *Solid State Commun.* **89**, 619 (1994).

[24] Y. Kanemitsu, K. Suzuki, K. Kondo, S. Kyushin, and H. Matsumoto, *Phys. Rev. B* **51**, 10666 (1995).

[25] Y. Kanemitsu, K. Suzuki, Y. Nakayoshi, and Y. Masumoto, *Phys. Rev. B* **46**, 3916 (1992).

[26] Y. Kanemitsu, K. Suzuki, Y. Masumoto, T. Komatsu, K. Sato, S. Kyushin, and H. Matsumoto, *Solid State Commun.* **86**, 545 (1993).

[27] H. Matsumoto, H. Miyamoto, N. Kojima, and Y. Nagai, *J. Chem. Soc., Chem. Commun.* 1316 (1987).

[28] H. Matsumoto, K. Higuchi, Y. Hoshino, H. Koike, Y. Naoi, and Y. Nagai, *J. Chem. Soc. Chem. Commun.* 1083 (1988).

[29] S. Saito, S. Ohnishi, C. Satoko, and S. Sugano, *Z. Phys. D* **14**, 237 (1989).

[30] P. A. Bianconi, F. S. Schilling, and T. W. Weidman, *Macromolecules* **22**, 1697 (1989).

[31] I. Hirabayashi, K. Morigaki, and S. Yamasaki, *J. Phys. Soc. Jpn.* **52**, 671 (1983).

[32] H. Tachibana, M. Matsumoto, Y. Tokura, Y. Moritomo, A. Yamaguchi, S. Koshihara, R. D. Miller, and S. Abe, *Phys. Rev. B* **47**, 4363 (1993).

[33] Y. Kanemitsu, K. Suzuki, S. Kyushin, and H. Matsumoto, *Phys. Rev. B* **51**, 13103 (1995).

[34] K. Takeda and K. Shiraishi, *Solid State Commun.* **85**, 301 (1993).

[35] K. Takeda and K. Shiraishi, *Phys. Rev. B* **39**, 11028 (1989).

[36] K. Takeda, in *Light Emission from Novel Silicon Materials* (Ref. [1]), p.1.

[37] W. L. Wilson and T. W. Weidman, *J. Phys. Chem.* **95**, 4568 (1991).

[38] T. Ohno, K. Shiraishi, and T. Ogawa, *Phys. Rev. Lett.* **69**, 2400 (1992).

[39] T. Takagahara and K. Takeda, *Phys. Rev. B* **46**, 15578 (1992).

[40] J. R. Proot, C. Delerue, and G. Allen, *Appl. Phys. Lett.* **61**, 1948 (1992).

[41] S. Y. Ren and J. D. Dow, *Phys. Rev. B* **45**, 6492 (1992).

[42] M. Hirao and T. Uda, *Surf. Sci.* **306**, 87 (1994).

[43] L. W. Wang and A. Zunger, *J. Chem. Phys.* **100**, 2394 (1994).

[44] Y. Kanemitsu, *Phys. Rev. B* **48**, 12357 (1993).

[45] Y. Kanemitsu, *Phys. Rev. B* **49**, 16845 (1994).

[46] R. Tsu, H. Shen, and M. Dutta, *Appl. Phys. Lett.* **60**, 112 (1992).

[47] Y. Kanemitsu, H. Uto, Y. Masumoto, T. Matsumoto, T. Futagi, and H. Mimura, *Phys. Rev. B* **48**, 2827 (1993).

[48] S. Schuppler, S. L. Friedman, M. A. Marcus, D. L. Adler, Y. H. Xie, F. M. Ross, T. D. Harris, W. L. Brown, Y. J. Chabal, L. E. Brus, and P. H. Citrin, *Phys. Rev. Lett.* **72**, 2648 (1994).

[49] A. Nakajima, Y. Ohshima, T. Itakura, and Y. Goto, *Appl. Phys. Lett.* **62**, 2631 (1993).

[50] A. Uhlir, *Bell Syst. Tech.* **35**, 333 (1956).

[51] V. Lehmann, H. Cerva, and U. Gosele, *Mater. Res. Soc. Symp. Proc.* **256**, 3 (1992).

[52] K. Barla, R. Herino, G. Bomchil, J. C. Pfister, and A. Freund, *J. Cryst. Growth* **69**, 726 (1984).

[53] H. Mimura, T. Matsumoto, and Y. Kanemitsu, *Mater. Res. Soc. Symp. Proc.* **358**, 635 (1995).

[54] Y. Kanemitsu, T. Ogawa, K. Shiraishi, and K. Takeda, *Phys. Rev. B* **48**, 4883 (1993).

[55] S. Miyazaki, K. Sakamoto, K. Shiba, and M. Hirose, *Thin Solid Films* **255**, 99 (1994).

[56] F. Koch, V. Petrova-Koch, and T. Muschik, *J. Lumin.* **57**, 271 (1993).

[57] T. Matsumoto, T. Futagi, H. Mimura, and Y. Kanemitsu, *Phys. Rev. B* **47**, 13876 (1993).

[58] J. C. Vial, A. Bsiesy, F. Gaspard, R. Herino, M. Ligeon, F. Muller, R. Romestain, and R. M. Macfarlane, *Phys. Rev. B* **45**, 14171 (1992).

[59] M. Kondo, *J. Non-Cryst. Solids* **164-166**, 641 (1993).

[60] N. Ookubo, N. Hamada, and S. Sawada, *Solid State Commun.* **92**, 369 (1994).

[61] T. Matsumoto, H. Mimura, and Y. Kanemitsu, in *Light Emission from Novel Silicon Materials* (Ref. [1]), p.182.

[62] Y. Kanemitsu and T. Matsumoto, *Proc. Int. Conf. Physics of Semiconductors, Vancouver, 1994*, p.2153.

[63] V. M. Dubin, J. N. Chazalviel, and F. Ozanam, *J. Lumin.* **57**, 61 (1993).

[64] K. H. Li, C. Tsai, J. Sarathy, and J. C. Campbell, *Appl. Phys. Lett.* **62**, 3192 (1993).

[65] C. Tsai, H. Li, J. Sarathy, S. Shin, J. C. Campbell, B. K. Hance, and J. M. White, *Appl. Phys. Lett.* **59**, 2814 (1991).

[66] V. Petrova-Koch, T. Muschik, A. Kux, B. K. Meyer, F. Koch, and V. Lehmann, *Appl. Phys. Lett.* **61**, 943 (1992).

[67] Y. Kanemitsu, T. Matsumoto, T. Futagi, and H. Mimura, *Mater. Res. Soc. Symp. Proc.* **298**, 205 (1993).

[68] Y. Kanemitsu, T. Futagi, T. Matsumoto, and H. Mimura, *Phys. Rev. B* **49**, 14732 (1994).

[69] D. I. Kovalev, I. D. Yaroshetzkii, T. Muschik, V. Petrova-Koch, and F. Koch, *Appl. Phys. Lett.* **64**, 214 (1994).

[70] F. Namavar, H. P. Maruska, and N. M. Kalkhoran, *Appl. Phys. Lett.* **60**, 2514 (1992).

[71] T. Futagi, T. Matsumoto, M. Katsuno, Y. Ohta, H. Mimura, and K. Kitamura, *Jpn. J. Appl. Phys.* **31**, L616 (1992).

[72] H. P. Maruska, F. Namavar, and N. M. Kalkhoran, *Appl. Phys. Lett.* **61**, 1338 (1992).

[73] T. Futagi, T. Matsumoto, M. Katsuno, Y. Ohta, H. Mimura, and K. Kitamura, *Appl. Phys. Lett.* **63**, 1209 (1993).

[74] P. Steiner, F. Kozlowski, and W. Lang, *Appl. Phys. Lett.* **62**, 2700 (1993).

[75] N. Koshida, H. Koyama, Y. Yamamoto, and G. J. Collins, *Appl. Phys. Lett.* **63**, 2655 (1993).

[76] K. H. Li, D. C. Diaz, Y. He, J. C. Campbell, and C. Tsai, *Appl. Phys. Lett.* **64**, 2394 (1994).

[77] H. Mimura, T. Matsumoto, and Y. Kanemitsu, (to be published, 1995).

[78] A. Bsiesy, F. Muller, M. Ligeon, F. Gaspard, R. Herino, R. Romestain, and J. C. Vial, *Phys. Rev. Lett.* **71**, 637 (1993).

[79] T. Matsumoto, N. Hasegawa, T. Tamaki, K. Ueda, T. Futagi, H. Mimura, and Y. Kanemitsu, *Jpn. J. Appl. Phys.* **33**, L35 (1994).

[80] T. Matsumoto, T. Futagi, N. Hasegawa, H. Mimura, and Y. Kanemitsu, in *Semiconductor Silicon 1994*, ed. by H. R. Huff, W. Bergholz, and K. Sumino (Electrochemical Society, Pennington, 1994), p.545.

[81] A. Fejfar, I. Pelant, E. Sipek, J. Kocka, G. Juska, T. Matsumoto, and Y. Kanemitsu, *Appl. Phys. Lett.* **66**, 1098 (1995).

[82] T. Matsumoto, H. Daimon, H. Mimura, and Y. Kanemitsu, *J. Electrochem. Soc.* (1995).

[83] Y. Kanemitsu, *J. Spectro. Soc. Jpn.* **43**, 385 (1994).

# *CHAPTER 6*

# OPTICAL PROPERTIES OF Pb-BASED INORGANIC-ORGANIC PEROVSKITES

Teruya Ishihara
*Department of Physical Electronics, Hiroshima University*

## 6.1. INTRODUCTION

Extensive studies on semiconductor quantum wells have resulted in better understanding of the physics involved in low-dimensional electronic systems. Before the advent of quantum wells, for example, the band gap was generally accepted to be inherent to the material. We now know that it can be tuned by changing the thickness of a quantum well. Doped semiconductors used to never get high mobility because the dopants themselves played a role of scatterers as well. The modulation doping technique was developed to circumvent such a difficulty, leading to an extremely high mobility. Studies of quantum Hall effect has flourished on such a background. Better understanding of physical principles may follow after one controls samples at will.

One of the most important strategies in material science at this stage of evolution is to design materials for intriguing physics and/or useful applications. It is always difficult, however, to design the system from the very beginning. The guiding principle for designing an artificial structure is to divide the system functions into several blocks. For this purpose, it is advantageous to start with low-dimensional systems, such as molecular crystals, quantum wells and multi-layered structures, because the blocks can be designed rather separately. Let us take an example of an exciton binding energy, which is determined by a balance between the kinetic energy and the Coulomb interaction. By using a low-dimensional structure, one may control the dielectric environment while the kinetic energy intact. It is easier to tailor the properties if parameters can be controlled independently. By tailoring physical properties, one can setup a system in order to observe interesting physical phenomena.

Low-dimensional electronic systems have been mainly studied in GaAs/AlGaAs and related quantum well structures fabricated by Molecular Beam Epitaxy or Metal-Organic Chemical Vapor Deposition techniques. One-dimensional (1D) and zero-dimensional (0D) structures have been tried to be fabricated by state-of-the-art techniques. Even with these technique, nanometer structures are still difficult to realize because of the inevitable mono-layer fluctuation.

On the other hand, there are some compounds which self-organize into low dimensional structures. Two-dimensional (2D) compounds $(C_nH_{2n+1}NH_3)_2MX_4$ ($n = 3, 4, \cdots, 18$; M=Cd, Mn, Fe, Cu, $\cdots$; X=Cl and Br) are known in the field of dielectrics and magnetism. Intensive studies of these compounds have been carried out, focusing on the structural phase transition [1, 2], and 2D magnetism [3]. Discussions from the chemical point of view were given in [4]. Since most of the compounds are transparent except for the wavelengths related to the d-electron transitions, only a few optical properties due to electronic transitions had been reported [5, 6]. The first report on alkylammonium leadiodide was given by Dolzhenko *et al.* [7] in 1986. Optical properties of these compounds had been investigated by Ishihara since 1989, in relationship to their low dimensional structures [8]. In this article, we review the present status of the research field on physics and chemistry of alkylammonium leadiodide layered perovskites and the related compounds, focusing mainly on their optical properties. First we illustrate the crystal structures in Sec. 6.2 and sample preparation in Sec. 6.3. In Sec. 6.4, fundamental optical properties of representative compounds will be surveyed. Interesting phenomena observed with these compounds will be discussed in Sec. 6.5. Finally, concluding remarks will be given in Sec. 6.6.

## 6.2. MATERIALS AND STRUCTURES

In this section we review chemical and crystallographical properties which are essential to understand the exotic optical properties found in these compounds.

### 6.2.1. *Perovskite Compounds*

Perovskites are compounds with the chemical formula $AMX_3$. Some examples are found in ferroelectrics such as $BaTiO_3$. The unit cell is cubic and the cation M is located at the body center, while the anions X are located at the face centers, as shown in Fig. 6.1. The cation A resides at the eight corners of the cube, being shared by eight unit cells. Thus, one A-cation contributes to a unit cell. The structure can be regarded as a 3D network of the octahedral complexes of $[MX_6]$, sharing six corners. It is known that the crystal structure undergoes several phase transitions at lower temperatures.

Layered perovskites have the chemical formula $A_2MX_4$, which is sometimes called the $K_2NiF_4$ structure. One of the examples is the so-called 2-1-4 high-$T_c$ copper oxide superconductor, $(La_{2-x}Sr_x)CuO_4$. In this structure, $[MX_6]$ octahedra share four corners to form a 2D network.

### 6.2.2. *Inorganic-Organic Perovskites*

An important class of the layered perovskites is *bis*-(alkylammonium) metal (II) tetrahalide $(C_nH_{2n+1}NH_3)_2MX_4$, where $n = 2, 3, 4, \cdots, 18$, M=Cd, Cu, Fe, Mn, Pd or Pb, and X=Cl, Br or I.

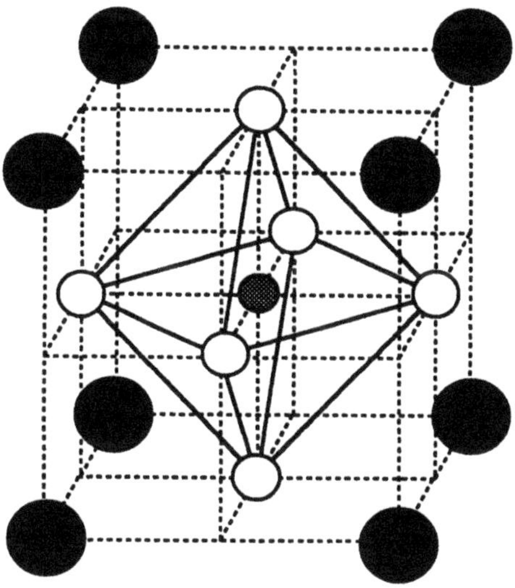

Figure 6.1. Unit structure of perovskites, closed circle: cation A, shaded circle: cation M, open circle: anion X.

The structure of these compounds is schematically shown in Fig. 6.2. The positively charged head of an alkylammonium is located in a void (X positions in the $A_2BX_4$ structure) of the network, within the same plane of the axial (not shared) X atoms. Here the attractive interaction between N and X originates from the hydrogen bonding. The alkylammoniums sandwich the two-dimensional plane of $[PbI_6]^{4-}$ to form a neutral layer. A multi-layer structure is organized with these neutral layers by the van der Waals interaction. The alkyl chains of the adjacent layers may or may not overlap, depending on the size of the ions in the inorganic layer and the A-cationic molecule. Thus this structure has three different bonding schemes: Ionic bond between metal ions and halogen ions, hydrogen bond between the halogens and the ammoniums, and van der Waals bond between the sandwiched layers. In this article, we will abbreviate $(C_nH_{2n+1}NH_3)_2MX_4$ as $C_n$-$MX_4$ for simplicity.

Compounds with these structures have been known to exist for many years, but no interest had been paid to their electronic structures until recently. Instead there has been a lot of research in the structural instabilities, on which a variety of experimental techniques such as X-ray scattering, dielectric responses, differential scanning calorimetry, and NMR have been utilized [1]. de Jongh performed a systematic study of the antiferromagnetic behavior of $C_n$-$MX_4$, where X is divalent magnetic ions [3].

The first report on the alkylammonium leadiodide compounds was made on the intercalation of some organic molecules into the "van der Waals gap" of this layered crystal [7]. But from the present viewpoint, it seems that the observed change of the lattice constant was not due to intercalation of hexane molecules but due to the

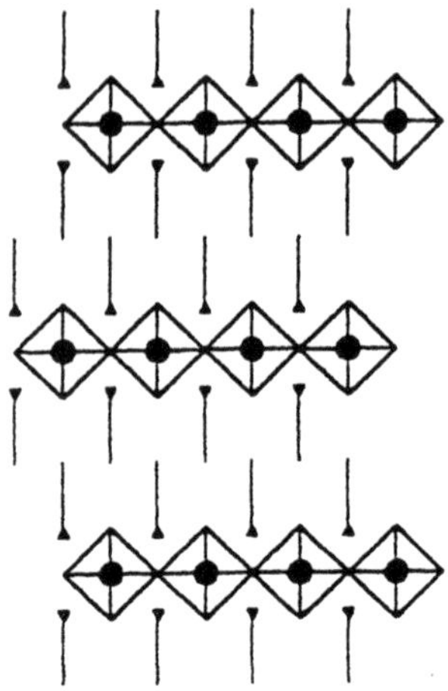

Figure 6.2. Schematic structure of $C_n$-$MX_4$.

structural phase transition induced by the indothermo reaction of hexane evaporation.

After the first report on an alkylammonium leadiodide, Ishihara *et al.* initiated the study of the optical properties of the compounds, regarding the compounds as naturally formed quantum wells (self-organized quantum wells). They found a huge exciton binding energy, and attempted to explain it in terms of the "dielectric confinement" effect for the excitons [8].

In 1991, a group at Du Pont expanded the field by showing that $C_n$-$PbI_4$ is only a subset of the generic class $(C_nH_{2n+1}NH_3)_2(CH_3NH_3)_{m-1}Pb_mI_{3m+1}$, $C_n$-$Pb_mI_{3m+1}$ for short, where m is the number of Pb-ion monolayer sheets within a well layer. This modification is in a sense a natural expansion of the strategy on high-$T_c$ superconducting materials: increase the thickness of the two-dimensional layer for higher $T_c$.

Figure 6.3(b) shows the schematic structure of $C_n$-$PbI_4$, where 2D sheets of $[PbI_6]^{4-}$ are separated by organic layers (not shown). Figure 6.3(c) shows the structure of $C_n$-$Pb_2I_7$, of which the quantum well layer is twice as thick as in $C_n$-$PbI_4$. Methylammonium ions (not shown) work as countercations to glue two layers. The extreme member of this class of compounds is $C_1$-$PbI_3$ for $m = \infty$. This structure lacks barrier layers as shown in Fig. 6.3(d) and can be considered as a 3D variation of the generic compounds. $C_1$-$PbI_6{\cdot}H_2O$ may be considered as a 0D analog, as is shown in (a), where $[PbI_6]^{4-}$ octahedra share no corners and are isolated. Thus the octahedra may share 0, 2, 4 or 6 corners to have 0, 1, 2, or 3 dimensional structures. So far the one-dimensional (1D) structure whose octahedra shares 2 corners has not been known. Instead, by sharing two triangular faces on the opposite side of the

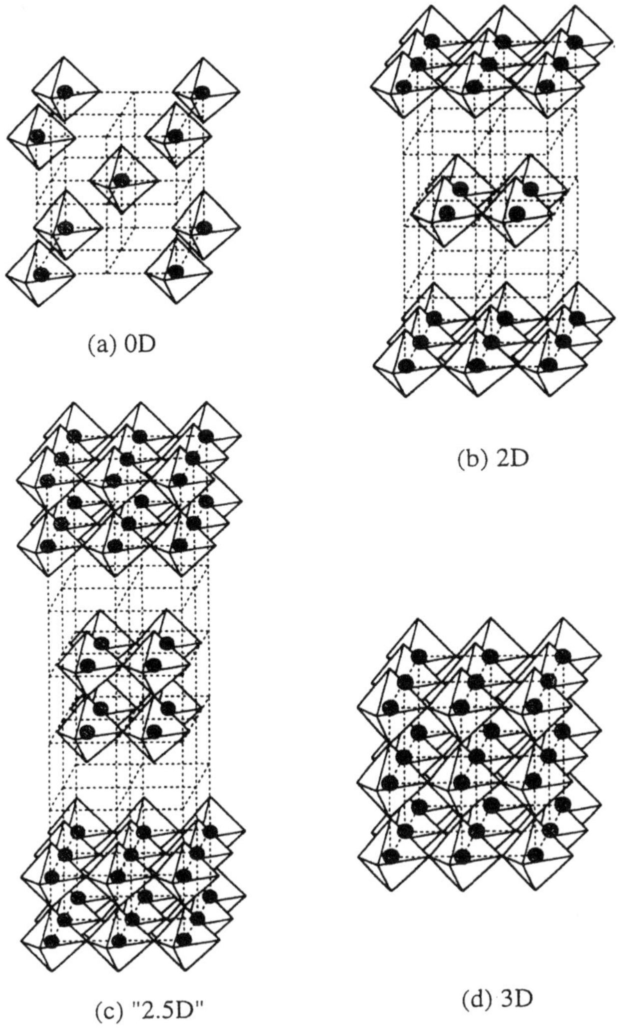


Figure 6.3. Structures based on $[PbI_6]^{4-}$ octahedra.

octahedra, a 1D structure can be realized. Hereafter we will refer these compounds with the dimensions of their $[PbI_6]^{4-}$ networks whenever it is necessary.

How can the compound family based on $[PbI_6]^{4-}$ octahedra exhibit such a variety of structures? It is important that the building block of $[PbI_6]^{4-}$ is an anion. In order to form a crystal out of these building blocks, cations are required to neutralize the electric charge. A key role is played by the variety of size and shape of the alkylammonium cations. When the octahedra combine with small cations such as $NH_4^+$ (or $K^+$, $Cs^+$), the cations are embedded into the voids to form a 3D structure. When they combine with a cation having a long tail such as *n*-decylammonium ion ($C_{10}H_{21}NH_3^+$), they sandwich a 2D plane to form a 2D structure.

Thus, PbI-based perovskite compounds have an enormous variety, which may be utilized for studying the physical properties inherent to these materials. Since the photon energy required for the electronic transition under consideration is smaller than that of the alkylammoniums, these compounds may be treated as self-organized quantum wells, wires and dots; these new materials have opened a new field for physics in low-dimensional systems.

### *6.2.3.* $C_{10}$-$PbI_4$ *(2D)*

Figure 6.4 shows the crystal structure of $C_{10}$-$PbI_4$ at room temperature determined from X-ray diffraction [9]. The top view (a) shows that the octahedra are rotated along their axes and the voids have the diamond shape. Thus the unit cell contains two formula units per layer. Two adjacent octahedra are connected by the glide-reflection operation. Octahedra are also tilted in the *ac*-plane to have a wavy structure in a two-dimensional plane along *a*-axis. The *b*-axis projection (b) shows that the chains from adjacent layers are overlapping to some extent. This is not the case for bromide and chloride compounds, where the lattice constant of *a*- and *b*-axis is smaller because of the smaller ionic radii of the halogen ions. It is not the case either for some iodide compounds with more bulky cation such as phenethylammonium [10]. The crystal structure of $C_9$-$PbX_4$ is essentially the same [11]: the lattice is tetragonal and the space group is $D_{2h}^{15}$ (Pbca).

It is known that the perovskite structures have an instability in their lattices and that causes structural phase transitions. The crystal structures of $C_{10}$-$PbI_4$— at the other phases were studied by Kondo *et al.* [12]. According to their results, the length of c-axis changes discontinuously at the phase transition temperatures as shown in Fig. 6.5.

It is clear that the crystal has four different structural phases. The first phase above 65 °C is tetragonal in terms of the heavy atoms; the alkylammonium chain seems to be "melting," judged from the large diffracted patterns. The second phase (the room temperature phase) is rhombohedral and the space group is $D_{2h}^{15}$ (Pbca). The alkylammonium chains are fairly disordered. The third phase reportedly has the same space group with the alkylammoniums more ordered. But since some spectral features splits in this phase, it seems that the space group is different from the second phase. The fourth phase is monoclinic with $P2_1/c$, where the crystal were found to

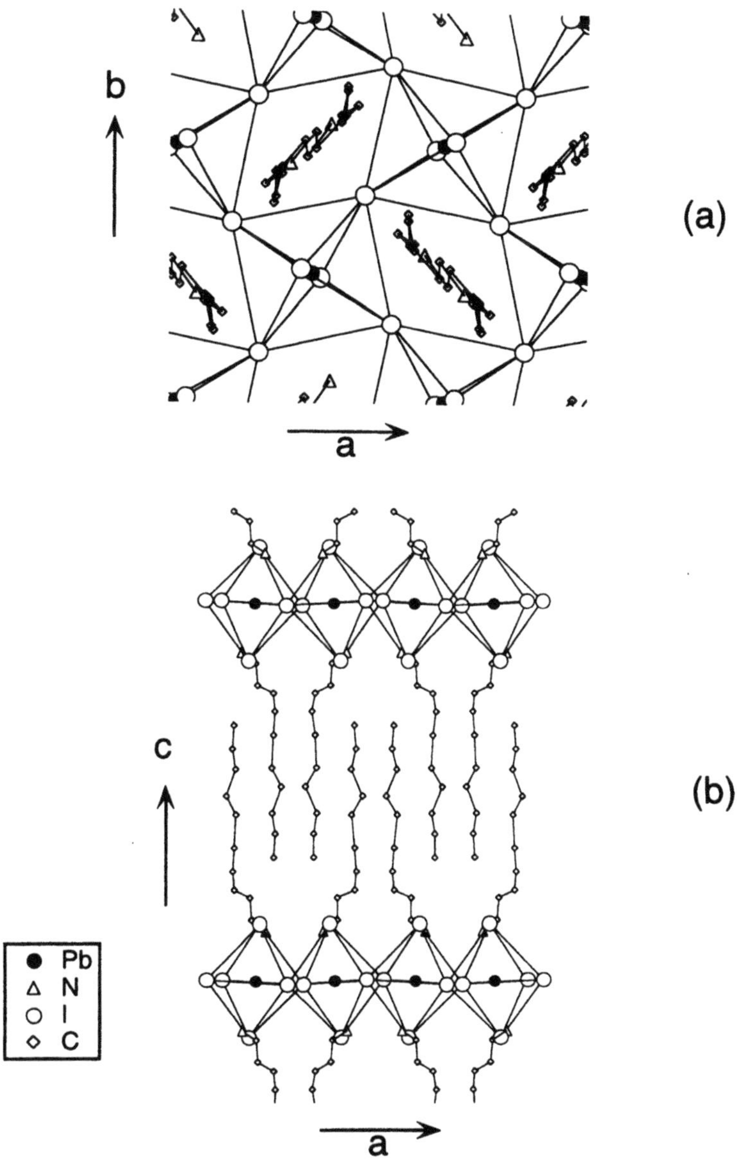


Figure 6.4. Crystal structure of $C_{10}$-$PbI_4$ at room temperature. (a) Top view. (b) Side view. (After [9]).

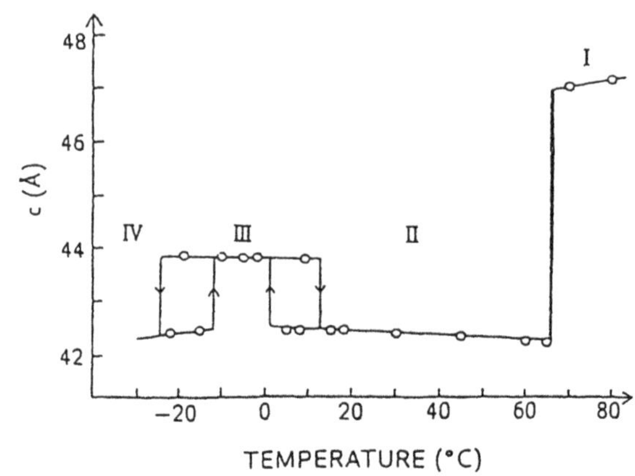


Figure 6.5. $c$-lattice constant of $C_{10}$-$PbI_4$ as a function of temperature. (After [17]).

have twin structures, and it was impossible to identify the atomic positions.

From the structural point of view, the most remarkable change between phase II and phase III is the angle between the octahedra. At this phase transition, color of the crystal changes from orange to yellow, as will be discussed later. The structures of the inorganic layer in phase II, III, and IV are shown in Fig. 6.6.

### 6.2.4. PhE-$PbI_4$ *(2D)*

$(C_6H_5C_2H_5NH_3)_2PbI_4$, bis-(phenethylammonium) tetraiodoplumbate (abbreviated as PhE-$PbI_4$) has been synthesized and the crystal structure was determined [10]. The space group is monoclinic C2/m. The lattice constants are $a = 32.508$ Å, $b = 6.131$ Å, $c = 6.185$ Å, and $\beta = 93.80°$. (Note that the choice of the axes is different.) Because of the bulky phenyl ring in the organic layer, the adjacent layers don't overlap each other, unlike the n-alkylammonium tetraiodoplumbates. Probably from the same reason, this compound does not have any structural phase transitions.

### 6.2.5. PhE-$Pb_2I_7$ *("2.5D")*

$(C_6H_5C_2H_5NH_3)_2(CH_3NH_3)Pb_2I_7$, (abbreviated as PhE-$Pb_2I_7$) has been reported in [10]. This compound can be regarded as a "bilayer" material. The space group is found to be triclinic $P\bar{1}$. Lattice constants are $a = 8.794$ Å, $b = 8.792$ Å, $c = 22.766$ Å, and $\alpha = 94.02°$, $\beta = 97.02°$, $\gamma = 90.18°$. This compound does not have any structural phase transition either.

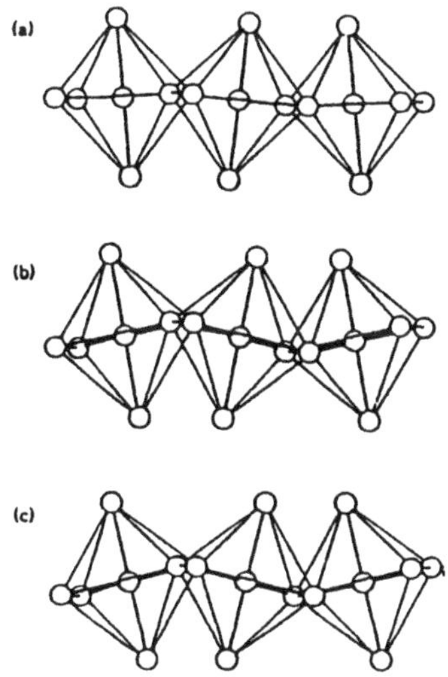


Figure 6.6. Inorganic layer structures of $C_{10}$-$PbI_4$ in phase II, III, and IV. (After [17]).

### 6.2.6. $C_1$-$PbI_3$ *(3D)*

The structure of $(CH_3NH_3)PbI_3$ ($C_1$-$PbI_3$ for short) was determined by Weber [13]. It is cubic with a lattice constant $a = 6.27$ Å and $Z = 1$ at room temperature.

### 6.2.7. $C_5$-$PbI_3$ *(1D)*

The structure of $C_5H_{10}NH_2PbI_3$, (abbreviated as $C_5$-$PbI_3$) has been reported in [14]. The organic part in this compound is piperidinium $C_5H_{10}NH_2^+$ which has a 6-membered ring structure. The $[PbI_6]^{4-}$ units are connected by sharing triangle faces to form a 1D structure. Thus this compound can be regarded as a "quantum wire" material, where the quantum wires of 4 Å in diameter are arranged with a spacing of 9 Å. The space group is found to be orthorhombic $C222_1$. Lattice constants are $a = 8.691$ Å, $b = 18.392$ Å, $c = 8.125$ Å, and the number of formula units in a unit cell is $Z = 4$. This compound does not have any structural phase transitions either.

### 6.2.8. $C_1$-$PbI_6{\cdot}H_2O$ *(0D)*

A $(CH_3NH_3)_4PbI_6{\cdot}2H_2O$ (abbreviated as $C_1$-$PbI_6{\cdot}H_2O$ hereafter) crystal contains isolated $[PbI_6]^{4-}$, $CH_3NH_3^+$ and $H_2O$ molecules [15]. In the context of the $[PbI_6]^{4-}$ networks, this structure can be considered to be quantum dot arrays or a 0D crystal. The space group of this yellow crystal is monoclinic $P2_1/n$. Lattice constants are $a = 10.421$ Å, $b = 11.334$ Å, $c = 10.668$ Å, $\beta = 91.73°$, and $Z = 2$.

## 6.3. SAMPLE PREPARATION

In this section, we review how to prepare nanostructures in a beaker with the aid of chemistry.

### 6.3.1. *Synthesis, Purification, and Crystal Growth*

In the first phase of the research on the compounds, single crystals of alkylammonium lead halides are obtained from a gel mixture [7, 16]. But with this method the crystals are easily contaminated, the surfaces are stained by gels and it is rather difficult to control the hardness of gel for appropriate growing condition. Fukuda developed the KI-acetone method [17], which was gel-free. And Hirasawa [18] developed a new solution method, which is contamination free. Here we introduce a typical solution method to make $C_{10}$-$PbI_4$ single crystals.

#### *6.3.1.1.* $C_{10}$-$PbI_4$ *(2D)*

Leadiodide ($PbI_2$, 99.9% $\sim$ 99.99%), decylamine ($C_{10}H_{21}NH_2$), and hydroiodic acid (HI) (aqueous solution, typically 57 wt%) are commercially available. Leadiodide should be distilled in vacuum ($10^{-2} \sim 10^{-3}$ Torr) by a rotary pump with a liquid nitrogen cold trap) at 400°C several times as it contains some residuals even if it claims to be 99.99%). Stoichiometric amounts of hydroiodic acid and decylamine are mixed in a beaker cooled with ice in order to take away the reaction heat. The reaction in oxygen-free atmosphere is preferable to prevent the oxidization of HI. After the reaction, water should be evaporated by elevating the temperature. The material is sublimed under vacuum at 140 °C to get rid of the residual water and impurities such as iodine molecules. The final product of alkylammonium iodides should be white.

In order to synthesize alkylammonium lead iodides, purified decylammoniumiodide and $PbI_2$ are reacted in acetone heated up to 40 °C. Although $PbI_2$ itself is not soluble in acetone, it dissolves as the reaction proceeds. After the reaction, the solution is filtered to get rid of the unreacted $PbI_2$. The compound is obtained as a precipitation, when a poor solvent such as nitromethane is added. Single crystals can be obtained from an acetone-nitromethane mixed solution. As the acetone evaporates, the nitromethane (the poor solvent) concentration increases and single crystals of $C_{10}$-$PbI_4$ precipitates. After about 2 to 4 weeks, single crystals with typical size of $4 \times 10 \times 0.1$ mm$^3$ are obtained.

#### *6.3.1.2. Other 2D compounds*

In order to have different organic materials for the barrier layer, their amine-derivatives are used as the starting materials.

For example, in order to make PhE-$PbI_4$, phenethylamine is used for decylamine in the procedure above. Orange platelets with a typical size of $2 \times 5 \times 0.1$ mm$^3$ are obtained. Besides the orange crystals, needle-like colorless transparent crystals can be grown in the presence of moisture. The structure of this form of the crystal is not known yet.

The method to make "two-layered" compounds such as PhE-$Pb_2I_7$ is just to use stoichiometric amounts of PhE-$NH_3I$ and $CH_3NH_3I$. The crystals are red in color with a typical size of $2.4 \times 4 \times 0.5$ mm$^3$.

Bromide and chloride compounds can be grown by similar methods but with different solvents. Since $C_{10}$-$PbBr_4$ crystals have poor solubility in acetone, it is better to use water instead. The typical size is $3 \times 3 \times 0.05$ mm$^3$.

As for $C_{10}$-$PbCl_4$, the evaporation method gives only powder crystals. Single crystals can be grown by cooling down the aqueous solution slowly (from 60 °C to 30 °C in two weeks). The typical size of these crystals is $1 \times 1 \times 0.05$ mm$^3$.

#### *6.3.1.3.* $C_1$-$PbI_3$ *(3D) and* $C_1$-$PbI_6 \cdot H_2O$ *(0D)*

Vincent [15] used saturated $Pb(NO_3)_2$ to precipitate $C_1$-$PbI_3$ from a $CH_3NH_3I$ aqueous solution. An alternative way for synthesizing $C_1$-$PbI_3$ is to react $PbI_2$ with $CH_3NH_3I$ in water. Black $C_1$-$PbI_3$ is obtained as precipitation when the solution is heated up to 60 °C. Single crystals of $C_1$-$PbI_3$ can be grown from a saturated acetonitrile solution by elevating the temperature slowly (20°C/hour) from −20 °C to 60 °C. The crystal is a black cube with well developed facets and its size is typically $0.1 \times 0.1 \times 0.1$ mm$^3$.

The 0D crystals $C_1$-$PbI_6 \cdot H_2O$ can be obtained from the stoichiometric solution of $CH_3NH_3I$ and $PbI_2$ in acetonitrile, allowing some moisture in the air to be incorporated. Evapolation of the solvent leads to colorless crystals of $1 \times 1 \times 2$ mm$^3$ that are grown in two weeks at room temperature.

#### *6.3.1.4. Spin coating*

Spincoating results in a film of a good quality on a substrate (glass or quartz) [19]. For $C_n$-$PbI_4$ and PhE-$PbI_4$ acetone or methylethylketone gives good films. The presence of moisture tends to give poor crystals (especially for PhE-$PbI_4$). The film is polycrystalline but the X-ray diffraction shows that it is highly oriented with the c-axis perpendicular to the substrate surface.

Caution should be taken with the spin-coated film, when it is used to study the phase transition. Thin spin-coated films tend to have no structural phase transition, which may be ascribed to the strain in the film. It is known that these compounds deteriorate under light. Therefore the crystals should be kept in the dark. The surface of the crystal gets foggy after a few weeks in the air, and it does not give any luminescence. Since the bromide crystal (which is colorless) survives longer than iodide compounds, the photochemical reaction seems to be responsible for this deterioration. This effect is much smaller at low temperatures.

In order to make a good spin-coated film of $C_n$-$PbBr_4$, it is necessary to clean the substrate with KOH in ethanol before coating.

*6.3.2. Cleavage*

Cleavage is an important technique for measurement of absorption spectrum with a single crystal, because the absorption coefficient of the compounds is of the order of $10^6$ $cm^{-1}$ at the lowest exciton peak.

For cleavage,
(i) glue one side of a crystal to a substrate (glass or quartz) (with hard and transparent glue such as cyanoacrylate or epoxy),
(ii) cleave a crystal with a piece of transparent adhesive tape,
(iii) take the thin flake on the tape or the counterpart on the substrate for the absorption measurement and repeat cleaving several times until the thickness is appropriate,
(iv) by using a stereomicroscope, choose a good uniform part with a pinhole.

## 6.4. FUNDAMENTAL OPTICAL PROPERTIES

### *6.4.1.* $C_{10}$-$PbI_4$ *(2D)*

$C_{10}$-$PbI_4$ is the most extensively investigated compound in the Pb-based inorganic-organic perovskite family [8, 9]. We look at its optical properties rather in detail, since most of the properties are common among other materials.

#### *6.4.1.1. Electronic levels in the visible and ultraviolet region*

Figure 6.7 shows the optical density spectra of $C_{10}$-$PbI_4$ at several temperatures. The wave vector of the unpolarized incident light is parallel to the *c*-axis of the crystal. The sample is cleaved several times by means of adhesive tapes. Note that there is a prominent absorption peak in the fundamental absorption edge region even at room temperature. The flat topped shape of these absorption peaks is due to reflection losses (more than 80%) for which we did not attempt any correction. The large spectral width of the absorption band (about 100 meV) is ascribed not to the inhomogeneous broadening of the level but to the high reflectivity in the longitudinal-transverse (LT) splitting region. The absorption peak at 2.4 eV at 293 K is located in essentially the same position as that at 275 K. At 261 K, it discontinuously shifts to the higher energy side by 100 meV.

Figure 6.8 shows the absorption peak energy together with the PL peak energy as a function of temperature [9]. The discontinuous change of the exciton transition energy is clearly seen at around 270 K. The color of the crystal at the lower temperature phases (phase III, IV) is yellow, while it is orange at the higher temperature phases (phase I, II). Let's call the phase transition as YO transition hereafter in this article. Such a behavior is commonly seen in the most of the $C_n$-$PbI_4$ crystals, although the transition temperature depends on the number of carbons in a chain, $n$, as shown in Fig. 6.15 later in this section.

Corresponding to this energy shift, a band at 3.2 eV at 293 K splits into two bands at 3.1 and 3.3 eV at low temperatures. The splitting of the band suggests lower symmetry in the yellow phase.

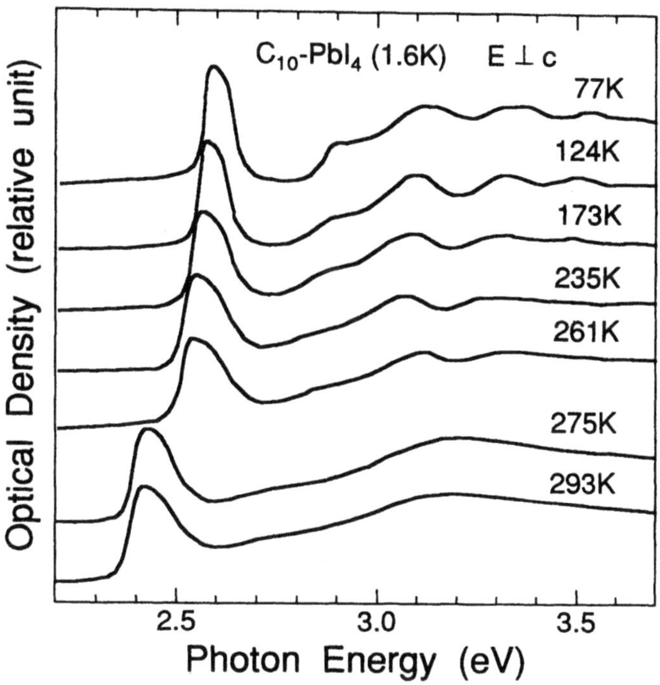


Figure 6.7. Absorption spectra of $C_{10}$-$PbI_4$ at several temperature.

As the temperature decreases further, the absorption structures become clearer, and at 77 K, a well-defined step-like structure at 2.87 eV is observed. As can be seen later in this section, this step represents a 2D density of states. Thus the excitonic binding energy is found to be 320 meV, which is very large compared to the binding energies in conventional semiconductors, 4 meV for GaAs, 28 meV for CdS, and 213 meV for CuCl. As for the higher members of the exciton series (2s, 3s, $\cdots$), they may form a quasi-continuum band and merge into the step-like continuum.

Figure 6.9 shows the reflection spectra of an as-grown side face of a $C_{10}$-$PbI_4$ single crystal [9]. The upper spectrum is taken for incident light with its polarization vector $\mathbf{E}$ perpendicular to the $c$-axis, while the lower spectrum is for the parallel configuration. The wave vector of the light, $\mathbf{k}$ is perpendicular to the $c$-axis for both of the spectra.

There is a large difference between the two spectra: The exciton structure at 2.55 eV is conspicuous for the $\mathbf{E} \perp \mathbf{c}$ configuration, while for the $\mathbf{E} \parallel \mathbf{c}$ configuration, the structure is by far narrower and less intense. The peak reflectivity amounts to more than 80 % for the $\mathbf{E} \perp \mathbf{c}$ configuration. From the absorption spectrum obtained by the Kramers-Kronig transformation, the oscillator strengths of the exciton for the two configurations are found to be $0.7 \pm 0.1$ and $0.03 \pm 0.01$, respectively, per formula

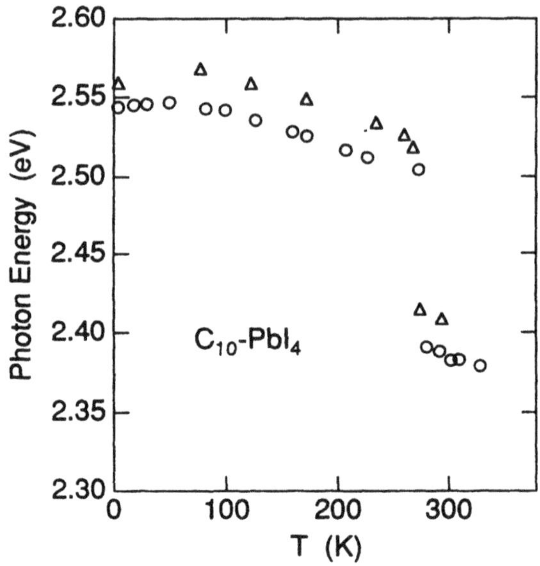


Figure 6.8. Absorption and PL peak of $C_{10}$-$PbI_4$ at several temperature.

unit. Such large oscillator strengths imply that the peak at 2.55 eV is due to a dipole-allowed direct exciton. Nevertheless it is exceptionally large compared to common semiconductors. The large enhancement will be explained in Sec. 6.5.4 in terms of the image charge effect (ICE).

At 2.85 eV, there is a cusp structure for the $\mathbf{E} \perp \mathbf{c}$ configuration, which corresponds to the step edge in the absorption spectrum.

The structures in the reflection spectra are classified into four groups with respect to their polarization dependence. The structures around 2.55 eV (group I) are mainly allowed for the $\mathbf{E} \perp \mathbf{c}$ configuration. At 3.1 and 3.3 eV, there are strong bands for the $\mathbf{E} \perp \mathbf{c}$ configuration. They belong to group II. Complicated structures between 3.54 and 4 eV (group III) are allowed for both polarizations. Broad bands at 4.3 and 4.7 eV (group IV) are stronger for $\mathbf{E} \parallel \mathbf{c}$. As will be discussed in Sec. 6.5.2, it is found that the valence band consists of Pb(6$s$) and I(5$p$) and the conduction band consists of Pb(6$p$) split into three bands by transfer anisotropy and the spin-orbit interaction. The polarization dependence of the reflection spectra is well explained with the symmetry consideration of the bands.

Figure 6.10 shows reflection spectra for $\mathbf{k} \parallel \mathbf{c}$ and $\mathbf{E} \parallel \mathbf{a}', \mathbf{b}'$ configurations [9]. Here the **a**' and **b**' axes refer to the **E** directions for which the largest and the smallest LT splittings are observed for the lowest single exciton state, respectively. The relation between these axes and the crystallographic axes are not confirmed yet. These two spectra are rather similar except for the structures around 3.2 eV. This anisotropy

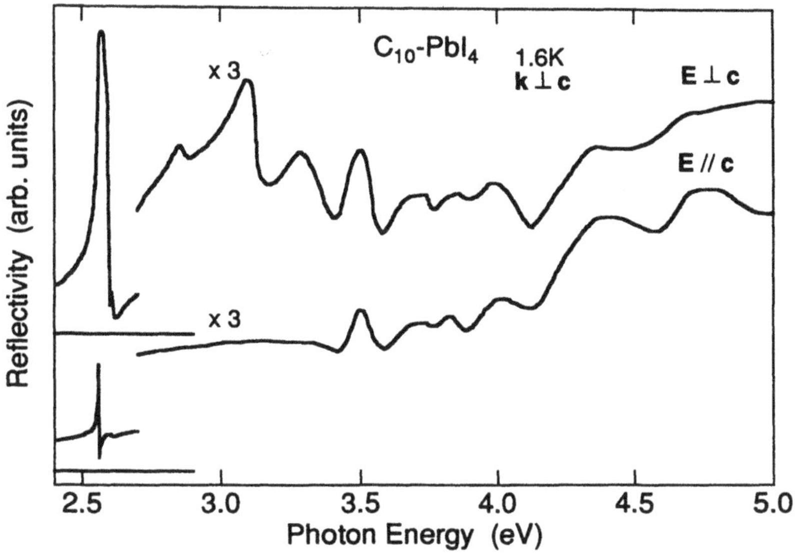


Figure 6.9. Reflection spectra of a single crystal of $C_{10}$-$PbI_4$ for **k**⊥ c configuration at 1.6 K.

vanishes in the orange phase crystal. As for the lowest exciton we may regard the crystal as uniaxial. The B band at 3.1 eV for **E** ⊥ **c** splits two for **E** || **c** configuration. Similarly the X band splits for **E** || **b**' configuration. The splitting should be interpreted in terms of the symmetry of the wave functions in these bands. The breaking of the symmetry may be caused by the tilting of the octahedra. In this sense, the *b*'-axis may correspond to the crystallographic *b*-axis, to which the octahedra incline in the yellow phase.

### *6.4.1.2. Luminescence properties of* $C_{10}$-$PbI_4$

The photoluminescence (PL) spectra of $C_{10}$-$PbI_4$ at several temperatures are shown in Fig. 6.11 [9]. Even at room temperature, fairly strong green luminescence is observed (easily seen with naked eye under UV illumination of few mW/cm$^2$). Since the absorption peak is at 2.4 eV at room temperature, the luminescence band overlaps with the absorption band. Thus it is a resonant luminescence of the exciton.

As the temperature decreases, the spectral position of the PL peak changes discontinuously at 268 K corresponding to the shift of the lowest exciton observed in the absorption spectrum. Below 124 K, a new band shows up at 2.45 eV. Since the shape and the intensity of the band is sample dependent, the band may be ascribed to a deep-trapped bound exciton. The strongest line at 2.53 eV at 4.2 K decreases its intensity rapidly as the temperature increases. Therefore, this line seems to be due

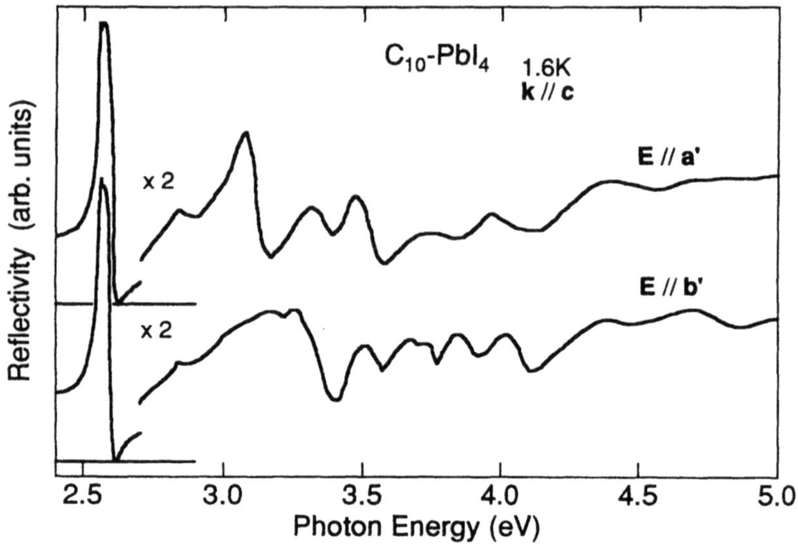


Figure 6.10. Reflection spectra of a single crystal of $C_{10}$-$PbI_4$ for $\mathbf{k}\|c$ configuration at 1.6 K.

to a shallow bound exciton. The luminescence band due to the free exciton is located at 2.55eV and is rather weak at 4.2 K. It becomes dominant at higher temperatures where the bound excitons ionizes. The band observed clearly at 10 and 20K, about 25 meV below the free-exciton band shifts to the lower energy side quickly as the temperature rises. Its origin is yet unknown at this stage but might be due to the acoustic phonon side band of the free exciton PL band.

PL intensity for the lowest exciton band is plotted as a function of 1000/T in Fig. 6.12 [9]. The intensity is almost constant up to 250 K, while it decreases exponentially at the higher temperatures. It can be shown that the slope gives the activation energy of the luminescence (see Sec. 6.5.5.2). The activation energy of $C_{10}$-$PbI_4$ is estimated to be 360 meV, which is comparable to the binding energy of the lowest exciton. Since a strong irradiation of the sample deteriorates the crystal, caution has to be taken for carrying out these experiments in order to keep the effect negligible.

### *6.4.2.* $C_n$-$PbI_4$ *(2D)*

The extensive study on $C_n$-$PbI_4$ family was reported in [20]: Although the distance between the adjacent leadiodide layers changes from 14.89 Å for $n = 4$ to 24.51 Å for $n = 12$, all the qualitative features in the spectra are essentially the same, which demonstrates that the interaction between the layers are weak enough to treat them independent. One of the small differences is the energy of the minimum in the reflec-

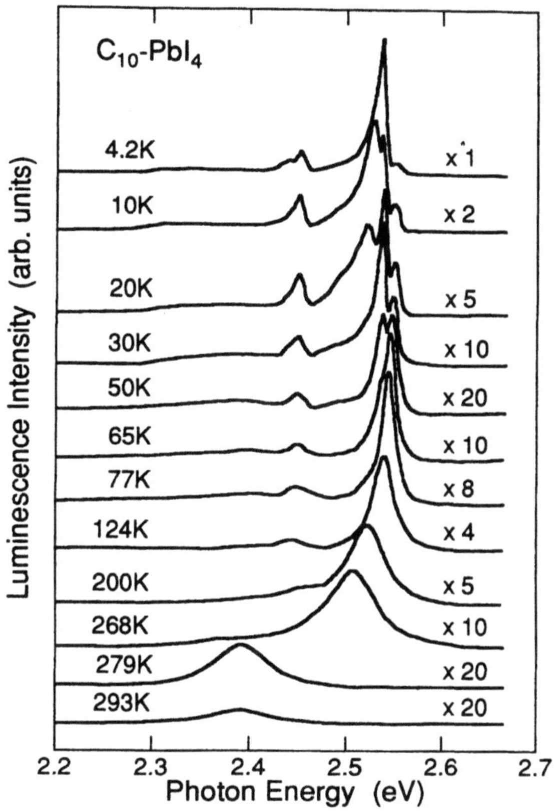


Figure 6.11. Photoluminescence spectra of a single crystal of $C_{10}$-$PbI_4$ at several temperatures.

tion spectrum around 2.6 eV.

Figure 6.13 manifests that the longitudinal exciton energy decreases as $n$ increases, while the transverse exciton energy is kept to be almost constant [9]. This tendency is reasonable because the oscillator strength decreases for thicker barrier layer as the density of the inorganic well layers (which supply electrons for the optical responses) becomes smaller.

Figure 6.14 indicates that the activation energy (determined from the Arhenius plot of PL intensity) $E_a$ and the "binding energy" (determined from the energy difference between the lowest exciton and the step) $E_b$ tends to increase as $n$ increases, while the oscillator strength $f$ (per formula unit) is almost constant [9]. This is consistent with the fact that the dielectric effect is important primarily for the band gap, as will be discussed later in Sec. 6.5.4.2.

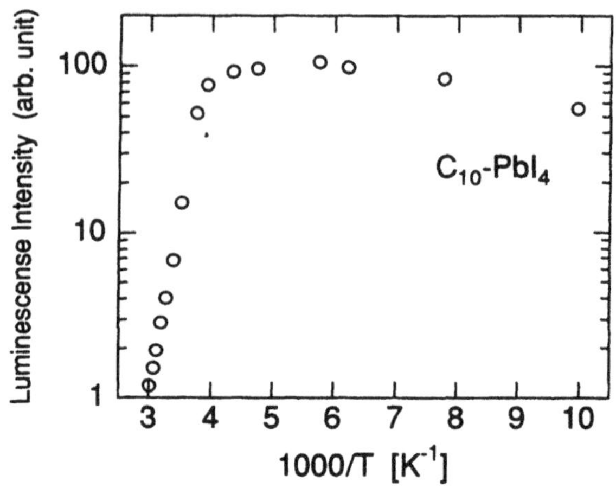


Figure 6.12. PL intensity on logarithmic scale as a function of inverse temperature.

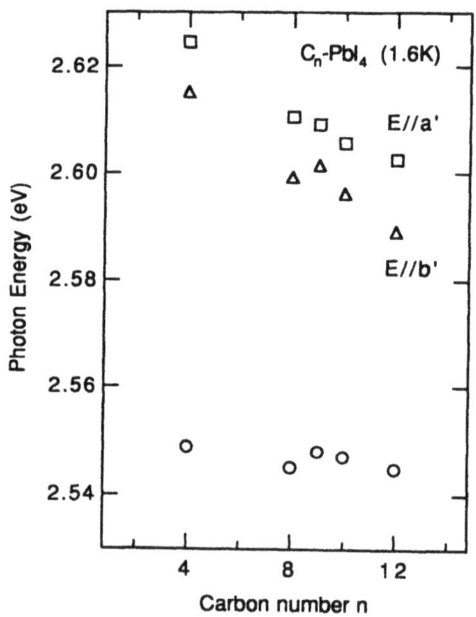


Figure 6.13. Exciton longitudinal-transverse splitting for $C_n$-$PbI_4$.

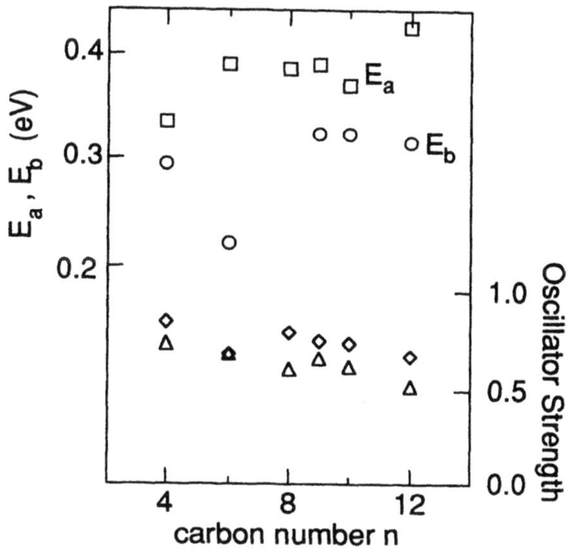


Figure 6.14. Exciton binding energy $E_b$, activation energy $E_a$, and oscillator strength $f$ as a function of carbon number in the barrier chain $n$.

It is the phase transition temperature that is the most remarkably affected by the number of carbons in an alkylammonium chain, $n$: Figure 6.15 shows YO transition temperatures of materials with different $n$. The melting temperatures for the corresponding alkylamines are also plotted as a comparison. It is clear that YO phase transition is related to the melting of the chain, which changes the ammonium motion in the cavity of the inorganic layer. The melting temperatures determined for alkylamines may be different from the phase transition temperatures, because the packing density is different.

### 6.4.3. $C_6$-$PbI_4$ (2D)

For some reason to be investigated, $C_6$-$PbI_4$ is eccentric in the $C_n$-$PbI_4$ family: It does not have YO phase transition and is in the orange phase even at low temperatures. Having no phase transitions, $C_6$-$PbI_4$ is ideal for studying optical spectra at low temperatures in detail [21]: The absorption band due to the lowest exciton state splits into five bands at low temperature. (Similarly PhE-$PbI_4$ has no phase transition, yet the absorption band width of the lowest exciton at low temperatures is wide, reflecting a large LT splitting.) The space group is Pbca, which is the same as the other $C_n$-$PbI_4$ compounds at room temperature. The chains are ordered at room temperature.

The energy splitting between the lowest exciton peak and the step structure is about 250 meV, which is considerably smaller than the binding energy for other $C_n$-$PbI_4$ compounds. From this fact, Kataoka *et al.* claim that the "dielectric con-

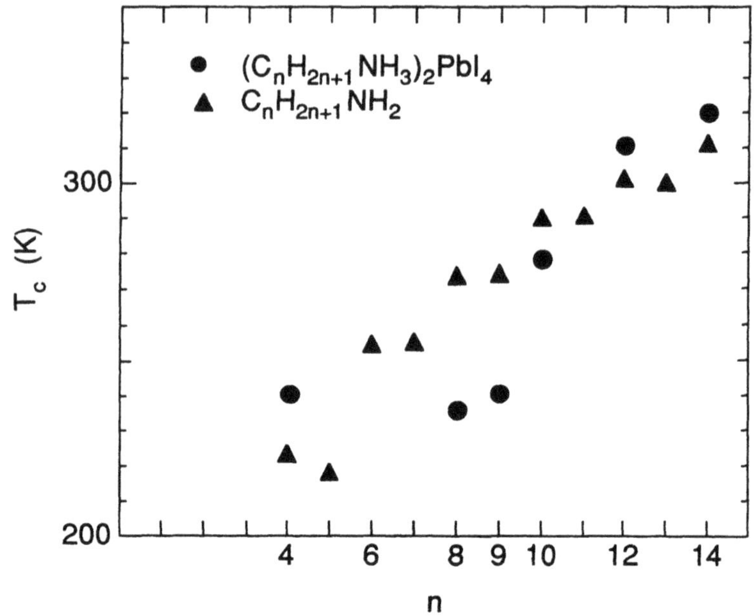


Figure 6.15. YO transition temperature of $C_n$-$PbI_4$ and the melting point temperature of $C_nH_{2n+1}NH_2$ as a function of the carbon number in the chain $n$.

finement effect" is not the dominating factor in determining the binding energy. They have measured magneto-absorption spectra both under the Faraday and Voigt configurations to investigate the Bohr radius and the electronic structure [21]. From the energy shift under the magnetic field, they found the diamagnetic shift coefficient for $C_6$-$PbI_4$ to be $3.53 \times 10^{-7}$ eV/ $T^2$ and determined 10 Å and 386 meV for the 2D Bohr radius and the exciton binding energy, respectively. For the related compound with doubled well thickness $C_6$-$Pb_2I_7$, the diamagnetic shift is $7.57 \times 10^{-7}$ eV/ $T^2$ [22], which is considerably larger than $C_6$-$PbI_4$, reflecting a smaller confinement in the well. Thus the exciton state in these compounds has a nonvanishing diamagnetic shift coefficient, which demonstrates the finite extent of the excitonic relative motion. It has a large binding energy not because it is of Frenkel type, but because it is in the unusual dielectric environment. The puzzling small binding energy observed for $C_6$-$PbI_4$ may be ascribed to its anomalous lattice structure among the $C_n$-$PbI_4$ family.

### 6.4.4. PhE-$PbI_4$ *(2D)*

#### *6.4.4.1. Electronic structure of* PhE-$PbI_4$

In this compound, phenethylammonium is used as the barrier material. Because of the presence of $\pi$-electrons in the phenyl ring, Hong *et al.* expected a smaller "enhance-

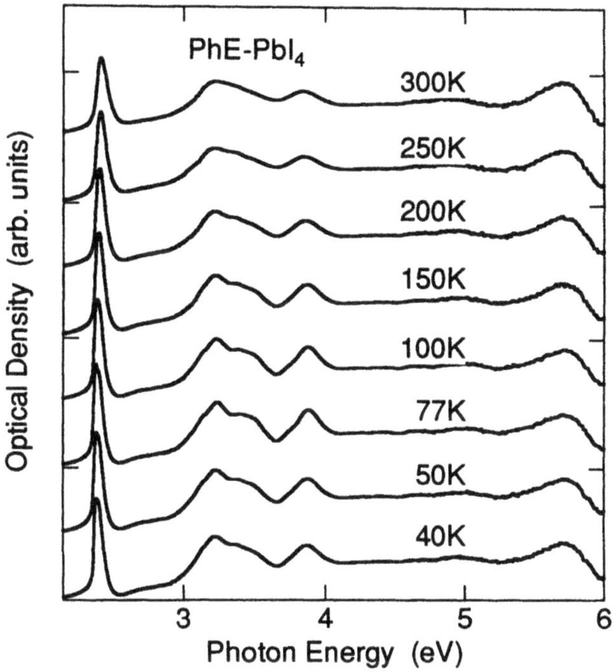


Figure 6.16. Absorption spectra of PhE-$PbI_4$ spin-coated film at several temperatures.

ment effect" on the exciton binding energy for this dielectric well structure. From the interference fringes in the optically transparent region, the dielectric constant in the optical region is determined to be 4.41 for PhE-$PbI_4$ and 3.39 for $C_{12}$-$PbI_4$. The refractive index is 1.83 for $C_{12}$-$NH_3I$ [23]. These values represent the average dielectric constant of the barrier and the well materials. Assuming a dielectric constant for the well of 6.1, the dielectric constant in the barrier is estimated to be 3.32 for PhE-$PbI_4$ and 2.44 for $C_n$-$PbI_4$.

Figure 6.16 shows the absorption spectra of a PhE-$PbI_4$ film spin-coated on a quartz substrate at several temperatures [57]. As can be seen in the figure, this material does not have a YO phase transition, which has been confirmed also with single crystals of the same material [see Fig. 6.17(b)]. The lowest exciton absorption peak is located at about 2.4 eV (orange phase) even at low temperatures and has slightly positive temperature coefficient of + 0.2 meV/K, in contrast to $C_n$-$PbI_4$ (−0.2 meV/K). In general, the temperature coefficient is determined by competition between the lattice thermal expansion and the electron-phonon interaction. As has been shown for $C_{10}$-$PbI_4$, the A-bandgap shrinks with hydrostatic pressure (see Sec. 6.5.6). Therefore the temperature coefficient should be positive if the electron phonon interaction is

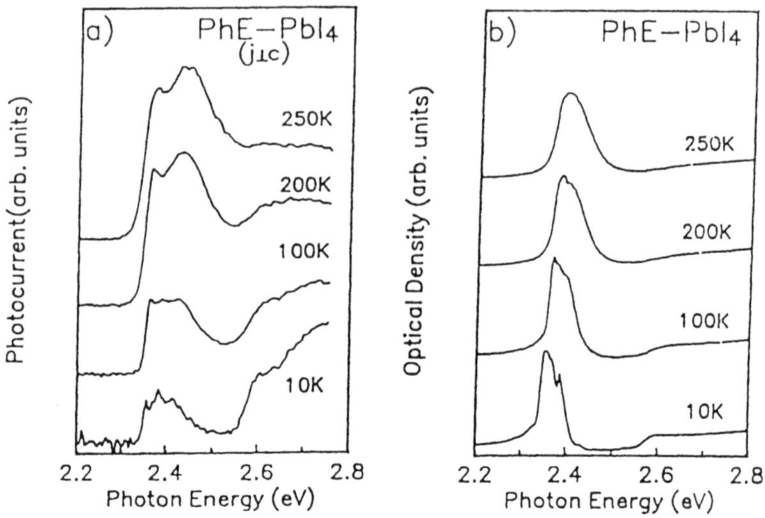


Figure 6.17. (a) Photoconductivity and (b) absorption spectra for a PhE-$PbI_4$ single crystal at different temperatures. (After [54]).

weak. In other words, the positive temperature coefficient suggests a weaker electron-lattice coupling in this compound. For a cleaved single crystal, the step like feature is clearly seen 220 meV above the lowest exciton peak [23], while this is less clear for the spin-coated film. This may be ascribed to the polycrystaline nature of the spin coated film. Note that the energy separation between the lowest exciton state and the step edge (220 meV) is fairly smaller than in $C_{10}$-$PbI_4$ (320 meV), which is ascribed to the smaller image charge effect on the exciton binding energy (see Sec. 6.5.4).

In a reflection spectrum, no polarization dependence is observed at low temperatures in the $\mathbf{k} \parallel \mathbf{c}$ configuration. Although the lattice structure is anisotropic in the *ab*-plane (orthorhombic), the electronic structure looks isotropic, which suggests some information on its symmetry. In the $\mathbf{k} \perp \mathbf{c}$ configuration, large anisotropy is observed as in $C_{10}$-$PbI_4$. Thus the uniaxial behavior (no polarization dependence in the *ab*-plane) is specific to all alkylammonium leadiodide layered compounds in the orange phase.

### *6.4.4.2. Photoconductivity*

Photoconductivity measurements at low temperatures can be reliably carried out in PhE-$PbI_4$. This is not the case for the most of the other compounds in this family, which have several structural phase transitions that may cause cracks in the single crystals.

The photocurrent excitation spectra of PhE-$PbI_4$ at several temperatures are shown in Fig. 6.17(a), while the corresponding absorption spectra are given in Fig. 6.17(b). At $T = 10$ K the absorption spectrum in (b) displays a strong exciton peak at 2.36 eV and a clear step at 2.58 eV. The step has been interpreted as a band-to-band transition, reflecting the 2D density of states. The step structure becomes less clear as the temperature increases.

A photocurrent flows when the compound is excited by a photon with energy above its band gap. Figure 6.17(a) shows that at 10 K the photocurrent is observed above the band gap at 2.6 eV. In principle no current is expected when the excitation is at the exciton state, because an exciton is a neutral entity. In reality, the photocurrent is observed even with photo-excitation directly into the exciton resonance, which could be explained by a partial ionization of the exciton state into free electron-hole pair states. This effect is rather less important at low temperatures but becomes substantial at elevated temperatures. At 250 K the amplitude of the photocurrent at the exciton region is much larger than those above the step.

Electroluminescence under a high electric field has been observed at low temperatures with this material as will be shown in Sec. 6.5.9.

### *6.4.5.* PhE-$Pb_2I_7$ *("2.5D")*

This compound has well layers twice as thick as in PhE-$PbI_4$. Thus it may be considered as a "2.5D" compound. The structure is reported in [10]. The lowest exciton peak is located at 2.15 eV [23]. Following a weaker potential confinement, the exciton states are located by 200 meV lower in energy as compared to PhE-$PbI_4$. The characteristic feature of this compound is the absorption peak at 2.62 eV, which is assigned to the second confined state (one node in the center) in the well. This spectral position corresponds to the step energy for PhE-$PbI_4$.

Since the exciton is less confined in this compound as compared to PhE-$PbI_4$, even a smaller binding energy is expected. In fact, the binding energy is reported to be 170 meV. The smaller bandgap was beneficial for the observation of electroluminescence in this compound (see Sec. 6.5.9).

### *6.4.6.* $C_1$-$PbI_3$ *(3D)*

A $C_1$-$PbI_3$ single crystal looks black suggesting a rather narrow band gap. The absorption spectrum of $C_1$-$PbI_3$ at 4.2 K obtained with the Kramers-Kronig method is shown in Fig. 6.18. The lowest exciton state is sharp and located in 1.64 eV. The LT splitting energy and the oscillator strength of this compound are estimated to be $5 \pm 1$ meV and $0.020 \pm 0.004$, respectively, which are much smaller than those in the low-dimensional compounds in the family. Above this sharp exciton band (band I), there are three broader bands: the band II at 2.77 eV, the band III at 3.6 eV and the band IV at 4.8 eV.

The spectrum is quite similar to those of other cubic perovskites $CsPbBr_3$ and $CsPbCl_3$, with rigid energy shift 0.7 eV ad 1.3 eV, respectively. Hirasawa *et al.* [36] assigned the band I and the band III to the direct transitions at the R([111]) and

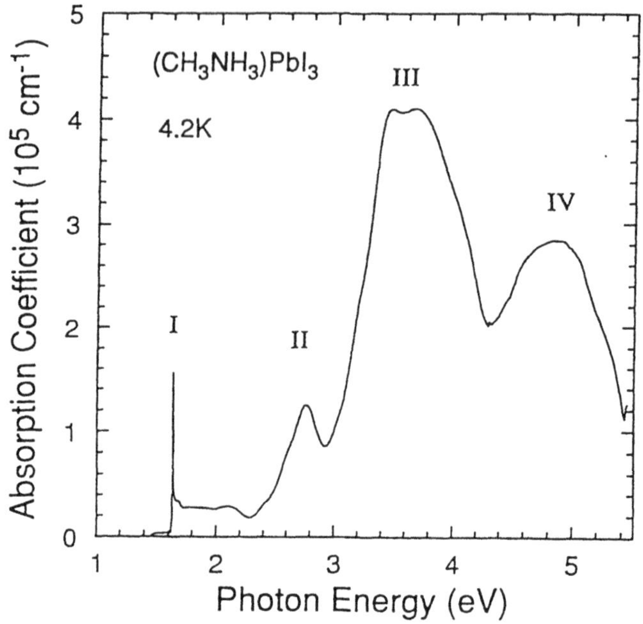


Figure 6.18. Absorption spectrum obtained from reflection data with Kramers-Kronig relation. (After [36]).

M([110]) points, respectively. The band II and the band IV are the spin-orbit partners of the band I and band III, respectively. This assignment is different from that given by Heidrich *et al.* [24] for $CsPbBr_3$ and $CsPbCl_3$, where the three upper bands are ascribed to the transitions at the R point. The comparison of absorption spectra for nD leadiodide perovskites (see Sec. 6.5.3) favors Hirasawa's assignment.

Absorption spectra of $C_1$-$PbI_3$ dispersed in PMMA have been measured in the presence of a magnetic field under the Faraday configuration [25]. The peak energy is plotted in Fig. 6.19. In addition to the Zeeman splitting, the exciton energy clearly shifts quadratically with a magnetic field, from which the diamagnetic shift coefficient is estimated to be $(2.7 \pm 0.1) \times 10^{-6}$ eV/T$^2$. Using an optical dielectric constant of 6.5, the Bohr radius, binding energy, and reduced mass are estimated to be 28 Å, 37 meV and 0.12 $m_0$, respectively. With these values, it is legitimate to treat this exciton as a typical Wannier exciton. The small reduced mass suggests large dispersions of the valence and the conduction bands in this compound.

### 6.4.7. $C_1$-$PbI_6$·$H_2O$ *(0D)*

The absorption spectrum obtained from reflection spectra with the Kramers-Kronig method are shown in Fig. 6.20. The lowest exciton is located at 3.4 eV, which is the highest in the leadiodide perovskite family, comparable to the absorption peak of

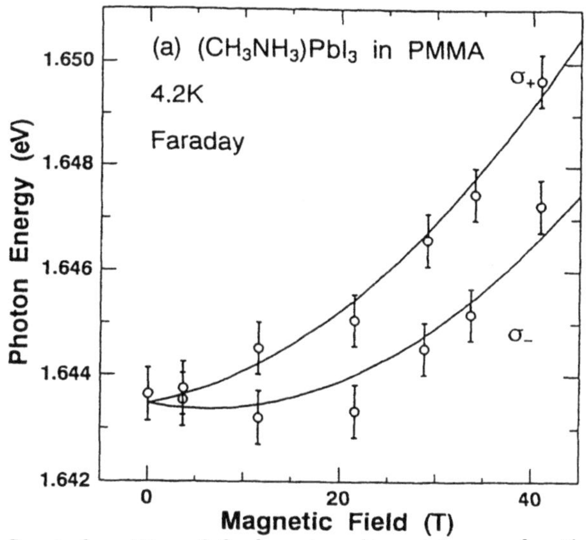


Figure 6.19. Spectral position of the lowest exciton state as a function of magnetic field. (After [25]).

the isolated $Pb^{2+}$ ion in KI crystal [26]. Furthermore the absorption spectra of the ion shows similar features as in Fig. 6.20. By comparing these spectra, the A and C bands are ascribed to the transitions from the highest occupied molecular orbital (HOMO) to the lowest unoccupied molecular orbital (LUMO) transition, with the angular momentum configuration $(j_e, j_h)_J = (1/2, 1/2)_1$ and $(3/2, 1/2)_1$, respectively. The D band is assigned to the second highest occupied molecular orbital (SOMO)-LUMO transition. SOMO can originate from the non-bonding I(5p) state. See the discussion in Sec. 6.5.1.

Papavassiliou and Koutselas [27] pointed out that the absorption spectrum of $C_1$-$PbI_6 \cdot H_2O$ is similar to $C_n$-$PbI_4$ solution in acetonitrile, where the complex ion $[PbI_6]^{4-}$ exists.

### *6.4.8.* $C_n$-$PbX_4$ (X=Br,Cl) *(2D)*

Figure 6.21 shows the absorption spectra of $C_{10}$-$PbI_4$ (X=I, Br, and Cl) obtained from the reflection spectra for **E** $\|$ **b**' by the Kramers-Kronig analysis. Each compound has a prominent absorption peak due to the lowest exciton state. The absorption coefficient at the exciton peak is of the order of $10^6$ $cm^{-1}$. Note that the features in the absorption spectrum blue shift as a whole, as the halogen proceeds upper in the periodic table. This fact suggests that the halogen wave functions participate the fundamental transitions in these compounds, which excludes one of the models for the electronic structure in these compounds (see Sec. 6.5.1).

$C_n$-$PbBr_4$ are important, because the spectral position of the lowest exciton falls in the wavelength region of a doubled Ti:sapphire laser, which can easily generate

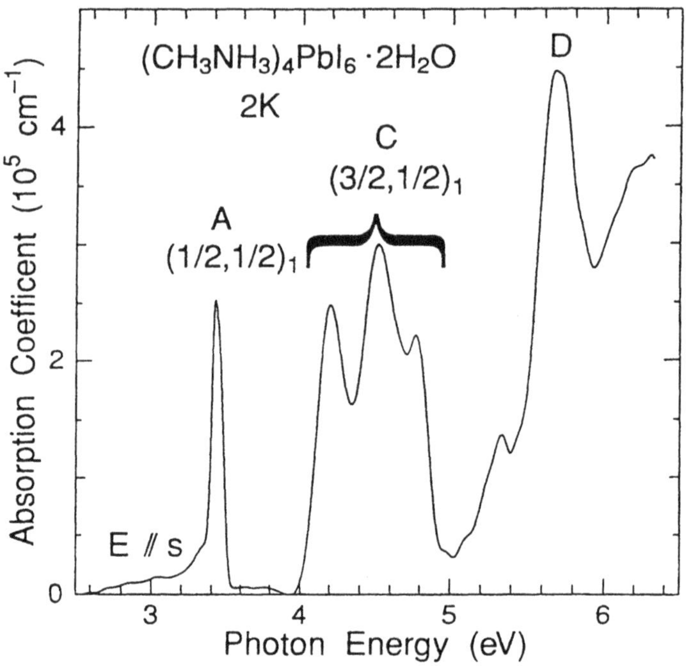


Figure 6.20. Absorption spectrum of $C_1$-$PbI_6{\cdot}H_2O$ obtained from reflection spectrum for $E \parallel c$ configuration. (After [18]).

temporally short (100 fs) light pulses. Time-resolved nonlinear spectroscopy will be applied in order to characterize ultrafast dynamics in these compounds.

### *6.4.9.* $C_n$-$PbX_{4-x}Y_x$ (X,Y=Cl,Br,I) *(2D)*

Absorption spectra of mixed halogen crystals $C_{10}$-$PbX_{4-x}Y_x$ (X,Y=Cl,Br,I) are reported by Papavassiliou *et al.* [28]. It was found that the mixed crystal is of congruent type and, there is a linear dependence of the wavelength of the band position on the $x$-value. Just by controlling the concentration of halide, one obtains samples whose absorption and/or luminescence bands can occur in any required position in the spectral region from about 330 nm for chloride to about 520 nm for iodide. Furthermore, by controlling the dimensionality of the perovskite networks, the tunable range expands to cover whole visible spectrum.

### *6.4.10.* $C_5$-$PbI_3$ *(1D)*

The optical data of this 1D compound were somewhat diversed: Muljarov *et al.* reported a broad luminescence band at 2.1 eV at 4.2 K [45], while Papavassiliou *et al.*

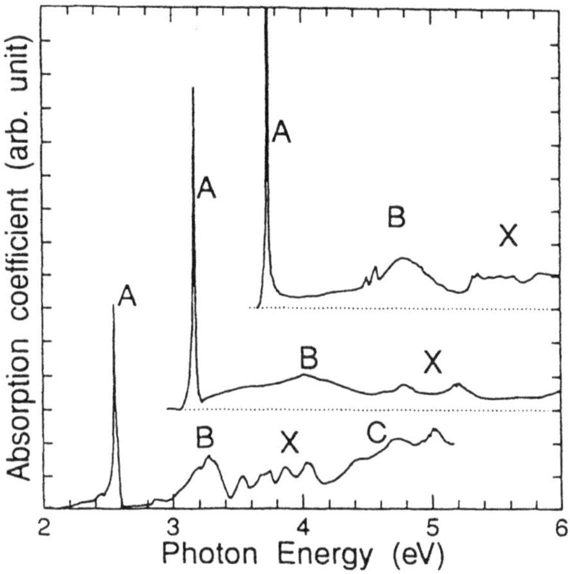


Figure 6.21. Absorption spectra of $C_{10}$-$PbX_4$ (X=I, Br, Cl) obtained from the reflection spectra. (After [34]).

[27] reported resonant emission at 410 nm (3.0 eV) at room temperature. Recently Nagami *et al.* [30] found that the exciton luminescence has a large Stokes shift of 1.2 eV (the exciton state at 3.2 eV while the luminescence band at 2.1 eV). This large Stokes shift may be ascribed to the self-trapped exciton characteristic to the 1D band structure.

### 6.4.11. *Related Compounds*

The Pb atom in $C_n$-$PbX_4$ can be replaced by Sn or Ge. Optical properties of $C_6$-$SnI_4$ were reported in [31, 27]: The crystal shape is a rectangular platelet and its color is brown-golden in reflected light or red in transmitted light. The excitonic absorption band is at 2.0 eV, and the PL band is resonant centered at 1.98 eV. The exciton band is found in the corresponding bilayer compound at 1.78 eV and in the 3D compound at 1.55 eV. The conductivity as a function of temperature shows semiconducting behavior in PhE-$SnI_4$ while metallic in the 3D compound.

Mitzi *et al.* [32] measured the dc conductivity of the $m$-layered perovskite $(C_4H_9NH_3)_2(CH_3NH_3)_{m-1}SnI_{3m+1}$, as a function of temperature, and found that they show a transition from semiconducting to metallic behavior with increasing $m$. The observation can be justified, when we consider the discussion on the band width in Sec. 6.5.2. The band gap collapses if the band widths are large enough for the valence and conduction bands to overlap with each other.

## 6.5. TOPICS IN Pb-BASED INORGANIC-ORGANIC PEROVSKITES

### 6.5.1. *Models for Electronic Structure*

In order to explain the optical properties of the Pb-based inorganic-organic perovskites, several models have been proposed. Some of them turned out to be inappropriate after experiments which were recently carried out.

#### 6.5.1.1. *Cationic model*

At an early stage, Ishihara ***et al.*** [20] proposed the cationic model to explain the electronic structure of the lead iodide alkylammonium perovskites. This model was originally proposed for $PbI_2$ in order to account for the polarization dependence of the reflection spectra [33]. It assumes that the optical transition takes place in a $Pb^{2+}$ cation. Katakoka *et al.* [21] claimed that the small diamagnetic coefficient observed in the experiments proves the cationic nature of the transition. But the rigid shift of the spectrum observed in the halogen-replaced compounds suggests that the halogen wave functions participate in the fundamental optical transition. The large quantum size effect in $n$D $[PbI_6]^{4-}$ networks shows that the electronic transfer in these compounds is rather large and that the wave functions are not localized in these materials [34]. Furthermore the dimension-dependent, finite diamagnetic shift is difficult to explain in the cationic model. In order to circumvent these difficulties, the charge transfer model has been proposed.

#### 6.5.1.2. *Charge transfer model*

The cationic model could not explain the p-like character in the photoemission spectra of $PbI_2$. Robertson carried out a tight-binding band calculation and showed that the highest valence band consists of 59 % iodine p-function. Thus the charge transfer nature is indispensable for understanding the electronic properties of $PbI_2$. The same argument applies to the alkylammonium leadiodide, because the essential point in Robertson's argument was the relative position of the one-electron levels of Pb($6s$) and I($5p$): Since the I($5p$) level is higher than the Pb($6s$) one, both of which are occupied in the ground state, the highest valence band consists mainly of I($5p$). The discussion of the cationic model on the polarization dependence of the reflection spectra is still valid in this charge transfer model, because the overall symmetry at the top point in the valence band has $s$-like symmetry.

Table 6.1. LCAO parameters used. (units in eV)

| | $E_{Xp}$ | $V_{sp}$ | $V_{pp\sigma}$ | $V_{pp\pi}$ | $3\lambda_X$ |
|---|---|---|---|---|---|
| I | −10.95 | 1.47 | 2.20 | −0.41 | 0.7 |
| Br | −12.41 | 1.75 | 2.62 | −0.49 | 0.4 |
| Cl | −13.76 | 1.90 | 2.85 | −0.53 | 0.1 |

### 6.5.2. *Band Calculations*

Tight-binding calculations of the electronic structure for these compounds were performed for a 2D square lattice by Hirasawa *et al.* [18, 34]. Neglecting all alkylammonium ions, they used two Pb(6*s*), 3 × 2 Pb(6*p*), and 4 × 3 × 2 halogen *p* atomic functions as the bases with spin functions for the band calculations. The spin-orbit interaction for Pb(6*p*) ($3\lambda = 0.9$ eV) and halogen are taken into account.The parameters used in the calculation are given in Table 6.1. $E_{Xp}$ is the energy for the atomic level, $V_{sp}$ the nearest neighbor transfer between Pb(6*s*) and halogen *p*. $V_{pp\sigma}$ and $V_{pp\pi}$ are the nearest neighbor transfer energies between Pb(6*p*) and halogen *np*. The transfer energies were assumed to have an inverse square dependence on the distance.

Figure 6.22 shows the calculated band structures. The band gap increases as the halogen goes from I to Cl. The energy increase in the neutral atomic level is partially compensated by the increasing band width. The lowest two interband transitions occur at the M(11) point in the square lattice, which corresponds to the A and B transitions in Fig. 6.21. The X band is ascribed to the direct transitions at the X point. The C band is again the direct transition at the M point. The top of the valence band at the M point consists of 57% equatorial $I_1$ ($5p_{xy}$), 28% Pb($6p_{xy}$), 4% Pb($6p_z$). The second conduction band at the M point consists of 99% Pb($6p_{xy}$). The third band consists of 81% Pb($6p_z$), 14% $I_2$($5p_z$) and 5% Pb($6p_{xy}$). These ratio readily explain the polarization dependence observed in Fig. 6.21. Note that in the real crystal, the octahedra rotate in the *ab*-plane; the unit cell is larger than the one in the square lattice. Therefore folding of the Brillouin zone takes place. The band calculation for PhE-$SnI_4$ is reported in [31].

The photoemission spectrum (PES) has been reported by Nagapetyan *et al.* [11]. The shallowest peak in PES lie at −3 eV which corresponds to the large density of states due to the dispersionless non-bonding I($5p_z$). Since the PES spectra looks quite similar among $PbI_2$, $C_{10}$-$PbI_4$ and $C_5$-$PbI_3$, they conclude that the valence band of these compounds is dominated by the lead and iodine wave functions but not by those of the organic cations.

### 6.5.3. *Unified Model for an nD Network of* $[PbI_6]^{4-}$

As was shown in the introduction, $PbI_6$ octahedra can be arranged in various ways. Let's take a look at how these crystals behave in the optical region. Figure 6.23 shows

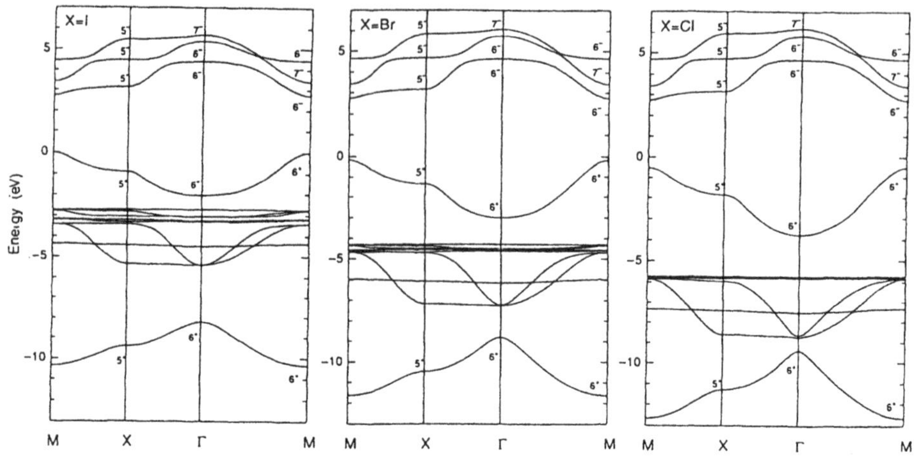


Figure 6.22. Tight binding band structure of alkylammonium leadhalogenides with square. (After [34]).

absorption spectra of $C_1$-$PbI_6 \cdot H_2O$ (0D), PhE-$PbI_4$ (2D), PhE-$Pb_2I_7$ ("2.5D"), and $C_1$-$PbI_3$ (3D) at liquid helium temperature [35]. They may be considered as 0D, 2D, "2.5D", and 3D networks of $PbI_6$ octahedra, respectively. The lowest exciton lies at 3.42 eV, 2.35 eV, 2.15 eV and 1.633 eV for the 0D, 2D, "2.5D" and 3D compounds respectively: It shifts towards a higher energy side as the exciton state is more confined.

This shift should be explained in terms of a transfer restriction, which is another way of describing the quantum confinement effect. Figure 6.24 illustrates this concept. Imagine a 3D semiconductor. Because of the electron transfer, the electron and hole levels form bands. As the $z$-direction is confined by barrier layers, the band width shrinks, while the band gap opens up. Neglecting the Coulomb interaction between electrons and holes for the time being, the band gap shrinks as the dimensionality increases from 0 to 3.

In order to treat the problem semiquantitatively, a unified model was introduced for the $n$D network of $[PbI_6]^{4-}$ octahedra [36]. In that model, an $s$-like HOMO and a $p$-like LUMO are assumed for the molecular orbitals in an octahedon. Then transfer interactions are introduced when two octahedra are connected by sharing an $I^-$ ion.

The Hamiltonian for the electron system is expressed as

$$H = \sum_{\vec{R}} H_0 \left( a^\dagger_{\vec{R}} a_{\vec{R}} + b^\dagger_{\vec{R}} b_{\vec{R}} \right)$$

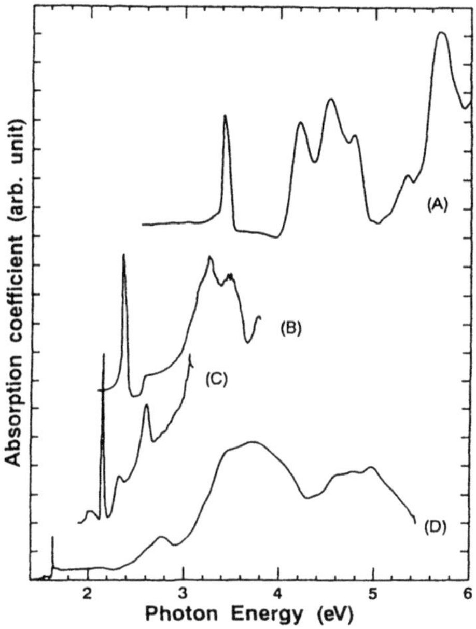


Figure 6.23. Absorption spectra of (A) $C_1$-$PbI_6 \cdot H_2O$ (0D), (B) PhE-$PbI_4$ (2D), (C) PhE-$Pb_2I_7$ ("2.5D") and (D) $C_1$-$PbI_3$ (3D) at liquid helium temperature. (Ishihara [35])

$$+ \sum_{\vec{R}_1,\vec{R}_2,j} \left[T_{cj}(\vec{R}_1,\vec{R}_2)a^\dagger_{\vec{R}_1}a_{\vec{R}_2} + T_v(\vec{R}_1,\vec{R}_2)b^\dagger_{\vec{R}_1}b_{\vec{R}_2}\right]$$
$$+ \sum_{\vec{R}_1,\vec{R}_2} V\left(\left|\vec{R}_1-\vec{R}_2\right|\right) a^\dagger_{\vec{R}_1}a_{\vec{R}_2}b^\dagger_{\vec{R}_1}b_{\vec{R}_2},$$

where $a^\dagger_{\vec{R}}(b^\dagger_{\vec{R}})$ are the creation operators for the electrons (holes). Transfer energy is anisotropic for the $p$-like electron orbital, while it is isotropic for the $s$-like hole orbital.

An exciton is an eigenstate of the Hamiltonian, with the creation operator,

$$B^\dagger_{\nu,\vec{K}} = \sum_{\vec{r}} U_{\nu,\vec{K}}(\vec{r})B^\dagger_{\vec{r},\vec{K}}, \tag{6.1}$$

$$B^\dagger_{\vec{r},\vec{K}} = \frac{1}{\sqrt{N}}\sum_{\vec{R}} e^{i\vec{K}\cdot\vec{R}}a^\dagger_{\vec{R}+\vec{r}}b^\dagger_{\vec{R}}. \tag{6.2}$$

Then the eigenenergy for the exciton state at $K = 0$ is

$$E_\nu = E_0 + \sum_{\vec{r}_1,\vec{r}_2} U^*_\nu(\vec{r}_1)[T_c(\vec{r}_1,\vec{r}_2) + T_v(\vec{r}_1,\vec{r}_2)]U_\nu(\vec{r}_2) + \sum_{\vec{r}} V(|\vec{r}|)|U_\nu(\vec{r})|^2. \tag{6.3}$$

Since $V_{eh}$ vanishes for a free electron-hole pair, the bandgap energies $E_g$ are obtained from this equation. Table 6.2 summarizes $E_g$'s at the R([1,1,1]) and M ([1,1,0])

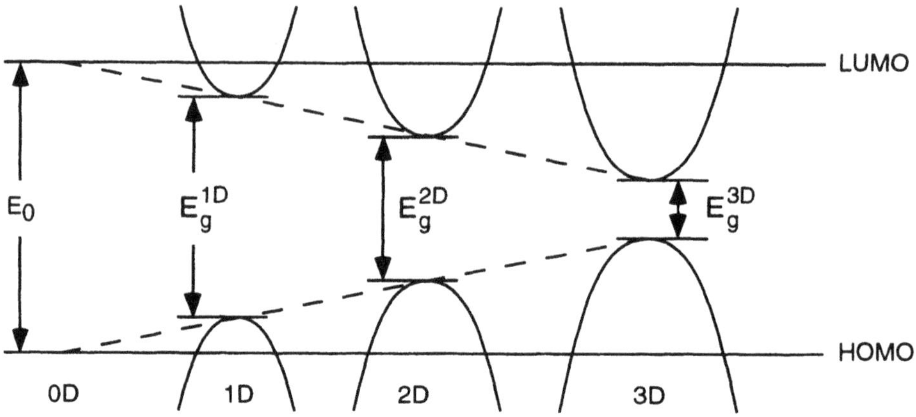


Figure 6.24. Concept of Quantum Confinement Effect: Band gap shrinks as the dimension increases.

points in 3D, the T point ([1,1]) in 2D and "2.5D", and the 0D transition energies, with corresponding experimental values. Here $T = T_c + T_v$ is the sum of the nearest neighbor transfer energies for the valence and conduction bands.

By comparing these energies to the singularities in the optical spectra, one can estimate $T$. We neglect the dielectric effect (the image charge effect) for the moment. By comparing $E_g$ of $C_1$-$PbI_3$ with $C_{10}$-$PbI_4$ and PhE-$PbI_4$, we see that $0.9 < 2T < 1.2$ eV. Thus electrons in these compounds are not localized but highly itinerant. This fact shows that the exciton in these compounds cannot be understood in terms of the Frenkel exciton model. The reason for the large binding energy of the exciton will be explained later in terms of the image charge effect. If the reduced mass were estimated

Table 6.2. Exciton and bandgap energies (in eV) for alkylammonium leadiodides of various dimensions.

| Name | Dimension | $E_{ex}^{exp}$ | $E_g^{exp}$ | $E_g^{model}$ |
|---|---|---|---|---|
| $C_1$-$PbI_6 \cdot H_2O$ | 0D | 3.4 | – | $E_0 - Q$ |
| $C_{10}$-$PbI_4$ | 2D (T) | 2.55 | 2.87 | $E_0 - 4T$ |
| PhE-$PbI_4$ | 2D (T) | 2.4 | 2.62 | $E_0 - 4T$ |
| PhE-$Pb_2I_7$ | "2.5D" (T) | 2.2 | 2.37 | $E_0 - 5T$ |
| $C_1$-$PbI_3$ | 3D (R) | 1.633 | 1.671 | $E_0 - 6T$ |
| $C_1$-$PbI_3$ | 3D (M) | 3.6 | 3.6 | $E_0 - 2T$ |

from the curvature at the bandgap, it would be 0.2 $m_0$, which is considerably larger than that estimated from the diamagnetic shift for $C_1$-$PbI_3$, 0.12 $m_0$. This suggests that the band shape is deviated from a cosine band structure.

This model predicts that $E_g$ at the T point in 2D crystal should coincide with the average energy of those for the R and M points in the 3D crystal, which is approximately satisfied for $C_{10}$-$PbI_4$ and PhE-$PbI_4$ when the third band of $C_1$-$PbI_3$ at 3.6 eV is ascribed for the M-point transition. Similarly the second confined transition energy in PhE-$Pb_2I_7$ ("2.5D") coincides with the band gap of PhE-$PbI_4$ (2D). Thus alkylammonium leadiodides of various dimensions can be understood in this simple model. The minor differences (of the order of 0.1 eV) between the spectra for $C_{10}$-$PbI_4$ and PhE-$PbI_4$ may be attributed to the details of the structure, which is beyond the scope of the model.

For $C_1$-$PbI_6 \cdot H_2O$, the octahedra are isolated and it can be considered as a 0D case. The exciton formation energy is $E_0 - Q$, where $Q$ is "the on-site Coulomb energy", which accounts for the kinetic energy of the relative motion and the Coulomb energy. From $E_0 - Q = 3.42$ eV, we find that $0.9 < Q < 1.8$ eV. This large on-site Coulomb energy may affect the exciton binding energy for low dimensional systems. Detailed discussion including the anisotropy of the transfer energies was given in [36].

### *6.5.4. Image Charge Effect*

#### *6.5.4.1. Image charge effect on exciton binding energy*

The aim of this section is to explain the large exciton binding energy in layered perovskite crystals.

Table 6.3 summarizes the exciton binding energies, diamagnetic shift and other parameters observed in the experiments. The binding energy is 320 meV for the 2D compound $C_{10}$-$PbI_4$ while it is only 37 meV for the 3D compound $C_1$-$PbI_3$. The binding energy in the 2D compound is enhanced more than 8 times as compared to that of the 3D compound. The increase in the dielectric constant of the barrier layer by the substitution of the barrier material gives a reduction of the binding energy from 320 meV in $C_{10}$-$PbI_4$ to 220 meV in PhE-$PbI_4$. Here we review the attempts to explain this experimental evidence.

First let's think about the Coulomb interaction between an electron and a hole involved in an exciton confined in a 2D thin layer. Because of the smaller kinetic energy in 2D system, the binding energy of a 2D exciton is four times larger than the one of a 3D exciton [39]. Is it the largest enhancement factor one can get in a 2D system? The answer is no.

In 1979, Keldysh [40] derived an analytic expression for the binding energy of an exciton in a thin semiconductor film surrounded by a medium of smaller dielectric constant. Because the surrounding medium screens the field less, the Coulomb attractive interaction is enhanced. It makes the qualitative estimation of the effect easier to consider an image charge potential produced by the difference of the dielectric constant of the three layers. Figure 6.25 illustrates the image charges for an exciton

Table 6.3. Exciton parameters in Pb-based perovskites: Experiments and theories.

| | $C_{10}$-$PbI_4$ | PhE-$PbI_4$ | PhE-$Pb_2I_7$ | $C_6$-$PbI_4$ | $C_1$-$PbI_3$ |
|---|---|---|---|---|---|
| Dimension | 2D | "2.5D" | 2D | 2D | 3D |
| $E_b^{exp}$(meV) | 320 [a] | 220 [b] | 170 [b] | – | 37 [c] |
| $f_{ex}^{exp}$ | 0.7 | 0.5 | 0.3 | – | 0.014 |
| $c_0^{exp}$($10^{-7}$ eV/T$^2$) | 2 $\sim$ 6 [c] | – | – | 3.5 [d] | 27 [e] |
| $L_w$(Å) | 6.36 | 6.36 | 12.72 | – | $\infty$ |
| $\epsilon_b$ | 2.44 | 3.22 | 3.22 | 2.44 | 6.1 |
| $a_B^{SDW}$(Å) [f] | 15.5 | 17.0 | 20.5 | – | 28 |
| $E_b^{SDW}$(meV) [f] | 315 | 230 | 175 | – | 33 |
| $c_0^{SDW}$($10^{-7}$ eV/T$^2$) [f] | 8.5 | 10.5 | 15.0 | – | 61 |
| $a_B^{MDW}$(Å) [g] | 13.0 | 14.2 | 16.5 | 13.1 | 33.0 |
| $E_b^{MDW}$(meV) [g] | 321 | 250 | 190 | 299 | 63 |
| $c_0^{MDW}$($10^{-7}$ eV/T$^2$) [g] | 2.18 | 2.60 | 3.50 | 2.21 | 14.2 |

[a] Reference [20].
[b] Reference [23].
[c] Reference [37].
[d] Reference [38].
[e] Reference [25].
[f] Reference [53].
[g] Reference [45].

confined in a "dielectric quantum well," which is sometimes used to emphasize that the barrier layer has much smaller dielectric constant than the quantum well.

Hanamura *et al.* [41] adopted the image charge method to obtain a Hamiltonian for an exciton in the well and gave a numerical estimate of the excitonic binding energy for a simple trial wave function.

Figure 6.26 shows the calculated enhancement of the binding energy normalized to the bulk value as a function of the dielectric constant ratio. As a possible correction to the binding energy in GaAs quantum wells, several treatments have been carried out. But the effect turned out to be very small in conventional systems [42, 43, 44]. In order to demonstrate the effect, large contrast in the dielectric constant and thin film thickness are required. If one could use NaCl as a barrier material to GaAs quantum well of 40 Å, the exciton binding energy should be enhanced 6 times. Since the bulk binding energy of GaAs is 5 meV, it would be higher than the thermal energy at room temperature. But it is rather difficult to grow a high quality GaAs quantum well sandwiched by NaCl layers. Thus it is difficult to fulfill these requirements with conventional systems. Such a structure, however, can be readily found in the inorganic-organic layered perovskite crystals.

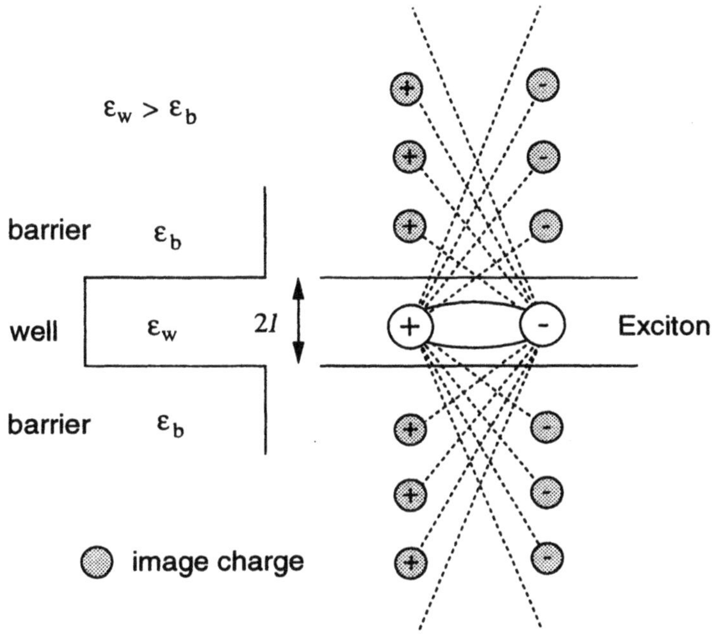


Figure 6.25. Image charges for an exciton in a dielectric quantum well.

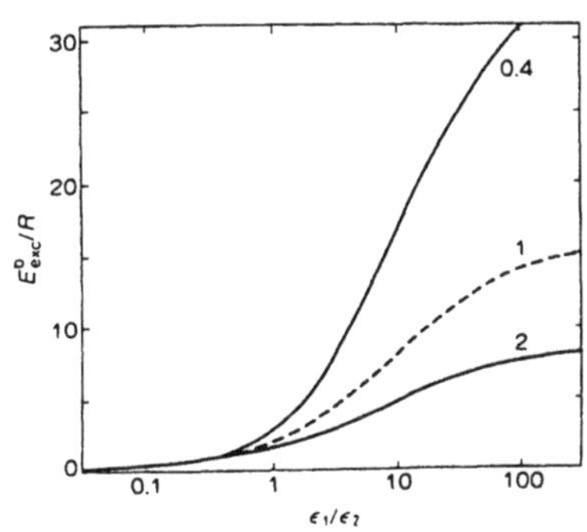


Figure 6.26. Calculated exciton binding energy in a dielectric quantum well. (After [41]).

Following the image charge potential formalism [41], the excitonic binding energy is estimated by minimizing the exciton formation energy [23]. The Hamiltonian for an exciton is

$$H = -\frac{\hbar^2}{2M}\nabla_R^2 - \frac{\hbar^2}{2\mu}\nabla_r^2 - \frac{\hbar^2}{2m_e}\frac{\partial^2}{\partial z_e^2} - \frac{\hbar^2}{2m_h}\frac{\partial^2}{\partial z_h^2} + V_{eh} + V_{\text{self}}, \tag{6.4}$$

where $V_{eh}$ is the potential due to the images of the electron (hole) to a hole (an electron),

$$V_{eh} = -\frac{e^2}{\epsilon_w}\sum \frac{[(\epsilon_w - \epsilon_b)/(\epsilon_w + \epsilon_b)]^{|n|}}{[(r_e - r_h)^2 + (z_e - (-1)^{|n|}z_h - 2nl)^2]^{1/2}}, \tag{6.5}$$

and $V_{\text{self}}$ is the potential due to the self images,

$$V_{\text{self}} = \frac{e^2}{2\epsilon_w}\sum \left(\frac{\epsilon_w - \epsilon_b}{\epsilon_w + \epsilon_b}\right)^{|n|} \left[\frac{1}{|z_e - (-1)^{|n|}z_e - 2nl|} + \frac{1}{|z_h - (-1)^{|n|}z_h - 2nl|}\right]. \tag{6.6}$$

Here the well thickness is $2l$. Assuming a simple trial wave function with one variational parameter $a$,

$$\Psi = \frac{1}{la}\left(\frac{2}{\pi}\right)^{1/2} \cos\left(\frac{\pi z_e}{2l}\right) \cos\left(\frac{\pi z_h}{2l}\right) \exp\left(-\frac{r}{a}\right), \tag{6.7}$$

the exciton binding energy is obtained by maximizing

$$E_b = -\langle\Psi| - \left(\frac{\hbar^2}{2\mu}\right)\nabla^2 + V_{eh}|\Psi\rangle. \tag{6.8}$$

Note that all the other terms are independent of $a$. The theoretically estimated Bohr radii, binding energies and diamagnetic shifts are summarized in Table 6.3 as $a_{\text{B}}^{\text{SDW}}$, $E_{\text{b}}^{\text{SDW}}$ and $c_0^{\text{SDW}}$, respectively. The observed binding energies are reproduced fairly well by choosing $\mu = 0.09m_0$, which is rather small. In this estimate the system is modeled as a single dielectric quantum well, while in reality the object is a multiple quantum well. Thus the image charge effect is overestimated in this work.

Multiple quantum well effects were considered in [45] by solving the Schrödinger equation was solved numerically, rather than estimated with a variational approach. The reduced mass was found to be $\mu = 0.2m_0$ in this model. The values estimated in this theory is summarized in Table 6.3 as $a_{\text{B}}^{\text{MDW}}$, $E_{\text{b}}^{\text{MDW}}$ and $c_0^{\text{MDW}}$.

The dielectric constant of the well layer in the layered perovskite crystal was estimated at 6 from the similar compound available, $PbI_2$. It may be more appropriate to use the dielectric constant of $C_1$-$PbI_3$ though. Since the dielectric constant is frequency dependent, we need to know the appropriate frequency for our problem. Usually the guideline is as following: the optical dielectric constant should be used if the binding energy is larger than the optical phonon energy. As the binding energy is 320 meV and the optical phonon is only 5 meV, the optical (high frequency) dielectric constant seems to be appropriate in our case.

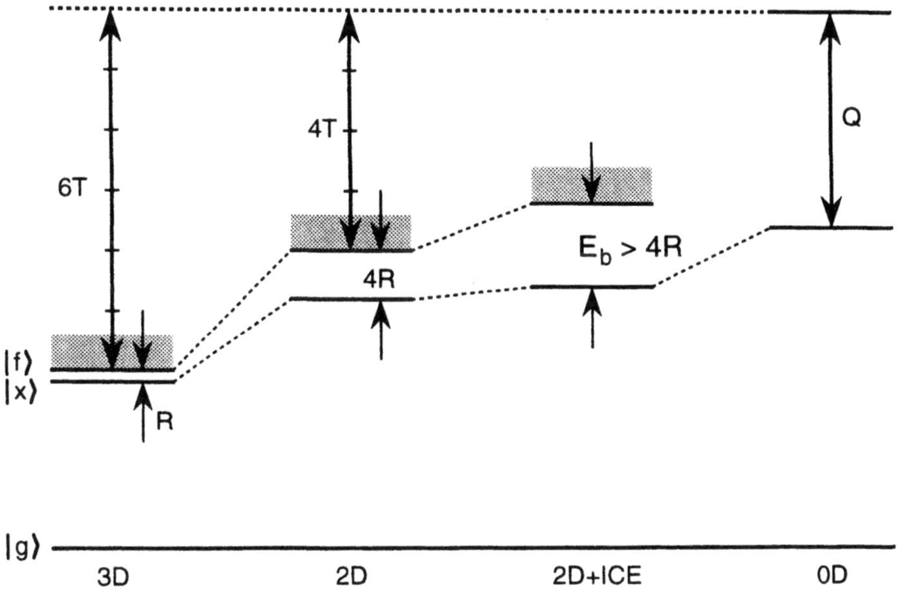


Figure 6.27. Exciton formation energy and band gap energy in $n$D system.

*6.5.4.2. Image charge effect on the spectral position*

So far we have discussed only the binding energy of an exciton in a dielectric quantum well. Now let us consider the formation energy of the exciton.

Imagine an ultrathin quantum well with a dielectric constant of $\epsilon_1$ sandwiched with barrier layers of $\epsilon_2$. If $\epsilon_1 = \epsilon_2$, the excitonic binding energy is four times larger than in 3D and the band gap increases by $2T$, where $T$ is the band width of the exciton band as is depicted in Fig. 6.27 (3D and 2D). Suppose $\epsilon_2$ decreases. Then the band gap energy increases, because of the larger electrostatic energy, while the exciton formation energy increases in much slower way, because it is charge neutral (Fig. 6.27, 2D+ICE). Thus the dielectric constant actually affects the band gap rather than the exciton. A similar situation is encountered in semiconductors under high density excitation, known as the band gap renormalization effect. The image charge effect on the exciton formation energy is only a minor correction. The energy difference between the band gap and the exciton is the enhanced binding energy which we calculated in Sec. 6.5.4. After the discussion in this section, it is clear that the image charge effect in a dielectric quantum well is important in order to understand its binding energy, even for an exciton with small radius. If $\epsilon_2$ were as large as $\epsilon_1$ in $C_{10}$-$PbI_4$, the binding energy would be $37 \times 4 = 148$ meV at most, while it is 320 meV in reality. The difference is attributed to the image charge effect.

Figure 6.28 shows the absorption spectra of cleaved single crystals of $C_n$-$PbI_4$ ($n$

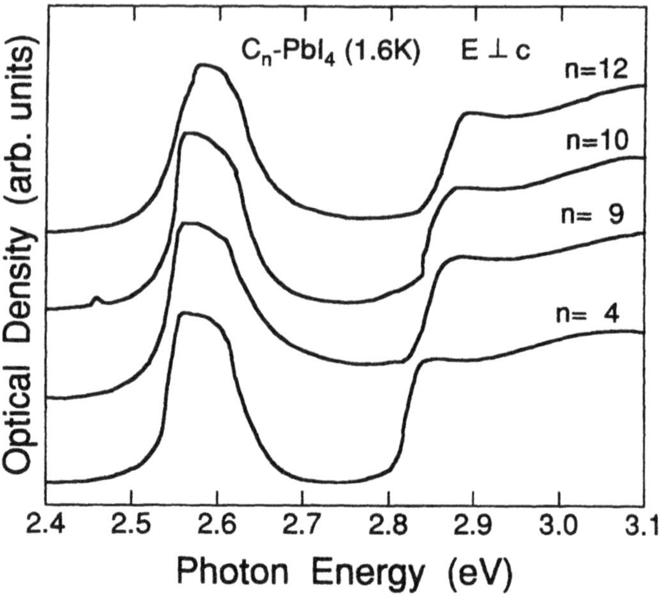


Figure 6.28. Optical density spectra of cleaved single crystals of $C_n$-$PbI_4$ with $n = 4, 9, 10$ and 12 at 1.6 K.

= 4, 9, 10 and 12). The absorption peak around 2.6 eV is due to the exciton state while the step like feature at 2.85 ~ 2.90 eV corresponds to the band gap. The band gap energy increases as $n$ increases, because the effective dielectric constant in the barrier layer decreases: the exciton energy is essentially constant, on the other hand. This data clearly demonstrates the discussion above.

### *6.5.5. Polariton Luminescence*

The luminescence band width for 2D PbI-based perovskites at room temperature is 50 meV which is fairly broad compared to the exciton luminescence bands of the familiar semiconductors. It is important, however, to keep in mind that this width is not a inhomogeneous broadening due to sample quality.

The oscillator strength for the lowest exciton state of $C_n$-$PbI_4$ is so high that the coupling to light is inevitable to form a mixed state, a polariton. The longitudinal-transverse splitting of the polariton amounts to about 60 meV which is larger than the thermal energy at room temperature. So all optical process should be explained in terms of polaritons. The characteristic feature in these compounds is that the exciton state is of Wannier type along the well and of Frenkel type along the c-axis. In other words, this polariton has a spatial dispersion along the well direction but none along the $c$-axis.

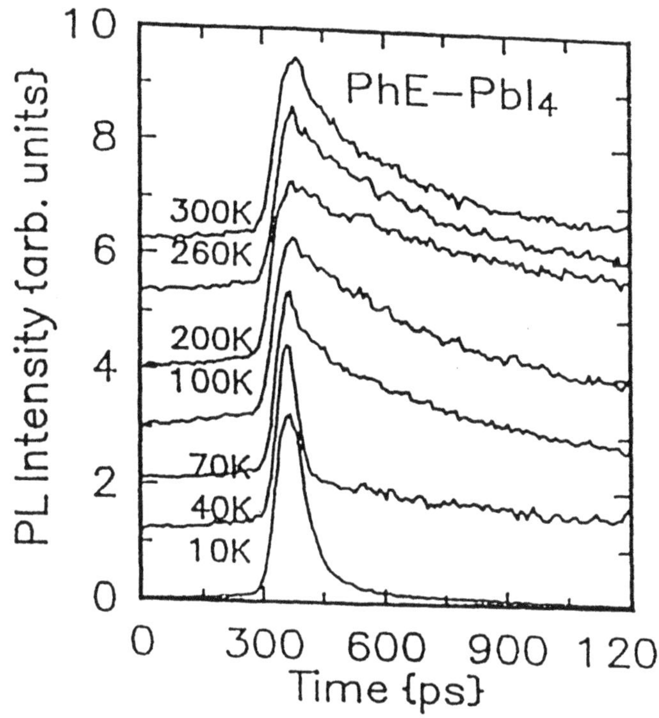


Figure 6.29. Time-resolved spectrally integrated PL of a PhE-$PbI_4$ single crystal in the backward scattering geometry at different temperatures. (After [46]).

The small energy shift of the PL peak from the absorption peak can be ascribed to the reabsorption effect (in the exciton picture) but it is more appropriate to ascribe it to the distribution of polaritons on the dispersion curve.

### *6.5.5.1. Time-resolved luminescence of* PhE-$PbI_4$

Figure 6.29 shows the polariton luminescence decay of PhE-$PbI_4$ in the back scattering geometry at several temperatures [46]. The excitation photon energy was above the band gap where the excitation depth is about 1000 Å. At 10 K, the response was pulse-like with half-life of 50 ps. Half-life time was measured because the decay is not exponential. At 40 K, a longer component emerges and dominates the kinetics above 100 K. This decay time increases to 1 ns at 200 K and decreases again, corresponding to the drop in the time-integrated PL intensity. Such a behavior can be explained in terms of polaritons [47]. Because the absorption coefficient for the excitation (above the bandgap) is fairly large, the excited region is essentially the surface (1000 Å) of the crystal. The electron-hole pair immediately relaxes to a polariton with a wave

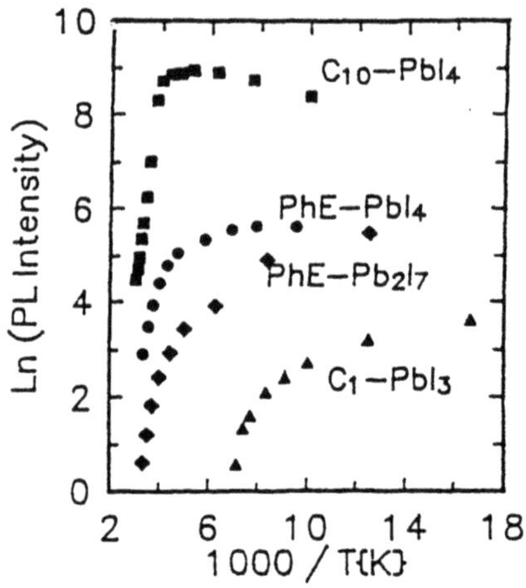


Figure 6.30. PL intensity of $C_{10}$-$PbI_4$ (2D), PhE-$PbI_4$ (2D), PhE-$Pb_2I_7$ ("2.5D") and $C_1$-$PbI_3$ (3D) on logarithmic scale as a function of inverse temperature. (After [46]).

vector in any direction. At low temperatures, only the polariton that happened to have the wave vector back to the illuminated surface can be observed. All the other polaritons can never be observed. Even though polaritons can immediately emit optical phonons, the probability is scarce, because the final density of state is small. As the temperature increases, some of the polaritons can absorb a phonon to be scattered back to the surface. That is why the luminescence has a longer tail in higher temperatures. Above 200 K, the lifetime becomes shorter, because the nonradiative decay channel through the ionized state is open (see Sec. 6.5.5). In this temperature range, PL efficiency becomes considerably smaller, while the photoconductivity is larger than that at low temperatures.

The decay time of the "bound exciton" state is rather long, about 1 ns at 1.6 K, which reflects a smaller spatial extent of the bound exciton compared to the free exciton state that extends over the crystal volume.

### *6.5.5.2. Temperature dependence of PL intensity*

Figure 6.30 shows the PL intensity of the dominant free exciton band as a function of inverse temperature for $C_{10}$-$PbI_4$ (2D), PhE-$PbI_4$ (2D), PhE-$Pb_2I_7$ (2.5D) and $C_1$-$PbI_3$ (3D). The temperature dependence of the PL intensity is distinct for the different compounds, although all excitons involve the same transition between electronic states of Pb(6*s*)+I(5*p*) to Pb(6*p*). In $C_1$-$PbI_3$, where the dielectric confinement effect does not apply, the PL intensity decreases rapidly with increasing temperature, and was too weak to be detected above 140 K. On the other hand, in $C_{10}$-$PbI_4$ and PhE-$PbI_4$, the PL can be seen at room temperature with a naked eye at several $mW/cm^2$ of UV

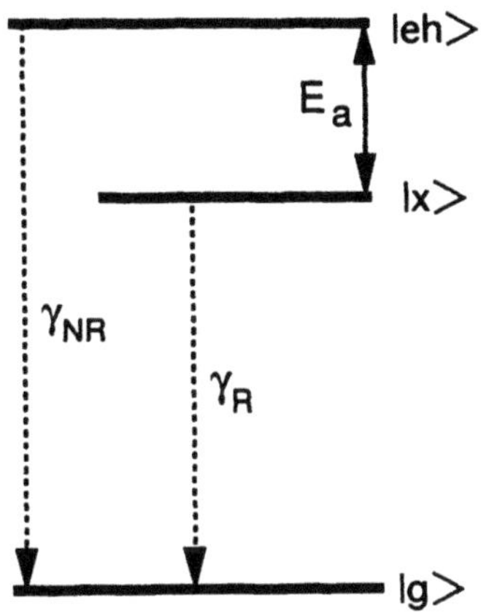


Figure 6.31. Three level model to describe temperature dependence of PL intensity.

excitation. The PL intensity for $C_{10}$-$PbI_4$ and PhE-$PbI_4$ remains almost constant up to 250 K and 200 K, respectively, and decreases as $\exp(E_a/k_BT)$ with $E_a$= 360 and 220 meV, where $E_a$is the activation energy described below. Note that these values are close to the exciton binding energy estimated directly from the absorption spectrum. Thus it is reasonable to ascribe this drop of PL intensity to an ionization process of an exciton to a free electron-hole state. This fact demonstrates again that the exciton binding energies can be tailored by varying the dielectric environment. One might ask the question why the exciton becomes unstable at temperatures which in energy equivalent terms are much smaller than the one expected from the exciton binding energy. The temperature dependence for the PL intensity is determined by the competition of radiative and nonradiative processes. In order to discuss this point semi-quantitatively, let's have an equation which describes this phenomenon.

Imagine a three-level system, with the ground, exciton, and free electron-hole pair states as in Fig. 6.31, where the nonradiative decay channel for the exciton state is to ionize into the free electron-hole pair state. Assuming thermal equilibrium between the exciton and the free electron-hole pair state, which are separated by $E_a$, the relative PL intensity $L(T)$ and the PL decay time $\tau(T)$ are expressed as

$$L(T) \propto \frac{\gamma_R(T)}{\gamma_R(T) + \gamma_{NR} e^{E_a/k_BT}}, \tag{6.9}$$

where $\gamma_{NR}$ and $\gamma_R(T)$ are the nonradiative and radiative decay rates, respectively. As can be seen from this expression, the PL intensity is constant if $\gamma_{NR} \exp(-E_a/k_BT)$ is negligible compared to $\gamma_R$. If it is the other way around, the PL intensity behaves as $\exp(-E_a/k_BT)$, as is observed experimentally above 250 K. Thus the steep drop of the PL intensity gives a large activation (or binding) energy.

The knee of the curve is given by

$$k_B T = E_a / \ln(\gamma_R / \gamma_{NR}). \tag{6.10}$$

Since $(\gamma_R)^{-1}$ is of the order of 500 ps, $\gamma_{NR}$ is $10^{-4}$ smaller, which is 50 fs. Such a behavior is often seen in wide-gap ionic semiconductors. For example, the binding energy of an exciton in CuCl is 213 meV and the activation energy is 150 meV, while the PL intensity decreases rapidly above 100 K (9 meV) [48]. The extremely rapid relaxation of free electron-hole pair might be ascribed to the quick relaxation of the polarizable ionic lattice.

By assuming $\gamma_{NR}^{-1} = 0.02$ ps and $\gamma_R^{-1} = 4.25T$ ps/K, we can fairly reproduce the lifetime and PL intensity data [46]. Note that if $\gamma_R$ were temperature independent and had a value 1/60 ps of 10 K, the knee of the curve would occur at a much higher temperature. Hence the reason for this decrease in the PL intensity is that at higher temperatures the polaritons cannot escape the crystal quickly before being trapped into the nonradiative channel.

### *6.5.6. Pressure Effect*

Hydrostatic pressure is one of the most powerful perturbations used to identify the electronic structure. Figure 6.32 shows absorption peak energies of $C_{10}$-$PbI_4$ under hydrostatic pressure at room temperature [18]. The energy of the lowest exciton absorption decreases as the pressure increases, while the peak energies for the B and C bands increase, which is consistent with the model of the electronic states: The large pressure enhances the transfer energy of the conduction band and crystal anisotropy between $xy$ and $z$ directions accordingly.

When the measurements are carried out at 77 K, a pressure induced phase transition is observed: Figure 6.33 shows absorption peak energies of $C_{10}$-$PbI_4$ at 77 K under hydrostatic pressure [18]. At ambient pressure, the lowest exciton is located at 2.57 eV and the step like feature is observed 320 meV above the lowest exciton. As the pressure is applied, these features shift together and reveal a discontinuous low energy shift of 0.2 eV at about 0.4 GPa. They red-shift above the critical pressure. This fact clearly shows the same electronic origin for the wave functions of the lowest exciton and the step structure: They are different only in terms of the relative motion.

### *6.5.7. Raman Scattering and Phase Transition*

#### *6.5.7.1. Raman scattering*

Raman scattering spectra give information on the microscopic symmetry of the crystal. Figure 6.34 shows non-resonant Raman scattering spectra of $C_{10}$-$PbI_4$ at several temperatures [8]. It is clearly seen that the crystal has at least four structural phases. The Raman shift is plotted in Fig. 6.35. Stronger peaks are plotted in larger marks in the figure. The peaks shift discontinuously at the structural phase transitions. In the phase IV(low temperature monoclinic phase), 25 $cm^{-1}$ and 52 $cm^{-1}$ are the dominant

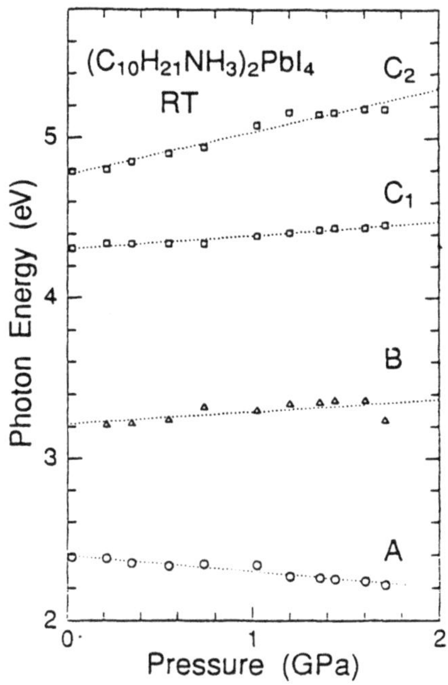


Figure 6.32. Absorption peak energies of $C_{10}$-$PbI_4$ under hydrostatic pressure. (After [18]).

lines. Because these lines shift to the higher energy side for $C_{10}$-$PbX_4$(X=Br, Cl), they are assigned to the vibration of a $[PbX_6]^{4-}$ octahedron. In the phase III (low temperature orthorhombic phase), a 48 $cm^{-1}$ line is found instead of the 52 $cm^{-1}$ line. In the phase II (room temperature orthorhombic phase) a 20 $cm^{-1}$ line is seen instead of the 25 $cm^{-1}$ line. In the phase I (high temperature tetragonal phase), a broad band is observed at 25 $cm^{-1}$, corresponding to the melt of the decylammonium chains.

#### *6.5.7.2. Structural phase transition and electronic states*

As was shown in Sec. 6.4, in most of $C_n$-$PbI_4$ the lowest exciton peak position shifts to the blue abruptly as the temperature decreases. We have seen that the transition temperature changes as a function of $n$, the number of carbons in the alkylammonium chain. As was shown in Fig. 6.15, the transition temperatures show a V shape while there is no phase transition for $n = 6$. For $n > 6$, the transition temperature increases very similarly to the melting temperature of corresponding amines. It clearly shows that the phase transition is related to the motion of the chain. On the other hand, as we have already discussed, the excitonic structure is mainly determined by the quantum well. As is known for structural phase transitions in other layered

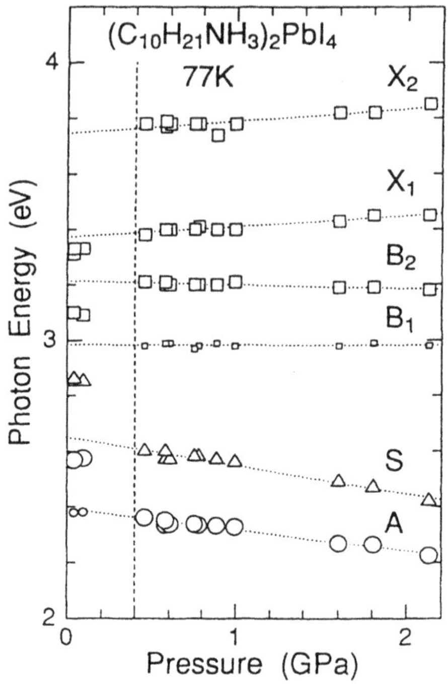


Figure 6.33. Absorption peak energies of $C_{10}$-$PbI_4$ at 77 K under hydrostatic pressure. (After [18]).

perovskites, the motion of the alkylammonium triggers a structural distortion of the two-dimensional network of $[PbI_6]^{4-}$ octahedra [1]. From the structural analysis in each phase, it is suggested that the YO transition in $C_{10}$-$PbI_4$ is caused by the wiggling of the 2D layer [17]. Note that the wiggling brings a perturbative anisotropy in the *ab*-plane of the $[PbI_6]^{4-}$ 2D plane. As a result, the B band splits into two subbands, while the A band exciton is essentially the same (with some change of the oscillator strength).

### 6.5.8. *Nonlinear Optical Properties*

#### 6.5.8.1. *Third harmonic generation*

Third harmonic generation of light has been demonstrated by Xu *et al.* [50]. By using a $LiNbO_3$ difference-frequency generator with a Q-switched Nd:YAG laser and a tunable dye laser, they measured third harmonic intensity of $C_{10}$-$PbI_4$ spin-coated film compared to quartz film. The magnitude of $\chi^{(3)}$ reaches $10^{-9}$ esu, which is comparable to the largest reported value on organic materials with one-dimensional conjugated $\pi$-electron structures.

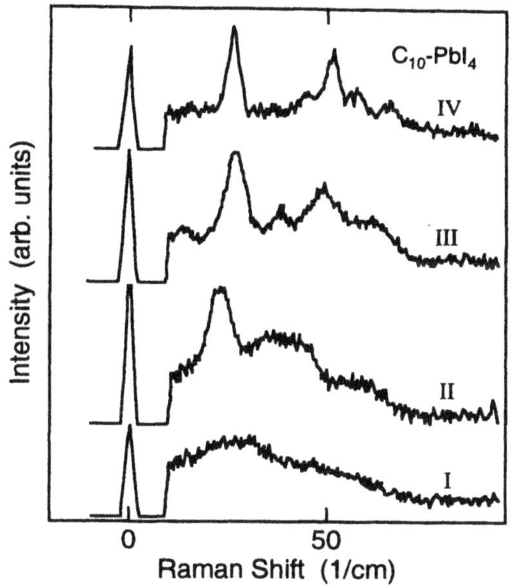


Figure 6.34. Raman spectra of $C_{10}$-$PbI_4$ for different structure phases.

### *6.5.8.2. High density optical excitation*

Under high-density optical excitation (about 10 kW/cm$^2$, with 10 ns pulse width) above the bandgap, a new luminescence band shows up 50 meV below the free exciton luminescence of $C_{10}$-$PbI_4$ at 4.2K, which grows as a square of the excitation intensity [51]. Time resolved luminescence spectroscopy reveals that the decay time is less than 100 ps. These observations suggests that the band is due to the radiative recombination of a biexciton leaving an exciton behind. Hence the binding energy of the biexciton state is 50 meV, which is enormously large compared to the values for conventional semiconductors, a few meV at most. The ratio of binding energies of a biexciton to an exciton is 50 meV / 320 meV = 0.16, which is plausible in 2D system [52].

Figure 6.36 shows the PL spectra of PhE-$PbI_4$ at 1.6 K excited by pulse laser with 10 ns with various intensities [53]. A band at about 2.31 eV, which is not observed with a weak light source, grows superlinearly and dominates at the highest excitation. Assuming it as the biexciton recombination, one can estimate the biexciton binding energy to be about 40 meV. The smaller biexciton binding energy compared to $C_{10}$-$PbI_4$ corresponds to the smaller exciton energy in PhE-$PbI_4$, where weaker dielectric effect is expected. As for the fine structures observed in the spectra, further investigation is required in the future.

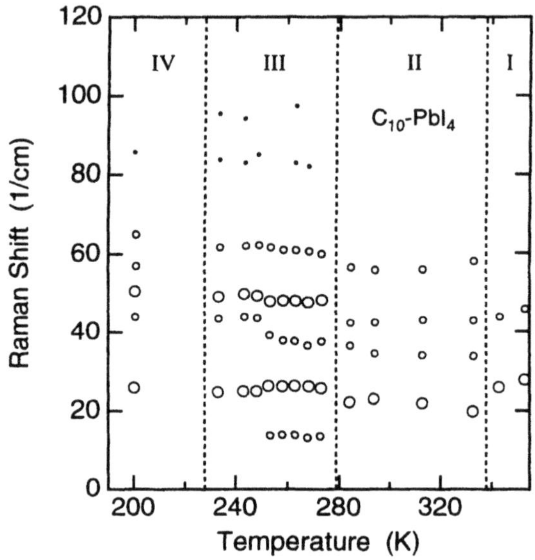


Figure 6.35. Raman shift of $C_{10}$-$PbI_4$ as a function of temperature.

*6.5.8.3. Two-photon absorption*

By applying laser pulses with photon energy $\hbar\omega > E_{ex}/2$, luminescence from the lowest exciton state is observed [55]. It may be ascribed to two-photon excitation of the lowest exciton state. Since the symmetry does not allow two-photon absorption to the exciton state, an odd-parity phonon is believed to be involved to the process. Further investigation is necessary to identify the mechanism.

*6.5.9. Transport and Electroluminescence*

In general, alkylammonium leadhalide crystals are good insulators. So far there have been no reports on doping to control carriers. By shining a little light, Hong *et al.* [54] measured photoconductivity in PhE-$Pb_2I_7$ as a function of electric field and showed the anisotropy of the electronic band structure. They observed much larger photocurrent along the well and a plateau for current through the wells. The plateau is ascribed to Wannier localization, from which the miniband width was calculated to be about 20 meV. Furthermore they observed electroluminescence (EL) for a single crystal of PhE-$Pb_2I_7$ with silver paint electrodes below 200 K. The current and luminescence was nonlinear to the applied field (they increase as $F^n$ with $n = 7 \sim 11$ and fluctuates). When the current increased to approximately 100 nA, EL from free exciton recombination was easily detected visually in a darkened room. As

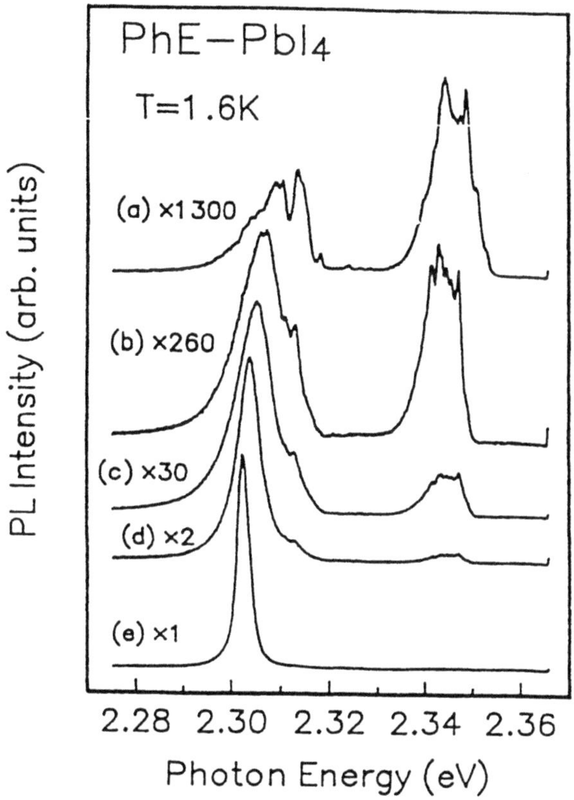


Figure 6.36. PL spectra of PhE-PbI$_4$ at 1.6 K under high density excitation, 0.3%, 2%, 15%, 40%, and 100% (about 100 kW/cm$^2$). (After [53]).

for PhE-PbI$_4$ single crystals, such EL was observed at a field of $F > 25$ kV/cm at 77 K. The current fluctuation and the temperature dependent threshold suggest an avalanche mechanism in this case.

Recently Era *et al.* [56] reported bright EL out of PhE-PbI$_4$ with a multi-layer EL devices. On account of the optimized organic electrodes, the EL was 10000 cd/m$^2$ at 2 A/cm$^2$ at 77 K, comparable to the brightest multilayer organic EL films. By utilizing the enormous variety of the compounds, it may be possible to optimize the conditions for the operation at room temperature in the near future.

## 6.6. CONCLUDING REMARKS

The inorganic-organic perovskites, such as alkylammonium leadiodides have an enormous flexibility in the structure. The electronic structure in the visible region is characterized by the inorganic part, which can be tailored by utilizing the quantum confinement effect, while the organic part modifies the dielectric environment. These materials can be grown as single crystals and/or film with excellent quality.

Explore the exotic quantum wells
Wires and dots as well
In order to study physics in low dimension
And to make them serve us well.

*All's well that ends well.* [58]

The author would like to express his best gratitude to Prof. Masamichi Yamanishi and Dr. Igal Brener for reading the manuscript critically and giving precious comments.

## REFERENCES

[1] R. Kind, *Ferroelectrics* **24**, 81 (1980).

[2] H. Arend, W. Huber, F. H. Mischgofsky, and G. K. Richter-van Leeuwen, *J. Cryst. Growth* **43**, 213 (1978).

[3] L. J. de Jongh, A. C. Botterman, F. R. de Boer, and A. R. Miedema, *J. Appl. Phys.* **40**, 1363 (1969).

[4] P. Day, *Phil. Trans. R. Soc. London* **A314**, 145 (1985).

[5] N. Watanabe, N. Kojima, T. Ban, and I. Tsujikawa, *J. Phys. C* **21**, 4795 (1988).

[6] T. Yoshinari, T. Namba, S. Shimanuki, M. Fujisawa, T. Matsuyama, M. Ikezawa, and K. Aoyagi, *J. Phys. Soc. Jpn.* **58**, 2276 (1989).

[7] Yu. I. Dolzhenko, T. Inabe, and Y. Maruyama, *Bull. Chem. Soc. Jpn.* **59**, 563 (1986).

[8] T. Ishihara, J. Takahashi, and T. Goto, *Solid State Commun.* **69**, 933 (1989).

[9] J. Takahashi, *Master Thesis*, Tohoku University (1989).

[10] J. Calabrese, N. L. Jones, R. L. Harlow, N. Herron, D. L. Thorn, and Y. Wang, *J. Am. Chem. Soc.* **113**, 2328 (1991).

[11] S. S. Nagapetyan, Yu. I. Dolzhenko, E. R. Arakelova, V. M. Koshkin, Yu. T. Struchkov, and V. E. Shklover, *Russian J. Inorg. Chem.* **33**, 1614 (1988).

[12] T. Kondo, H. Ogawa, Y. Sato, and R. Ito, *Report of Intense X-ray Lab, Univ. Tokyo*, (1989) [in Japanese].

[13] D. Weber, *Z. Naturf.* **33b**, 1443 (1978).

[14] G. V. Gridnova, E. A. Ziger, V. M. Koshkin, S. V. Lindeman, Yu. T. Struchkov, and V. E. Shklover, *Daklady Akademii Nauk* **278**, 414 (1984).

[15] B. R. Vincent, K. N. Robertson, T. S. Cameron, and O. Knop, *Can. J. Chem.* **65**, 1042 (1987).

[16] H. K. Henisch, *Crystal Growth in Gels* (The Pennsylvania State University Press, 1970).

[17] R. Ito, C. Xu, and T. Kondo, in *Solid State Materials*, ed. by S. Radhakrishna and A. Daud (Narosa Publishing, House, New Delhi, 1991).

[18] M. Hirasawa, *Ph.D. Thesis*, Tohoku University (1994).

[19] Y. Takahashi, T. Fukuzawa, K. Kumata, S. S. Kano, F. Schellenberg, V. Lee, Y. Shiraki, and R. Ito, *Extended Abstracts of the Japan Society of Applied Physics, 1990*, **3**, 1063 (1990).

[20] T. Ishihara, J. Takahashi, and T. Goto, *Phy. Rev. B* **42**, 11099 (1990).

[21] T. Kataoka, T. Kondo, and R. Ito, *Phys. Rev. B* **47**, 2010 (1993).

[22] T. Kataoka, T. Kondo, R. Ito, S. Sasaki, K. Uchida, and N. Miura, *Physica B* **201**, 423 (1994).

[23] X. Hong, T. Ishihara, and A. V. Nurmikko, *Phys. Rev. B* **45**, 6961 (1992).

[24] K. Heidrich, W. Schäfer, M. Schreiber, J. Söchtig, G. Trendel, J. Treusch, T. Grandke, and H. J. Stolz, *Phys. Rev. B* **24**, 5642 (1981).

[25] M. Hirasawa, T. Ishihara, T. Goto, K. Uchida, and N. Miura, *Physica B* **201**, 427 (1994).

[26] T. Tsuboi, *phys. stat. sol. (b)* **96**, 321 (1979).

[27] G. C. Papavassiliou and I. B. Kouteslas, *Synth. Metals* (in print).

[28] G. C. Papavassiliou, A. P. Patsis, D. J. Lagouvardos, and I. B. Koutselas, *Synth. Metals* **55–57**, 3889 (1993).

[29] E. A. Muljarov, N. A. Gippius, A. V. Kvit, G. Ya. Zueva, G. N. Mikhailova, A. M. Prokhorov, and S. G. Tikhodeev, *2nd Int. Conf. Nanostructures.* (St. Petersburg, 1994).

[30] A. Nagami, K. Okamura, T. Kuitani, and T. Ishihara, (in print).

[31] G. C. Papavassiliou, I. B. Koutselas, A. Terzis, and M. -H. Whangbo, *Solid State Commun.* **91**, 695 (1994).

[32] D. B. Mitzi, C. A. Feild, W. T. A. Harrison, and A. M. Guloy, *Nature* **369**, 467 (1995).

[33] G. Harbeke and E. Tosatti, *Phys. Rev. Lett.* **28**, 1567 (1972).

[34] T. Ishihara, M. Hirasawa, and T. Goto, *Jpn. J. Appl. Phys. Suppl.* **34-1**, 71 (1995).

[35] T.Ishihara, *J. Lumin.* **60&61**, 269 (1994).

[36] M. Hirasawa, T. Ishihara, and T. Goto, *J. Phys. Soc. Jpn.* **63**, 3870 (1994).

[37] M. Hirasawa, T. Ishihara, T. Goto, S. Sasaki, K. Uchida, and N. Miura, *Solid State Commun.* **86**, 479 (1993).

[38] T. Kataoka, T. Kondo, R. Ito, S. Sasaki, K. Uchida, and N. Miura, *Physica B* **184**, 132 (1993).

[39] M. Shinada and S. Sugano, *J. Phys. Soc. Jpn.* **21**, 1936 (1966).

[40] L. V. Keldysh, *Pis'ma Zh. Eks. Teor. Fiz.* **29**, 716 (1979).

[41] E. Hanamura, N. Nagaosa, M. Kumagai, and T. Takagahara, *Mater. Sci. Eng. B* **1**, 255 (1988).

[42] M. Kumagai and T. Takagahara, *Phys. Rev. B* **40**, 12359 (1990).

[43] L. Wendler and B. Hartwig, *J. Phys. Matter* **3**, 9907 (1991).

[44] D. B. Tran Thoai, R. Zimmermann, M. Grundmann, and D. Bimberg, *Phys. Rev. B* **42**, 5906 (1990).

[45] E. A. Muljarov, S. G. Tikhodeev, N. A. Gippius, and T. Ishihara, *Phys. Rev. B* (in print).

[46] X. Hong, T. Ishihara, and A. V. Nurmikko, *Intern. Conf. Phys. Semiconductors*, 1152 (1992).

[47] M. Kuwata, T. Kuga, H. Akiyama, T. Hirano, and M. Matsuoka, *Phys. Rev. Lett.* **61**, 1226 (1988).

[48] T. Goto, T. Takahashi, and M. Ueta, *J. Phys. Soc. Jpn.* **24**, 314 (1968).

[49] T. Ishihara, (unpublished).

[50] C. Xu, T. Kondo, H. Sakakura, K. Kumata, Y. Takahashi, and R. Ito, *Solid State Commun.* **79**, 245 (1991).

[51] T. Ishihara, X. Hong, and A. V. Nurmikko, *Surf. Sci.* **267**, 323 (1992).

[52] D. A. Kleinman, *Phys. Rev. B* **15**, 871 (1983).

[53] X. Hong, *Ph.D. Thesis*, Brown University (1993).

[54] X. Hong, T. Ishihara, and A. V. Nurmikko, *Solid State Commun.* **84**, 657 (1992).

[55] K. Okamoto, H. Ohmaki, T. Ishihara, and M. Yamanishi, *Abstracts of the Meeting of the Physical Society of Japan, Sectional Meeting Pt. 2*, 241 (1994).

[56] M. Era, S. Morimoto, T. Tsutsui, and S. Saito, *Appl. Phys. Lett.* **65**, 676 (1994).

[57] K. Okamura, *Master Thesis*, Hiroshima University (1995).

[58] W. Shakespeare.

# CHAPTER 7

# SOLID STATE PROPERTIES OF $C_{60}$ AND ITS RELATED MATERIALS

Yoshihiro Iwasa

*Japan Advanced Institute of Science and Technology*



## 7.1. INTRODUCTION

The cagelike carbon cluster, fullerene, has achieved wide interdisciplinary interest in materials science. $C_{60}$ first appeared in the laser desorption mass spectroscopy of carbon in 1985 [1]. Although calculations of many physical properties were made, it had not been possible to produce sufficient quantities of the material to permit measurements of physical properties to test the theoretical calculations or to evaluate the possible applications. The historical discovery of mass production of fullerenes by Krätchmer and Huffman in 1990 [2] opened the field to anyone who is interested in this class of materials. Macroscopic amount of $C_{60}$ and higher fullerenes such as $C_{70}$

and $C_{76}$ can now be made with an arc furnace, which consists of a vacuum chamber and a power supply for an arc welding.

Fullerene is the third solid form of pure carbon following diamond and graphite. Here, we call a group of fullerene based materials the fullerene family. This family increases year after year since 1990. One of the reasons for this increase is that the cage-like carbons have both sufficient stability and considerable chemical activity. One of the classification of fullerenes is given as follows:

(1) fullerene ($C_{60}$ and higher fullerenes),
(2) host-guest compounds (intercalation compounds),
(3) exohedral fullerenes,
(4) endohedral fullerenes,
(5) carbon nanotubes,
(6) carbon nanocapsules,
(7) heterofullerenes,
(8) fullerene polymer.

The first group (1) is the so-called fullerene which has an empty hollow inside the cage. This class involves not only $C_{60}$ and $C_{70}$ but also higher fullerenes such as $C_{76}$, $C_{82}$, and $C_{84}$,... The second group (2) is the compounds composed of fullerenes and other molecules or elements, such as alkali metal intercalated $C_{60}$. (3) is the exohedral fullerenes, in which foreign atoms or molecules are covalently bonded on the outside of the carbon cage. These are synthesized by organic chemical routes. The fourth endohedral fullerenes (4) contain one or a few metal atoms inside the carbon cage, for example La@$C_{82}$ where @ indicates that La is inside the cage. This compound is synthesized by an arc discharge of metal containing carbon rods. (5) is the carbon nanotubes which are graphite tubes with diameters of nanometer dimensions. They are synthesized by similar techniques to that for fullerene synthesis. (6) is the multilayered graphite cages with typically several tens of nanometers in size. The cage is air-tight and single-domain crystals of metal carbides such as $LaC_2$ are usually entrapped inside. (7) is the fullerene cages in which some carbon atoms are exchanged with boron or nitrogen atoms. (8) is a new type of $C_{60}$ solids, in which the molecules are chemically bonded forming covalent solids rather than usual molecular crystals.

Amongst various kinds of fullerenes, $C_{60}$ is quite exceptional, because of three reasons. First of all, $C_{60}$ is fascinating and beautiful because of its high symmetry. 12 pentagons and 20 hexagons are arrayed in a soccer-ball-like structure (Fig. 7.1) that belongs to the icosahedral point group, $I_h$. Fullerene molecules with the same symmetry has never been synthesized yet. The second reason is that the abundance of $C_{60}$ is extremely larger than that of the other fullerenes. This makes the research of $C_{60}$ much easier. The third is that $C_{60}$ displays a rich variety of solid state properties in intercalation compounds (group 2), for example, superconductivity [3] and ferromagnetism [4]. The third point invites special attention to fullerene compounds by chemists, physicists, materials scientists.

The diameter of $C_{60}$ is 7.1 Å. One of characteristic features of the soccer ball is that pentagons are isolated from each other by hexagons. It is believed that the

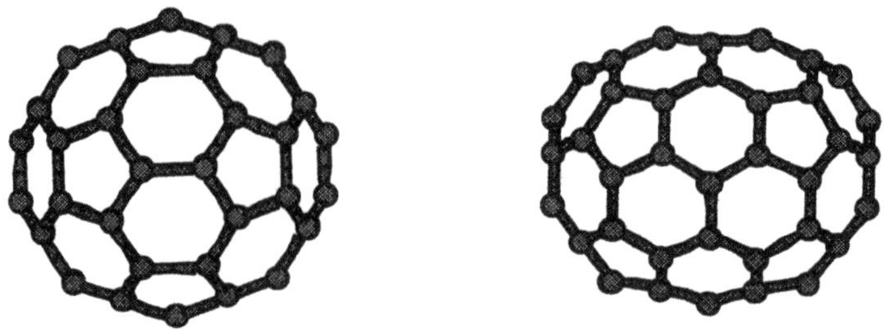

Figure 7.1. $C_{60}$ (left) and $C_{70}$ (right) molecules.

high stability of fullerene molecules is caused by the topological constraint that the less stable pentagons are isolated from each other by more stable hexagons. This constraint is called "isolated pentagon rule" (IPR). All found fullerene molecules satisfy the IPR. It is noted that $C_{60}$ is the smallest cage-like molecule that satisfies the IPR and that $C_{70}$ is the second smallest. Although $C_{60}$ molecules consist of sixty carbon atoms, there exist only two kinds of carbon-carbon bonds: That is sixty C–C single bonds and thirty C=C double bonds. The C-atoms in the pentagonal rings are connected by single bonds (C–C, 1.455 Å) and the C-atoms shared by adjacent hexagons are connected by double bonds (C=C, 1.391 Å). All carbon atoms are equivalent in $C_{60}$, while there are several inequivalent carbon sites in higher fullerenes. Thus higher fullerenes contain many kinds of carbon-carbon bonds.

$C_{60}$ is attracting much attention not only due to its beautiful and unique molecular structure, but also due to its chemical possibilities. A variety of derivatives has been synthesized starting from $C_{60}$ [5]. In the classification presented in the first page of this chapter, the group (2), (3) and (8) are chemical derivatives from $C_{60}$. In the present article, I shall be concerned with the groups (1), (2) and (8). I will review our work on the structural and electronic properties of solid $C_{60}$ and its derivatives; alkali-intercalated, and polymerized $C_{60}$'s.

$C_{60}$ molecules form a close-packed solid with a face-centered cubic (fcc) structure due to their spherical shape [6]. Since the center-to-center distance between neighboring molecules is about 10 Å and the van der Waals radius of carbon is 1.70 Å, the $C_{60}$ solid is regarded as a molecular (van der Waals) crystal in which the molecular characters are well maintained even in a solid form. In Sec. 7.2, I will focus on the optical spectroscopy for solid and molecular $C_{60}$, and explain unique optical properties of $C_{60}$.

Research into solid fullerenes was accelerated considerably by the discovery of conductivity [7] and superconductivity [3] in alkali intercalated $C_{60}$. Superconducting transition temperatures of $K_3C_{60}$ [3, 8] and $Rb_3C_{60}$ [9] are 19 K and 29 K, respectively. The highest $T_c$ reached 33 K in a binary compound, $Cs_2RbC_{60}$ [10]. An important constraint for the occurrence of superconductivity is the chemical composition. Although many kinds of alkali intercalated $C_{60}$ compounds, $A_xC_{60}$, have been synthesized, the superconductivity is observed only in materials with an essential chemical formula of $A_3C_{60}$. Here A denotes alkali metals. Intercalation of alkali metals into $C_{60}$ induces a dramatic change in the electronic structures of these compounds. Optical spectroscopy is a useful means to investigate the electronic structures. In Sec. 7.3, I present optical data for $A_xC_{60}$, and discuss on the variation of electronic structures with changing the x value.

Both doped and undoped $C_{60}$ are essentially molecular crystals in which molecules interact with each other very weakly. However, different kinds of $C_{60}$ solids were found recently. The fcc form of undoped $C_{60}$ is rather stable in an inert and dark condition. However, light-irradiation induces a formation of chemical bonds between $C_{60}$ molecules. Later on, similar chemical bonds were found to be formed by high pressure and temperature. Polymerization also takes place in some alkali intercalated $C_{60}$ compounds. These findings indicate that the connected $C_{60}$ is another metastable phase of carbon solids. In Sec. 7.4, I describe three kinds of polymerized forms of $C_{60}$ with emphasis on pressure-induced polymerization.

## 7.2. OPTICAL PROPERTIES OF UNDOPED FULLERENES

In this section, I review spectroscopic properties of fullerenes, and discuss about characteristic features of fullerenes. Fullerene molecules are composed of double and single bonds of carbons, indicating that $\pi$ electrons plays a leading role in the low energy electronic properties. Comparison of spectroscopic data for several allotropes of solid carbon is helpful for a survey of the electronic structures of fullerenes.

Figure 7.2 shows optical reflection spectra of diamond, graphite, $C_{60}$, and $C_{70}$ in the range of 0.1–35 eV, covering all the transitions of valence electrons (2s and 2p of carbon). Single crystals of $C_{60}$ and $C_{70}$, were grown by sublimation of purified and dried powders. The reflection measurements have been made at room temperature using a synchrotron radiation source at the Institute for Solid State Physics, University of Tokyo.

The crystal structure of $C_{60}$ at room temperature is fcc in which molecules spin freely. The free rotation of molecules is frozen at about 250 K accompanied with a structural transition to simple cubic in which molecules are orientationally ordered [11]. The spectra were taken at room temperature for the as-grown (111) surface which has an excellent quality for the optical reflection measurements. The surface dimension was $1 \times 2$ mm$^2$ being sufficient for the accurate determination of reflectivity even in the high energy region.

$C_{70}$ is the second abundant fullerene. The structure of $C_{70}$ is obtained from $C_{60}$ by the addition of a ring of hexagons around the equator (Fig. 7.1). It is no longer spherical, but has a slightly elongated rugby-ball shape; the molecular symmetry is $D_{5h}$. Due to the reduced symmetry and elongated form, the orientational transition and crystal phases are more complicated in $C_{70}$. Fischer and his coworkers reported that although as-prepared $C_{70}$ generally had hcp (hexagonal-close-packed) or modified hcp structures, purified, dry, and annealed $C_{70}$ generally had fcc or modified fcc structures [12]. The hcp and fcc type structures differ in the stacking sequence of close-packed triangular planes: ABAB in the former case and ABCABC in the latter. In the fcc-type stacking, the orientational ordering transition occurs in two stages. Above 345 K, the structure is fcc with complete orientational disorder (free rotation) of molecules. Between 345 K and 295 K, only the orientation of the long axis is ordered, and disorder persists about this axis. The lattice is distorted into rhombohedral. This structure is obtained by the elongation of the fcc lattice along a unique $[111]_{fcc}$ direction, along which the long axis of $C_{70}$ molecules are aligned. Below 295 K, the rhombohedral lattice undergoes a further distortion, resulting in a monoclinic structure in which molecules are presumed to be essentially static. The structural properties of $C_{70}$ is still in controversy because of the poor crystallinity that results mainly from two reasons: One is the stable supercooled fcc phase and the other is the contamination of the other metastable phases such as the hcp phase [13].

The single crystal of $C_{70}$ for the reflection measurements were grown by a sublimation following a careful purification and a drying procedure. The structure of $C_{70}$ crystal was hcp, in contradiction to the statement of [12]. The crystallinity of $C_{70}$ used for the reflection measurement was not as good as that of $C_{60}$. The reflectivity was measured for the as-grown (101) which was good in optical quality but small in size ($0.5 \times 1$ mm$^2$). The small size slightly reduces the reliability of the absolute reflectivity in the vacuum ultraviolet region.

Diamond displays a single broad reflection band with three Van Hove singularities. Since there are no $\pi$ electrons in diamond, all the observed excitations are attributed to the $\sigma$–$\sigma^*$ interband transitions. The spectra suggest that the $\sigma$ electrons form a broad band. In sharp contrast, the reflection spectra of the other three forms are separated into two bands. In general, the first band below 8 eV is attributed to a $\pi$–$\pi^*$ transition and the second band above 10 eV is the $\sigma$–$\sigma^*$ transition. Since the $\pi$ electrons are weakly bound in solids, the transition energies are smaller than that of $\sigma$ electrons. In the case of fullerenes, since the $\pi$ and $\sigma$ electrons are mixed due to their nonplaner shapes, the structures above 8 eV involve the $\sigma$–$\pi^*$ and $\pi$–$\sigma^*$ transitions as well. These spectra are consistent with a simple picture that graphite and fullerenes are $\pi$-electron systems.

However, the low energy part of the spectra (below 8 eV) for graphite and fullerenes are significantly different. In graphite the reflectivity increases with decreasing photon energy below 2 eV. This Drude-like feature agrees with the semimetallic nature of graphite. The reflection spectra shows a single peak at 5 eV that is attributed to the $\pi$–$\pi^*$ interband transition. The single broad peak suggests that a broad $\pi$ band is formed in graphite. On the other hand, the reflectivity for $C_{60}$ and $C_{70}$ decreases with photon energy at low energy region ($h\nu < 1$ eV), indicating that these materials are insulators. The spectra of the fullerenes display multi-peak-structures which are attributed to the intramolecular electronic transitions. The sharp peaks indicate the molecular character is well maintained in the fullerene solids that and the bandwidth is very narrow.

Small anisotropy was observed in the reflection spectra of $C_{70}$ in contrast to the isotropic spectra of $C_{60}$. The reflection spectra below 6 eV are shown in Fig. 7.3. The crystal structure of the $C_{70}$ sample measured is hcp, in which the long axes of

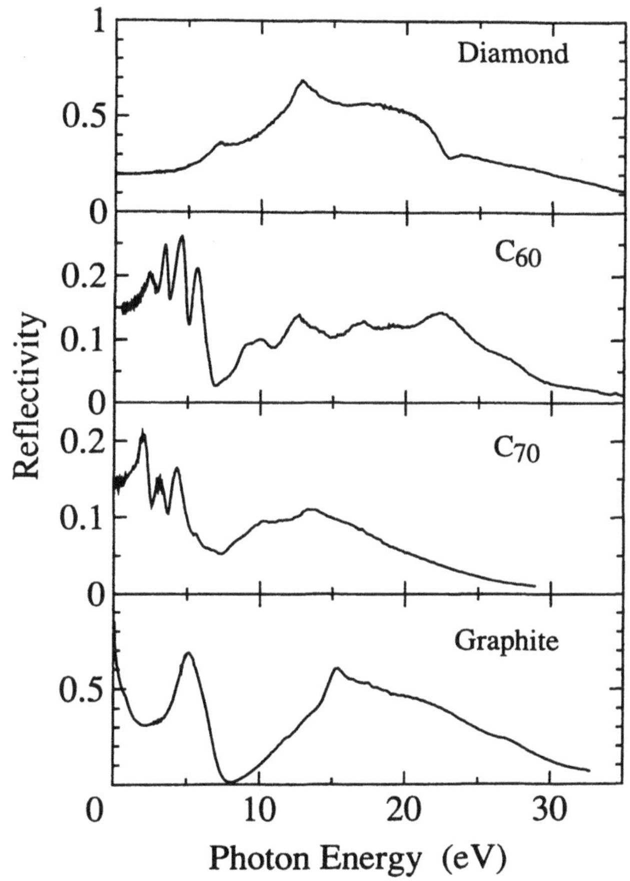


Figure 7.2. Reflectivity spectra of four carbon allotropes; diamond (top), $C_{60}$ (second), $C_{70}$ (third), and graphite (bottom). For $C_{70}$, the spectra is shown for E $\parallel$ a which is parallel to the short axis of molecule. The spectra for graphite is for the polarization E $\perp$ c which is parallel to the two-dimensional sheet.

molecules are aligned along the c-direction. The [100] direction is parallel to the short axis of the molecules in this structure. The E $\parallel$ and $\perp$ [100] polarizations correspond to the E $\parallel$ and $\perp$ the short axis of $C_{70}$, respectively. Figure 7.3 shows that the lowest absorption peak energy is smaller for E $\parallel$ the short axis. The lower transition energy along the short axis is possibly explained by the local electronic transition inside the ring of hexagons inserted around the equator. Since the $\pi$ electrons are more delocalized in hexagons than in pentagons, the transition energy becomes smaller for

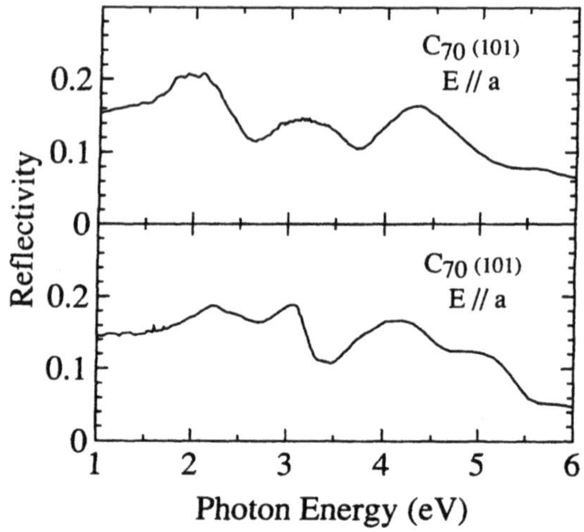


Figure 7.3. Anisotropic reflectivity spectra of hcp $C_{70}$ for the (101) plane. The top is for the polarization $\parallel$ a and the bottom is for $\perp$ a.

the polarization along the hexagon rings.

Let us discuss about precise features of the optical absorption spectra ($< 6$ eV) of fullerenes. Figure 7.4 shows the optical absorption spectra of molecular and solid forms of $C_{60}$. The former spectra was taken for the hexane solution and the latter was for evaporated thin films. The data represents the spectra for isolated molecules and condensed solids, respectively. The peak structures correspond to the ones observed in the low energy side in Fig. 7.2, and are assigned to intramolecular transitions of $C_{60}$. The absorption spectra for molecules and solids are very similar. Particularly the spectral features above 3 eV are almost identical except for the red shift and slight broadening of the peaks in solids. The similar optical spectra for molecules and solids indicate that the molecular character of $C_{60}$ is well maintained in the solid form. Accordingly, we interpret the optical absorption spectra for solid $C_{60}$ in terms of the intramolecular (Frenkel-type) excitons. The red shift and broadening in solids are characteristics of Frenkel excitons commonly observed in various molecular solids.

These transitions are compared with the molecular orbital calculations in Fig. 7.5 [14]. Since the peaks below 6 eV are attributed to $\pi$–$\pi^*$ transitions, only the $\pi$ orbitals are shown. Due to the high symmetry, molecular orbitals have high degeneracy. The icosahedral group is close to spherical symmetry and one can clearly recognize the parentage of the low-lying molecular orbitals in spherical harmonics ($Y_{LM}$ where $L$ is the orbital angular momentum quantum number). In particular, the S, P, and D ($L =$ $1, 2,$ and 3) spherical harmonics become $a_g$, $t_{1u}$, and $h_g$ representations in $I_h$. These

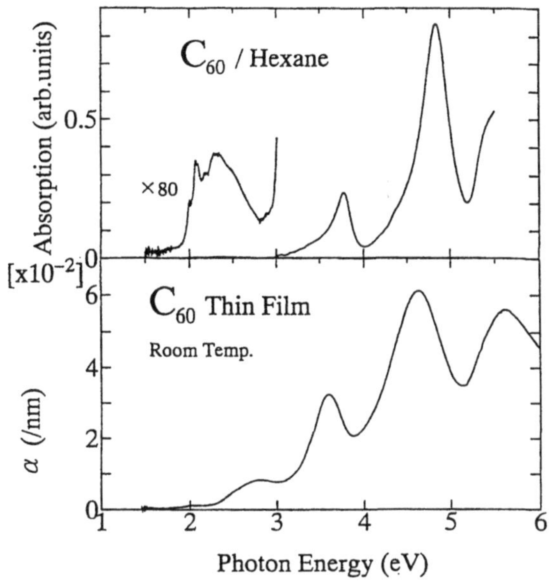


Figure 7.4. Absorption spectra of molecular (top) and solid (bottom) forms of $C_{60}$. They were taken for hexane solution and evaporated thin films, respectively.

three levels are seen at the bottom of Fig. 7.5. The $L = 3$ functions split into $t_{2u}$ and $g_u$. The two components ($g_g$ + $h_g$) of the $L = 4$ level remain accidentally degenerate. Above $L = 4$, the various components mix strongly, reflecting the $I_h$ symmetry. Sixty $\pi$ electrons occupy the levels up to $h_u$ which is the highest occupied molecular orbital (HOMO). The $t_{1u}$ is the lowest unoccupied molecular orbital (LUMO).

One of the characteristic features of the energy scheme (Fig. 7.5) is that the lowest optical transition from the HOMO to the LUMO is forbidden. Correspondingly, we observe an extremely weak absorption band peaking at about 2.3 eV in the molecular form of $C_{60}$ (upper panel of Fig. 7.4). This forbidden transition is sometimes activated by a coupling with the intramolecular vibrations. In fact, weak but definite phonon side band structures are observed in the peak. Thus, we tentatively assign the weak peak at 2.3 eV is the transition from the HOMO to LUMO activated by the coupling with intramolecular phonons. Accordingly, the first strong peak at 3.8 eV corresponds to the optical allowed transition from $h_u$ (HOMO) to $t_{1g}$ (the second LUMO). The peak at 4.8 eV is attributed to a superposition of the two transitions: that is ($g_g$+$h_g$ $\rightarrow$ $t_{1u}$) and ($h_u \rightarrow t_{1g}$).

In solid forms, a new peak appears at about 2.8 eV, which is slightly larger than the peak position (2.3 eV) in molecular forms. The weak absorption tail reaches down to the edge at around 1.7 eV. These features significantly differ from those usually

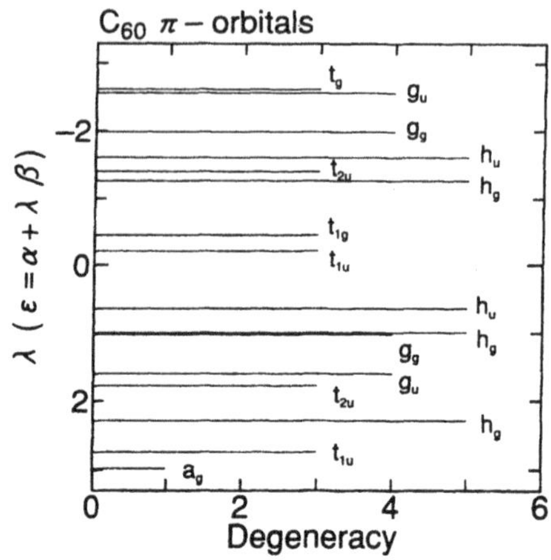


Figure 7.5. Hückel energy levels of $C_{60}$.

observed in molecular materials, such as $C_{70}$. The absorption spectra of $C_{70}$ are shown in Fig. 7.6 for molecular and solid forms. In this case, both spectra are very similar to each other except for the red shift and broadening of peaks in the film form in all photon energy region.

Unusual difference in absorption spectra of molecular and solid forms of $C_{60}$ is possibly attributed to the forbidden transition between the HOMO and LUMO. In solid state, the forbidden transition is weakly activated not only by the exciton-molecular vibration coupling but also by crystal fields. We attribute the weak absorption with the onset energy of 1.7 eV to the weakly allowed HOMO–LUMO transitions. The absorption band peaking at 2.8 eV that is absent in the molecular form is possibly attributed to another type of exciton called a charge transfer (CT) exciton. The CT exciton is schematically depicted in Fig. 7.7 with a Frenkel (molecular) exciton. In CT excitons, photogenerated electrons and holes exist separately on neighboring molecules. The CT exciton is characteristic of condensed matters and dominates the lowest electronic excitations in semiconductor or insulators. Even in molecular crystals, such as anthracene and phthalocyanines [15], CT excitons are known to play an important role particularly in the photoconductivity and third-order nonlinear optical properties.

The contribution of CT excitons is supported by theoretical calculations. Tsubo and Nasu simulated optical spectra of $C_{60}$ taking account of the exciton effects [16]. Figure 7.8 displays calculated optical absorption spectra of $C_{60}$ for molecular (a) and solid forms (b). The horizontal axis is scaled by the overlap integral ($T_0$) of $\pi$ electrons on the C–C bond. They estimated that $T_0 \sim 2.3$ eV. The calculation failed to reproduce the spectral shapes, but succeeded to explain the difference between the solid and molecular spectra. Since they neglected electron-molecular vibration cou-

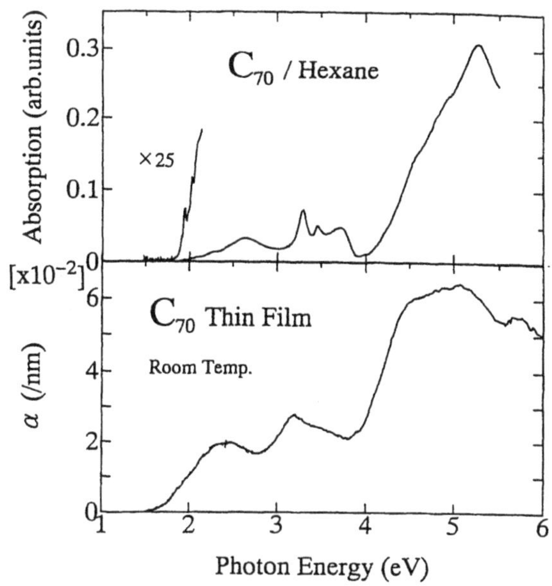


Figure 7.6. Absorption spectra of molecular (top) and solid (bottom) forms of $C_{70}$. They were taken for hexane solution and evaporated thin films, respectively.

Figure 7.7. Schematic drawings of CT-type (left) and Frenkel-type (right) excitons in $C_{60}$ solids.

pling, the HOMO–LUMO absorption is absent in the molecular form (a). In sharp contrast to (a), additional bands indicated by F1, F2, C1, and C2 appear in the low energy spectra of solid $C_{60}$ (b). F1 and F2 are attributed to the Frenkel exciton transitions between HOMO and LUMO, and C1 and C2 are the CT excitons. The formers become allowed by a combined effect of crystal fields and intermolecular interactions. Although the calculations need to be refined, the contribution of forbidden HOMO–LUMO transition and CT exciton is very clear in the low energy spectra of solid forms.

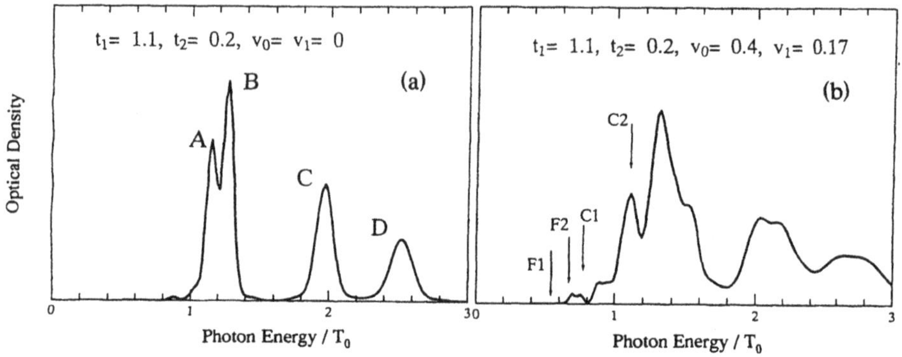


Figure 7.8. Calculated absorption spectra of molecular (a) and solid (b) forms of $C_{60}$. In the latter form, the charge transfer effect between neighboring molecules are included [16].

Although the CT exciton is known for many other molecular crystals, the role of CT excitons is enhanced in the low energy absorption spectra of $C_{60}$ because the HOMO–LUMO transition is parity-forbidden. The CT exciton plays another important role in the photoinduced polymerization of $C_{60}$ which is briefly explained in Sec. 7.4.

The relaxation path of the highly photoexcited species to the CT excitons competes with that to the lower lying molecular excitons. The latter states can be investigated by means of photoluminescence spectroscopy. Although the HOMO–LUMO intramolecular transition is forbidden in $C_{60}$, the luminescence is observed possibly due to the interaction with the intramolecular vibrations.

Photoluminescence spectra of fullerene solids are significantly affected by the crystallinity: For example, the spectra of single crystals and evaporated films are different. Fine structures are lost in the latter. The contamination of oxygen molecule results in a similar degradation even in single crystals. Figure 7.9 shows the luminescence spectra for $C_{70}$ single crystals at 2K. The excitation source was a 514.5 nm Ar-ion laser. The lower panel represents the spectra for the air-free crystal and the upper is for the crystal exposed to air. Air-free single crystal samples are mounted onto a sample holder inside the He-glove box just after taking out from the quartz tube for growth. Exposing to air smears out fine structures in the spectra. The same degradation effect by air was also observed in $C_{60}$ crystals. The broadening by air is possibly attributed to the energy transfer to the intercalated $O_2$ molecule. Assume that the width of the fine peaks in the spectra is determined by the lifetime of the lowest exciton state. An effective energy transfer to $O_2$ reduces the lifetime, resulting in a broadening of fine peaks. In fact it is reported that the energy transfer rate to $O_2$ is reported to be considerably high in toluene solutions of $C_{60}$ and $C_{70}$ [17]. Contamination of oxygen also results in a significant change in the transport properties

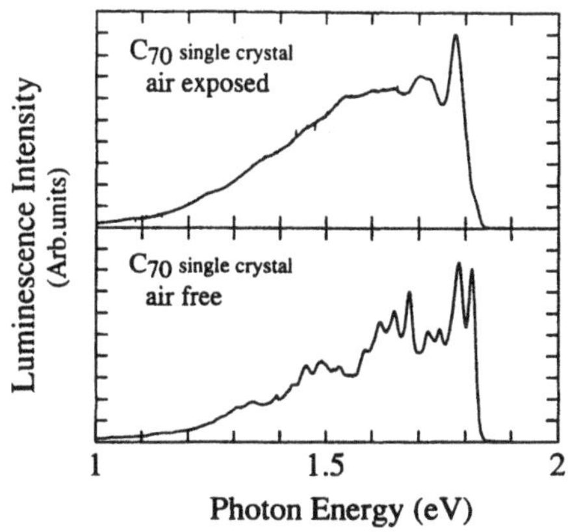


Figure 7.9. Photoluminescence spectra of $C_{70}$ single crystals at 2 K. The top panel displays the spectra for crystals exposed to air and the bottom is for the air-free crystals.

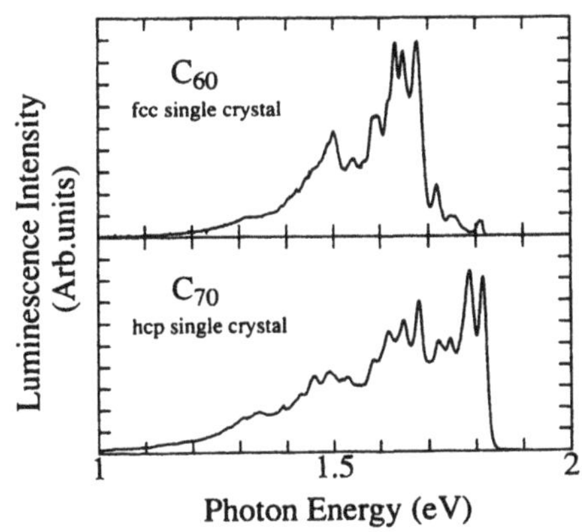


Figure 7.10. Photoluminescence spectra for air-free single crystals of $C_{60}$ (top) and $C_{70}$ (bottom) at 2 K.

on undoped $C_{60}$ single crystals [18].

Figure 7.10 shows the photoluminescence spectra for air-free crystals of $C_{60}$ and $C_{70}$. Both spectra are broad extending from 1.2 eV to 1.8 eV. Some periodic series of fine peaks are observed: They are assigned to 1440cm$^{-1}$ and 1300cm$^{-1}$ phonon side bands for $C_{60}$ and $C_{70}$, respectively. These high wavenumbers are attributed to tangential vibration of fullerene skeletons. These vibrations originate from the C–C or C=C stretching.

In Fig. 7.11, we show luminescence spectra at 2 K for molecular forms of $C_{60}$ and

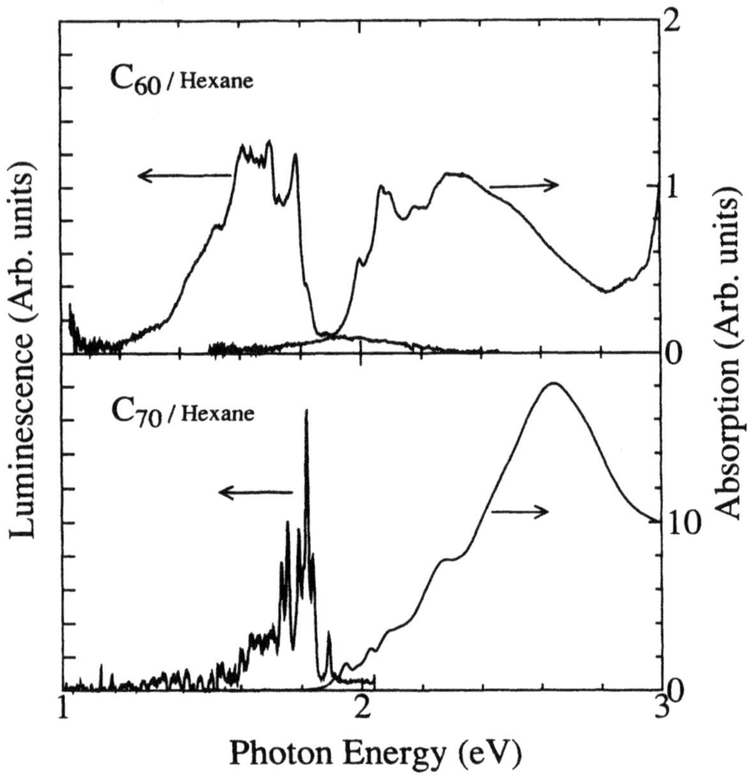


Figure 7.11. Luminescence and absorption spectra of molecular forms of $C_{60}$ (top) and $C_{70}$ (bottom).

$C_{70}$ dispersed in hexane matrix. Room temperature absorption spectra are superimposed. In contrast to the broad luminescence spectrum of $C_{70}$ solids, $C_{70}$ molecules display a relatively sharp spectrum. The mirror image relation between absorption and luminescence approximately holds in $C_{60}$, while does not in $C_{70}$. Such spectral shape differences may suggest that the interaction between excitons and intramolecular vibrations is stronger in $C_{60}$ than in $C_{70}$. We propose a Jahn-Teller-type exciton-molecular vibration coupling in $C_{60}$. Since the HOMO and LUMO are fivefold and triply degenerate, respectively, the lowest exciton states have high degeneracy as well. Such degenerate states are unstable to Jahn-Teller distortions. $C_{70}$ molecules does not have degeneracies in HOMO and LUMO, they are free from this type of interaction. The Jahn-Teller effect in $C_{60}$ solids is also discussed in the following section.

To summarize this section, fundamental optical spectra of pure $C_{60}$ and $C_{70}$ have been presented. The optical transitions in $C_{60}$ are classified into two bands; $\pi$–$\pi^*$ transition below 8 eV and $\pi$–$\sigma^*$, $\sigma$–$\pi^*$, and $\sigma$–$\sigma^*$ transitions above 8 eV. The sharp absorption peaks below 8 eV are attributed to the intramolecular transitions (Frenkel exciton). This is consistent with the crystal structure in which $C_{60}$ molecules are only weakly coupled with each other by Van der Waals forces. Anomalous absorption spectra below 3 eV are qualitatively explained taking account of the CT interaction between neighboring molecules. Small subpeaks in the luminescence spectra of $C_{60}$ and $C_{70}$ are smeared out by exposing to air. This degradation is possibly explained by the effective energy transfer to contaminated oxygen molecules. The luminescence spectra of molecular forms indicate that there exists a strong exciton-molecular vibration interaction particularly in $C_{60}$.

## 7.3. SPECTROSCOPIC STUDY OF ALKALI $C_{60}$ FULLERIDES

### *7.3.1. Introduction*

Intercalation of alkali or alkali-earth metals induces a dramatic reconstruction of electronic structures in fullerene solids. Evolution of electronic properties of intercalation compounds is one of the central topics in the fullerene research. The electronic structures of alkali metal (A) fullerides ($A_xC_{60}$) have been a subject of considerable interest because a high $T_c$ superconductivity has been discovered in this class of materials [3].

A number of theoretical and experimental studies have been made on these materials to clarify the electronic structures. Table 7.1 summarizes structurally identified $A_xC_{60}$'s. Various compounds are obtained for $x = 1, 2, 3, 4, 6$ and A=Na, K, Rb, and Cs. In most of these materials, the solubility limit is very narrow, forming line phase compounds rather than solid solutions. Although many compounds stay in fcc structure, the intercalation of large alkali ions sometimes induces structural transitions [19].

For fcc structure there are two kinds of interstitial sites. One is an octahedral site surrounded by an octahedron formed by eight $C_{60}$ molecules and the other is a tetrahedral site surrounded by an tetrahedron formed by four $C_{60}$ molecules. There are one octahedral site and two tetrahedral sites per $C_{60}$ molecule. The role of the two kinds of interstitial sites is very contrasting. The radius of octahedral and tetrahedral sites are 2.06 Å and 1.12 Å, respectively. Intercalation into the large octahedral site results in a lattice contraction due to the Coulomb interaction between metal cations and $C_{60}$ anions. On the other hands, intercalation into tetrahedral sites expands lattice. This is a good means to increase lattice constants of $A_3C_{60}$ and at the same time the onset temperature of superconductivity. However, intercalation of too many ions makes the fcc structure unstable.

For K, Rb, and Cs, there exists an $A_1C_{60}$ phase where alkali ions are intercalated into the octahedral site [20]. This structure is so called NaCl structure and is stable

Table 7.1. Structurally identified alkali metal intercalated compounds $A_xC_{60}$.

| $x$ | 1 | 2 | 3 | 4 | 5 | 6 |
|---|---|---|---|---|---|---|
| Na | fcc | fcc | fcc | | | fcc |
| K | fcc | | fcc | bct | | bcc |
| Rb | fcc | | fcc | bct | | bcc |
| Cs | fcc | | | bct | | bcc |

only at higher temperatures than about 100°C. Below this temperature, $C_{60}$ polymers are formed as shown in the following section. In contrast, Na ion is intercalated into the tetrahedral site due to its small ionic radius [21]. There exist solid solutions in the range of $1 < x < 3$ [22]. The $x = 3$ phase involves well-known superconductors $K_3C_{60}$, $Rb_3C_{60}$, and other ternary compounds designated as $A_xA'_{3-x}C_{60}$ [23, 24]. The crystal structure of $A_3C_{60}$ is fcc in which both octahedral and tetrahedral sites are occupied [25]. For alkali metals with large ionic radii than K, the $x = 4$ and 6 materials are in bct [26] and bcc [27] structures, respectively. For Na, even $x = 6$ and 11 compounds stay in fcc [28].

Basically most of the alkali intercalated $C_{60}$ compounds can be synthesized by a direct reaction of alkali metals and $C_{60}$ powders. Detailed procedures depend on materials and are shown in literatures. However, these compounds are extremely sensitive to air. Even brief exposure to air results in serious degradation of all physical properties. Samples should be maintained in an inert atmosphere at all times.

As to the electronic properties, the superconducting $A_3C_{60}$ has been a main target. Extensive experimental and theoretical studies show that both the normal and superconducting properties of $A_3C_{60}$ can be approximately understood by conventional theories in contrast to the high $T_c$ cuprates. At low temperatures most of the electronic properties seems to be explained in the framework of the Fermi liquid picture [29]. The superconducting properties seems to be explained by the BCS theory [30, 31, 32]. However, due to the its degenerate and narrow bands, high frequency phonons, and strong on-molecule Coulomb interactions, the quantitative understanding is left to be achieved.

In contrast to the superconducting $A_3C_{60}$, the nature of nonsuperconducting compounds such as $A_1C_{60}$, $A_2C_{60}$, and $A_4C_{60}$ are poorly understood. The molecular energy scheme of $C_{60}$ and the band structure of $K_3C_{60}$ are shown in Fig. 7.12 [33]. The LUMO forms a conduction band into which doped electrons are introduced [34]. Since the LUMO is triply degenerate, the conduction band can be occupied by six electrons. Therefore, according to the band picture, $A_xC_{60}$ compounds can be metallic for $x = 1, 2, 3, 4$, and 5. In fact band calculations predict that the compounds with intermediate x become at least metallic [35]. If materials are metallic, they might superconduct. However, in reality, the superconductivity occurs only in $A_3C_{60}$. Then a question arises why only $A_3C_{60}$ is superconducting. This question induces

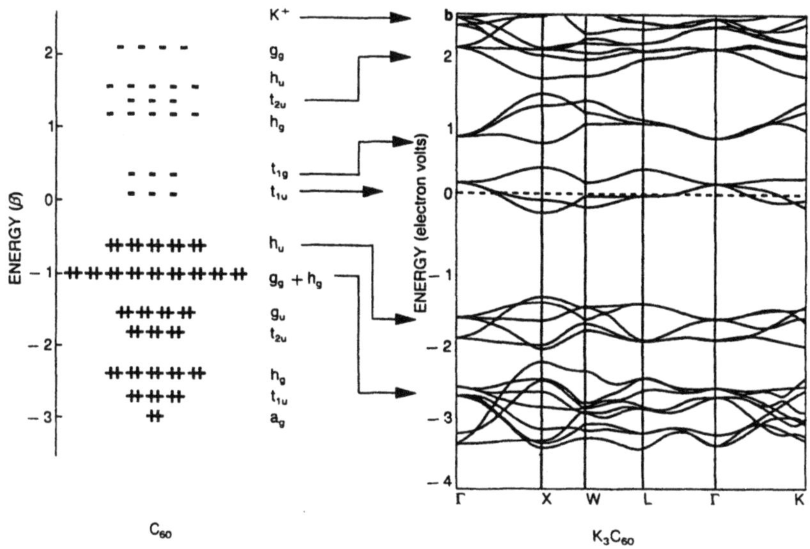


Figure 7.12. Relation between $C_{60}$ molecular levels (left) and band structure of $K_3C_{60}$ (right).

other questions such as what is the electronic states of $A_xC_{60}$ with $x$ other than 3. Therefore, the electronic properties of nonsuperconductors are still important issues even in a view point of superconductivity. In this section I review our recent results of optical spectra for $A_xC_{60}$ materials. These studies indicate that the simple band model which was approximately applicable to $A_3C_{60}$ is not necessarily correct for other $A_xC_{60}$ materials [37, 36].

### 7.3.2. *Optical Spectra*

$A_xC_{60}$ samples for optical measurements were grown by a cosublimation method in which alkali metal and $C_{60}$ are simultaneously vaporized [38]. The advantage of this method over *in situ* measurements on thin films is that the samples can be characterized both by x-ray diffraction and Raman spectra. An evacuated quartz tube is placed in a furnace having two temperature zones; a hot end (550–590°C) for the sublimation of $C_{60}$ and a cooler end (200–450°C) for the vaporization of alkali metals. Simultaneously vaporized $C_{60}$ and alkali metals react in a gas phase. After a few days reactions, flake-like samples were formed in the middle part of the tube. The $x = 4$ phase samples ($K_4C_{60}$ and $Rb_4C_{60}$) were usually obtained as a mixture with $x = 6$ phases. A deintercalation process was necessary to get pure samples of $A_4C_{60}$.

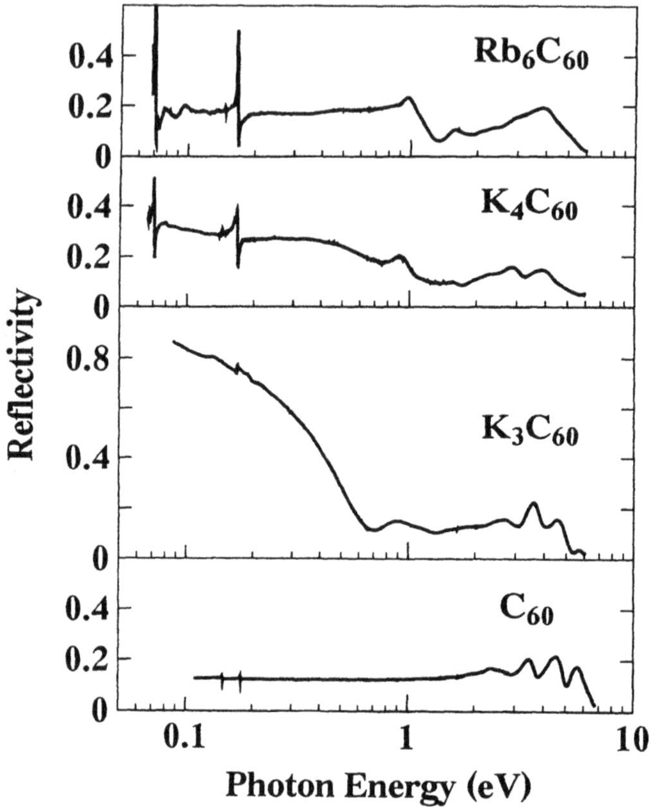


Figure 7.13. Reflectivity spectra of undoped $C_{60}$, $K_3C_{60}$, $K_4C_{60}$, and $Rb_6C_{60}$. (After [36]).

For a characterization of the cosublimated samples, we measured magnetization, x-ray diffraction patterns, and Raman spectra. Magnetic measurements were made on superconducting $K_3C_{60}$ and $Rb_3C_{60}$. The shielding fraction for both compounds reached more than 50%. The x-ray data confirmed that the samples were polycrystalline but in a single phase. A contamination of impurity phases was ten percent at most, which does not seriously affect the quality of the optical reflection data. We also measured Raman spectra of the pentagonal pinch $A_g(2)$ mode since this mode is known to be a very sensitive probe for the valence state of $C_{60}$ molecules [39]. In contrast to the x-ray measurements, the Raman spectra probes for the surface in the scale of the penetration depth of light. Accordingly, the Raman measurement is an

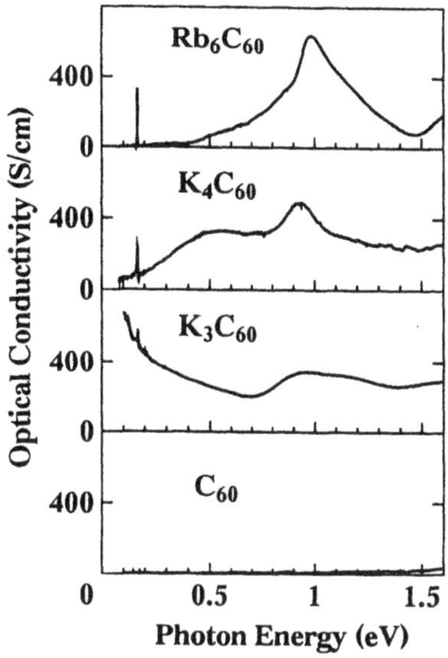


Figure 7.14. Infrared optical conductivity spectra for undoped $C_{60}$, $K_3C_{60}$, $K_4C_{60}$, and $Rb_6C_{60}$. (After [36]).

important characterization of the samples for reflection measurements. The single peak of $A_g(2)$ mode is another evidence for the absence of phase separation in the surface of cosublimated samples.

Optical reflection measurements were made at room temperature only for the samples which were confirmed to be in a single phase by these characterizations. The samples were sealed in a hand-made vacuum cell having an appropriate optical window in order to prevent the degradation by air. The sample surfaces were slightly round since it is a trace of the inner wall of the quartz tube. Therefore an optical microscope was used for the measurement of the optical reflection spectra. In the scale of the microscope, the sample surfaces were flat and shiny, showing sufficient optical qualities.

Figure 7.13 shows the reflection spectra of $C_{60}$, $K_3C_{60}$, $K_4C_{60}$, $Rb_6C_{60}$ [36]. The spectrum for $C_{60}$ was taken for the (111) as-grown surface of single crystals. The reflectivity in the mid-IR region is small for $C_{60}$ and $Rb_6C_{60}$, indicating that these materials are insulators. For $K_3C_{60}$ and $K_4C_{60}$, the reflectivity at low energy is relatively high, indicating that there exist excitations in the lower energy region. The small structures in the IR region in pristine $C_{60}$ are attributed to the intramolecular vibration modes which corresponds to the tangential motion of $C_{60}$ skeletons. One of the modes $[T_{1u}(4)]$ shows a drastic increase in intensity upon doping.

Figure 7.14 shows the conductivity spectra in the infrared region. The optical conductivity was obtained by the Kramers-Kronig transformation of reflectivity spectra. In the doped materials, the peak near 1 eV is commonly observed. According to the calculation of molecular orbital and band structure of $C_{60}$, this peak is assigned to the transition from the LUMO (lowest unoccupied molecular orbital, $t_{1u}$) to the next LUMO ($t_{1g}$).

In $Rb_6C_{60}$, the mid-IR region becomes silent, because the $t_{1u}$ band is fully occupied by the electrons transferred from the alkali metals. The lowest electronic excitation is the $t_{1u} \rightarrow t_{1g}$ interband (intramolecular) transition. These features are in a reasonable agreement with the previous results [40].

For insulating materials such as undoped $C_{60}$ and $Rb_6C_{60}$, transmission measurement provides reliable data. However, for relatively conducting materials such as $K_3C_{60}$ and $K_4C_{60}$, quantitatively reliable spectra are obtained only by the reflection measurements. In addition to the 1 eV transition, we observe a Drude-like conductivity spectra in $K_3C_{60}$. Detailed features of the spectra are discussed later in comparison with the spectra of $Rb_3C_{60}$ and $K_4C_{60}$.

In contrast to the metallic features of $K_3C_{60}$, the conductivity of $K_4C_{60}$ at low energy decreases with the photon energy, indicating that $K_4C_{60}$ is an insulator. The lowest absorption peak is located at around 0.5 eV. This behavior is not understood in the framework of the simple band picture which predicts that the $x = 4$ compounds are metallic since the triply degenerate $t_{1u}$ LUMO is still partially filled. In fact, the band calculation [35] shows that the metallic state is maintained in $K_4C_{60}$ although the symmetry of the bct structure is low in comparison with the fcc structure.

When we compare the present result with the other experimental results, on the other hands, we find a fairly good agreement. The $\mu$SR measurements found that both $K_4C_{60}$ and $K_6C_{60}$ are insulators [41]. It is interesting to point out that the energy gap ($\sim$ 0.3 eV) of $K_4C_{60}$ estimated from the temperature dependence of the relaxation time $T_2$ of $\mu$SR [41] is close to the lowest peak energy in the optical conductivity spectra. The $T_1$ of NMR displays a non-Korringa-like behavior in $Rb_4C_{60}$, supporting an insulating state of the $x = 4$ phase [19]. A small Curie-like spin susceptibility [42, 37] and photoemission spectra [43, 44] also confirms the insulating state of $K_4C_{60}$. Recently measured susceptibility data for $Rb_4C_{60}$ supports that this material is a nonmagnetic insulator [45].

### 7.3.3. *Discussion on* $A_4C_{60}$

The most serious problem in Fig. 7.14 is the absence of metallic behavior in $K_4C_{60}$. The absence of superconductivity in $A_4C_{60}$ results from the insulating ground state of this compounds. However, no reasonable explanation has been given to the insulating state of $A_4C_{60}$. To answer this question several mechanisms have been proposed.

(1) Electron localization due to the strong electron correlation.
(2) A possibility of charge density wave (CDW) formation along c-axis of bct structure.

(3) The orientational disorder may result in the localization of electrons.
(4) Jahn-Teller distortion; the triply degenerate LUMO is split into two.

While the model (1) predicts a magnetic ground state, the other three mechanisms result in nonmagnetic ground states. Magnetic measurements are good tests for these models. An electron spin resonance (ESR) measurements show that only a Curie-like spin susceptibility is observed with a small spin density. (in the order of 1/10 per $C_{60}$) [42, 37] Other experiments such as NMR and $\mu$SR failed to find any magnetic ordering above 4 K [19, 41]. Therefore, the observed Curie-like susceptibility is attributed to the localized spins due to the nonstoichiometry of alkali metals, indicating that the ground state of $K_4C_{60}$ is a nonmagnetic insulator. A recent magnetometry on $Rb_4C_{60}$ also shows that it is a nonmagnetic insulator [45]. Thus it seems unlikely that only the strong correlation effect can explain the insulating state of $A_4C_{60}$.

The model (2) was proposed by Pennsylvania group in the early stage of the $C_{60}$ research based on a band calculation for $K_4C_{60}$ which showed that there exists a one-dimensional Fermi surface [35]. This model predicts a period-doubling along the c-axis. Since the observed optical gap in Fig. 7.14 is considerably large ($\sim$ 0.5 eV), the superstructure is expected to be strong enough for experimental detection. Nonetheless, the superlattice reflection has never been observed in the x-ray diffraction (XRD) measurements [46]. Thus the CDW model (2) is excluded.

The model (3) starts from a question why the structure of $K_4C_{60}$ can be tetragonal. Since $C_{60}$ does not have a four-fold axis, there must exist some kind of orientational disorder of molecules in the tetragonal $K_4C_{60}$. Two kinds of orientational disorder have been proposed; that is a free rotation of $C_{60}$ molecules and a random occupation of two stable orientations. Zhou *et al.* showed that the latter model better explains the XRD pattern. The latter is called merohedral disorder in which the stable alignment of two-fold axes of $C_{60}$ is along the [100] and [010] and the occupation is random [46]. This merohedral disorder in $A_4C_{60}$ is the same as that observed in $A_3C_{60}$, in which the metallic state is maintained [25]. Accordingly, the orientational disorder model (3) can not be a dominant mechanism for the gap-opening in $A_4C_{60}$.

Now let me consider about the Jahn-Teller model (4). As mentioned earlier, the molecular energy levels of $C_{60}$ are highly degenerate due to the high symmetry of the molecule. For instance, the LUMO is triply degenerate, and as a result, doped molecules are unstable to the Jahn-Teller distortion. The distortion occurs when the electronic energy gain by the degeneracy lifting overcomes the energy loss due to the molecular deformation. All $C_{60}$ radicals $C_{60}^{x-}$ is unstable to this distortion if $0 < x < 6$. In fact, it is reported that the Jahn-Teller distortion occurs in an electrochemically produced monoanion state $C_{60}^{-}$ [47]. Speculated molecular energy levels in Jahn-Teller distorted states are displayed in Fig. 7.15 for $A_xC_{60}$ ($x = 2, 3, 4$). In the case of $K_4C_{60}$, $C_{60}$ molecule is deformed like a pancake. The $t_{1u}$ LUMO splits into two and only the lower two levels are occupied by four electrons. This splitting might result in a nonmagnetic insulating state in $A_4C_{60}$. The theoretically estimated splitting of the LUMO depends on authors but is in the range of 0.3–1 eV [48], which is not far from the optical data.

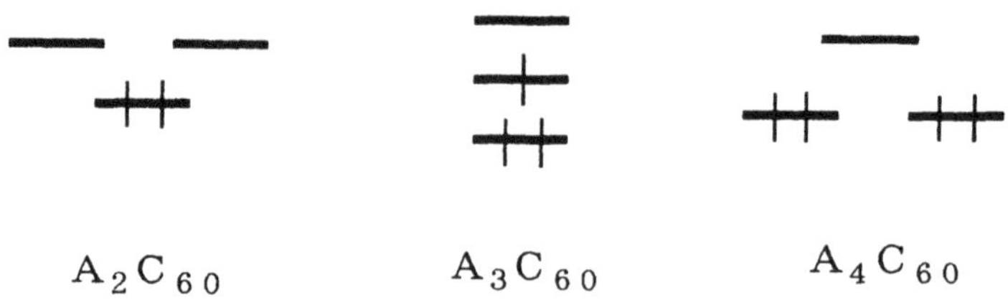


Figure 7.15. Speculative LUMO ($t_{1u}$) levels in the presence of Jahn-Teller instability for $A_2C_{60}$, $A_3C_{60}$, and $A_4C_{60}$.

Since the Jahn-Teller model predicts molecular distortions, a careful x-ray analysis can detect them in theory. However, this experiment might be very difficult because of the following reason. $C_{60}$ has fifteen two-fold axes, ten three-fold axes, and six five-fold axes. This means that there are as many distortion patterns as the symmetry axes. It is expected that since the energy of these patterns are very similar to each other and the barriers are very small, the molecules change the distortion patterns dynamically even at low temperatures. This situation might hinder the experimental observation of Jahn-Teller effect in $A_4C_{60}$.

Another experimental method to detect the Jahn-Teller distortion is the phonon splitting due to the symmetry reduction. Figure 7.16 shows conductivity spectra of the IR active intramolecular vibration modes for $C_{60}$, $K_3C_{60}$, $K_4C_{60}$, and $Rb_6C_{60}$ [36]. The peaks at 1182cm$^{-1}$ and 1428cm$^{-1}$ for $C_{60}$ are assigned to the $T_{1u}(3)$ and $T_{1u}(4)$ modes, respectively. These modes are tangential vibration of $C_{60}$ skeletons. The drastic increase in peak intensity and decrease in wavenumber is observed in the $T_{1u}(4)$ mode. In contrast, the $T_{1u}(3)$ mode does not show a remarkable change between $C_{60}$ and $Rb_6C_{60}$, although we did not observe any well-defined peak of $T_{1u}(3)$ in $x = 3$ and $x = 4$ phases. These results indicate that the $T_{1u}(4)$, having more double-bond-like character, has an extraordinarily strong coupling constant with the electrons introduced in the $t_{1u}$ conduction band.

In the spectra of $K_4C_{60}$, we see a clear splitting of 30cm$^{-1}$ in the $T_{1u}(4)$ mode. The peak of the lower component is at 1320cm$^{-1}$ which is lower than that for the $Rb_6C_{60}$. This is a strong indication that the lower component observed in $K_4C_{60}$ is not attributed to extrinsic effects such as a phase separation in the sample, but intrinsic in origin.

If $C_{60}$ molecules undergo a static Jahn-Teller distortion in $A_4C_{60}$, the degeneracy of the $T_{1u}$ phonon is lifted. The 30cm$^{-1}$ splitting in the $T_{1u}(4)$ mode observed in $A_4C_{60}$ (Fig. 7.16) might be an evidence for the Jahn-Teller distortion. The intensity ratio of the two peaks is expected to be 2:1, which is not far from the observation (Fig. 7.16). However, there is an alternative interpretation. The crystal structure of $A_4C_{60}$ is bct. In the tetragonal crystal field, the $T_{1u}$ mode also splits. This type of crystal field splitting of phonons was observed in the fivefold degenerate $H_g$ mode in the Raman spectra of bcc $Rb_6C_{60}$ [49]. To identify the origin of the phonon splitting,

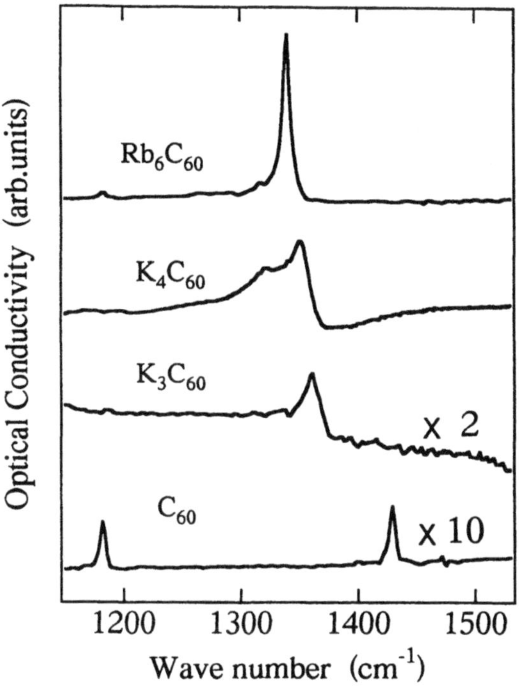


Figure 7.16. Intramolecular vibration spectra for undoped $C_{60}$, $K_3C_{60}$, $K_4C_{60}$, and $Rb_6C_{60}$ derived by the Kramers-Kronig transformation of reflectivity spectra. (After [36]).

a quantitative argument is necessary.

### 7.3.4. *Discussion on* $A_3C_{60}$

We have shown in the previous subsection that a significant deviation from the band model occurs in $A_4C_{60}$, although the mechanism is not clear yet. Then the next question comes out: How about the neighboring $A_3C_{60}$ phase? This question about the electronic structures of $A_3C_{60}$ is essential also from the view point of superconductivity with considerably high $T_c$. As mentioned in the previous section, the band model approximately works in $A_3C_{60}$. However, recent careful experiments show that the band model is not necessarily sufficient for a quantitative understanding.

A well-known example is an enhancement of Pauli-paramagnetism [50]. Ramirez *et al.* found that $K_3C_{60}$ and $Rb_3C_{60}$ display temperature-independent susceptibilities and showed that Pauli-paramagnetic contribution is $8.5 \times 10^{-3}$ emu/mole and $1.25 \times 10^{-3}$ emu/mole for $K_3C_{60}$ and $Rb_3C_{60}$, respectively. A direct application of free electron model results in a density of states at Fermi energy $N = \chi/2\mu_B^2 = 14$

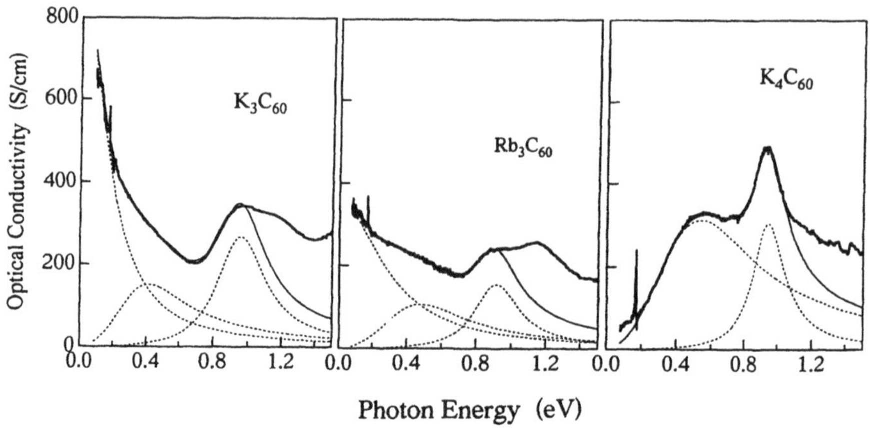


Figure 7.17. Optical conductivity spectra for $K_3C_{60}$, $Rb_3C_{60}$, and $K_4C_{60}$. The best fit curves by the equation in the text are shown by thin solid lines. Contributions from the Drude and two Lorentz oscillators are shown by thin broken lines. The parameters are summarized in Table 7.2. (After [36]).

Table 7.2. Parameters derived from the fit of the optical conductivity spectra in Fig. 7.17.

| | $\omega_P$ (eV) | $\gamma_P$ (eV) | $f_d$ | $\omega_1$ (eV) | $\gamma_1$ (eV) | $\omega_L$ (eV) | $\omega_2$ (eV) | $\gamma_2$ (eV) |
|---|---|---|---|---|---|---|---|---|
| $K_3C_{60}$ | 1.34 | 0.18 | 0.65 | 0.42 | 0.56 | 0.82 | 0.96 | 0.34 |
| $Rb_3C_{60}$ | 1.14 | 0.30 | 0.61 | 0.48 | 0.62 | 0.60 | 0.91 | 0.31 |
| $K_4C_{60}$ | 1.34 | - | 1 | 0.54 | 0.77 | 0.73 | 0.95 | 0.24 |

states/eV·spin·$C_{60}$ for $K_3C_{60}$, and 19 states/eV·spin·$C_{60}$ for $Rb_3C_{60}$. For this value a local density approximation and extended Hückel band structure calculations yield N=6.6 and 9.6 for $K_3C_{60}$, respectively [51, 52, 53]. This means that the observed N is enhanced by a factor of 1.4-2.3. The large spin susceptibilities of $K_3C_{60}$ are also found in the NMR [54] and ESR measurements [55]. Such an enhancement is frequently observed in strongly correlated electron systems and there are two possible explanations; one is the effective mass enhancement and the other is the spin fluctuation enhancement. Combining five independent experiments, Ramirez *et al.* proposed the latter mechanism. However, due to the small enhancement (in the order of 2), quantitative argument is very limited.

Deviation from the simple band model is more clearly observed in a high energy spectroscopy. Let us go back to Fig. 7.14. The conductivity spectra of $K_3C_{60}$ is

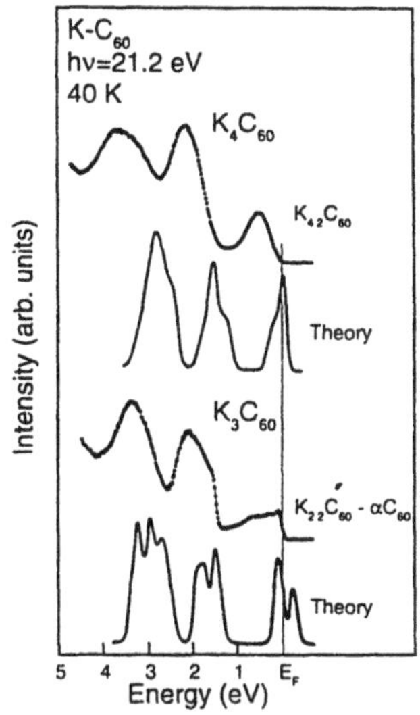


Figure 7.18. Photoemission spectra of $K_3C_{60}$ and $K_4C_{60}$. Local density calculations of the density of states are shown in comparison. The discrepancy between theory and experiment is seen near the Fermi edge. For both compounds, the spectral weight is shifted away from the Fermi edge. (After [43]).

approximated by the Drude formula below 0.7 eV. Similar Drude-like conductivity is observed in $Rb_3C_{60}$ (Fig. 7.17). Note that the spectral weight below 0.7 eV for $K_3C_{60}$ is larger than that for $Rb_3C_{60}$. The Drude-spectral weight is proportional to $n/m^*$, where $n$ is the carrier density and $m^*$ is the effective mass. The observed difference in the spectral weight is consistent with the simple band model that predicts a larger effective mass in $Rb_3C_{60}$ with a larger intermolecular spacing. However, strictly speaking, the simple Drude formula can not fully explain the low energy spectra of these superconductors. In order to fit the data completely, we must assume another Lorentzian oscillator centered at 0.4–0.5 eV. Results of analysis is presented in Fig. 7.17 for $K_3C_{60}$, $Rb_3C_{60}$, and $K_4C_{60}$ [36]. In this analysis we used a following dielectric function:

$$\epsilon(\omega) = \epsilon_\infty + \omega_P^2 \left[ -\frac{f_d}{\omega(\omega + i\gamma_P)} + \frac{1 - f_d}{\omega_1^2 - \omega^2 - i\omega\gamma_1} \right] + \frac{\omega_L^2}{\omega_2^2 - \omega^2 - i\omega\gamma_2}. \tag{7.1}$$

Here $\epsilon_\infty$ is the high frequency dielectric constant. The second term is the mid-IR excitation, and the last term corresponds to the 1 eV transition. The second term is divided into Drude and Lorentz parts. A fair agreement between experimental data (thick solid lines) and best fit calculations (thin solid lines) are obtained below 1 eV. The three component (one Drude and two Lorentz oscillators) are shown by dotted lines. The parameters used in the fit are summarized in Table 7.2. The

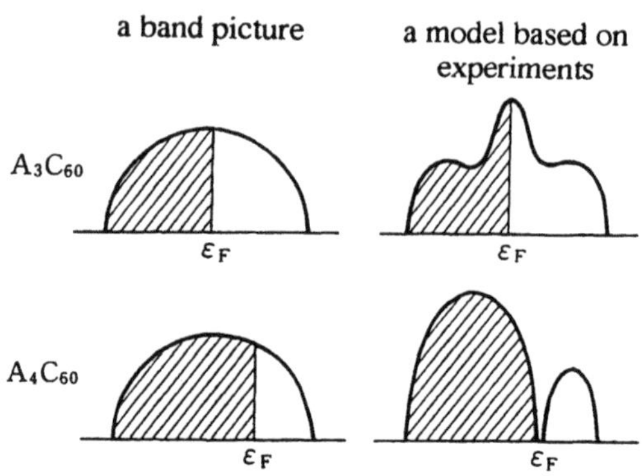


Figure 7.19. Density of states for $A_3C_{60}$ (upper) and $A_4C_{60}$ (lower). The left graphs show the naive band models, and the right show the experimentally derived models combining the susceptibility enhancement and the spectroscopic anomaly.

Lorentzian contribution centered at 0.9 eV is assigned to the intramolecular electronic transition from $t_{1u}$ to $t_{1g}$, while the lower energy contributions are attributable to the intermolecular or intra-$t_{1u}$ band transitions. The mid-IR absorption peaking at about 0.4 eV was crucial for a sufficient fit for the data both for $K_3C_{60}$ and $Rb_3C_{60}$. Moreover, it is noted that the energy of mid-IR absorption in $A_3C_{60}$ is close to the absorption peak in $K_4C_{60}$. This result suggests that the peak at 0.4 eV in $A_3C_{60}$ is a pseudo-gap which has the same origin as the real gap in $A_4C_{60}$. The Deviation of the conductivity spectra from the simple Drude model is an example of the failure of the band picture.

Another anomalous behavior is observed in the photoemission spectra. While a joint density of states is measured by optical spectra, the photoemission spectra directly correspond to the density of states of valence bands in a first-order approximation. Figure 7.18 displays 40 K photoemission spectra for $K_3C_{60}$ and $K_4C_{60}$ in comparison with a band model calculation of the density of states [43]. In the experimental spectra of $K_4C_{60}$ there is no density of states at the Fermi edge $\epsilon_F$ in sharp contrast with the calculation, but in agreement with other kinds of experiments shown in the previous subsection. Of particular interest is the photoemission spectra of $K_3C_{60}$. In addition to a sharp Fermi edge, we find a broad component expanding to 1 eV binding energy. The latter is not explained by the band model, since the width of $t_{1u}$ band is less than 0.5 eV. Similar spectra have been observed by other groups.[56, 44] The broad photoemission spectra are qualitatively consistent with the optical data (Fig. 7.17). The optical transition around 0.5 eV is possibly related to the broad component in the photoemission spectra.

Taking account of the two kinds of spectroscopy and the enhancement of the density of states at Fermi energy, experimentally derived density of states are drawn in Fig. 7.19. They dramatically differ from the naive band models both for $A_3C_{60}$ and $A_4C_{60}$. In contradiction with the band picture, a gap is formed at $\epsilon_F$ in $A_4C_{60}$.

Even in $A_3C_{60}$, the three-peak structure is necessary to account for the experimentally observed features. The DOS spectra seem more complicated than the band theory. The rich energy structure reminds us the Jahn-Teller splitting for single molecules shown in Fig. 7.15.

It is noted that in the far-infrared reflection measurements, Degiorgi *et al.* found another structure which is not explained by the simple band model either [57]. The observed rich structures in the low energy response in superconducting $A_3C_{60}$ suggest that electron-phonon interaction or electron correlation effect plays a crucial role in $A_3C_{60}$.

If there exists the Jahn-Teller effect in $A_3C_{60}$, the superconductivity mechanism mediated by the Jahn-Teller type electron-phonon interaction must be reconsidered [58, 59]. Although the deviation from the simple band model is not so dramatic, the above argument suggests that more precise discussion is necessary for full understanding the electronic properties and superconductivity of $A_3C_{60}$.

## 7.4. $C_{60}$ POLYMERS

### 7.4.1. *Introduction*

It is well-known that the third form of carbon, $C_{60}$, forms an fcc molecular crystal, in which molecular characters are well maintained. This is also confirmed by optical spectra as shown in Sec. 7.2. However, new forms of $C_{60}$ have been recently discovered. That is $C_{60}$ polymer in which chemical bonds are formed between molecules. $C_{60}$ polymer is a covalent solid rather than a molecular crystal. From the very beginning of fullerene research, synthesis of fullerene polymers has been one of the attractive goals to be achieved [5, 60]. In addition to the beautiful and elegant structures, $C_{60}$ polymers are expected to exhibit interesting electronic properties. Since 1993, several kinds of $C_{60}$ polymers have been discovered. Schematic drawings of $C_{60}$ polymers are given in Fig. 7.20. Not only one-dimensional but also two-dimensional polymers have been found and the $C_{60}$ polymer is now forming a new subfamily. In all polymers found, molecules are connected by four-membered rings between molecules. This type of reaction mechanism is called [2+2] cycloaddition, in which double bonds on adjacent molecules convert to the four-membered rings. Three kinds of polymerization of $C_{60}$ have been reported:

(1) Photo-induced polymerization.
(2) Polymerization in alkali-intercalated compounds.
(3) Polymerization under high pressure and temperature.

Although the photoinduced polymerization was reported relatively earlier, it was not a bulk phenomenon: polymerization does not take place uniformly in a macroscopic scale. In fact, no clear structural change was found under light irradiation. On the other hands, the latter two involve dramatic structural changes, showing that the polymerization of $C_{60}$ can occur as macroscopic bulk phenomena. In 1994–1995 several kinds of bulk $C_{60}$ polymers were found and identified. The new polymer phases of

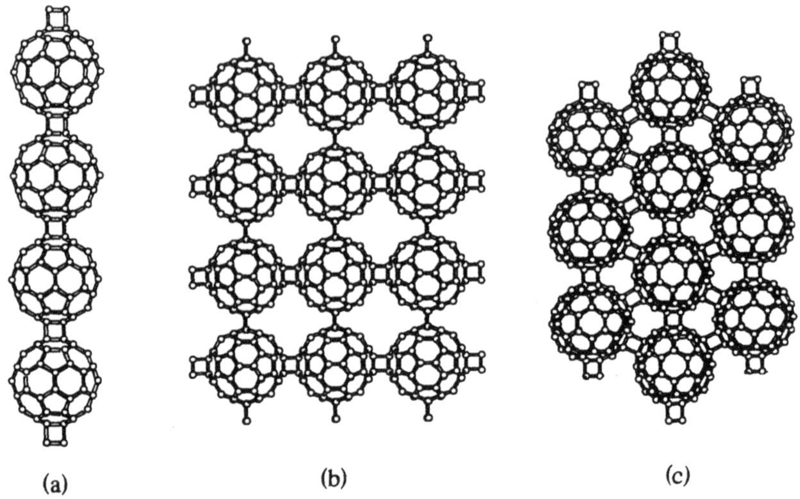

Figure 7.20. $C_{60}$ polymers involving four membered rings as intermolecular chemical bonds. (a)one-dimensional polymer and two-dimensional polymers with (b) tetragonal and (c) rhombohedral structures. Four membered rings in (a) and (b) are either parallel or perpendicular to the text plane. In contrast the four membered rings in (c) is inclined by 30°.

$C_{60}$ are not fcc any more, having lower crystal symmetries, reflecting the anisotropic structures of polymers. In this section, a short review is given on the polymerization of $C_{60}$ with an emphasis of our discovery of the pressure-induced polymerization.

### *7.4.2. Photoinduced Polymerization*

The polymerization was first reported by Rao *et al.* in evaporated thin films of $C_{60}$ [61]. They are first motivated by a fact that the Raman shift of the pentagonal pinch mode depends on several research groups. They made a hypothesis that unknown changes are induced by a strong irradiation of laser light, and made various kinds of characterization on the film. It has been well-known that $C_{60}$ dissolves into organic solvent such as toluene or benzene. (This solubility was essential for successful isolation of fullerenes from soot.) Surprisingly, Rao *et al.* found that the UV irradiated $C_{60}$ films did not dissolve into the solvent any more. This remarkable property can be applied for lithography and one micron wide lines with good edge definition has already been demonstrated [62].

Rao *et al.* found that UV irradiated $C_{60}$ films display richer IR absorption and Raman spectra than those for pristine $C_{60}$. The richer phonon spectra strongly indicate that the high molecular symmetry of $C_{60}$ ($I_h$) is reduced. They also observed

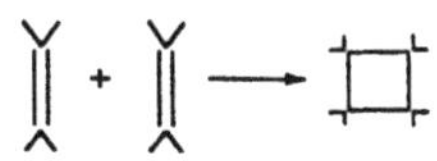

Figure 7.21. A model for polymerization of $C_{60}$; [2+2] cycloaddition.

$(C_{60})_n$ oligomers up to n=20 in the mass spectra of UV exposed $C_{60}$ films. They concluded that UV exposure induces a cross-linking or polymerization of $C_{60}$. Particularly, observation of oligomers is a convincing evidence, and the loss of solubility is consistent with the polymerization. Since the UV-exposure was performed in a high vacuum free from solvents and oxygen, the polymerization is regarded as an intrinsic phenomenon to $C_{60}$. Other experiments such as photoemission spectroscopy also supported an occurrence of photoinduced polymerization [63].

They proposed [2+2] cycloaddition as a mechanism of polymerization (Fig. 7.21). Parallel double bonds in adjacent $C_{60}$ molecules, one from each molecule, are broken and reform as single bonds between the molecules. To test this proposal, extensive efforts have been made.

Zhou *et al.* found that the photoinduced polymerization is significantly suppressed below 250 K, which is close to the orientational ordering transition temperature of solid $C_{60}$ [64]. This result indicates that the free rotation of molecule is crucial for the polymerization and supports the [2+2] cycloaddition model. In the low temperature simple cubic phase, the orientation of neighboring molecules is not preferable for the [2+2] cycloaddition. The polymerization efficiency is proportional to the incident light flux, indicating that photoinduced polymerization is a single photon process [65]. Wang *et al.* proposed a polymerization mechanism *via* a triplet CT exciton (Fig. 7.7).

Despite these notable findings in spectroscopy, structural information is extremely limited. For example, the XRD pattern showed that the photopolymerized film stays in fcc except for a very small shrinkage. This result seems contrary to the cycloaddition model that predicts a significant reduction of intermolecular distances. The connectivity of intermolecular bonds still remains unsolved. The difficulty in the structural study possibly results from the lack of structural coherence in the photoinduced polymerization process. It is likely that the light-induced polymerization does take place in a microscopic scale without a large scale coherence. This situation hinders detailed structural studies of photopolymerized materials. For a precise study of polymerization, a different synthesis route was necessary. It was found a year after.

### 7.4.3. *Polymerization in Alkali $C_{60}$ Fullerides*

A breakthrough to bulk polymerization was made by two independent and totally different processes. One is the polymerization in alkali-intercalated materials, and

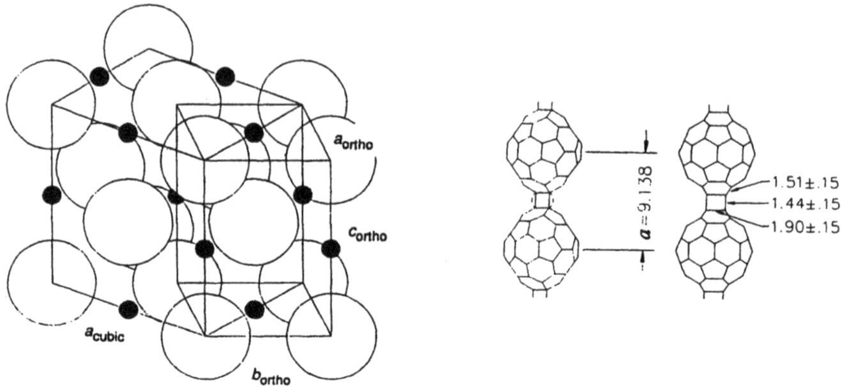


Figure 7.22. (Left) relation between the fcc and orthorhombic cells. The polymer chain is parallel to the a-axis of the orthorhombic cell. (Right) the structure of $C_{60}$ polymer derived from the Rietveld refinement.

the other is the high pressure and temperature-induced polymerization.

The former is $A_1C_{60}$, where A is alkali metal. While $A_3C_{60}$ (A=K or Rb) is a well-known superconductor, $A_1C_{60}$ is not superconducting above 1.8K. This phase was first identified by Winter and Kuzmany by Raman spectra [66]. Zhu *et al.* found that $A_1C_{60}$ have an fcc structure at higher temperatures than 400 K (Table 7.1) [20]. The structures of pure $C_{60}$, fcc $A_1C_{60}$, $A_3C_{60}$ are very similar. The arrangement of $C_{60}$ is essentially the same in these materials. While in $A_1C_{60}$ the alkali ions occupy the octahedral interstitial sites, both octahedral and tetrahedral sites are occupied in $A_3C_{60}$. The fcc structure of $A_1C_{60}$ is nothing but an NaCl-type structure.

At about 400 K, $A_1C_{60}$ undergoes a dramatic phase transition to a non-cubic structure. Based on a careful sample preparation and structural analysis, Chauvet *et al.* showed that the slow cooling of $A_1C_{60}$ results in an orthorhombic distortion with the lattice parameters of $a = 9.12$ Å, $b = 10.10$ Å, and $c = 14.26$ Å [67]. Note that the high temperature fcc phase can be viewed as a bct lattice with dimensions of $a = b = 9.95$ Å and $c = 14.072$ Å. Thus the low temperature orthorhombic phase is a compression of this bct lattice along the the $[100]_{bct}$ or $[110]_{fcc}$ direction. Along the $[110]_{fcc}$ or $[100]_{ortho}$, the molecular spacing was found to be 9.12 Å, which is extremely small comparing to the value of 10.02 Å for the normal fcc $C_{60}$. The relation between the cubic and orthorhombic cells is described in Fig. 7.22. Pekker *et al.* proposed a one-dimensional $C_{60}$ polymer model for orthorhombic $Rb_1C_{60}$ [68]. This model is schematically shown in Figs. 7.20(a) and 7.22. Later on, better orthorhombic-$A_1C_{60}$

samples with less phase separations have been prepared, and a Rietveld refinement of powder XRD data has been performed by Stephens et al [69]. They confirmed the one-dimensional polymeric structure and clarified the distorted structure of $C_{60}$ as a result of the formation of intermolecular covalent bonds (Fig. 7.22). Molecules are elongated along the polymer chain to form intermolecular chemical bonds. Since the molecular symmetry is reduced in the orthorhombic phase, the infrared absorption spectra show rich structures, similar to the photoinduced polymer and pressure-induced polymers [71]. Pekker *et al.* synthesized single crystals of $K_1C_{60}$ by the cosublimation technique [70], and found that the orthorhombic polymer phase is stable in the air. This is the first air-stable alkali-doped $C_{60}$. They estimated the degree of polymerization to be more than 100,000.

Chauvet *et al.* performed an ESR (electron spin resonance) study for $Rb_1C_{60}$, and found that this compound is a one-dimensional conductor [67]. Since one electron is transferred from alkali to $C_{60}$, the metallic state is consistent with the band picture. $Rb_1C_{60}$ exhibits following features : The ESR linewidths are considerably narrow comparing to the broad lines of $Rb_3C_{60}$ [72], and the width decreases with temperature following the Elliot law. These features are characteristic of one-dimensional metals, being consistent with the polymeric structure of $C_{60}$ in $Rb_1C_{60}$. At about 50 K, $Rb_1C_{60}$ undergoes a metal-insulator transition. Chauvet *et al.* claimed that most possible low temperature state is the spin density wave.

Although orthorhombic phase is the most stable at low temperatures, a rapid cooling across the transition temperature results in other new types of $C_{60}$ linkings. For example, Chauvet *et al.* first reported that a new phase is obtained by temperature-quenching. Very recently, Zhu *et al.* clarified that the quenched phase has another orthorhombic structure in which $C_{60}$ molecules are dimerized [73]. Other new structures are now being identified. Cross-linking of $C_{60}$ results in a rich variety in structures and electronic properties of $A_1C_{60}$ [74].

### *7.4.4. Pressure-Induced Polymerization*

An alternative method to synthesize the bulk $C_{60}$ polymer is the pressure-induced polymerization [75]. In contrast to the one-dimensional polymer in $A_1C_{60}$, the pressure treatment of $C_{60}$ results in higher dimensional polymers, such as two-dimensional polymers. Pressure effect has been extensively studied on pure $C_{60}$: for example, non-hydrostatic compression around 20 GPa induces a transformation to diamond [76]. Spectroscopic study showed that unknown structures are formed above 15 GPa [77]. Samara et el., on the other hand, reported the pressure-dependence of orientational transition temperature of $C_{60}$ [78, 79]. All these high pressure experiments were made near room temperature. The crystal structure and the stability of $C_{60}$ cage at high temperature and pressure were open questions until 1994.

Synthesis at high pressure and high temperature is now well-known for the growth of new families of cuprate high $T_c$ superconductors [80]. A wedge-type cubic anvil apparatus was used for the synthesis [81]. Several tens of mg of $C_{60}$ was pressed into a pellet and sandwiched between two Ti-powder oxygen getters. They were loaded

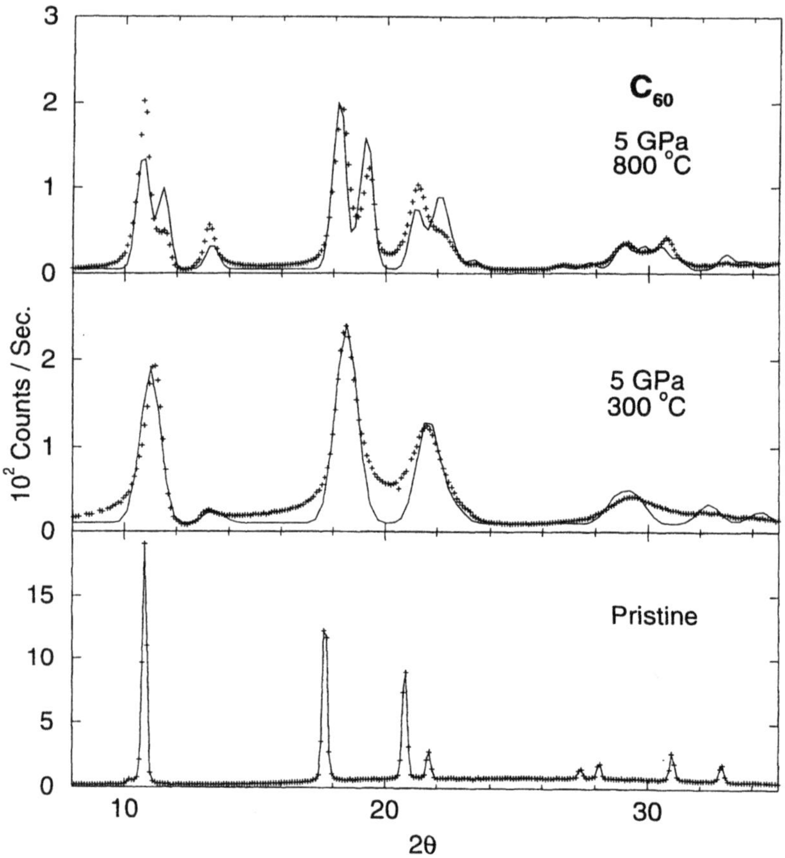


Figure 7.23. The x-ray diffraction patterns for pristine fcc $C_{60}$ and high pressure forms synthesized at the indicated conditions. Crosses and solid lines are experimental points and patterns calculated with the spherical shell model, respectively.

into a gold capsule 4.5mm in diameter and 10mm in length. The capsule was loaded in a dry box to minimize the exposure of $C_{60}$ to air. This capsule was surrounded by a cylindrical graphite heater enclosed inside an insulating NaCl spacer. Pyrophyllite was used as a pressure medium.

The pressure and temperature were applied in steps. First, the pressure was increased to 5 GPa over a 1-hour period. The temperature was then increased for 20 min and held for 1 hour. Subsequently, the temperature was quenched in 5 s, followed by a gradual release of pressure over 1 hour. The sample was not exposed to air until

Table 7.3. Lattice parameters of pristine and the pressure-induced forms of $C_{60}$.

| Phase | Cubic Cell | Hex Cell | | $Vol/C_{60}$ |
|---|---|---|---|---|
| | $a$ (Å) | $a$ (Å) | $c$ (Å) | Å$^3$ |
| pristine fcc | 14.17 | 10.02 | 24.54 | 711 |
| fcc($pC_{60}$) | 13.6 | 9.62 | 23.6 | 629 |
| rh($pC_{60}$) | — | 9.22 | 24.6 | 603 |

after the reaction. The reacted sample formed a solid ingot with a clear interface and there was no observable reaction of $C_{60}$ with Ti.

To rule out the effect of solvent or impurities, we repeated the experiments with three different starting materials, which were (1) chromatographically purified, (2) purified and vacuum-dried at 200°C for several hours, or (3) purified, dried, and sublimated. The same results were obtained for all preparations. To check for possible oxygen effects after high pressure treatment, we placed the sample in a sealed quartz capillary within a He glove box. The subsequent XRD data were the same as those taken in air. Based on these results, we conclude that the observed pressure-induced effects are intrinsic to $C_{60}$.

High pressure synthesis was performed at 100°C steps from room temperature to 800°C at fixed pressure $P = 5$ GPa. CuK$\alpha$ XRD powder patterns are shown in Fig. 7.23 for pristine $C_{60}$, a sample treated at $P = 5$ GPa and 300°C, and a sample treated at $P = 5$ GPa and 800°C. Whereas ground powders of pristine $C_{60}$ typically show relatively sharp lines, pressurized materials show gross broadening, indicating that crystalline order extends over 50 to 100 Å in these samples. Despite the broadening, both of the pressure-treated samples can be indexed as single phase.

The obtained patterns are classified into two. The 300°C sample is typical of materials synthesized between 300°C and 400°C. Between 600°C and 800°C, we obtained a different pattern, and the 500°C treatment results in a mixture of these two patterns. The pattern at lower temperatures can be indexed as fcc with a lattice parameter of 13.6 Å, a contraction of about 3.7% from the pristine $C_{60}$. This lattice parameter is larger than the *in situ* value at 5 GPa at room temperature [82]. On the other hand, the pattern at 800°C can not be indexed as fcc any more. We found that this is indexed as rhombohedral with $a = 9.77$ Å and $\alpha = 56.3°$. The hexagonal cell parameter of this structure is $a = 9.22$ Å, and $c = 24.6$ Å.

The thin solid line in Fig. 7.23 is the result of a calculation in which the $C_{60}$ molecule was replaced by a spherical shell of charge with a radius of 3.5 Å. The calculation for the 800°C sample was made for the space group $R\bar{3}m$. This model contains a single sphere at the origin in the primitive rhombohedral cell. In the hexagonal cell, there are three equivalent molecules at (0,0,0), $(\frac{1}{3}, \frac{2}{3}, \frac{1}{3})$ and $(\frac{2}{3}, \frac{1}{3}, \frac{2}{3})$.

The rhombohedral structure can be thought of as a distortion of the fcc structure resulting from an elongation along $[111]_{fcc}$. The model qualitatively describes the

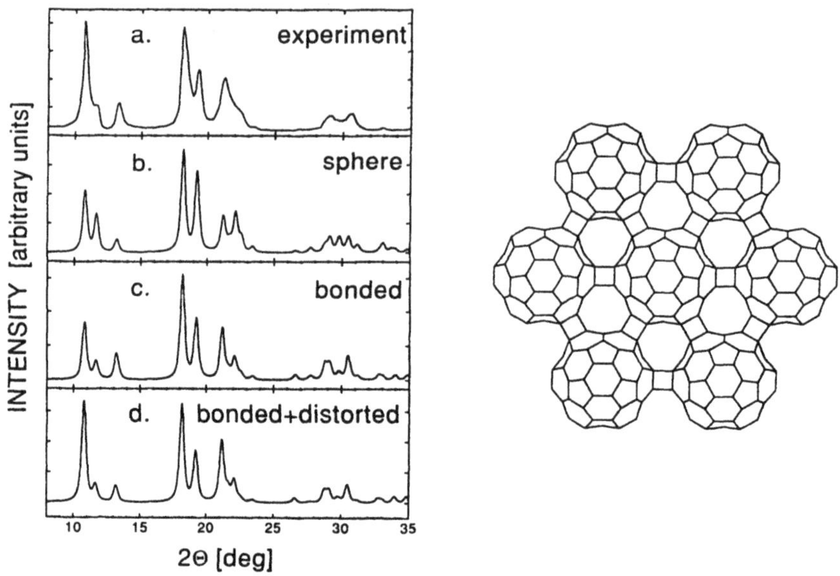


Figure 7.24. (Left) a. Experimental x-ray diffraction data. b. Model calculation using spherical shell model. c. Model calculation using bonded but undistorted molecule. Bonding is approximated by pulling out carbon atoms by 0.3 Å along the bonds. d. Model calculation using polymerized $C_{60}$ with a 5% uniaxial compression along the c direction. (Right) A model of polymerized ab plane of the rhombohedral phase. (After [83]).

data and correctly predicts the systematic absences but fails to predict the detailed features of the data even in the lower angle. The difference between the experiment and calculation means that the spherical shell model is not exactly correct. In the conventional $C_{60}$ solids, either doped or undoped, the spherical shell model is appropriate for the case where $C_{60}$ freely rotates. Even in the phases where the rotation of molecules is frozen, the spherical shell model works at least for the first few peaks in the XRD patterns. This is not the case here: Even the relative intensities of the main peak can not be explained by the simple model. This indicates that the soccer ball-like $C_{60}$ cage is significantly deformed in the rhombohedral phase.

Table 7.3 summarizes lattice parameters for pristine and pressure-induced phases of $C_{60}$. In the pristine and the two pressure-induced structures, the nearest intermolecular distance is given by the hexagonal cell parameter a. The two close-packed cubic structures have isotropic nearest neighbor distances with 12 nearest neighbors (Table 7.3). The rhombohedral structure is anisotropic with six nearest neighbors

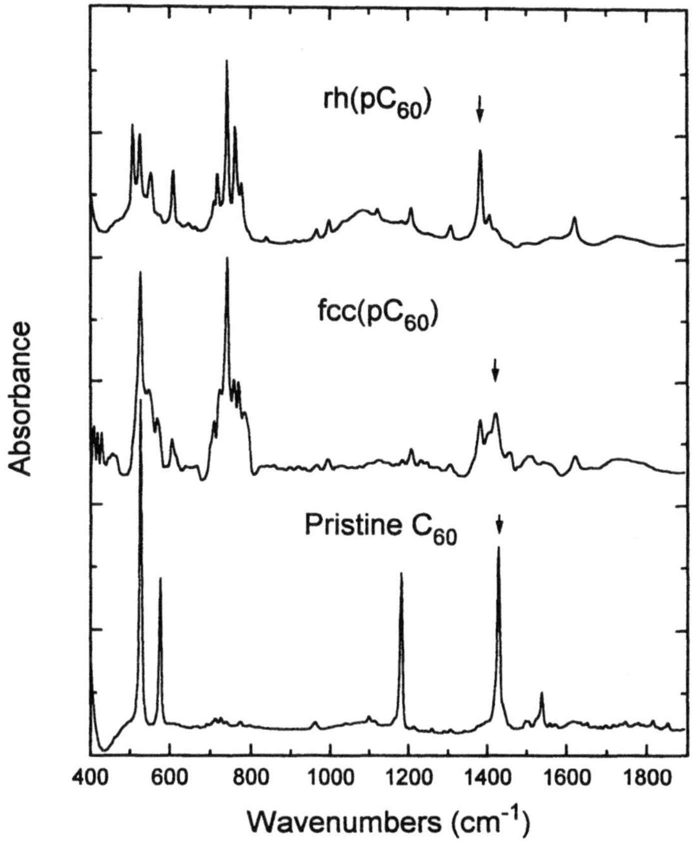


Figure 7.25. Infrared absorption of powders dispersed in KBr pellets for pristine $C_{60}$, fcc($pC_{60}$), and rh($pC_{60}$). Arrows indicate the strongest peak in the C=C stretching vibration region.

(9.22 Å) in the "waist" [the old fcc (111) plane] and six next nearest neighbors (9.76 Å) above and below the waist. Although the volume per molecule monotonically decreases as the synthesis temperature increases (Table 7.3), the hexagonal c axis goes through a minimum in the pressure-induced fcc structure. It is noted that the intermolecular distance of 9.22 Å in the ab-plane of the rhombohedral phase is very small and close to that in the one-dimensional polymer in $A_1C_{60}$. The small nearest neighbor distance in the ab-plane suggests the two-dimensional bonding of molecules within the layer.

Later, Oszlanyi and Forro proposed a structural model to explain the observed XRD pattern for the rhombohedral phase [83]. The right panel of Fig. 7.24 shows the

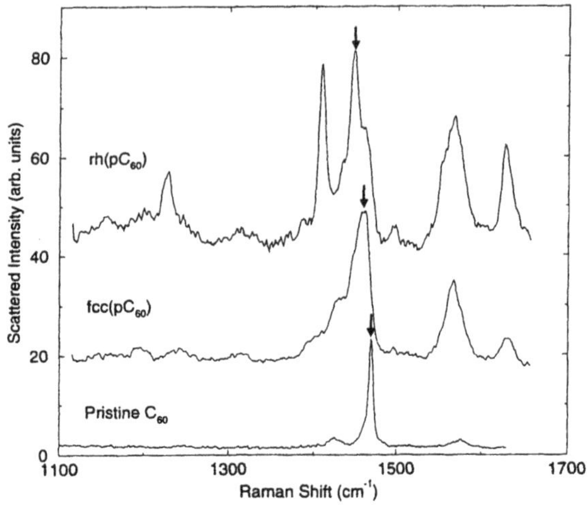


Figure 7.26. Raman scattering spectra of pristine $C_{60}$, fcc($pC_{60}$), and rh($pC_{60}$). Arrows mark the $C_{60}$ $A_g$ mode.

structural model of the rhombohedral phase. The structural model shows that all the $C_{60}$ molecules are oriented the same way with hexagons facing the c-direction of the hexagonal cell. This orientation allows the simultaneous formation of polymer bonds along the [100], [010], [110] directions. The neighboring molecules are bonded by four membered rings in the same manner as the one-dimensional polymer in $A_1C_{60}$. The bond formation is approximated with pulling out the carbon atoms of the double bonds toward each other by 0.3 Å. Molecules are significantly deformed due to the bond formation in sharp contrast with schematic diagram of two-dimensional polymer [Fig. 7.20(c)]. The left panel of Fig. 7.24 shows the experimental XRD data and three different model calculations. The calculation using the above model is shown in Fig. 7.24(c) and it describes the experiment [Fig. 7.24(a)] better than the spherical shell model [Fig. 7.24(b)] used in the analysis in Fig. 7.23. However, in both cases the intensity of the first peak is seriously underestimated. Oszlanyi and Forro found that the intensity of this peak can not be increased without distorting the whole molecule along the c-axis. A 5% uniaxial compression of molecules increases the intensity of the first peak without affecting the rest of the pattern [Fig. 7.24(d)]. This model is reasonable considering that the carbon atoms near the equatorial which are involved in the bond formation are pulled out, leaving space to the rest of the molecule to distort. If the bond lengths within the molecules are kept constant with the exception of those carbons which form the intermolecular bonds, the molecule is compressed along the c direction, in other words, expanded in the ab plain.

While pristine $C_{60}$ is well known for its good solubility to organic solvent such as toluene or 1,2-dichlorobenzene [84], the both pressure-induced phases do not dissolve

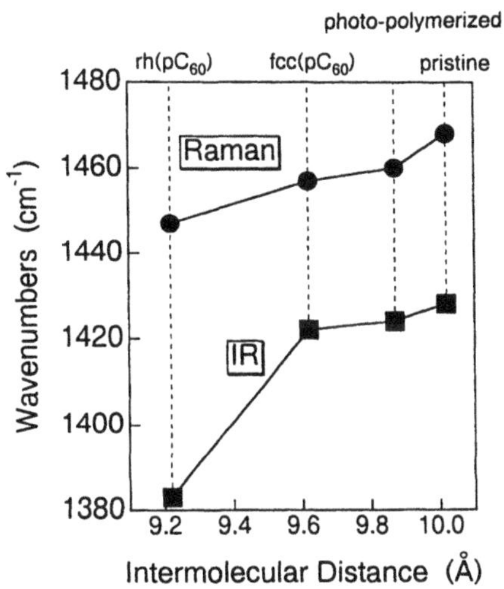


Figure 7.27. Wavenumbers of infrared- and Raman-active C=C stretching modes indicated by arrows in Figs. 7.25 and 7.26 against the nearest intermolecular distance.

in either solvent. In the same manner as the photopolymerization, the loss of solubility is a convincing evidence for the crosslinking of $C_{60}$ suggested by the XRD analysis.

Spectroscopic studies on the pressurized materials also suggest a formation of intermolecular bonds. Figure 7.25 shows the powder infrared absorption spectra for pristine, pressure-induced fcc and rhombohedral phases. The latter two are designated as fcc($pC_{60}$) and rh($pC_{60}$), respectively. Only four strong absorption peaks are observed in pristine $C_{60}$. This is because of the high symmetry of molecules and weak intermolecular interactions in molecular crystal form. In sharp contrast, the spectra for two pressurized materials have substantially richer structures, indicating a significant reduction of the $I_h$ symmetry. Particularly, the region from 700cm$^{-1}$ to 800cm$^{-1}$ is featureless for pristine $C_{60}$, while intense peaks are observed for both pressure-induced forms. Similar IR spectra are observed when the $C_{60}$ skeleton is partly modified, for instance, an epoxide $C_{60}O$ [85]. Photo-polymerized $C_{60}$ and $Rb_1C_{60}$ also shows a similar spectral change [61, 71].

Activation of extra bands are also observed in the Raman spectra (Fig. 7.26). Although the peak assignment for the pressure-induced polymer phases were not performed yet, the rich structure in the vibration spectra is a common feature of the $C_{60}$ polymers.

Another notable feature in the vibration spectra is the softening of the modes originating from the C=C stretching motion, which appear around 1400cm$^{-1}$. The strongest lines in this region are indicated by arrows both in IR (Fig. 7.25) and Raman (Fig. 7.26) spectra. The wavenumbers of these modes are plotted in Fig. 7.27 against the intermolecular distance determined by the XRD analysis. The data for the photopolymerized $C_{60}$ are also shown since it exhibits the same trend as the

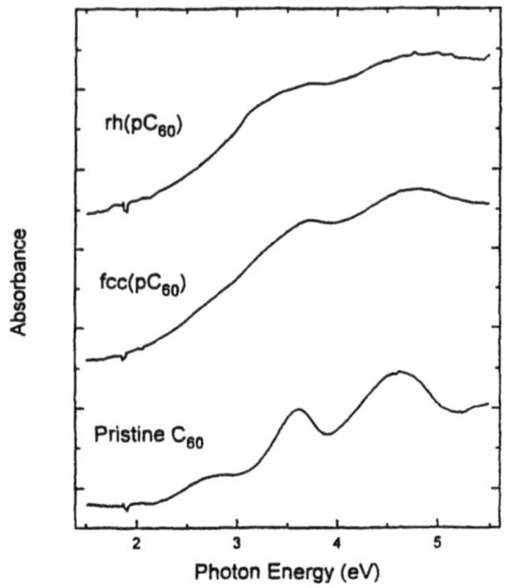


Figure 7.28. Optical absorption spectra of pristine $C_{60}$, fcc($pC_{60}$), and rh($pC_{60}$).

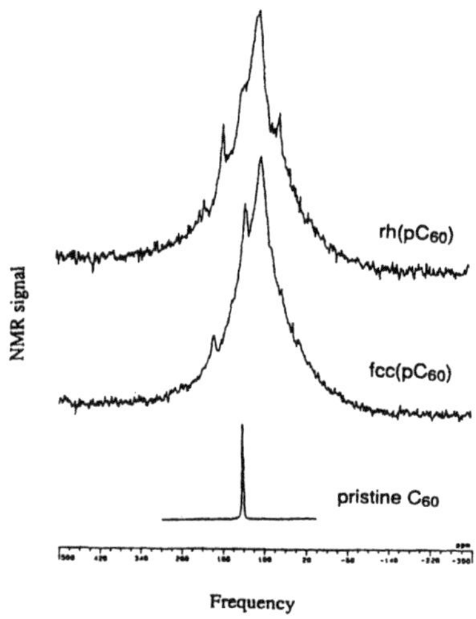


Figure 7.29. Solid state $^{13}C$ NMR spectra of pristine $C_{60}$, fcc($pC_{60}$), and rh($pC_{60}$).

pressure-induced forms. Although the IR active mode is observed at 1428cm$^{-1}$ in pristine $C_{60}$, it is shifted to 1422cm$^{-1}$ and 1383cm$^{-1}$ for pressure-induced fcc and rhombohedral phases, respectively. The position of this mode, as well as Raman active mode is known to be very sensitive to the degree of charge transfer on the $C_{60}$ molecule. For example, the IR active ($t_{1u}$) mode is observed at 1341cm$^{-1}$ for $K_6C_{60}$ [40]. The observed shift of 45cm$^{-1}$ in the rhombohedral phase is significantly large,

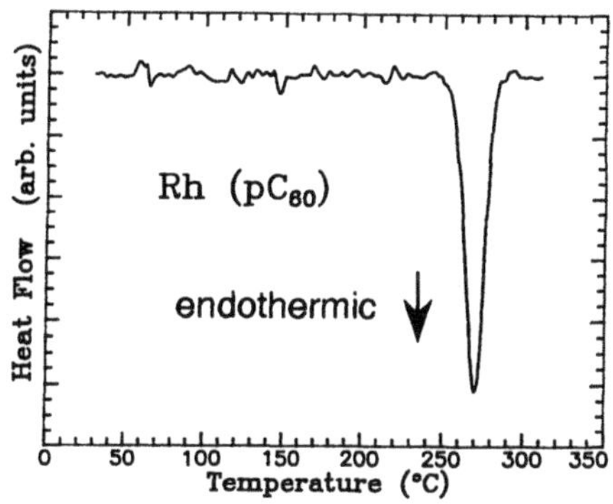


Figure 7.30. Differential scanning calorimetry data for the heating process of rh($pC_{60}$).

especially when one considers that the material is undoped.

The trend in Fig. 7.27 can not be understood if $C_{60}$ molecules are not crosslinked. *In situ* pressure-dependence of the IR and Raman active modes have been measured at room temperature where bulk scale polymerization is prohibited. In sharp contrast to Fig. 7.27, the C=C vibration (IR active $t_{1u}$ and Raman active $A_g$) modes shifts to higher wavenumber under pressure at room temperature [86]. In other words, in a conventional molecular crystal of $C_{60}$ the C=C stretching mode becomes hard as the intermolecular distance is smaller. The opposite trend observed in the pressure-induced phases indicates that these forms are not conventional molecular crystals but the chemical bonds are reconstructed in pressure-induced forms.

The softening of C=C stretching mode is qualitatively understood based on the polymer model. In general, the C=C double bond is harder than the single bond. Since then, the softening of the band means that the single-bond-character increases in the pressure treated materials. This is consistent with the [2+2] cycloaddition model, in which double bonds on neighboring molecules are converted to single-bond four membered rings.

Other spectroscopic data provide more evidence for the symmetry reduction of $C_{60}$ molecules. Fig. 7.28 shows the absorption spectra in the visible to ultraviolet region, in which intramolecular electronic absorption are observed. The data were taken on the powder samples dispersed in KBr pellets. In contrast to the sharp absorption peaks in pristine $C_{60}$, the two pressure-induced forms exhibit considerably broad spectra. This broadening is possibly explained by the degeneracy-lifting due to the deformation of molecules. The absorption edge is shifted to lower energy comparing to that of pristine $C_{60}$, being consistent with the recent calculation [87].

Figure 7.29 shows the solid-state $^{13}C$ nuclear magnetic resonance (NMR) spectra taken at 100.46 MHz with magic angle spinning. In a marked contrast with a sharp single peak for pristine $C_{60}$, a broad peak centered at 125 parts per million was observed accompanied with sub-peaks in both pressure-induced phases, providing a strong evidence for an existence of several inequivalent carbon sites. The R$\bar{3}$m structural model shown in Fig. 7.24 predicts that there exist six inequivalent site in rh(p$C_{60}$). We observe four well-defined peaks in the NMR spectra of this phase. This discrepancy between the model and the experiment is possibly explained by unresolved peaks around 120 ppm frequency shift.

The pressure-induced $C_{60}$ polymers revert back to $C_{60}$ upon heating. Figure 7.30 shows a differential calorimetry curve for the heating scan of the rhombohedral $C_{60}$. An endothermic peak with an enthalpy of 23 J/g was observed, while no structure was found in the cooling scan. This indicates an occurrence of irreversible transition. Bashkin *et al.* reports the same heat value in the recovery of the pressure-induced phase [88]. By means of XRD and IR spectroscopy, we confirmed that the rhombohedral $C_{60}$ reverts back to the conventional fcc $C_{60}$. This result indicate that the pressure-induced polymers are metastable phases of $C_{60}$. The transition back to monomer $C_{60}$ is a commonly observed phenomenon in the other two kinds of $C_{60}$ polymers [89, 70]. All three kinds of polymers are unstable at high temperatures. Only $A_1C_{60}$ polymer undergoes a reversible transition between monomer and polymer states. The observed heats of transition are 2.4–9.4 J/g, $\sim$ 12 J/g, and $\sim$ 15 J/g for $K_1C_{60}$, $Rb_1C_{60}$, and $Cs_1C_{60}$, respectively being smaller than that for rhombohedral $C_{60}$ polymer [20]. This result suggests that the enthalpy change increases with the number of intermolecular bonds per molecule.

Recently, Núñez-Regueiro *et al.* found that different conditions produce other types of $C_{60}$ polymers [90]. At 3 GPa and 600°C, a tetragonal phase [Fig. 7.20(b)] is obtained as a mixture with the rhombohedral phase. The 8 GPa and 300°C treatment results in an orthorhombic phase involving one-dimensional polymer [Fig. 7.20(a)]. These observations indicate that variation of synthesis conditions might produce a rich variety of $C_{60}$ polymers with potentially interesting properties.

### 7.4.5. *Discussion*

One of the necessary conditions for the occurrence of solid state polymerization is the decrease in the lattice constant. In this sense, the pressure-induced polymerization is a reasonable phenomenon, since the pressure reduces the lattice constants. In the other two cases, the lattice contraction is not necessarily trivial.

In the case of $A_1C_{60}$, the key is the position of alkali ions. The large alkali ions such as $K^+$, $Rb^+$, and $Cs^+$ occupy the octahedral site, since they are larger than the tetrahedral site. In this case, the Coulomb interaction reduces the lattice constant slightly. This contraction may be a trigger for the polymerization.

In the photoinduced polymerization, the charge transfer exciton (CT) plays an important role. Photoexcited species rapidly relaxes to the ground state emitting the excess energy to the lattice. During this process the excited species sometimes stay in several stable excited states. In the case of $C_{60}$, one of them is the lowest exciton state which is the center of photoluminescence (Sec. 7.2) and the other is the triplet charge transfer (CT) exciton state [65]. Since the lifetime of triplet CT states is considerably large, the Coulomb interaction between the electron and hole on the adjacent molecules induces a slight decrease in the intermolecular distance. This is a kind of self-trapped exciton, and triggers a formation of the covalent intermolecular bond.

These are the proposed mechanisms as the trigger of polymerizations. The reduction of intermolecular distance is a bulk scale phenomenon both in the pressure effect and $A_1C_{60}$. In the case of photopolymerization, on the other hand, the contraction is a local phenomenon in a sharp contrast to the other two. This might explain the absence of bulk scale structural coherence in photoinduced polymerization.

However, only the lattice contraction is not sufficient for the occurrence of polymerization. In fact, the pressurization at room temperature does not result in a polymerization. In addition to the lattice contraction, some other conditions are necessary. In the case of photopolymer and $A_1C_{60}$, the existence of electron or hole seems crucial. The high temperature is another necessary condition for the pressure-induced polymerization.

When these conditions are satisfied, the polymerization occurs relatively easily, This fact indicates that the monomer $C_{60}$ phase (the molecular crystal phase) is not a unique stable state. Thus the problem of $C_{60}$ polymer is an important issue related to the variety of solid phase of carbon. As to the pressure-induced phases, they are understood as metastable states that is produced just before destruction of $C_{60}$ cage. In fact, $C_{60}$ transforms into diamond under extremely high pressure and decomposes into amorphous carbon at high temperature. Moderate and appropriate conditions produce unique polymer phases. The structure of $C_{60}$ polymer is understood as a partial breakdown of double bonds in $C_{60}$ cages. In this sense, polymer phases are interpreted as intermediate states between $C_{60}$ and diamond.

In this section I reviewed a recent progress in the new solid form of $C_{60}$. $C_{60}$ have been recognized to form typical molecular crystals. However, it has been clarified that several covalent forms of $C_{60}$ do exist. They are so called one-dimensional and two-dimensional $C_{60}$ polymers, in which neighboring molecules are connected by four membered rings. These polymers have common properties. The polymeric forms are not stable at high temperatures at ambient pressure, and reverts to the normal molecular solids.

## 7.5. SUMMARY

In Chapter 7, I have reviewed the investigation of physical properties of $C_{60}$ and its related compounds, and exploration of new fullerene family, $C_{60}$ polymer. In the optical properties of $C_{60}$, Frenkel and CT excitons play an essential role in the optical processes. Since the lowest optical transition is parity-forbidden, contribution of CT excitons is enhanced near the absorption edge.

The electronic structure of $C_{60}$ is significantly modified by doping of alkali metals. While a naive band model predicts that all $A_xC_{60}$'s are metallic for $0 < x < 6$, spectroscopic studies show that $A_4C_{60}$ is an insulator with the lowest optical transition peak at 0.5 eV. This is in contrast with $A_3C_{60}$ in which the band model is approximately correct. However, a careful analysis shows that a pseudo-gap behavior is observed in $A_3C_{60}$. This result indicates that, even in the metallic $A_3C_{60}$, there exists a mechanism that makes the naive band model inapplicable. The insulating state of $A_4C_{60}$ is possibly related to the same mechanism that forms a pseudo-gap in $A_3C_{60}$. We suggest the Jahn-Teller effect for the failure of the band model. The Jahn-Teller effect might play an important role in the moderately high $T_c$ superconductivity of fullerenes.

Covalent forms of $C_{60}$, so called $C_{60}$ polymers are synthesized. Among several synthesis routes, the application of high pressure can produce not only one-dimensional but also two-dimensional polymers. These $C_{60}$ polymers are providing an promising opportunity to synthesize new kinds of intercalation materials.

The fullerene research is an extremely interdisciplinary project covering the astronomy, chemistry and solid state physics. In fact, $C_{60}$ is synthesized by an arc discharge of carbon which was totally far from conventional materials scientists. It is clear that an interdisciplinary exploration of new materials will open a new field in solid state chemistry and physics.

I thank my co-workers T. Koda, S. Koda, T. Yasuda, K. Tanaka, T. Kaneyasu, S. Watanabe, T. Arima, and T. Yagi at University of Tokyo, and R. M. Fleming, O. Zhou, L. J. Rothberg, T. Siegrist, K. B. Lyons, R. Tycko, and R. C. Haddon at AT&T Bell Laboratories. I am grateful to A. F. Hebard and G. A. Thomas for their hospitality and enlightening discussions. I thank T.Tsubo for providing me with Fig. 7.8.

## REFERENCES

[1] H. W. Kroto, J. R. Heath, S. C. O'Brien, R. F. Curl, and R. E. Smalley, *Nature* **318**, 162 (1985).

[2] W. Krätchmer, L. D. Lamb, K. Fostiropoulos, and D. R. Huffman, *Nature* **347**, 354 (1990).

[3] A. F. Hebard, M. J. Rosseinsky, R. C. Haddon, D. W. Murphy, S. H. Glarum, T. T. M. Palstra, A. P. Ramirez, and A. R. Kortan, *Nature* **350**, 600 (1991).

[4] P. -M. Allemand, K. C. Khemani, A. Koch, F. Wudl, K. Holczer, S. Donovan, G. Grüner, and J. D. Thompson, *Science* **253**, 301 (1991).

[5] R. Taylor and D. R. M. Walton, *Nature* **363**, 685 (1993).

[6] R. M. Fleming, T. Siegrist, P. M. March, B. Hessen, A. R. Kortan, D. W. Murphy, R. C. Haddon, R. Tycko, G. Dabbagh, A. M. Mujsce, M. L. Kaplan, and S. M. Zahurak, in *Clusters and Cluster-Assembled Materials* ed. by R.S.Averback, D. L. Nelson, and J. Bernholc, MRS Symposium Proceedings No.206, Materials Research Society, Pittsburgh (1991).

[7] R. C. Haddon, A. F. Hebard, M. J. Rosseinsky, D. W. Murphy, D. J. Duclos, K. B. Lyons, B. Miller, J. M. Rosamilia, R. M. Fleming, A. R. Kortan, S. H. Glarum, A. V. Makhija, A. J. Muller, R. H. Eick, S. M. Zahurak, R. Tycko, G. Dabbagh, and F. A. Thiel, *Nature* **350**, 320 (1991).

[8] K. Holczer, O. Klein, S. -M. Huang, R. B. Kaner, K. -J. Fu, R. L. Whetten, and F. Diederich, *Science* **252**, 1154 (1991).

[9] M. J. Rosseinsky, A. P. Ramirez, S. H. Glarum, D. W. Murphy, R. C. Haddon, A. F. Hebard, T. T. M. Palstra, A. R. Kortan, S. M. Zahurak, and A. V. Makhija, *Phys. Rev. Lett.* **66**, 2830 (1991).

[10] K. Tanigaki, T. W. Ebbesen, S. Saito, J. Mizuki, J. S. Tsai, Y. Kubo, and S. Kuroshima, *Nature* **352**, 222 (1991).

[11] P. A. Heiney, J. E. Fischer, A. R. McGhie, W. J. Romanow, A. M. Denenstein, J. M. McCauley Jr., A. B. Smith III, and D. E. Cox, *Phys. Rev. Lett.* **66**, 2911 (1991).

[12] J. E. Fischer and P. A. Heiney, *J. Phys. Chem. Solids* **54**, 1725 (1993).

[13] G. B. M. Vaughan, P. A. Heyney, J. E. Fischer, D. E. Luzzi, D. A. Ricketts-Foot, A. R. McGhie, Y. -W. Hui, A. L. Smith, D. E. Cox, W. J. Romanow, B. H. Allen, N. Coustel, J. P. McCauley Jr., and A. B. Smith III, *Science* **254**, 1350 (1991).

[14] R. C. Haddon, L. E. Brus, and K. Raghavachari, *Chem. Phys. Lett.* **125**, 459 (1986).

[15] Y. Tokura, T. Koda, Y. Iyechika, and H. Kuroda, *Chem. Phys. Lett.* **102**, 174 (1983).

[16] T. Tsubo and K. Nasu, *J. Phys. Soc. Jpn.* **63**, 63 (1994); *Solid State Commun.* **91**, 907 (1994).

[17] J. W. Arbogast, A. P. Darmanyan, C. S. Foote, Y. Rubin, F. N. Diederich, M. A. Alvarez, S. J. Anz, and R. L. Whetten, *J. Phys. Chem.* **95**, 11 (1991).

[18] T. Arai, Y. Murakami, H. Suematsu, K. Kikuchi, Y. Achiba, and I. Ikemoto, *Solid State Commun.* **84**, 827 (1992).

[19] D. W. Murphy, M. J. Rosseinsky, R. M. Fleming, R. Tycko, A. P. Ramirez, R. C. Haddon, T. Siegrist, G. Dabbagh, J. C. Tully, and R. E. Walstedt, *J. Phys. Chem. Solids* **53**, 1321 (1992).

[20] Q. Zhu, O. Zhou, J. E. Fischer, A. R. McGhie, W. J. Romanow, R. M. Strongin, M. A. Cichy, and A. B. Smith III, *Phys. Rev. B* **47**, 13948 (1993).

[21] M. J. Rosseinsky, D. W. Murphy, R. M. Fleming, R. Tycko, A. P. Ramirez, T. Siegrist, G. Dabbagh, and S. E. Barrett, *Nature* **356**, 416 (1992).

[22] T. Yildrim, J. E. Fischer, A. B. Harris, P. W. Stephens, D. Liu, L. Brard, R. M. Strongin, A. B. Smith III, *Phys. Rev. Lett.* **71**, 1383 (1993).

[23] R. M. Fleming, A. P. Ramirez, M. J. Rosseinsky, D. W. Murphy, R. C. Haddon, S. M. Zahurak, and A. V. Makhija, *Nature* **352**, 787 (1991).

[24] K. Tanigaki, I. Hirosawa, T. W. Ebbesen, J. Mizuki, Y. Shimakawa, Y. Kubo, J. S. Tsai, and S. Kuroshima, *Nature* **356**, 419 (1992).

[25] P. W. Stephens, L. Mihaly, P. L.Lee, R. L. Whetten, S. -M. Huang, R. Kaner, F. Diederich, and K. Holczer, *Nature* **351**, 632 (1991).

[26] R. M. Fleming, M. J. Rosseinsky, A. P. Ramirez, D. W. Murphy, J. C. Tully, R. C. Haddon, T. Siegrist, R. Tycko, S. H. Glarum, P. Marsh, G. Dabbagh, S. M. Zahurak, A. V. Makhija, and C. Hampton, *Nature* **352**, 701 (1991).

[27] O. Zhou, J. E. Fischer, N. Coustel, S. Kycia, Q. Zhu, A. R. McGhie, W. J. Romanow, J. P. McCauley Jr., A. B. Smith III, and D. E. Cox, *Nature* **351**, 462 (1991).

[28] T. Yildrim, O. Zhou, J. E. Fischer, N. Bykovetz, R. M. Strongin, M. A. Cichy, A. B. Smith III, C. L. Lin, and R. Jelinek, *Nature* **360**, 568 (1992).

[29] T. T. M. Palstra, A. F. Hebard, R. C. Haddon, and P. B. Littlewood, *Phys. Rev. B* **50**, 3462 (1994).

[30] L. Degiorgi, P. Wachter, G. Grüner, S. -M. Huang, J. Wiley, and K. B. Kaner, *Phys. Rev. Lett.* **69**, 2984 (1992); L. Degiorgi, G. Briceno, M. S. Fuhrer, A. Zettl, and P. Wachter, *Nature* **369**, 541 (1994).

[31] A. P. Ramirez, A. R. Kortan, M. J. Rosseinsky, S. J. Duclos, A. M. Mujsce, R. C. Haddon, D. W. Murphy, A. V. Makhija, S. M. Zahurak, and K. B. Lyons, *Phys. Rev. Lett.* **68**, 1058 (1992).

[32] R. F. Kiefl, W. A. MacFarlane, K. H. Chow, S. Dunsiger, T. L. Duty, T. M. S. Johnston, J. W. Schneider, J. Sonier, L. Brard, R. M. Strongin, J. E. Fischer, and A. B. Smith III, *Phys. Rev. Lett.* **70**, 3987 (1993).

[33] A. F. Hebard, *Physics Today* **45**, 26 (1992).

[34] S. Saito and A. Oshiyama, *Phys. Rev. Lett.* **66**, 2637 (1991); A. Oshiyama, S. Saito, N. Hamada, and Y. Miyamoto, *J. Phys. Chem. Solids* **53**, 1457 (1992).

[35] S. C. Erwin, in *Buckminsterfullerenes* ed. by W.E.Billups and M.A.Ciufolini (VCH, New York, 1992).

[36] Y. Iwasa and T. Kaneyasu, *Phys. Rev. B* **51**, 3678 (1995).

[37] Y. Iwasa, in *Advanced Materials '93, I/B: Magnetic, Fullerene, Dielectric, Ferroelectric, Diamond and Related Materials* ed. by H. Homma *et al.* (Elsevier, Amsterdam, 1994), p. 1085.

[38] Y. Iwasa, K. Tanaka, T. Yasuda, T. Koda, and S. Koda, *Phys. Rev. Lett.* **69**, 2284 (1992).

[39] P. C. Eklund, P. Zhou, K. -A. Wong, G. Dresselhaus, and M. S. Dresselhaus, *J. Phys. Chem. Solids* **53**, 1391 (1992).

[40] K. -J. Fu, W. L. Karney, O. L. Chapman, S. -M. Huang, R. B. Kaner, F. Diederich, K. Holczer, and R. L. Whetten, *Phys. Rev.* **B46**, 1937 (1992).

[41] R. F. Kiefl, T. L. Duty, J. W. Schneider, A. MacFarlane, K. Chow, J. W. Elzey, C. Niedermayer, D. R. Noakes, C. E. Stonach, B. Hitti, and J. E. Fischer, *Phys. Rev. Lett.* **69**, 2005 (1992).

[42] M. Kosaka, K. Tanigaki, I. Hirosawa, Y. Shimakawa, S. Kuroshima, T. W. Ebbesen, J. Mizuki, and Y. Kubo, *Chem. Phys. Lett.* **203**, 429 (1993).

[43] P. J. Benning, F. Stepniak, and J. H. Weaver, *Phys. Rev.* **B48**, 9086 (1993).

[44] T. Morikawa and T. Takahashi, *Solid State Commun.* **87**, 1017 (1993).

[45] I. Lukyanchuk, N. Kirova, F. Rachdi, C. Goze, P. Molinie, and M. Mehring, *Phys. Rev. B* **51**, 2978 (1995).

[46] O. Zhou and D. E. Cox, *J. Phys. Chem. Solids* **53**, 1373 (1992).

[47] T. Kato *et al.*, *Chem. Phys. Lett.* **186**, 35 (1991).

[48] S. Suzuki and K. Nakao, preprint.

[49] Y. Wang, J. M. Holden, A. M. Rao, W. -T. Lee, X. -X. Bi, S. L. Ren, G. W. Lehman, G. T. Hager, and P. C. Eklund, *Phys. Rev. B* **45**, 14396 (1992).

[50] A. P. Ramirez, M. J. Rosseinsky, D. W. Murphy, and R. C. Haddon, *Phys. Rev. Lett.* **69**, 1687 (1992).

[51] S. C. Erwin and W. E. Pickett, *Science* **254**, 842 (1991).

[52] R. C. Haddon, *Acct. Chem. Res.* **25**, 127 (1992).

[53] M. P. Gelfand and J. P. Lu, *Phys. Rev. Lett.* **68**, 1050 (1992).

[54] R. Tycko, G. Dabbagh, M. J. Rosseinsky, D. W. Murphy, A. P. Ramirez, and R. M. Fleming, *Phys. Rev. Lett.* **68**, 1912 (1992).

[55] W. H. Wong, M. E. Hanson, W. G. Clark, G. Grüner, J. D. Thompson, R. L. Whetten, S. -M. Huang, R. B. Kaner, F. Diederich, P. Petit, J. -J. André., and K. Holczer, *Europhys. Lett.* **18**, 79 (1992).

[56] M. Knupfer, M. Merkel, M. S. Golden, J. Fink, O. Gunnarsson, and V. P. Antropov, *Phys. Rev. B* **47**, 13944 (1993).

[57] L. Degiorgi, G. Grüner, P. Wachter, S. -M. Huang, J. Wiley, R. L. Whetten, R. B. Kaner, K. Holczer, and F. Diederich, *Phys. Rev. B* **46**, 11250 (1992); L. Degiorgi, E. J. Nicol, O. Klein, G. Grüner, P. Wachter, S. -M. Huang, J.Wiley, and R.B.Kaner, *Phys. Rev. B* **49**, 7012 (1994).

[58] C. M. Varma, J. Zaanen, and K. Raghavachari, *Science* **254**, 989 (1991).

[59] O. Gunnarsson, H. Handschuh, P. S. Bechthold, B. Kessler, G. Ganteför, and W. Eberhardt, *Phys. Rev. Lett.* **74**, 1875 (1995).

[60] J. E. Fischer, *Science* **264**, 1548 (1994).

[61] A. M. Rao, P. Zhou, K. -A. Wang, G. T. Hager, J. M. Holden, Y. Wang, W. -T. Lee, X. -X. Bi, P. C. Eklund, D. S. Cornett, M. A. Duncan, and I. J. Amster, *Science* **259**, 955 (1993).

[62] A. F. Hebard, C. B. Eom, R. M. Fleming, Y. J. Chabal, A. J. Muller, S. H. Glarum, G. P. Pietsch, R. C. Haddon, A. M.Mujsce, M. A. Pazkowski, and G. P. Kochanski, *Appl. Phys. A* **57**, 299 (1993).

[63] A. Ito, T. Morikawa, and T. Takahashi, *Chem. Phys. Lett.* **211**, 333 (1993).

[64] P. Zhou, Z. -H.Dong, A. M. Rao, and P. C. Eklund, *Chem. Phys. Lett.* **211**, 337 (1993).

[65] Y. Wang, J. M. Holden, Z. -H. Dong, X. -X. Bi, and P. C. Eklund, *Chem. Phys. Lett.* **211**, 341 (1993).

[66] J. Winter and H. Kuzmany, *Solid State Commun.* **53**, 1321 (1992).

[67] O. Chauvet, G. Oszlányi, L. Forro, P. W. Stephens, M. Tegze, G. Faigel, and A. Jánossy, *Phys. Rev. Lett.* **72**, 2721 (1994).

[68] S. Pekker, L. Forró, L. Mihály, and A. Jánossy, *Solid State Commun.* **90**, 349 (1994).

[69] P. W. Stephens, G. Bortel, G. Faigel, M. Tegze, A. Jànossy, S. Pekker, G. Oszlanyi, and L. Forró, *Nature* **370**, 636 (1994).

[70] S. Pekker, A. Jánossy, L. Mihaly, O. Chauvet, M. Carrard, L. Forró, *Science* **265**, 1077 (1994).

[71] M. C. Martin, D. Koller, A. Rosenberg, C. Kendziora, L. Mihaly, *Phys. Rev. B* **51**, 3210 (1995).

[72] A. Jànossy, O. Chauvet, S. Pekker, J. R. Cooper, and L. Forro, *Phys. Rev. Lett.* **71**, 1091 (1991).

[73] Q. Zhu, D. E. Cox, and J. E. Fischer, *Phys. Rev. B* **51**, 3966 (1995).

[74] M. Kosaka, K. Tanigaki, T. Tanaka, T. Atake, A. Lappas, and K. Prassides, *Phys. Rev. B* (in print).

[75] Y. Iwasa, T. Arima, R. M. Fleming, T. Siegrist, O. Zhou, R. C. Haddon, L. J. Rothberg, K. B. Lyons, H. J. Carter Jr., A. F. Hebard, R. Tycko, G. Dabbagh, J. J. Krajewski, G. A. Thomas, and T. Yagi, *Science* **264**, 1570 (1994).

[76] M. Núñez Regueiro, P. Monceau, and J. -L. Hodeau, *Nature* **355**, 237 (1992).

[77] F. Moshary, N. H. Chen, I. Silvera, C. A. Brown, H. C. Dorn, M. S. de Vries, and D. S. Bethune, *Phys. Rev. Lett.* **69**, 466 (1992).

[78] G. Kriza, J. -C. Ameline, D. Jèrome, A. Dworkin, H. Szwarc, C. Fabre, D. Schütz, A. Rassat, P. Bernier, and A. Zahab, *J. Phys. I France* **1**, 1361 (1991).

[79] G. A. Samara, J. E. Schirber, B. Morosin, L. V. Hansen, D. Loy, and A. P. Sylwester, *Phys. Rev. Lett.* **67**, 3136 (1991).

[80] M. Azuma, Z. Hiroi, M. Takano, Y. Bando, and Y. Takeda, *Nature* **356**, 775 (1992).

[81] S. Akimoto, N. Hamaya, and I. Shirotani, in *Material Science of the Earth's Interior* ed. by I.Sunagawa (Terrapub, Tokyo, 1984) p. 131.

[82] S. J. Duclos, K. Brister, R. C. Haddon, A. R. Kortan, and F. A. Thiel, Nature 351, 380 (1991).

[83] G. Oszlanyi and L. Forro, *Solid State Commun.* **93**, 265 (1995).

[84] R. S. Ruoff, D. S. Tse, R. Malhotra, D. C. Lorents, *J. Phys. Chem.* **97**, 3379 (1993).

[85] K. M. Creegan, J. L. Robbins, W. K. Robbins, J. M. Millar, R. D. Sherwood, P. J. Tindall, D. M. Cox, A. B. Smith III, J. P. McCauley Jr., D. R. Jones, and R. T. Gallagher, *J. Am. Chem. Soc.* **114**, 1103 (1992).

[86] For example, H. Yamawaki, M. Yoshida, Y. Kakudate, S. Usuba, H. Yokoi, S. Fujiwara, K. Aoki, R. Ruoff, R. Malhotra, and D. Lorents, *J. Phys. Chem.* **97**, 11161 (1993).

[87] C. H. Xu and G. E. Scuseria, *Phys. Rev. Lett.* **74**, 274 (1995).

[88] I. O. Bashkin, V. I. Rashchupkin, A. F. Gurov, A. P. Moravsky, O. G. Rybchenko, N. P. Kobelev, Y. M. Soifer, and E. G. Ponyatovsky, *J.Phys. Condensed Matter* **6**, 7491 (1995).

[89] Y. Wang, J. M. Holden, Z. -H. Dong, X. -X. Bi, and P. C. Eklund, *Chem. Phys. Lett.* **217**, 417 (1994).

[90] M. Núñez Regueiro, L. Marques, J. -L.Hodeau, O. Béthoux, and M. Perroux, *Phys. Rev. Lett.* **74**, 278 (1995).

*CHAPTER 8*

# ARRAYED NANOCLUSTERS IN ZEOLITE CRYSTALS

Yasuo Nozue
*Department of Physics, Tohoku University*



## 8.1. SPECTROSCOPIC STUDY OF NEW MATERIALS

### *8.1.1. The Aim of Spectroscopic Study*

Generally, when we start the research on new condensed matter, we need a structural analysis followed by the characterization of electronic properties. A spectroscopic study plays an important role in the basic understanding of the electronic state, because it offers us precise and detailed information about the electronic configuration of materials by observing the optical transition between electronic states.

Usually we first try to understand observed electronic properties based on the one-electron model, such as the energy band model. In the ground state, however, electrons are interacting with each other. Hence, the true electronic configuration is different from that based on the one-electron model. As stated later, even in alkali metals, the electronic state is different from the free-electron Fermi gas model in a strict sense, and the difference appears as new magnetic properties in their clusters, such as ferromagnetism. Also in an excited state, the electronic state is different from that expected from the one-electron model. Sometimes, we create a lot of interest in the mutual interaction of electrons, and specify it as the special phenomenon. For example, the attractive interaction between electron and hole, which will constitute an exciton, is caused by the repulsive interaction between an excited electron in the conduction band and other many electrons in the valence band.

In the excited state, the relaxation process occurs due to the configuration rearrangement of many electrons, and sometimes followed by the lattice displacement. In the case of electronic rearrangement, the effect occurs during the optical excitation process, and a photon creates the rearranged excited state. For example, in the exciton state, the electric field of electron-hole pair rearranges core electrons as well as the lattice polarization. This is called the screening effect for the electron-hole Coulomb interaction. In the case followed by the lattice relaxation, the self-trapping of exciton, for example, occurs after the generation of exciton. As for the plasmon oscillation, the classical electromagnetic theory can deal with the collective motion of free carriers phenomenologically, because the long range part of electron-electron Coulomb interaction can be calculated in the classical electromagnetic theory. The RPA model is helpful to understand the phenomenon based on the quantum statistical mechanics.

We have already enriched our experience in the spectroscopy of traditional materials. Now, we should put our experience into the research of novel new-materials, because the optical measurement is increasingly playing an indispensable role as one of the complemental way of research for new materials. Especially for the low-dimensional material, we often need reliable information about the electronic state without detailed knowledge of the interatomic bonding, even though the optical measurement plays a subsidiary role.

When we classify a lot of materials, physicists will look at the periodic table. We can find a lot of element crystals, such as Si, Ge, etc. As the next class, we think about compound materials, such as CdS, GaAs, NaCl, etc. We already have a lot of knowledge on these traditional materials. The knowledge is helpful to understand new materials which is different from traditional ones, because we sometimes find a phenomenon similar to but somewhat different from that observed in traditional materials. As recent trends, materials with microstructure are increasingly important. Many different materials with the structure of nanometer scale are synthesized by using various technique. The properties of these materials are sometimes beyond those observed in traditional ones, because new quantum electronic states are generated by making microstructures. The optical study is one of the important methods to elucidate properties of such low dimensional materials. In other words, when we generate a new element of microstructure, such as quantum-well, molecules, clusters, etc., we can construct a new material based on these microstructural element, such as quantum-well superlattice, molecular crystal or complex, and cluster crystal in zeolite matrix, etc.

Material types are usually classified into 0-, 1-, 2- and 3-dimensions. Strictly speaking, real materials, however, can not be classified into them, because different dimensions are combined into the different type of dimension. For example, in a quantum-well superlattice, the two-dimensional electronic state of one quantum well can interact with that in different wells. This is different from the true two-dimensional system, but quasi-two-dimensional one. A molecular complex also has the array of 0-dimensional electronic states of molecule, but observed properties are classified to the dimension different from 0-dimension. An essential point may resides not in the structural dimension but in the scientific insight.

### *8.1.2. Spectroscopic Study of Low-Dimensional Materials*

Here, we classify these microstructural materials into four types according to the method of the synthesis. The first one is given by the artificial growth of thin crystals, such as MBE technique. The quantum well superlattices have brought us a lot of variety and have opened new field in science and technology. The second one is a natural growth of microstructural materials, such as copper oxide high-$T_c$ superconductors, clathrate compounds, quasi-crystals, etc.. The third one is the synthesis of stable clusters or molecules followed by the crystal growth of complex, such as $C_{60}$ complexes and organic superconductors, etc. The last one is the method detailed here; clusters and chains generated in the microporous crystals, such as zeolites.

When the material size is reduced into the nanometer scale, some part of the electronic properties are different essentially from those of bulk crystal. Usually, physicist prefers an advanced macroscopic concept in low-dimensional materials. If we can use a macroscopic model in low-dimensional materials, such as the energy band model and the effective mass approximation, the observed properties which are different from the bulk crystal can be ascribed to the boundary condition problem. If we can not find out any similarity with the bulk crystal concept, the materials are substantially new or may belong to a new field.

In some cases, however, we can analyze the optical response based on the classical electromagnetic theory without any change of the local dielectric function given in the original bulk material, because the shape and size effect can be partly calculated in the classical electromagnetic theory. Even though the size of material is reduced to the nanometer scale, the optical spectrum sometimes can be calculated without any change of the dielectric function. A typical case is the surface plasmon response of metal sphere with the size smaller than the wavelength of electromagnetic wave but larger than the cluster size, as stated later. The optical response of metal sphere is largely different from that of original bulk metal. The difference can not be ascribed to the difference in the quantum electronic state of material nor to the difference in the local dielectric function. According to the Mie theory based on the classical electromagnetic theory, we can calculate the absorption spectrum of metal sphere embedded in the matrix by the introduction of the boundary condition for the electromagnetic field. This is the benefit of the boundary condition in the classical electromagnetic theory. Another typical case is the surface exciton. We can calculate the reflection spectrum of surface exciton in anthracene crystal under the assumption of the surface dielectric layer with the mono-molecular thickness [1]. When we go into detail the microscopic feature of optical response, we can simplify it to the point dipole model.

The essential difference is not in the change in the quantum electronic state but in the change in the shape or thickness of molecular array. In inorganic semiconductors, essential difference occur, when the thickness is comparable or less than the effective Bohr radius of exciton, so called the electron-hole individual confinement region.

For the non-local optical response, we already have improved successfully the effective dielectric function of condensed materials by the introduction of the spatial dispersion; the k-dependence of the resonance energy of eigenstate. For example, we can define the equivalent exciton polariton state within the classical electromagnetic theory, except for the damping process [2]. The damping process occurs as the quantum mechanical phenomena. Hence, after the assumption of the additional boundary condition (ABC) and a dead layer for exciton [3], we can calculate the transmission or reflection spectrum of exciton polariton in thin semiconductor films without any change of the dielectric function derived from the bulk crystal. The ABC problem occurs due to the non-local dielectric function, but can not be solved easily. The thickness of the dead layer may be determined partly by the quantum mechanical electronic state of exciton near the crystal surface.

The classical electromagnetic theory is very convenient to calculate the optical response of materials, but has the limitation in the handling of the non-local response at the boundary. If we have a microscopic concept for the interaction between the electromagnetic wave and the material, we can solve the problem. The process to the solution, however, is not easy. The process is desired to be easy comparable with that in the classical electromagnetic theory. We already knows that we can easily solve the electromagnetic interaction with many atoms and/or molecules by the introduction of the dielectric function. From the quantum mechanical viewpoint, however, the classical electromagnetic theory involves the limitation; the amplitude and the phase of electromagnetic wave are given at the same time. Both of them, however, can not be determined at the same time in the quantum mechanical theory.

In any case, the phenomenon which requires a new concept is most interesting, such as a new optical phenomenon, an optical response of new quantum state, a new relaxation process of excited state, a new relation or combination with the different phenomenon. We can expect the development of advanced science in the field of optical properties of low dimensional materials. When we synthesize the new materials, however, we often obtain new materials unexpectedly. Their properties sometimes diverge beyond the originally intended optical properties. In such case, we should extend our research field outside of the spectroscopic research. The nature is usually beyond our knowledge. The research of optical properties is outstanding but one of the fields of the material science and technology.

## 8.2. ZEOLITE CRYSTALS AS THE CONTAINER OF CLUSTER

The most typical and easiest method for the preparation of small particles is the deposition in a glass matrix. This method has the advantage of the wide range control of particle size. However, as for the small particles with nanometer size, such as clusters, this method is not effective, because the chemical reaction with matrix

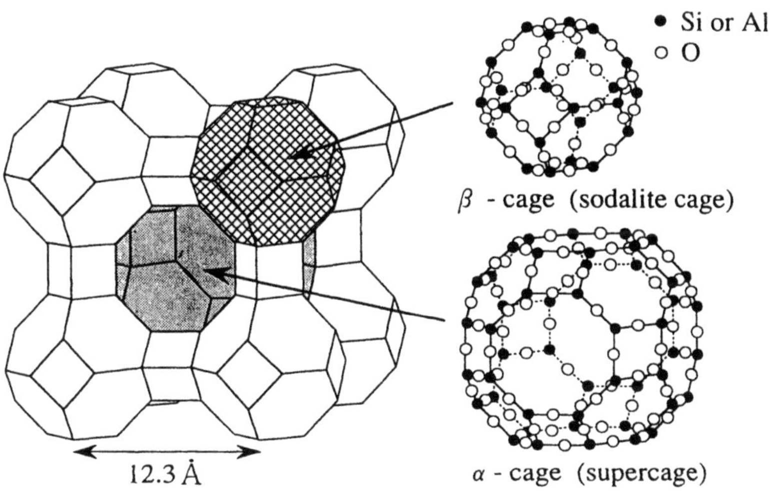


Figure 8.1. Framework structure of zeolite LTA. Closed and open circles indicate (Si, Al) and O, respectively. The $\beta$-cages are connected at the 4-membered rings in a simple cubic structure.

makes a serious problem at the boundary of the particle, and we can not generate particles with unique size. In order to solve these problems, the zeolite method has the advantages of the size control and of the reduction of the undesired chemical reaction with the matrix. Zeolite is a microporous crystal, and widely used for molecular sieve, catalysis, adsorbent, builder, etc. Zeolite research has been well developed already at certain level, although we are expecting a further development.

The most typical zeolite is alminosilicate. Various types of zeolite crystals are synthesized under various growth conditions. Some kinds of zeolites are found also in nature. A general chemical formula of alminosilicate zeolite is given by

$$\mathrm{M}_{m/r} \cdot \mathrm{Al}_m\mathrm{Si}_n\mathrm{O}_{2(m+n)} \cdot x\mathrm{H_2O}, \quad (m \leq n), \tag{8.1}$$

where $\mathrm{Al}_m\mathrm{Si}_n\mathrm{O}_{2(m+n)}$ constructs the framework of zeolite, and M is an $r$-valent cation and sitting at the site near the framework. A lot of water molecules are captured in the space of framework, when zeolite is kept in air. The framework network is constructed by the sharing of O atom of $SiO_4$ or $AlO_4$ tetrahedra. The most typical cation is $Na^+$. We can easily change $Na^+$ to many other cations, such as $K^+$, $Ca^{2+}$, $Cd^{2+}$, $Ag^+$, etc. More than one hundred kinds of zeolite framework structure types are already known. Various types of zeolite can be synthesized in the different chemical composition with the common framework structure, because zeolites are synthesized in different Si to Al ratios with different kinds of cation.

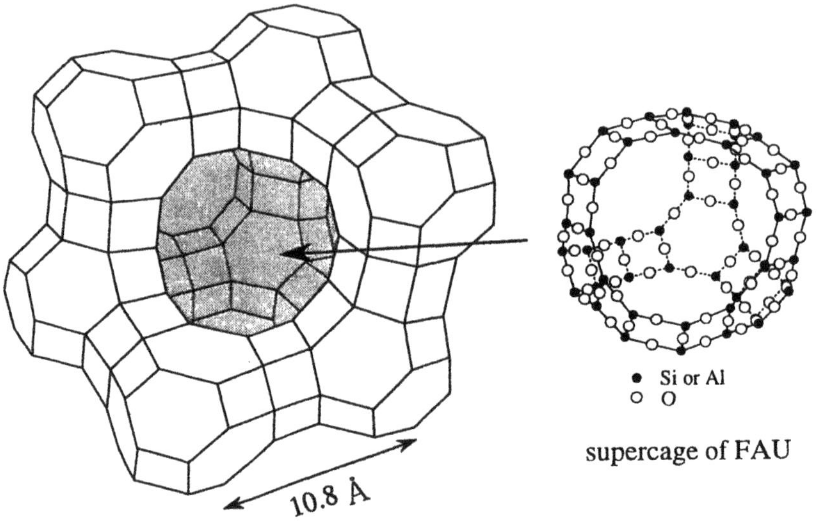


Figure 8.2. Framework structure of zeolite FAU. Closed and open circles indicate (Si, Al) and O, respectively. The $\beta$-cages are connected at the 6-membered rings in a diamond structure.

LTA is one of the most typical zeolites. In Fig. 8.1, LTA framework structure is shown schematically, where cations are neglected in the figure. Open and closed circles indicate O and (Si, Al) atoms, respectively. A basic unit cage is called the $\beta$-cage or the sodalite cage. The $\beta$-cages are connected at 4-membered rings, and arrayed in a simple cubic structure. Among $\beta$-cages, a new cage, called the $\alpha$-cage (or the supercage of LTA), exists. In other words, the $\alpha$-cages are connected by the sharing of 8-membered rings in a simple cubic structure. In LTA, the Si to Al ratio is usually unity, and Si and Al atoms are alternately ordered. We denote Na type zeolite LTA with Si to Al ratio of unity as Na-LTA(1), hereafter. Zeolite LTA is sometimes called a different name, A, and that of Na type is called NaA.

Besides LTA, there are FAU (faujasite) and SOD (sodalite) zeolites which have the $\beta$-cage as a basic unit cage. In Fig. 8.2, framework structure of FAU is shown schematically. In FAU, $\beta$-cages are connected at 6-membered rings, and arrayed in a diamond structure. In SOD, $\beta$-cages are connected by the sharing 4-membered rings, and arrayed in a body centered cubic structure. Zeolite MOR (mordenite) has a different type of framework, and has parallel channels with an inner diameter of about 7 Å.

Zeolite crystals have remarkable features of container in the fabrication of clusters. The framework material is transparent between infrared and ultraviolet regions. This point is very important in the optical study of guest materials. Zeolite is fully crystalline. The number density of $\alpha$-cage amounts to $5.4 \times 10^{20}$ cm$^{-3}$. We can estimate

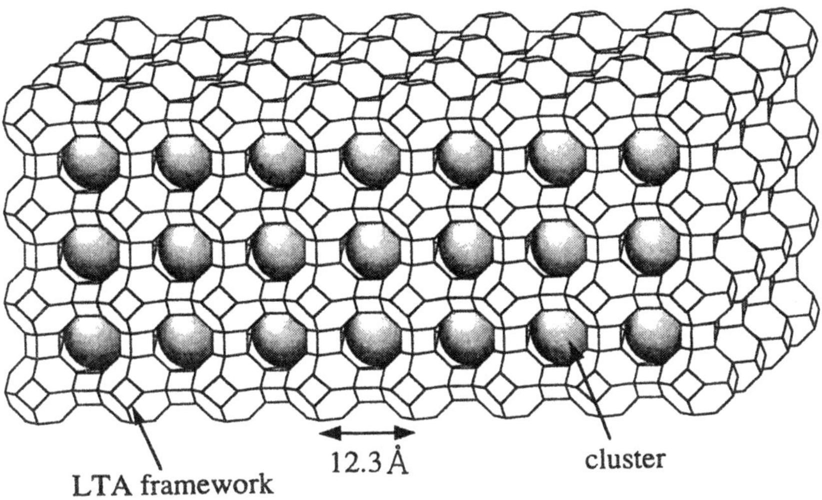


Figure 8.3. Schematic representation of arrayed clusters in zeolite LTA.

an average density of guest material per cage. Zeolite is stable at high temperatures. It is possible to introduce different materials into zeolite cages or channels. In Fig. 8.3, clusters arrayed in the space of LTA framework is schematically indicated. When we use nanometer size pores of zeolite crystals as the container of materials, structurally unstable clusters can be supported there. Clusters in zeolite can have a homogeneous size and are arrayed according to the arrangement of zeolite cages. A new type of low-dimensional material can be fabricated due to the intercluster interaction.

## 8.3. SEMICONDUCTOR CLUSTERS

### *8.3.1. Quantum Electronic States*

The quantum electronic state of isolated clusters is, of course, considered as one of the molecular orbital. Usually, molecules have a stable structure, and have a closed chemical bonding. Generally, however, the chemical bonding is not completed at crystal surface. This will lead to the dangling bond state at the crystal surface. In the case of clusters, this leads to an indistinct criterion of semiconductor. We should think whether the semiconductor is truly semiconductor like in the cluster. Usually, the dangling bond is terminated by the chemical reaction with other materials when we keep clusters in air. In free InP clusters, optical behavior is different from that found in usual semiconductor clusters [4]. If we neglect a dangling bond problem, we can overview the optical properties of semiconductor particles in terms of the size.

If the semiconductor film is much thicker than the effective Bohr radius of exciton,

$a_B$, the reflection and transmission spectra can be calculated by the electromagnetic theory by using k-dependent dielectric function and the additional boundary condition. In the case of spherical particle, we have no simple expression for the optical response. In the case of k-independent materials, the optical response of spheres with the radius smaller than the wave length can be calculated easily, as follows. According to the Mie theory, an absorption coefficient of spheres dispersed in a matrix is given by

$$A(\omega) = \frac{9\epsilon_m^{3/2} F}{c} \frac{\omega\epsilon_2(\omega)}{[\epsilon_1(\omega) + 2\epsilon_m]^2 + \epsilon_2^2(\omega)}, \tag{8.2}$$

where $A(\omega)$ is the absorption coefficient, $\epsilon(\omega)$ and $\epsilon_m$ are the dielectric function of particle material and matrix, respectively, $F$ is the volume fraction, and $c$ is the velocity of light. The dielectric function $\epsilon(\omega)$ is given by

$$\epsilon(\omega) = \epsilon_1(\omega) + i\epsilon_2(\omega). \tag{8.3}$$

Equation (8.2) tells us that the resonant absorption occurs at

$$\epsilon_1(\omega) = -2\epsilon_m. \tag{8.4}$$

This formula can be used for the case that the dielectric function is independent of wave vector $k$, which means that the sphere is made of material with rather heavy effective mass.

Equation (8.2) shows that the oscillator strength of each sphere is proportional to the volume of sphere. This is quite reasonable for the sum rule of the oscillator strength. With the increase of the radius, the oscillator strength sometimes becomes a huge value. A similar expression for ellipsoidal particles can be described in the extended Maxwell-Garnett theory [5]. The expression for cylindrical wires also has been given in the theory.

In a small sphere, the translational motion of carrier is quantized. The quantization problem is common in the calculation of the quantum levels of single carrier confined in the spherical-well potential. As shown in Fig. 8.4, the quantum levels of $1s$, $1p$, $1d$, $2s$, etc. are given in the increasing order of energy. The kinetic energy of $1s$ state is given as

$$E_{K,1s} = \frac{\hbar^2}{2m}\left(\frac{\pi}{r}\right)^2, \tag{8.5}$$

where $m$ is the mass of carrier and $r$ is the radius of the sphere. In actual semiconductors, we must consider the spin-orbit interaction as well as the deviation from the parabolic dispersion of carrier.

If the particle size is much larger than $a_B$, the translational motion of exciton is quantized, where $m$ is given by the reduced mass of exciton in Eq. (8.5). If the particle size is comparable with $a_B$, the quantum states are deviate from the exciton confinement scheme, because of the finite size of exciton [6, 7]. In the smallest limit, such as the cluster, the electron and hole are confined individually, because the mutual interaction between electron and hole no longer plays the most important role in

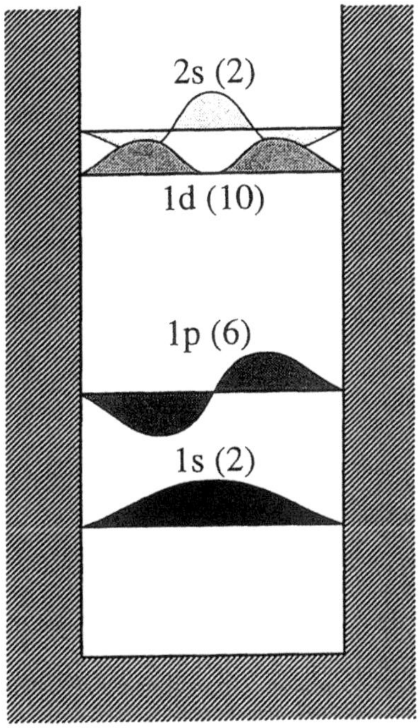


Figure 8.4. Quantum electronic levels of carrier confined in the infinite spherical quantum well. Levels $1s$, $1p$, $1d$, $2s$, etc. appear in the increasing order of energy.

the determination of their wave function [6, 7]. A little effect of the electron-hole interaction, however, still remains even in the smallest limit. In the intermediate size, such as the same size with $a_B$, usually an electron is confined directly by the wall of particle, but a hole is indirectly confined by the potential of electron, because the effective mass of electron is usually much smaller than that of hole. Hence, the effective size of hole wave function is much smaller than that of electron. This case is not in the region of the electron-hole individual confinement. In order to confine electron and hole, the particle size must be much smaller than $a_B$. This is usually the cluster size.

In the case of cluster, we can not explain the result by an effective mass approximation, because the particle size is comparable with the lattice constant. Furthermore, the basic feature of carriers are considerably different from the macroscopic feature of bulk crystal. We must consider that the structure of cluster is basically different from that of bulk crystal, because of the following two reasons that cluster has no translational symmetry and that the most of atoms are located at the surface. The latter point causes the change in the chemical bonding.

Although there are many problem in the basic feature of cluster, we can find a common or similar aspect in terms of the quantum confinement effect in the excitation energy and the oscillator strength. In aromatic hydrocarbons, such as benzene, naphthalene, anthracene etc., quantum electronic levels are roughly interpreted by assuming a perimeter free electron orbital [8]. We assume circular ring orbitals for $\pi$-electrons at the perimeter of aromatic hydrocarbon molecule. According to the length of circuit and the number of $\pi$-electrons, we can easily calculated the energy of the transition from HOMO to LUMO by the introduction of the quantization condition of momentum. In this calculation, we can use a free electron mass. It means that the electron travels in molecule as a free electron. This result is indicative of the common aspect of the quantization of carrier in a molecular scale. A similar result is found in the size dependence on the excitation energy of $ns$-electron trapped at the F-center in alkali halide, called the Mollwo-Ivey rule [9].

### 8.3.2. *Optical Properties of* $PbI_2$ *Clusters*

Bulk $PbI_2$ is well known layered type semiconductor. The excitonic spectra have been investigated in detail. The value of $a_B$ is estimated to be 19 Å [10], but the effective radii of the electron and hole in $PbI_2$ are estimated to be 6 and 14 Å, respectively. Hence, we need to generate the cluster with the size comparable with above radii, when we realize the electron-hole individual confinement.

$PbI_2$ clusters are well generated in the nanometer-size cage of zeolite. In Fig. 8.5, the absorption spectra are shown for $PbI_2$ clusters generated in Na-FAU(1.25), where the average molecular number of clusters are indicated in each curve [11]. $PbI_2$ clusters are generated in the FAU supercage which has the inner radius of 6.5 Å. In the most dilute sample with the average molecular number of 0.15, a single-molecular absorption band $B_1$ is observed at 5.5 eV. With increasing the average molecular number, new absorption bands, $B_2$–$B_5$, appear successively at lower energies. Band $B_n$ has been assigned to $n$-molecular clusters. The structure of clusters are not well known yet, but the excitation energy shifts systematically to lower energy with the increase in the size of $PbI_2$ cluster. In the upper most curve, the average molecular number coincides with the maximum molecular number of cluster. This means that the unique size of clusters are generated in each cage of Na-FAU(1.25).

If there is no electronic exchange interaction between clusters, an excitonic state of arrayed clusters is equivalent to that of the molecular crystal. In the molecular crystals, an excited state of molecule runs over many equivalent molecules, and this is called a Frenkel exciton [12]. The Frenkel exciton model is available even at the crystal surface [13], but there appears an important problem in case that molecules are randomly located at regular sites. When $PbI_2$ is loaded into zeolite Na-LTA(1), absorption bands similar to those observed in Na-FAU(1.25) appear depending on the average $PbI_2$ density. However, the reflectivity at the exciton resonance energy 3.25

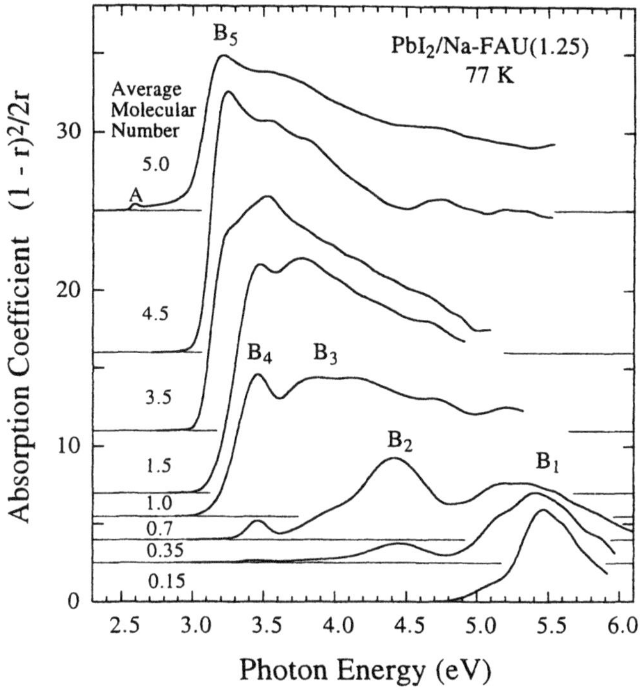


Figure 8.5. Absorption spectra of $PbI_2$ clusters generated in zeolite Na-FAU(1.25) [11]. The number in the figure indicates the average number of $PbI_2$ molecules per supercage of FAU.

eV quickly increases near the saturation limit of $PbI_2$, as shown in Fig. 8.6, where the number density of $(PbI_2)_5$ clusters increases in proportion to the loading density in the figure [14]. The broken line indicates the linear dependence of the reflectivity normalized at the maximum value. The reflectivity does not increase in proportion to the occupancy. According to the dynamical coherent-phase-approximation (D-CPA), a non-linear change of the effective damping parameter is expected as a function of the excitonic band width which linearly increases with the number density of clusters. With increasing the band width, the effective damping energy decreases nonlinearly, because of the motional narrowing effect. However, the observed reflectivity is beyond the calculation based on the D-CPA. If molecules are distributed randomly at regular sites, the Frenkel exciton state can not be solved easily. We must consider the percolation criterion as the function of the occupancy of cluster. According to the simulation based on the two-level lattice model, there appears a sudden change

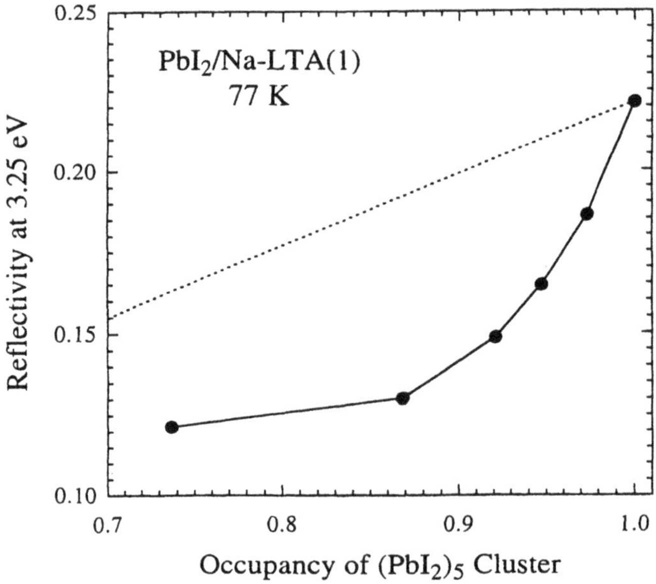


Figure 8.6. Reflectivity at the exciton resonance energy 3.25 eV of $PbI_2$ clusters generated in zeolite Na-LTA(1) as the function of the average number of $(PbI_2)_5$ molecules in each $\alpha$-cage [14]. Broken line indicates the linear dependence of the reflectivity normalized at the maximum value.

in the density of state of the exciton state [15].

According to the Mie theory, the oscillator strength of particle is in proportion to its volume. This do not violate the sum rule. In small sizes, however, electronic states are quantized. If we look at the specific excitonic transition, the sum rule may seem to be changed. In Fig. 8.7, the spectra of Fig. 8.5 are normalized by the average number of $PbI_2$ molecule [11]. The spectral intensity of $B_n$ band is rather enhanced in lower density samples. This result means that the oscillator strength of the lowest-energy excitation does not seem to obey the sum rule expected in the Mie theory.

According to the quantum mechanical theory, the oscillator strength f is generally given by

$$f = \frac{2}{\hbar\omega m_0} \left|\langle\Psi_e|P|\Psi_g\rangle\right|^2 , \tag{8.6}$$

where $\Psi_e$ and $\Psi_g$ are the wave functions of the excited and ground states, respectively. When the electron wave function fully overlaps with the hole wave function and the excitation is an allowed-transition type, the above value becomes size-independent

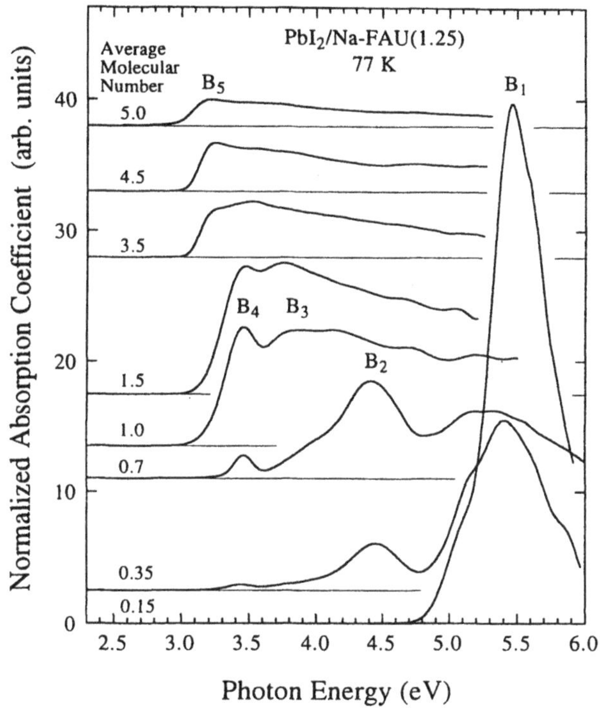


Figure 8.7. Normalized absorption spectra of $PbI_2$ clusters generated in zeolite Na-FAU(1.25). The absorption coefficient is normalized by the average number of $PbI_2$ molecules. The number in the figure indicates the average number of $PbI_2$ molecules per supercage of FAU. (After [11]).

[11]. On the other hand, the number of occupied quantum states decreases in proportion to the number of constituent atoms. Hence, the total oscillator strength is in proportion to the number of electrons included in the particle. When the absorption spectrum is normalized by the number of atoms, the smaller clusters show the larger intensity. The experimental result shown in Fig. 8.7 is manifestation of the quantum size effect on the oscillator strength.

The excitation energy of small clusters can not be explained on the bases of the effective mass approximation, but some systematic behavior is found in the experimental data [16]. As shown in Fig. 8.8, the excitation energy of $HgI_2$ and $PbI_2$ clusters shift in inversely proportion to the square of the effective diameter $d$, where the same volume sphere with the diameter $d$ is assumed for each cluster. The slope of

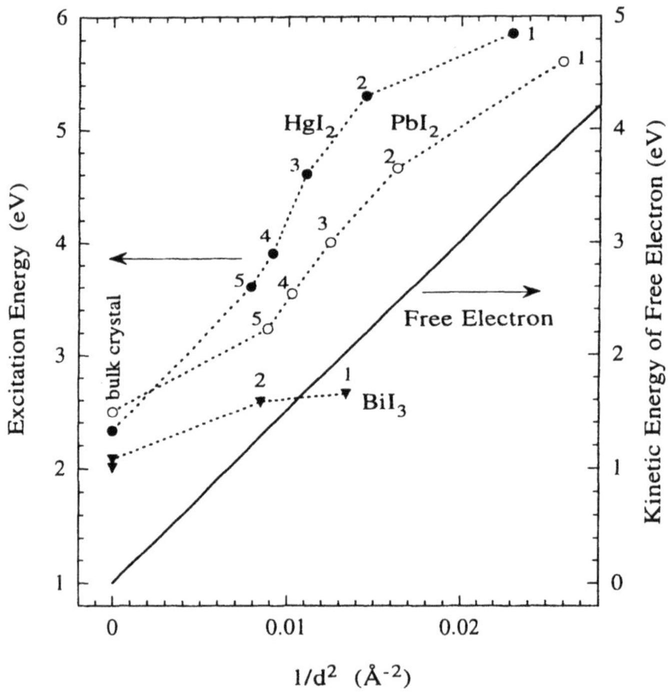


Figure 8.8. Excitation energy of $HgI_2$, $PbI_2$ and $BiI_3$ clusters and molecules and the bulk-crystal exciton energy as the function of $1/d^2$, where $d$ is the effective diameter of cluster. The kinetic energy of free electron confined in the sphere is also indicated by straight line in the figure.

the observed size dependence is quite similar to the kinetic energy of the free electron confined in the spherical potential, as shown by the straight line in the figure. This result is a good indication of the Mollwo-Ivey rule [9]. However, the result in $BiI_3$ clusters is different from others. In $BiI_3$, the electrons are strongly localized in $BiI_3$ molecule, and the electronic state may be quite similar to the molecular crystal.

## 8.4. ALKALI METAL CLUSTERS

### 8.4.1. *Quantum Electronic States*

In metal particles larger than cluster size, the optical properties are calculated based on the classical electromagnetic theory. In bulk metals, the long-range repulsive interaction between electrons leads to the plasmon oscillation as the collective motion of electrons. Especially in alkali metals, the most of oscillator strength is exhausted as the plasmon state. A small part of oscillator strength is shared with the interband transition. In noble metals and transition metals, however, the contribution of the interband transition is not small. The interband transition sometimes strongly modifies the color of metals, such as copper and gold. Usually, the oscillator strength of collective motion of metals is much larger than the excitonic oscillator strength of insulators.

Here, we restrict the theory within the simple metals, such as alkali metals. Alkali metals are called the free electron metals. The total volume of cations in bulk alkali metals is only 6.2, 10.1, 13.8, 15.6 and 17.6 % in Li, Na, K, Rb and Cs, respectively, when we assume the respective cationic radii of 0.68, 0.97, 1.33 1.48 and 1.67 Å. The first ionization energy of alkali metal atom is smaller than that of other metal atoms. Hence, $ns$ electrons of alkali metals can move almost freely as free electrons independent of the cation potential. The effective mass of electron is almost the same as the free electron.

The simplest dielectric function of free electron metal is given by

$$\epsilon(\omega) = 1 - \frac{\omega_p^2}{\omega(\omega + i\gamma)}, \tag{8.7}$$

where $\gamma$ is the damping energy mainly due to the electron-phonon interaction. The parameter $\omega_p$ is the bulk plasmon energy, and given as

$$\omega_p^2 = \frac{4\pi n_e e^2}{m_0}, \tag{8.8}$$

where $n_e$ is the number density of electron, $e$ the electron charge and $m_0$ the mass of free electron. In this formula, all oscillator strength of electrons is exhausted as the plasmon state. According to Eq. (8.4), the surface plasmon resonance absorption of metal sphere occurs at the energy $\omega_{sp}$:

$$\omega_{sp} = \frac{\omega_p}{\sqrt{1 + 2\epsilon_m}}. \tag{8.9}$$

The energy $\omega_{sp}$ is rather independent of radius in the scale of eV, because of a rather heavy mass of the plasmon state, as expected in RPA. The oscillator strength is proportional to the volume of sphere.

In bulk metals, usually the plasmon state is a good eigenstate of the collective excitation at the small $k$ region. At the large $k$ region, however, the plasmon state has the strong damping due to the individual electron excitation. This is called the Landau damping. Metal particles larger than nanometer scale show a significant absorption due to above surface plasmon state. However, some change occur in the spectral width due to the Landau damping with decreasing the size of metal particles [17].

When the size is reduced into the nanometer size, *i.e.* cluster size, the ***spillout*** of electron wave function causes the decrease of the resonance energy. The optical

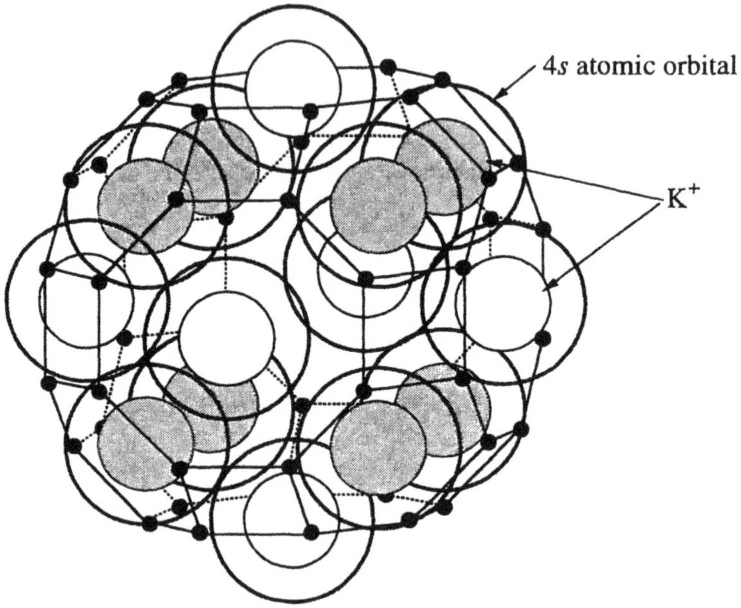


Figure 8.9. Schematic representation of $K^+$ cations in LTA and their 4*s* orbitals. The position of $K^+$ cations are simplified according to the structural data of dehydrated K-form LTA. Some other $K^+$ cations are not shown here.

properties of metal clusters are intensively calculated by several methods [18]–[31]. In alkali metal clusters, properties of *ns* electrons seem to be understood based on the model of free electrons confined in a simplified potential, such as a jellium potential of positive core background. The deviation from the spherical shape of cluster leads to the splitting of the surface plasmon energy. Before the cluster size approaches the atomic size, the surface plasmon oscillation must disappear, because the plasmon oscillation is due to the collective motion of many electrons. When the cluster has a few electrons, the surface plasmon picture in LDA calculation causes the self-interaction problem [31]. In Na clusters incorporated in zeolite, the surface plasmon state has been observed at the electron number of four or more than four [32].

In zeolite crystals, many alkali-metal-cations are distributed in the space of framework. For example, $K^+$ cations are located near the center of 6-membered ring and at the center of 8-membered ring, as shown schematically in Fig. 8.9, where several simplifications are introduced to the cation site according to the structural data of dehydrated K-form LTA [33]. There exist other cations at different sites. The effective outside radius of 4*s* orbital of K atom is 2.38 Å, and they are shown by thin circles in Fig. 8.9. These 4*s* orbitals are mutually overlapped with each other. These

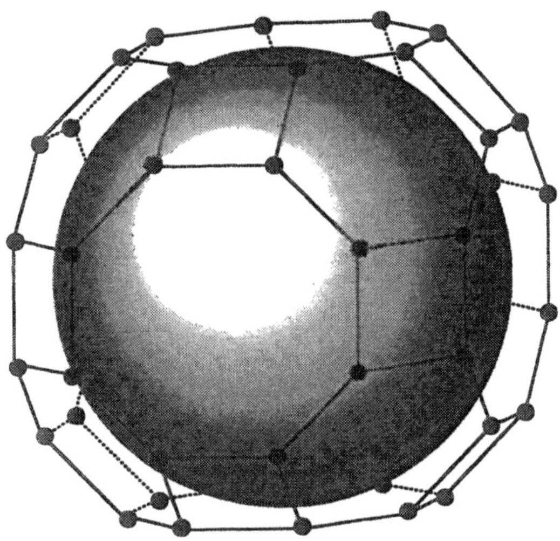

Figure 8.10. Expected $1s$ cluster-orbital of K-cluster confined in $\alpha$ cage of LTA.

orbitals are empty before the adsorption of guest K atoms. When a guest K atom is adsorbed into the space of zeolite, $4s$ electron can be resonantly transferred over many cations. The framework potential, however, is repulsive to electrons, because the framework is negatively charged. Hence, an electron wave function spreads over many cations but is confined in the $\alpha$-cage. These cations are located near the spherically symmetric sites. The sum of Coulomb potential of cations and framework has the flat bottom near the center of the $\alpha$-cage. Hence, we can simplify the effective potential for electron as the spherical one as shown in Fig. 8.4. The lowest quantum electronic state of cluster, $1s$ state, may be given as shown in Fig. 8.10. If we assume the effective diameter of the $\alpha$-cage, 11 Å, the $1p$ state is located at 1.2 eV higher than the $1s$ state.

When K metal is dilutely adsorbed in K-form LTA, a significant absorption band has been observed at 1.2 eV, as shown in Fig. 8.11 [34]. The resonance energy just coincides with that expected in the model mentioned above. This is a clear evidence for the quantum levels of K cluster with the effective diameter of 11 Å. When the kind of alkali metal is changed to Rb, a similar band has been observed [34]. As shown in Fig. 8.12, when Rb is dilutely adsorbed in Rb-form LTA, a quite similar band has been observed. This result suggests that the effective size of cluster is not determined by the kind of alkali metal but by the size of framework.

The spectra in Figs. 8.11 and 8.12 show some fine structures. The fine structure can not be interpreted by the model of the optical transition between $1s$ and $1p$ localized states, because no splitting occurs in the cubic symmetry. Each $\alpha$-cage has six windows shared with the nearest neighbor cage. Hence, the wave function of

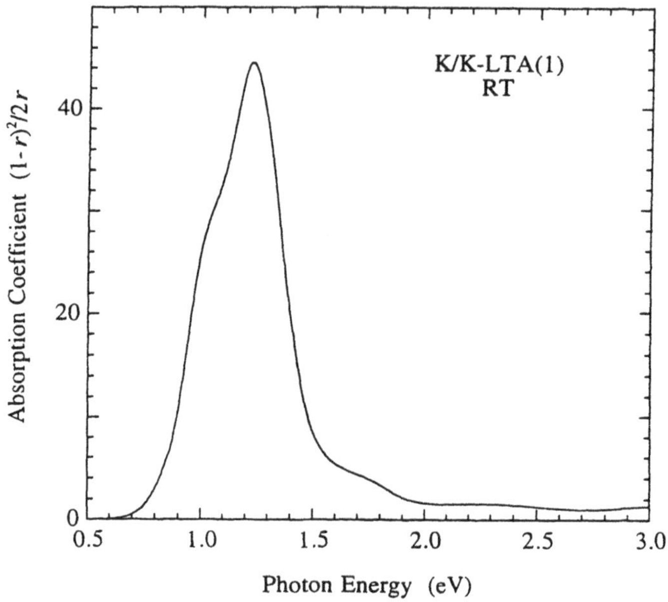


Figure 8.11. Absorption spectrum of dilutely K-adsorbed K-form LTA at room temperature.

molecular orbital of cluster is expected to overlap with that in clusters in adjacent cages. This leads to the energy band with the simple cubic symmetry. When we assume the nearest neighbor transfer, we can calculate the energy band of $1s$ and $1p$ states by using the tight-binding model. In Fig. 8.13, the Brillouin zone of simple cubic lattice and an example of energy bands are shown, where the $\sigma$ transfer energy of $1p$ orbital is assumed to have the opposite sign of the $\pi$ transfer energy. According to the theoretical calculation, the second nearest neighbor transfer energy of $1p$ orbitals plays an important role in the stability of ferromagnetism of K clusters [35]. When the second nearest neighbor transfer is included in the energy band calculation, the dispersion curve of Fig. 8.13 is varied and splits depending on the value of the parameters.

When the density of guest alkali metal increases, $ns$ electrons successively occupy $1s$ and $1p$ orbitals of cluster. Two and six electrons fill the $1s$ and $1p$ orbitals, respectively. Cluster with many electrons shows an intense band due to the surface plasmon excitation. This is one of the main topic of the optical study of cluster. In the classical electromagnetic theory shows that the surface plasmon occurs at the energy given by Eq. (8.9). In Na metal particle in vacuum, this energy is calculated to be 3.3 eV. In free Na clusters ($Na_3$–$Na_{20}$), many absorption bands have been observed

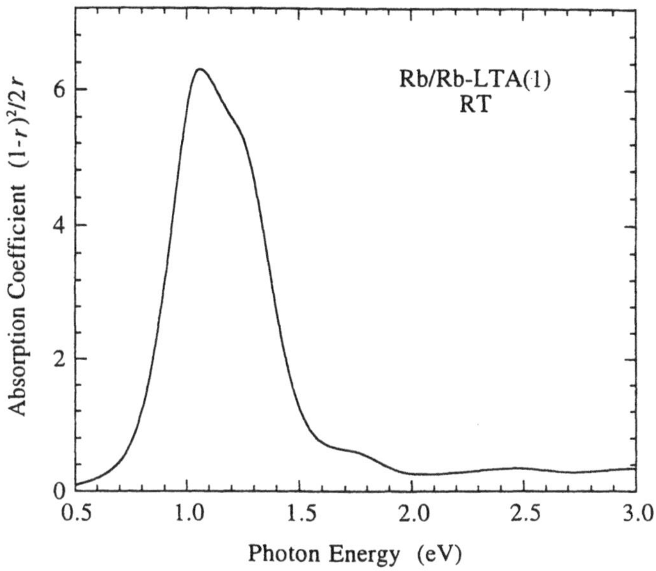


Figure 8.12. Absorption spectrum of dilutely Rb-adsorbed Rb-form LTA at room temperature. (After [34]).

by the photodepletion technique [36]. In larger clusters, a remarkable absorption bands appear due to the surface plasmon state. The resonance energy, however, is around 2.5 eV. The smaller energy is ascribed to the spillout of electrons and/or the difference in the inter-atomic distance. A numerical calculation by means of the *ab initio* large-scale configuration interaction procedure shows good agreement with the experimental results including the oscillator strength [29]. However, the physical meaning is not clear. On the other hand, the method based on the physical insight gives a good assignment for the phenomenon, but the coincidence of the numerical values are not so good [18]–[28], [30, 31]. In the cluster size, the difficulty exists in the criteria of the individual excitation and the collective excitation of electrons, because both excitation energies are very close to each other.

In alkali metal clusters in zeolite cage, a remarkable reflection band has been observed due to the surface plasmon resonance. Solid curve of Fig. 8.14 shows the reflection spectra of Na, K and Rb clusters incorporated in Na-, K- and Rb-form LTAs, respectively. The average electron density is 4.9, 5.4 and 5 per a cage in curves (a), (b) and (c), respectively. Dotted curve indicates the region where the transmission component of each powder particle can not be neglected in the total

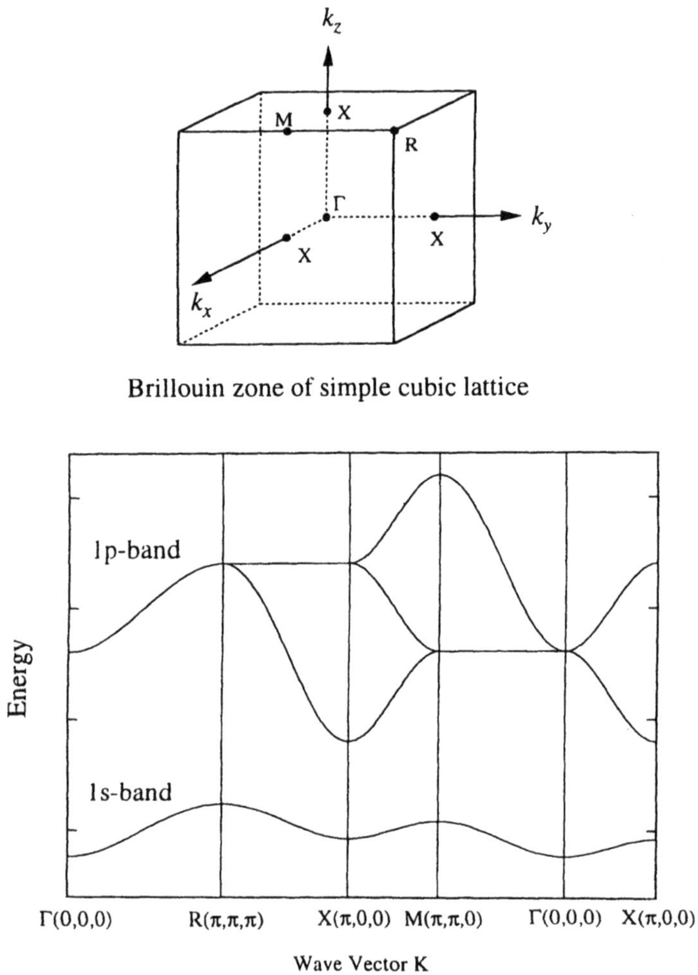


Figure 8.13. Brillouin zone of simple cubic lattice and an example of energy bands. The $\sigma$ transfer energy of $1p$ orbital is assumed to have the opposite sign of the $\pi$ transfer energy.

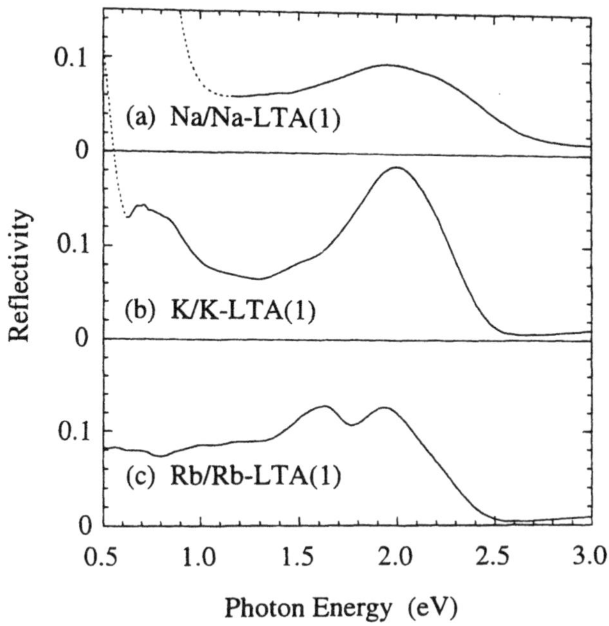


Figure 8.14. Solid curve shows the reflection spectra of Na, K and Rb clusters incorporated in Na-, K- and Rb-form LTAs, respectively. The average electron density is 4.9, 5.4 and 5 per $\alpha$-cage in curves (a), (b) and (c), respectively. Dotted curve indicates the region where the transmission component of each powder particle can not be neglected in the value of diffuse reflection. (After [40]).

value of the diffuse reflection. In curve (b), there appears the strongest band at 2 eV. This band has been assigned to the surface plasmon excitation of K clusters generated in the $\alpha$-cage [34]. Besides them, relatively weak bands due to the $1s$–$1p$ and $1p$–$1d$ interband transitions appear around 0.9 and 1.5 eV, respectively [34]. In curve (a), the similar surface plasmon band appears at 2 eV, but other bands are scarcely seen. The peak value of reflectivity of the 2 eV band in curve (a) is lower than that in curve (b). In curve (c), the band at 2 eV is weaker and narrower than that of curve (b), and the remarkable band appears at the peak energy of 1.6 eV. The lower energy bands appear at energies lower than that in curve (b).

From the analysis of the reflection spectra, the oscillator strength of the 2 eV band in Na clusters is estimated to be 4 per cluster, which indicates that at least 4 electrons contribute to this band. The value 4 is slightly larger than that of the similar band at 2 eV in K clusters, 3.5. The oscillator strength of 2 eV band in curve (c), however, is 1.6 per cluster. This value is much smaller than those in curves (a) and (b). The large oscillator strength of the 2 eV band in curve (a) is due to the strong surface plasmon enhancement effect. No remarkable enhancement, however, is seen for any bands in curve (c).

The difference in the appearance of the surface plasmon band is qualitatively explained by the difference in the potential depth of cluster. According to the theoretical calculation of the oscillator strength [28], the surface plasmon band becomes remarkable but the individual excitation becomes weak, with increasing the confinement potential depth for electrons. Conversely, with decreasing the confinement potential depth, the surface plasmon band becomes indistinct and the individual excitation dominates the spectrum. In Na clusters in LTA, the cluster potential is expected to be deeper than that of K clusters, because of the larger ionization energy of Na atom. Hence, the surface plasmon band reasonably dominates the spectrum. In Rb clusters, however, the cluster potential is expected to be shallower than that of K clusters, because of the smaller ionization energy of Rb atom. The surface plasmon band does not dominate the spectrum in Rb clusters. The 2 eV band as well as the 1.6 eV band in curve (c) may be ascribed mainly to the $1p$–$1d$ transition. The doublet structure of these bands is explained by the splitting of the $1d$ state, because the $1d$ state splits into two band in the cubic symmetry.

### *8.4.2. Electron-Phonon Interaction*

The electron-phonon interaction plays an important role in the structural stability of clusters in zeolite, because there are many equivalent cages with displaceable cations there. When cations shift to the center of cage, the potential for electron is deepened as the potential pocket, and an electron is trapped there. If the electron does not have an enough transfer energy to the nearest neighbor cage, the electron is self-trapped at a fixed cage. If the depth of the potential pocket is deepened by the second electron, two electrons can be trapped at the common cage in spite of the repulsive Coulomb interaction. In this case, the magnetic state is singlet when the level has no degeneracy. On the other hand, when the electron-phonon interaction is weak, electrons are mobile over many site due to the finite transfer energy. As stated later, the magnetic properties are dramatically changed due to the difference in the electron-phonon interaction, such as ferromagnetism and diamagnetism.

The number density of cations increases with increasing the loading density of alkali metal. Stable clusters are varied depending on the number density of cations, because of the finite number of cation sites. When Na is adsorbed dilutely into Na-form LTA, the firstly observed absorption band is different from that in K and Rb clusters generated in the $\alpha$-cage. Figure 8.15 shows the absorption spectra of Na loaded Na-form LTA [37]. The loading density increases in the order of curves 1 to 7. Dotted curve indicates the region where the value is somewhat saturated. In curves 1 and 2, bands $D_1$ and $D_2$ are observed around 3 eV. This band is assigned to the cluster generated in the $\beta$-cage, because the excitation energy well coincides with that calculated in the spherical potential with the inside diameter of the $\beta$-cage,

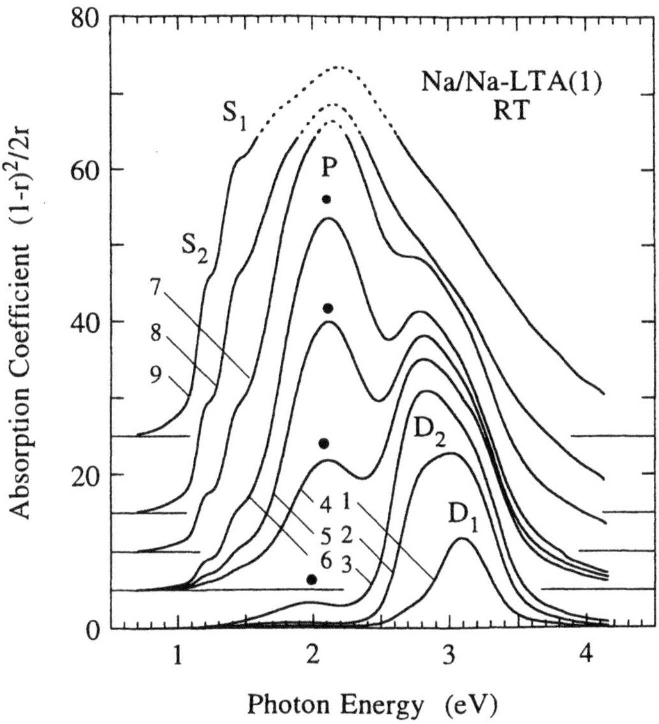


Figure 8.15. The absorption spectra of Na loaded Na-form LTA. The loading density increases in the order of curves 1 to 7. Dotted curve indicates the region where the value is saturated. (After [37]).

about 7 Å. With increasing the loading density, new absorption bands P, $S_1$ and $S_2$ are observed instead of $D_1$ and $D_2$ bands. These bands are assigned to the cluster generated in the $\alpha$-cage [37]. The band P is assigned to the surface plasmon band observed in the reflection spectrum of Fig. 8.14(a).

The oscillator strengths of 2 eV band in curves (a) and (b) in Fig. 8.14 are similar to each other, but the value of the reflectivity is different. From the analysis of the reflection spectrum, the damping energy of the 2 eV band in curve (a) is larger than that in curve (b). This result indicates that the electron-phonon interaction in Na clusters is stronger than that in K clusters. In Na clusters, the strong lattice displacement occurs at the photoexcitation of electrons. The difference in the electron-phonon interaction is qualitatively expected from the ionization potential difference and the atomic mass difference between Na and K atoms. Generally, the deep ionization po-

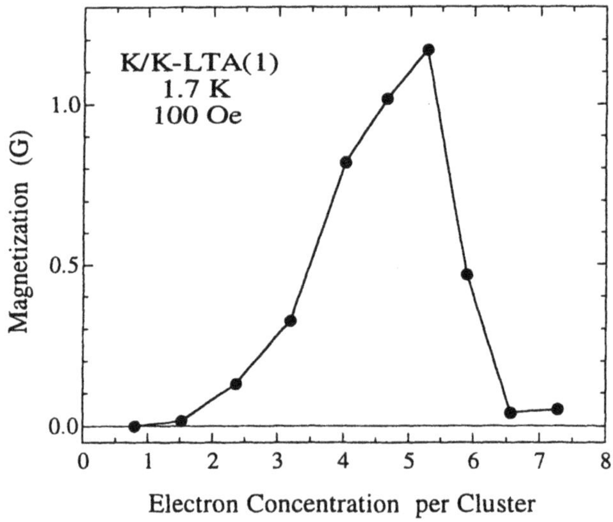


Figure 8.16. Magnetization at the external magnetic field of 100 Oe and at 1.7 K in K-loaded K-LTA(1) is plotted as a function of the average electron concentration per cluster. (After [39]).

tential of atom leads to the strong interaction between the electron wave function and the cation configuration. The light atomic mass enhances the cation motion in the photoexcitation process. In Rb clusters, the damping energy is found to be smaller than those in Na and K clusters from the analysis of curve (c) in Fig. 8.14. Hence, the electron-phonon interaction in Rb clusters is weaker than that in K clusters. This is due to the smaller ionization potential and the heavier atomic mass of Rb atom.

### *8.4.3. Magnetic Properties*

Magnetic properties are manifestations of quantum electronic states and mutual interaction of electrons localized at site and/or being itinerant over many sites. An optical study plays an important role in the research of the electronic levels, but less important role in the research of the spin configuration. On the other hand, magnetic properties are informative about the spin state of the ground state. The optical information sometimes plays a complimentary role in the whole of research. The material phase is expected to vary widely depending on the transfer energy, the on-site Coulomb repulsion energy, and the electron-phonon interaction energy [38]. In the alkali metal clusters in zeolite, we can control or change many effective parameters; the electron transfer energy between clusters $T$, the on-site Coulomb repulsion energy

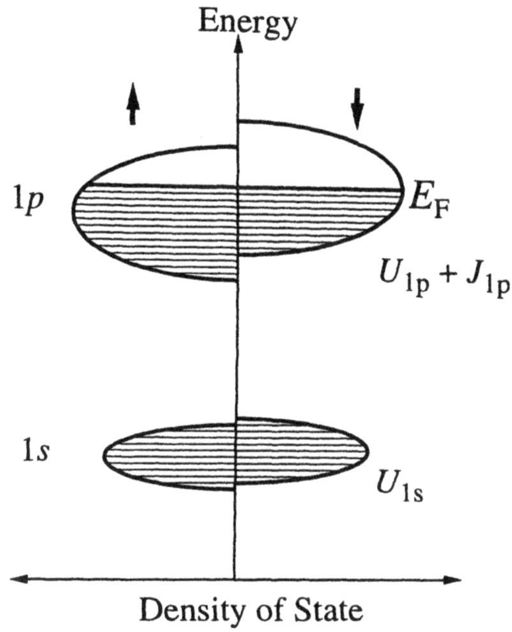


Figure 8.17. Density of state of the energy bands originating from $1s$ and $1p$ orbitals and the Fermi level in the ferromagnetic state. These bands have narrow widths. When the Fermi level is located near the center of the $1p$ energy band, the Stoner condition may be satisfied, and the ferromagnetism has been observed at low temperatures. In the $1p$ state, the sum of the Hund coupling energy $J_{1p}$ and on-site Coulomb energy $U_{1p}$ satisfies the ferromagnetic condition.

$U$, the electron-phonon interaction $S$, the degeneracy of quantum level, the density of $ns$-electrons and the cluster arrangement. In these materials, we need to extend the research to many fields.

In K clusters in K-form LTA, a ferromagnetism has been observed depending on the average electron density [39]. In Fig. 8.16, the magnetization at the external magnetic field of 100 Oe and at 1.7 K is plotted as a function of the average electron concentration per cluster. The magnetization increases when the average electron concentration per cluster exceeds about 2, and has the peak around 5 and decreases at higher concentrations. The observed magnetization is several orders larger than that observed in the ordinary paramagnetic materials. Hence, the observed magnetization around 5 electrons per cluster indicates a ferromagnetism. The maximum magnetic moment is estimated to be $0.24\mu_B$ per cluster. In this material, no magnetic element is contained. The observed ferromagnetism is ascribed to the $4s$ electrons provided by K atoms. The qualitative interpretation of ferromagnetism is given by the Stoner model. As shown in Fig. 8.17, energy bands originating from $1s$ and $1p$ orbitals have narrow widths. When the Fermi level is located near the center of the $1p$ energy band, the Stoner condition may be satisfied, and the ferromagnetism has been observed at low temperatures. In the $1p$ state, the sum of the Hund coupling energy $J_{1p}$ and on-site Coulomb energy $U_{1p}$ may satisfies the ferromagnetic condition.

At the just half-filled condition, such as the 5 electrons per cluster, an antiferromagnetic state is usually expected to be more stable than ferromagnetism. When the second nearest neighbor transfer of electron is introduced in addition to the first nearest one, the ferromagnetism is found to be stabilized even at the just half-filled condition [35].

In Na clusters, no ferromagnetism has been observed at any electron concentration. The resultant magnetic state is diamagnetism, indicating that even number of electrons are paired to have the singlet state. The essential difference is assigned to the stronger electron-phonon interaction than that in the K clusters. In the degenerated $1p$ level of Na cluster, the Jahn-Teller effect may overcomes the Hund coupling, and the degeneracy may be resolved, indicating that the spin singlet state is stabilized at any average electron density [40]. In Rb clusters, a ferromagnetism has been observed similar to that in K clusters [40, 41].

## REFERENCES


[1] Y. Nozue, M. Kawaharada, and T. Goto, *J. Phys. Soc. Jpn.* **56**, 2570 (1987).

[2] J. J. Hopfield, *Phys. Rev.* **112**, 1555 (1958).

[3] J. J. Hopfield and D. G. Thomas, *Phys. Rev.* **132**, 563 (1963).

[4] K. D. Kolenbrander and M. L. Mandich, *Phys. Rev. Lett.* **65**, 2169 (1990).

[5] R. W. Cohen, G. D. Cody, M. D. Coutts, and B. Abeles, *Phys. Rev. B* **8**, 3689 (1973).

[6] L. E. Brus, *J. Chem. Phys.* **80**, 4403 (1984).

[7] Y. Kayanuma, *Phys. Rev. B* **38**, 9797 (1988).

[8] M. Pope and C. E. Swenberg, *Electronic Process in Organic Crystals*, (Oxford University Press, 1982).

[9] W. B. Fowler, *Physics of Color Centers*, (Academic Press, New York, 1968).

[10] Y. Nagamune, S. Takeyama, and N. Miura: *Phys. Rev. B* **40**, 8099 (1989).

[11] Z. K. Tang, Y. Nozue, and T. Goto, *J. Phys. Soc. Jpn.* **60**, 2090 (1991).

[12] M. R. Philpott, *Adv. Chem. Phys.* **23**, 227 (1973).

[13] J. M. Turlet, Ph. Kottis, and M. R. Philpott, *Adv. Chem. Phys.* **54**, 303 (1983).

[14] Z. K. Tang, Y. Nozue, O. Terasaki, and T. Goto, *Mol. Cryst. Liq. Cryst.* **218**, 585 (1992).

[15] Y. Kayanuma, *J. Phys. Soc. Jpn.* **62**, 346 (1993).

[16] Z. K. Tang, Y. Nozue, and T. Goto, *J. Phys. Soc. Jpn.* **61**, 2943 (1992).

[17] A. Kawabata and R. Kubo, *J. Phys. Soc. Jpn.* **21**, 1765 (1966).

[18] W. Ekardt, *Phys. Rev. B* **31**, 6360 (1985).

[19] Z. Penzar, W. Ekardt, and A. Rubio, *Phys. Rev. B* **42**, 5040 (1990).

[20] W. Ekardt and Z. Penzar, *Phys. Rev. B* **43**, 1322 (1991).

[21] J.M. Pacheco and W. Ekardt, *Phys. Rev. Lett.* **68**, 3964 (1992).

[22] S. Tomiya and K. Kawamura, *Z. Phys.* **12**, 469 (1989).

[23] C. Yannouleas, J.M. Pacheco, and R.A. Broglia, *Phys. Rev. B* **41**, 6088 (1990).

[24] C. Yannouleas and R.A. Broglia, *Europhys. Lett.* **15**, 843 (1991).

[25] C. Yannouleas and R.A. Broglia, *Phys. Rev. A* **44**, 5793 (1991).

[26] M. Bernath, C. Yannouleas, and R.A. Broglia, *Phys. Lett. A* **156**, 307 (1991).

[27] C. Yannouleas, R.A. Broglia, M. Brack, and P.F. Bortignon, *Phys. Rev. Lett.* **63**, 255 (1989).

[28] C. Yannouleas, *Chem. Phys. Lett.* **193**, 587 (1992).

[29] V. Bonacic-Koutecky, M.M. Kappes, P. Fantucci, and J. Koutecky, *Chem. Phys. Lett.* **170**, 26 (1990).

[30] S. Saito, S.B. Zhang, S.G. Louie, and M.L. Cohen, *Phys. Rev. B* **42**, 7391 (1990).

[31] S. Saito, *Phys. Rev. B* **43**, 6804 (1991).

[32] T. Kodaira, Y. Nozue, S. Ohwashi, N. Togashi, and O. Terasaki, *Surf. Rev. Lett.* (in print).

[33] R. Y. Yanagida, A. A. Amaro, and K. Seff, *J. Phys. Chem.* **77**, 805 (1973).

[34] T. Kodaira, Y. Nozue, S. Ohwashi, T. Goto, and O. Terasaki, *Phys. Rev. B* **48**, 12245 (1993).

[35] A. Z. Chowdhury and K. Nasu, *J. Phys. Chem. Sol.* (in print).

[36] C. R. C. Wang, S. Pollack, T. A. Dahlseid, and G. Koretsky, *J. Chem. Phys.* **96**, 7931 (1992).

[37] T. Kodaira, Y. Nozue, and T. Goto, *Mol. Cryst. Liq. Cryst.* **218**, 55 (1992).

[38] Y. Shinozuka, *J. Phys. Soc. Jpn.* **56**, 4477 (1987).

[39] Y. Nozue, T. Kodaira, S. Ohwashi, T. Goto, and O. Terasaki, *Phys. Rev. B* **48**, 12253 (1993).

[40] Y. Nozue, T. Kodaira, S. Ohwashi, N. Togashi, T. Monji, and O. Terasaki, *Stud. Surf. Sci. Catal.* **84**, 837 (1994).

[41] Y. Nozue, T. Kodaira, S. Ohwashi, N. Togashi, and O. Terasaki, *Surf. Rev. Lett.* (in print).

www.ingramcontent.com/pod-product-compliance
Lightning Source LLC
LaVergne TN
LVHW051053210926
848486LV00002B/97
* 9 7 8 9 8 1 0 2 2 2 3 1 4 *